NANOFINISHING SCIENCE AND TECHNOLOGY

Basic and Advanced Finishing and Polishing Processes

MICRO AND NANO MANUFACTURING SERIES

Series Editor
Dr. V. K. Jain
Professor, Dept. of Mechanical Engineering
Indian Institute of Technology, Kanpur, India

Published Titles:

Nanofinishing Science and Technology: Basic and Advanced Finishing and Polishing Processes, by Vijay Kumar Jain

MICRO AND NANO MANUFACTURING SERIES

Nanofinishing Science and Technology

Basic and Advanced Finishing and Polishing Processes

edited by
V. K. JAIN

CRC Press
Taylor & Francis Group
Boca Raton London New York

CRC Press is an imprint of the
Taylor & Francis Group, an **informa** business

CRC Press
Taylor & Francis Group
6000 Broken Sound Parkway NW, Suite 300
Boca Raton, FL 33487-2742

First issued in paperback 2019

ISBN-13: 978-1-4987-4594-9 (hbk)
ISBN-13: 978-0-367-87521-3 (pbk)

Library of Congress Cataloging-in-Publication Data

Names: Jain, V. K. (Vijay Kumar), 1948- editor.
Title: Nanofinishing science and technology : basic and advanced finishing and polishing processes / edited by Vijay Kumar Jain.
Description: Boca Raton, FL : CRC Press is an imprint of Taylor & Francis Group, an Informa Business, [2017] | Series: Micro and nano manufacturing series | Includes bibliographical references and index.
Identifiers: LCCN 2016028023| ISBN 9781498745949 (hardback : alk. paper) | ISBN 9781315404103 (ebook)
Subjects: LCSH: Finishes and finishing. | Grinding and polishing. | Nanomanufacturing.
Classification: LCC TJ1280 .N36 2017 | DDC 667/.9--dc23
LC record available at https://lccn.loc.gov/2016028023

Visit the Taylor & Francis Web site at
http://www.taylorandfrancis.com

and the CRC Press Web site at
http://www.crcpress.com

Dedicated to Prof. P.C. Pandey (UOR, Roorkee, India), the

late Prof. J.P. Dwivedi (MACT Bhopal, India) and the late

Prof. Ranga Komanduri (OSU, Stillwater, USA)

Live and Let Live

Contents

Section IV   Magnetic Field Assisted
Nanofinishing Processes

Section V   Hybrid Nanofinishing Processes

Section VI   Miscellaneous

Foreword

Super surface finishing is an important requirement of high-value engineering products. There are many desirable properties that can be attributed directly to nanofinishing of products. Some examples are improved surface properties against wear, fatigue and corrosion and functional properties such as closer form and fitting tolerance, increased load bearing surfaces and sealing capability. Other qualities such as high aesthetics, smoothness and shine could project good image and add to the attractiveness of the product.

Traditionally, high surface finishing is achieved through a series of tedious processes such as fine grinding, polishing, honing and lapping. These processes are usually carried out by skilled craftsmen with many years of experience. Such processes often add to the cost of the products as well as a long lead time required to reach market. A product labelled 'hand polished' was usually perceived to be exclusive and prestigious in the past. Modern finishing methods have largely replaced manual processes with equal, if not better, results.

This book compiled by Prof. Vijay K. Jain of the Indian Institute of Technology, Kanpur, entitled *Nanofinishing Science and Technology: Basic and Advanced Finishing and Polishing Processes,* contains an impressive and comprehensive collection of nanofinishing processes, from traditional, to advanced, to magnetic field assisted, to hybrid nanofinishing processes, as well as miscellaneous measurement and optimisation techniques. Each technique is explained in depth both scientifically and with application examples. It is certainly one of the most complete books that I have come across in this field. This book complements an earlier book edited also by Prof. Vijay K. Jain, *Micromanufacturing Processes* (CRC Press, 2013), where various micromachining, microforming and microjoining processes are covered.

I applaud the great effort put together by Prof. Vijay K. Jain and many of his colleagues, and I believe this book will be of tremendous reference value for both academia and industry.

Andrew Y.C. Nee, DEng, PhD
Editor-in-Chief, International Journal of Advanced Manufacturing Technology
Professor of Mechanical Engineering
National University of Singapore

Preface

This unique book on *Nanofinishing Science and Technology: Basic and Advanced Finishing and Polishing Processes* is the result of the combined effort of 34 eminent professors and researchers in different fields of nanofinishing science and technologies. The main objective of this book is to acquaint the readers with both the science and technology of many newly developed nanofinishing processes as well as commonly practiced nanofinishing processes in a single volume. Such a collection of the articles will definitely help teachers, researchers, as well as shop floor engineers. This book encompasses not only the nanofinishing technologies but also the process optimisation methodologies that will help in extracting the best from the process under consideration. Knowledge of the application areas of different nanofinishing processes is equally important. Hence, almost every chapter deals with the specific applications and a separate section deals with general applications of different nanofinishing processes. Further, *you cannot finish a product unless you can measure it.* Hence, this book deals with topics such as optimisation, applications of nanofinishing processes and measurement techniques for nano-level finished parts. Most of the chapters have an extensive references list or bibliography, a set of review questions and in some cases, solved problems as well. All this material makes this book a suitable candidate for a textbook on nanofinishing science and technologies that can be offered to both undergraduate and postgraduate students, apart from being a reference book for researchers and practicing engineers.

This book is divided into six major sections. Section I gives an introductory overview of nanofinishing technologies, their working principles and in some cases, process-specific applications. Section II deals with two traditional nanofinishing processes, namely, honing and lapping. Section III discusses the advanced nanofinishing techniques requiring no magnetic field assistance. This section includes four processes, namely, abrasive flow finishing, elastic emission machining (EEM), elasto-abrasive finishing (EAF) and focused ion beam (FIB) nanofinishing for ultra-thin tunneling electron microscope (TEM) sample preparation. The beauty of the first process of this category is that it can finish, deburr and radius very large, three-dimensional (3D), complex and freeform surfaces at comparatively very high finishing rate. However, the best surface roughness value reported is around 20 nm. EEM removes the material almost atom by atom by using sub-micron-sized abrasive particles. In the case of the EAF process, elastomer balls embedded with the required type of abrasive particles are prepared as a finishing tool to achieve the desired surface roughness value on different kinds of workpieces. However, the last process (FIB) is useful only for dealing with submicron or a few micron-size working areas.

Section IV describes the nanofinishing processes that are assisted by magnetic field. In these processes, to some extent, you can control the forces on-line. This section includes the following processes: magnetic abrasive finishing (MAF) process, magnetorheological finishing (MRF), nanofinishing of freeform surfaces using ball end MRF and nanofinishing process for spherical components. These processes are also known as deterministic nanofinishing processes, in which the forces acting in the process can be controlled on-line through external means. These processes are useful for complicated 3D, complex and freeform surfaces. The last process of this category, known as magnetic float polishing, is mainly useful for nanofinishing spherical components, while others cannot finish spherical components efficiently.

Most of these previously mentioned processes come under the category of independent processes, which have some merits and some weaknesses as well. To minimise the effects of weaknesses and to exploit the effects of their merits, many times, two or more processes (two finishing or finishing and machining processes) are combined together, and such processes are known as *hybrid processes*. Section V deals with hybrid finishing/machining processes, which include the following processes: chemomechanical magnetorheological finishing, electrochemical grinding, electrochemical magnetic abrasive finishing (ECMAF), electric-discharge diamond grinding (EDDG) and electrochemical honing. The first is a combination of chemomechanical polishing and MRF, while the second combines electrochemical machining (ECM) and grinding to take advantage of the merits of the constituent processes. ECMAF combines ECM and MAF, while EDDG combines electric discharge machining (EDM) and diamond grinding. The last process of this section combines the ECM and honing processes. The main motive of all these hybrid processes is to enhance the finishing rate through enhancing material removal rate and to minimise the achievable final surface roughness value. Most of these processes have been able to attain their intended objective.

Section VI deals with the miscellaneous but important areas related to nanofinishing processes. Some of the chapters, which are the requirements of any nanofinishing technologies for their evaluation and performance enhancements, are as follows: measurement systems for characterisation of micro/nano-level finished surfaces, optimisation of nanofinishing processes for better output, molecular dynamic simulation to understand the science of nanoscale cutting processes and applications of nanofinishing processes in various fields such as biomedical implants.

I would definitely like to thank all the contributors for their efforts in writing their chapters. I will definitely like to put on record my sincere thanks to Taylor & Francis (CRC Press) in general and Cindy Carelli, in particular, who has helped in getting this book processed through different stages.

I will appreciate receiving suggestions from the readers of the book.

Vijay K. Jain
vkjain@iitk.ac.in

Editor

Vijay K. Jain, PhD, earned his BE (Mechanical) degree from Maulana Azad College of Technology Bhopal (under Vikram University Ujjain) in 1970 and ME (Production) and PhD degrees from University of Roorkee (now Indian Institute of Technology Roorkee). He has 43 years of teaching and research experience. He has served as a visiting professor at the University of California at Berkeley (USA) and University of Nebraska at Lincoln (USA). He has superannuated as a Professor in June 2016 (1983–2016) from Indian Institute of Technology Kanpur. He has also served on the faculty at other Indian institutions, namely, MR Engineering College Jaipur, BITS Pilani and MNR Engineering College Allahabad.

Dr. Jain has guided 15 PhD and more than 100 MTech/ME theses of postgraduate students. He has approximately 325 publications to his credit. He has published more than 200 research papers in refereed journals. Dr. Jain has won many medals and best paper awards in recognition of his research work. Dr. Jain has written nine books and many chapters for different books published by international publishers. He has been awarded a Lifetime Achievement Award by the AIMTDR (NAC). Dr. Jain has been appointed as an editor-in-chief of two international journals and associate editor of three other international journals. He has also worked as a guest editor for more than 15 special issues of different international journals. He has been chosen as a member of the editorial board of 12 international journals.

Contributors

R. Balasubramaniam
Bhabha Atomic Research Center
  Bombay
Mumbai, India

Satish Bukkapatnam
Department of Industrial and
  Systems Engineering
Texas A&M University
College Station, Texas

David Lee Butler
School of Mechanical and Aerospace
  Engineering
Nanyang Technological University
Singapore, Singapore

Saeed Zare Chavoshi
Mechanical Engineering Department
Imperial College London
London, United Kingdom

Saurav Goel
Precision Engineering Institute
School of Aerospace, Transport,
  and Manufacturing
Cranfield University
Cranfield, Bedfordshire,
  United Kingdom

Faiz Iqbal
Department of Mechanical
  Engineering
Indian Institute of Technology Delhi
New Delhi, India

Neelesh Kumar Jain
Discipline of Mechanical Engineering
Indian Institute of Technology Indore
Madhya Pradesh, India

Pramod Kumar Jain
Department of Mechanical
  and Industrial Engineering
Indian Institute of Technology
  Roorkee
Roorkee, India

Vijay K. Jain (Ret.)
Department of Mechanical
  Engineering
Indian Institute of Technology
  Kanpur
Kanpur, India

Sunil Jha
Department of Mechanical
  Engineering
Indian Institute of Technology Delhi
New Delhi, India

Jomy Joseph
Mechanical Engineering
  Department
Indian Institute of Technology
  Kharagpur
Kharagpur, India

K.B. Judal
Government Engineering College
  Patan
Gujarat, India

Deepu Kumar
Department of Mechanical
  Engineering
Indian Institute of Technology
  Guwahati
Guwahati, India

Rakesh G. Mote
Department of Mechanical
    Engineering
IIT Bombay
Mumbai, India

Adrian Murphy
School of Mechanical and
    Aerospace Engineering
Queen's University
Belfast, United Kingdom

Divyansh S. Patel
Department of Mechanical
    Engineering
Indian Institute of Technology
    Kanpur
Kanpur, India

Sunil Pathak
Discipline of Mechanical
    Engineering
Indian Institute of Technology
    Indore
Madhya Pradesh, India

Jinu Paul
Mechanical Engineering
    Department
Indian Institute of Technology
    Kharagpur
Kharagpur, India

V. Radhakrishnan
Indian Institute of Space Science
    and Technology
Kerala, India

Dhiraj P. Rai
Department of Mechanical
    Engineering
S. V. National Institute of
    Technology, Surat
Gujarat, India

J. Ramkumar
Department of Mechanical
    Engineering
Indian Institute of Technology
    Kanpur
Kanpur, India

Prabhat Ranjan
Bhabha Atomic Research Center
    Bombay
Mumbai, India

R. Venkata Rao
Department of Mechanical
    Engineering
S. V. National Institute of
    Technology, Surat
Gujarat, India

M. Ravi Sankar
Department of Mechanical
    Engineering
Indian Institute of Technology
    Guwahati
Guwahati, India

G.L. Samuel
Manufacturing Engineering Section
Department of Mechanical
    Engineering
Indian Institute of Technology Madras
Chennai, India

Ajay Sidpara
Mechanical Engineering Department
Indian Institute of Technology
    Kharagpur
Kharagpur, India

Ashif Iquebal Sikandar
Department of Industrial
    and Systems Engineering
Texas A&M University
College Station, Texas

Harpreet Singh
Department of Mechanical and
 Industrial Engineering
Indian Institute of Technology
 Roorkee
Roorkee, India

Sachin Singh
Department of Mechanical
 Engineering
Indian Institute of Technology
 Guwahati
Guwahati, India

V.S. Sooraj
Indian Institute of Space Science
 and Technology
Kerala, India

Arun Srinivasa
Department of Mechanical
 Engineering
Texas A&M University
College Station, Texas

Vinod K. Suri
Bhabha Atomic Research Center
 Bombay
and
MGM Institute of Health Sciences
Mumbai, India

Naveen Thomas
Department of Mechanical
 Engineering
Texas A&M University
College Station, Texas

Li Xiaomin
WinTech Nano-Technology Services
 Pte Ltd
Singapore, Singapore

Vinod Yadava
Mechanical Engineering
 Department
Motilal Nehru National Institute of
 Technology Allahabad
Uttar Pradesh, India

Section I

Introduction

1

Nanofinishing: An Introduction

Vijay K. Jain

Department of Mechanical Engineering, Indian Institute of Technology Kanpur, Kanpur, India

CONTENTS

1.1 Introduction

Finishing is usually the last operation in the manufacturing sequence of a part. It is an important operation to achieve certain surface properties on the part to fulfil its functional performance requirements. This operation is quite expensive and time consuming. In recent years, extensive technological developments have taken place in the field of nanofinishing. Traditional finishing processes (TFPs) have many constraints with reference to the size and shape of the parts that can be finished, the surface integrity of the finished parts and the level to which the surface finish can be achieved. To overcome some of the constraints of the TFPs, advanced finishing processes (AFPs) have been developed to the extent that they are being used on the shop floor of medium and large scale industries. There is a specific need for such processes especially in cases of free-form surfaces which need flexible finishing tool to achieve nanometre level surface finish without any surface and sub-surface defects. Some of the examples where we come across the free-form surfaces are human implants as shown in Figure 1.1.

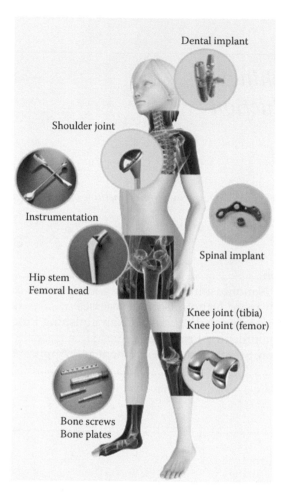

FIGURE 1.1
Human implants that would require high level of finish. (Courtesy of Rosler Metal Finishing USA, LLC.)

The AFPs would be required to finish some of the human implants without any surface and sub-surface defects. The nanofinishing processes can be classified in the following categories:

Classification of nanofinishing processes
Traditional nanofinishing processes
1. Grinding
2. Honing
3. Lapping

Advanced nanofinishing processes

General advanced nanofinishing process

4. Abrasive flow finishing (AFF)
5. Elastic emission finishing (EEF)
6. Elastic abrasive finishing (EAF)
7. Focused ion beam finishing (FIBF)

Magnetic field assisted advanced nanofinishing processes

8. Magnetic abrasive finishing (MAF)
9. Double-disk MAF (DDMAF)
10. Magnetorheological finishing (MRF)
11. Magnetorheological AFF (MRAFF)
12. Nanofinishing of free-form surfaces using ball end magneto-rheological tool (BEMRT)
13. Magnetic float polishing (MFP)

Hybrid advanced nanofinishing processes

14. Electrochemical grinding (ECG)
15. Electrochemical MAF (ECMAF)
16. Electrochemical honing (ECH) of gears
17. Electrolytic in process dressing (ELID) grinding
18. Chemomechanical polishing (CMP)
19. Chemomechanical MRF (CMMRF)
20. Electric discharge diamond grinding (EDDG)

In this chapter, a very brief introduction to the working principles of different nanofinishing processes is given. Detailed scientific analysis and working principles are discussed in the individual chapters in this book. The shape, size and finish requirements of a component play an important role in the selection of a particular type of nanofinishing process. For example, regular shapes such as cuboids, cylinders, spheres, etc. and their combinations can be comparatively easily finished by TFPs as well as AFPs. However, nanofinishing of free-form surfaces is comparatively difficult because it requires continuously varying relative motion between the workpiece and the finishing tool in three or more axes. For example, knee joints, hip joints, turbine blades, etc., have free-form surfaces and they cannot be finished so easily and uniformly by TFPs and some of the AFPs. In view of this requirement, engineers have tried to utilise the capabilities of computer numerical control (CNC) and robots to achieve a relative motion very close to the free-form surface coordinates. Keeping this in view, the nanofinishing processes for free-form surfaces can be classified as shown in Figure 1.2.

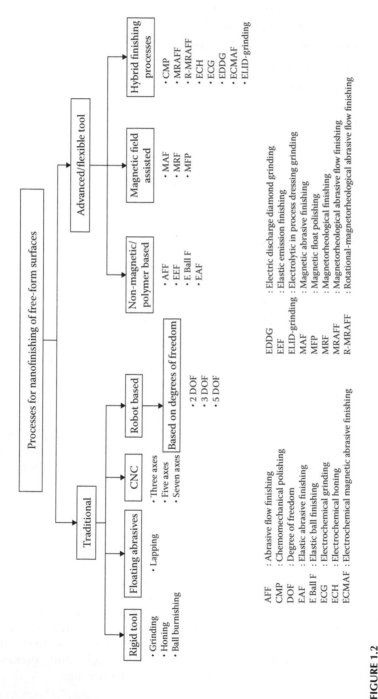

FIGURE 1.2

Classification of nanofinishing processes for free-form surfaces.

1.2 Traditional Nanofinishing Processes

This category consists of TFPs, mainly grinding, honing and lapping, that can produce nanometre level (or nano level) surface finish on different kinds of workpieces with certain constraints such as size, shape, complexity, workpiece material properties, etc.

Grinding: Grinding is a machining process that employs a grinding wheel having abrasive particles at its periphery. The grinding wheel rotates at high speed to remove material from a workpiece comparatively softer than the abrasive particles. Modern grinding machines may have computer-controlled feed drives and slide-way motions, allowing complex shapes to be machined through CNC programs without needing any human intervention during operation. Grinding machines equipped with adaptive controls would allow the machine to run for most of the time under optimum conditions. Some CNC grinding machines may have algorithms to compensate for wheel wear during the process.

Cooling and lubrication play an important role during the grinding process while aiming at a high material removal rate (MRR) and high quality of the ground surfaces. In this direction, many new grinding fluids and methods of delivering grinding fluid have been developed in the past. Minimum quantity lubrication provides an alternative to flood and jet delivery coolant and lubrication, aimed at environment-friendly manufacturing. In practice, four basic grinding processes are in use: (a) surface grinding, (b) peripheral cylindrical grinding, (c) face surface grinding and (d) face cylindrical grinding. However, other categories/classifications also have been mentioned in the literature.

Honing: Honing is an abrasive machining process that produces a precision surface by scrubbing abrasive stone against it along a controlled path on the pre-machined workpiece. It uses a bonded abrasive tool and it is recommended to finish hard materials and hardened surfaces. Honing is normally performed after precision machining such as grinding to achieve the desired surface characteristics. High-precision workpieces are usually first ground and then honed. Grinding determines the size, and honing improves the shape. Honing can give surface finish in the range 130–1250 nm compared to 900–50,000 nm in grinding. Honing is also known as a super finishing process.

There are many types of honing processes, but all of them consist of one or more abrasive stones that are held under pressure against the surface they are working on. During honing, the stone/stick (abrasive tool) should not leave the work surface any time during finishing and it must cover the entire work length. Some of the process input parameters that affect the process performance are revolutions per minute (RPM) of the honing tool, length and position of the honing stroke, honing stick pressure, etc. Thus, honing is an abrasive finishing process that is used to enhance the dimensional and geometrical accuracy of the functional surfaces of engineering parts. In this process, the abrasive particles apply a comparatively low cutting pressure.

Normal machining methods like turning, milling and classical grinding cannot meet these stringent requirements mainly due to process limitations.

Honing is primarily used to improve the geometric form of a surface; however, it also changes the surface texture. Typical applications are finishing of cylinders for internal combustion engines, air bearing spindles and gears. The results of the honing process are tight tolerances, high geometrical accuracy and good surface finish. A 'cross-hatch' pattern is used to retain oil or grease to ensure proper lubrication. It is possible to correct certain types of errors of the cylindrical components, as shown in Figure 1.3.

The honing process has been combined with other advanced machining process, namely, electrochemical machining (ECM), and developed a new hybrid process known as the ECH process, as discussed later in this book.

Lapping: Lapping is a machining process in which two surfaces are rubbed together with abrasive particles between them, by hand movement or usually by using a machine (Figure 1.4). The other form of lapping involves a softer

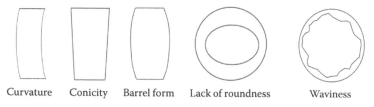

Curvature   Conicity   Barrel form   Lack of roundness   Waviness

FIGURE 1.3
Errors on the workpieces that can be corrected by the honing process.

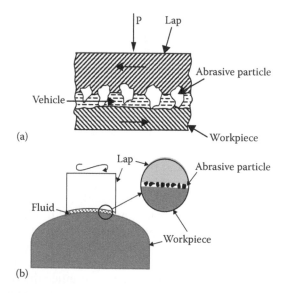

FIGURE 1.4
Lapping using (a) a charged lap and (b) free abrasive particles between the lap and workpiece.

material such as pitch, cast iron or a ceramic for the lap, which is 'charged' with abrasive particles. The lap is then used to cut a hard material – the workpiece. The abrasive particles (say, alumina, silicon carbide, etc.) embed within the soft material (Figure 1.4a), which holds it and permits it to score across and cut the hard material (that is workpiece). The carriers used in lapping operations are grease, olive oil, kerosene with other oil, etc. Lapping can be used to obtain a specific surface roughness and accurate surfaces, usually very flat surfaces. Surface roughness and surface flatness are two quite different concepts.

The mechanism of lapping on brittle material (brittle fracture) is different from the mechanism for metals that are deformed plastically (shear deformation). Lapping process is found be an ideal option to achieve the highest surface quality requirements.

1.3 General Advanced Nanofinishing Processes

General advanced nanofinishing processes can be divided into different types, as shown below. The following types of general advanced nanofinishing processes do not need the assistance of a magnetic field during finishing operation; hence, they have been kept in a category different from the category of magnetic field assisted advanced nanofinishing processes. This category includes the following four different types of nanofinishing processes:

a. AFF
b. EEF
c. EAF
d. FIBF

Abrasive flow finishing (AFF): AFF is a highly versatile finishing process that can finish components irrespective of their complexities, material properties, internal or external surfaces (including concave and convex), free-form surfaces and size. Starting from a simple to a highly complex feature, this process has been able to finish up to a few tens of nanometre centre line average, Ra value. According to the shape and size of the feature to be finished, one should design the medium viscosity, composition and finishing conditions. This process uses viscoelastic medium (polymer + abrasive particles + plasticiser + additives), which is passed through the passage/feature to be finished. The initial viscosity of the medium and the passage restriction for the medium to flow decide what forces will act on the workpiece surface to be finished. Figure 1.5a shows a complex workpiece (knee joint) that can be comparatively more easily finished by an AFF process. Figure 1.5b shows surface

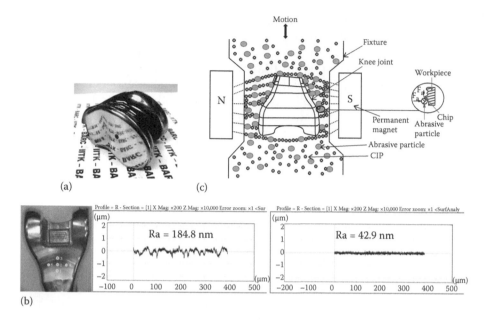

FIGURE 1.5
(a) Finished knee joint having a free-form surface. (From Sidpara, A., Magnetorheological fluid based nanofinishing of flat and freeform surfaces, Dissertation, Indian Institute of Technology Kanpur, India, 2013.) (b) Complex face of a knee joint finished using the AFF process. Initial surface roughness = 184.8 nm; final surface roughness = 42.9 nm; percentage reduction in surface roughness value = 72.86%. (From Sarkar, M., Nanofinishing of freeform surfaces by abrasive flow finishing (AFF) process, MTech Thesis, Indian Institute of Technology Kanpur, India, 2013.) (c) AFF of a knee joint. Enlarged view on the right side showing normal (Fn) and tangential (Ft) forces acting on the workpiece surface. (From Kumar, S., Nanofinishing of freeform surfaces by magnetorheological abrasive flow finishing process, MTech Thesis, IIT Kanpur, August 2013.)

roughness plots of one of the faces of a knee joint before and after finishing. Figure 1.5c shows the normal force (Fn) and axial force (Fa) acting on the workpiece to remove material in the form of micro/nano-chips. The size of the chip is governed by the magnitude of the forces acting on the workpiece. In this nanofinishing process, normal force is responsible for the penetration of an abrasive particle into the workpiece surface and axial/shearing force is responsible for removing the material in the form of micro/nano-chips. In this process, the forces acting on the workpiece cannot be varied once the medium composition and hydraulic pressure (which decides axial force, Fa) are fixed. The viscosity of the medium changes to some extent based on the passage shape and size through which it is extruded. Thus, it can be concluded that on-line control of forces in AFF is not feasible. The details of the different types of setup configurations (one-way AFF, two-way AFF and orbital AFF), force analysis, medium properties and different applications in high-tech industries and biomedical applications are discussed in different chapters. Some researchers have used sintered ferromagnetic abrasive particles in the viscoelastic carrier to control the movement of the ferromagnetic abrasive particles by the magnetic field during the finishing operation. It is claimed that this has improved the finishing rate (FR) of the process.

Figure 1.6a shows an AFF machine developed by CMTI Bangalore in collaboration with IIT Kanpur (India). Figure 1.6b shows a component that has been finished by this process efficiently in spite of its complexity. In this component, a large number of mini holes have been deburred and finished simultaneously.

This process has been employed for nanofinishing of aerospace components, gas turbine components, automobile components, etc. Lately, this

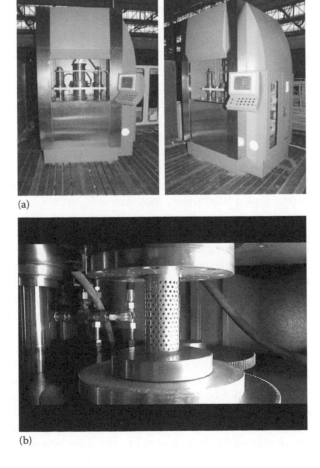

FIGURE 1.6
(a) AFF machine developed by CMTI Bangalore and IIT Kanpur (India). (b) A tube having a large number of mini holes deburred and finished simultaneously. (Courtesy of CMTI Bangalore, India.)

process has been employed for finishing free-form surfaces efficiently, namely, knee joints, in just one fourth time of what was taken by MRF process (Sarkar and Jain, 2015).

Elastic emission finishing (EEF): In this process, material is removed mechanically but atom by atom. Ideally, each nano-level abrasive particle removes material atom by atom from the top surface of the workpiece. This process is normally called *elastic emission machining*; however, in this chapter, it is called EEF because the objective of the process is to improve the surface characteristics, not to create any features on the part as done in case of machining process (Google Scholar: elastic emission machining). This process is truly capable of producing nano or sub-nano levels of surface finish but it is definitely a very slow process. Under certain conditions, this is the only mechanical but comparatively simple process that can provide solutions to some problems where a sub-nano level of surface finish is required.

In this process, abrasive particles are dragged by the flow of liquid over the workpiece surface (not pressed directly by the tool on the workpiece). There is a high probability that the nano-size abrasive particles detach some atoms from the stationary solid material that is the workpiece. If it continues for a long time, atom-by-atom material removal is possible without applying a large mechanical force (Mori et al., 1987). Workpiece material is one solid body and loose abrasive particle is another solid body in EEM process.

Elastic abrasive finishing (EAF): In this process, abrasive particles are embedded in the elastomeric beads in the form of spherical balls of meso/microscale dimensions. When these particles-embedded balls strike the peaks of the rough surface of a workpiece, they are able to remove a very small amount of material by shearing the peaks but at the same time the balls are flexible enough not to penetrate too deep to deteriorate the existing surface finish. These elasto-abrasive balls can also be made with magnetic characteristics by adding iron particles along with abrasive particles. By adding ferromagnetic particles, the flexibility of the balls reduces, but on-line control of force acting on the workpiece becomes feasible. In all finishing processes, the objective is to enhance the surface characteristic (including surface integrity) or improve surface roughness value, or both, in place of having higher material removal rate.

Focused ion beam finishing (FIBF): The last process of this class, FIBF is a slightly different process compared to the other three processes discussed earlier. This process cannot be used for finishing a large surface area, say, even in terms of a few millimetre square, because it is comparatively a very slow process. This process is mainly suitable for creating nano-features or some time micro-features. Here, a stream of ions (ions beam) hits the workpiece surface. Theoretically, it is assumed that the energy carried by an individual ion while hitting the workpiece surface is slightly higher than the bonding energy of the atoms on the top surface of the workpiece. As a result, when an ion hits an atom, the atom is knocked off (or, sputtered off) from

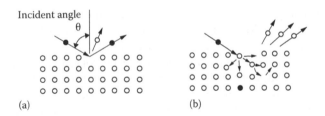

FIGURE 1.7
Focused ion beam machining/finishing. (a) Low and (b) high energy ion beam machining.

the surface of the workpiece and the ion is also knocked off (Figure 1.7a). However, if the energy carried by an ion is less than the bonding energy of the atom, then the ion will be knocked off without removing an atom from the workpiece surface. If the energy carried by an ion is much larger than the bonding energy of the atoms, then in place of one atom, it will knock off many atoms from the workpiece surface at a time, and this ion will be implanted inside the workpiece surface (and form a surface defect) (Figure 1.7b). This process, by its nature of material removal, gives surface roughness value in the range of sub-nanometre or nanometre but on a very small area (normally, in the μm^2 or nm^2). This process is also known as ion beam figuring (Google Scholar: ion beam figuring).

1.4 Magnetic Field Assisted Advanced Nanofinishing Processes

There is an important class of nanofinishing processes in which the magnetic field is used to control the forces acting on the workpiece to remove material from the workpiece in the form of nano-chips so that a nanometre level finish can be achieved on the targeted surface. This class of processes includes the following:

a. MAF

b. DDMAF

c. MRF

d. MRAFF

e. Nanofinishing of free-form surfaces using a ball end magnetorheological tool

f. MFP

Magnetic abrasive finishing (MAF): The previous list is not exhaustive in the sense that there could be other versions of such types of processes that are not listed. These processes can be further classified in two sub-categories: The first category includes first two processes (a and b) in which iron (or ferromagnetic) particles are mixed with the required abrasive particles. This mixture is brought close to the magnet and the workpiece surface to be finished. Due to the magnetic field effect, this mixture forms a flexible magnetic abrasive brush (FMAB), which can change its shape according to the shape of the workpiece, definitely, within certain limits of the size and shape. This brush, when brought in close proximity to the surface to be finished, applies a small normal force that is partly responsible for the indentation of abrasive particles of the FMAB, inside the workpiece surface. Usually, direct normal force is also applied to enhance the FR. When there is a relative motion between the FMAB (particularly, abrasive particles) and the workpiece, the abrasive particles shear off the peaks on the workpiece surface and the surface finish improves (Figure 1.5c). The shearing force is applied by rotating the FMAB (or the magnet) using an electric motor. Further, the MAF process can use either sintered ferromagnetic abrasive particles or a mixture of abrasive and ferromagnetic particles. It forms a series of chain-like structures. However, in many cases, in place of the mixture of ferromagnetic and abrasive particles, researchers and industrial users prefer the use of sintered ferromagnetic abrasive particles, which are claimed to give a higher MRR or higher FR. Using this process, less than 10 nm surface roughness value on the silicon nitride workpiece has been achieved (Komanduri, 1996). In this process and other processes of this category, electromagnets usually are used for producing the magnetic field cloud. However, the electromagnets are heavy and their rotation (or rotation of FMAB) becomes a bit difficult. Hence, some researchers have used permanent magnets to create the magnetic field cloud. Successful experiments have been carried out by different researchers using permanent magnets in place of electromagnets. However, it is achieved at the cost of reduced flexibility in terms of on-line/real-time changing magnetic flux density, which is possible in electromagnets.

Double-disk MAF (DDMAF): This second process of this category is a modified version of the MAF process. The modified version of the MAF process was developed so that the productivity of the process could be enhanced. In this case, the disc type of workpiece is finished on both sides (top and bottom surfaces) at the same time by producing FMAB on both sides by using two independent sets of magnets (or, electromagnets) so that the time taken to finish the workpiece is reduced to approximately half.

Magnetorheological finishing (MRF): The second sub-category of the processes encompasses the remaining four processes (c to f) listed earlier. In these processes, magnetorheological fluid (MR fluid) is used as a finishing medium. This fluid consists of carbonyl iron particles (CIPs) or iron particles as magnetic particles, abrasive particles, carrier fluid and some additives (to equip the fluid with specific properties, say, anti-corrosive nature, etc.). In place of a mixture of iron and abrasive particles, sintered ferromagnetic abrasive particles can also be tried for better performance. Again, when this

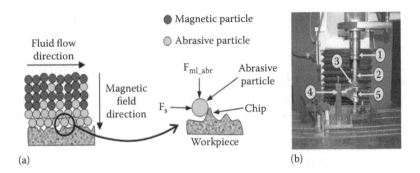

(a) (b)

FIGURE 1.8
(a) Close view of MR polishing fluid. (b) Photograph of MR finishing tool for free-form surfaces (1 – CNC milling machine head, 2 – MR finishing tool, 3 – MR polishing fluid [or FMAB], 4 – fixture for knee joint implant, 5 – knee joint implant). (From Sidpara, A., Magnetorheological fluid based nanofinishing of flat and freeform surfaces, Dissertation, Indian Institute of Technology Kanpur, India, 2013.)

medium is brought under the magnetic field zone, FMAB (Figure 1.8b) is formed whose strength depends on the composition of the medium and finishing parameters including magnetic field strength.

When MR fluid comes under the influence of the magnetic field, its viscosity increases. During the presence of the magnetic field, the CIPs arrange along the magnetic lines of force and the abrasive particles are entangled between or within the chains. When this fluid interacts with the workpiece and they have the relative motion, the normal force on the workpiece through the abrasive particles results in the abrasive penetration and shearing (tangential or axial) force removes the material in the form of micro/nano-chips (Figure 1.8a). These micro/nano-chips get mixed up with the medium (MR fluid) and reduce the machining efficiency if the same fluid is used for a long time. To minimise this effect, $X\%$ of the used MR fluid is replaced by the fresh $X\%$ of MR fluid at a definite interval of finishing time. A similar concept can be applied to the MAF (discussed earlier) process as well. Using this process, it is claimed (Jacobs et al., 2000) that one is able to achieve a surface roughness value of less than 1 nm on the glass surface. The performance of the processes in the second class (processes a to f) also depends upon whether the workpiece material is magnetic or non-magnetic in nature.

Magnetorheological AFF (MRAFF): This process is a modified version of the MRF process that has the limitation that it can mainly finish comparatively softer materials and external surfaces. To overcome these limitations of the process, a process was developed in which the setup of the AFF process was used and the MR fluid medium was modified to finish very hard and complex shaped components including 3D and free-form surfaces (Jha and Jain, 2004). This modified process has been used to finish free-form surfaces (knee joint implants) made of titanium alloy (Kumar et al., 2015). This process can be easily employed for internal as well as external free-form and other complex surfaces made of hard to finish materials by articulating the fixture, the medium constituents and magnetic field (Leeladhar et al., 2016).

Nanofinishing of free-form surfaces using ball end magnetorheological tool: This is again a modified version of the MRF process in which a ball-shaped brush is formed by using a straight magnet (or cylindrical or other shape), either permanent magnet or electromagnet. The brush (or FMAB) is given raster motion with reference to the surface of the workpiece to be finished. This process can be quite useful especially in cases of micro-sized complex features that are to be finished to a certain acceptable level of surface roughness value.

Magnetic float polishing (MFP): All the nanofinishing processes of class 2 except MFP are applicable for finishing flat, cylindrical and other simple and complex shaped workpieces but they cannot efficiently finish spherical shaped workpieces such as ceramic or stainless steel balls, say, for bearings. MFP is the process that can efficiently finish spherical workpieces. In this process, again, MR fluid is used along with a float, bank of magnets and a rod that pushes the fluid downward and rotates on its own axis. Here, the buoyancy force makes the abrasive particles to rub against the ceramic balls. Shear force removes the material in the form of micro/nano-chips. This is an efficient and deterministic process that has not been exploited to its fullest capabilities (Sidpara and Jain, 2014).

1.5 Hybrid Nanofinishing Processes

When two or more types of machining/finishing processes (traditional or/and advanced) are combined together to take advantage of their merits and to minimise the effect of their weaknesses, this new process is called a hybrid process. There are many hybrid nanofinishing processes that have been developed over the years as per industries' requirements. Some of them are as follows:

  a. ECG
  b. ECMAF
  c. ECH
  d. CMP
  e. ELID and grinding
  f. CMMRF
  g. EDDG

Electrochemical grinding (ECG): When a particular nanofinishing process independently is not capable to satisfy the job requirements, then two or more processes (traditional or/and advanced processes) can be combined together to get the desired results. For example, grinding as such gives reasonably good surface finish, but it results in thermal defects such as micro/nano-cracks, thermal residual stresses, etc. Some researchers combined the grinding process with ECM. Here, ECM helps in removing a small layer of material so that thermal defects, if any, due to grinding no longer remain in the workpiece. This process is known as ECG, which gives better quality of the finished component and higher FR as well as higher MRR. However, this also has a problem of corrosion of the finished component. But its potential has not been appropriately exploited by the shop floor engineers as well as researchers working in this field. It is believed that this process has a lot of potential for economic and highly efficient nanofinishing of difficult-to-finish materials. In the ECG process, material is simultaneously removed by the grinding process as well as by an anodic dissolution process (ECM). Here, the ECG wheel life increases as much as 10 times the life in traditional grinding, and MRR is higher than the individual MRR of grinding as well as ECM. Further, FR is also higher than in grinding process. Figure 1.9 shows the mechanism of material removal by grindings as well as by electrochemical (EC) dissolution (or ECM) simultaneously. Figure 1.9 shows an electrochemical surface grinding with three zones through which the material removal is taking place. In zone 1 and zone 3 (or stage 1 and stage 3), where the abrasive particles are not in contact with the workpiece, the material removal is taking place due to electrochemical dissolution only. In zone 2, where the abrasive particles are in contact with the workpiece surface, they are forming electrolytic cells between the bonding material (metal) of the grinding wheel and workpiece. The workpiece is the forming anode and the grinding wheel is the forming cathode. The material is removed due to electrolytic dissolution.

Since the abrasive particles are interacting with the workpiece material, the material is being removed by grinding process also. This is shown in Figure 1.9b under 'Grinding + EC dissolution'. Since the material removal by grinding phenomenon in ECG is very low, grinding wheel life during ECG is very high (say, up to 10 times).

Electrochemical magnetic abrasive finishing (ECMAF): ECM is normally used for shaping and sizing a component made of electrically conducting material but the level of surface finish achieved is not very high. Hence, many times after ECM, a finishing operation is performed on a separate setup. This makes the manufacturing process less efficient and comparatively more expensive and time consuming. However, if an ECM operation is continuously followed by a nanofinishing operation (in the present case, MAF operation), then in one setup and in one operation, one can get the machined as well as the finished component. This later combined operation is a hybrid operation (ECM + MAF), which is called the ECMAF operation (Judal and Yadav, 2013). This hybrid operation results in higher productivity. Thus, the material is simultaneously removed by both the mechanisms of electrochemical anodic dissolution and mechanical shearing of the workpiece material in the

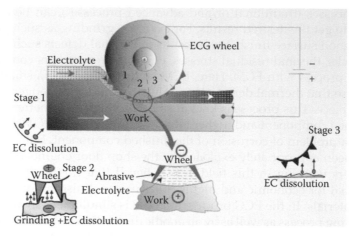

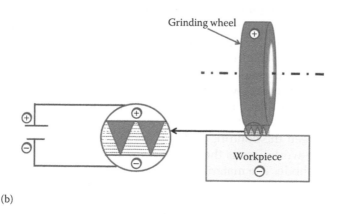

(b)

FIGURE 1.9
Mechanism of material removal during ECG. (a) Front view of surface grinding clearly show-ing three different zones and (b) side view of surface grinding showing mainly zone 2.

form of micro/nano-chips by FMAB. It definitely improves productivity by increasing both MRR and FR and by performing both operations (machin-ing and finishing) at the same time. ECMAF is very similar to ECG except that in ECMAF, the finishing operation is performed by MAF (or by a FMAB) process, while in ECG, it is done by a grinding operation (or a fixed or a non-flexible tool).

Electrochemical honing (ECH): Gears are extensively used in different mechanical machines, including automobiles, aerospace, heavy industries, machine tools, etc. The finishing of these gears takes a good amount of total production time of gears. For finishing gears, grinding, honing and lapping are used. However, all these conventional processes have their limitations in terms of productivity and cost. AFF is another advanced nanofinishing process that is also used in many industries because it can do three functions (finishing, deburring and radiusing), sometimes at the same time. But in this process, substantial material removal cannot take place to adjust the dimensional inaccuracies. There is another process that has been advantageously explored for finishing gears and for adjusting minor dimensional inaccuracies. This process is known as ECH of gears. In this process, ECM and mechanical honing are combined together, hence the name ECH. This hybrid process simultaneously overcomes the individual process limitations and makes the process more versatile by amalgamating the capabilities of both processes (ECM + honing). This process being practically non-thermal in nature, it is capable of a producing nano-level finished surface without any thermal defects.

Chemomechanical polishing (CMP): This process is also known as chemomechanical planarisation. This process uses a medium known as slurry that consists of carrier fluid + abrasive particles + additives. In this process, the chemical reaction between the slurry and the workpiece forms a layer comparatively much softer than the original workpiece material hardness. As a result, this layer of reaction products is easily removed by the comparatively softer abrasive particles in the slurry. This is the most commonly used process for finishing silicon wafers in microelectronics and other related industries.

Electrolytic in process dressing and grinding (ELID grinding): This hybrid process uses a hard-metal-bonded grinding wheel having embedded abrasive particles in it. In this process, two operations are simultaneously performed: grinding of a workpiece along with dressing of the grinding wheel. For the grinding operation, the workpiece can be electrically conductive or nonconductive. In this process, an electrode is used that covers a part of the grinding wheel (say, 15% to 25% of the wheel outer periphery area). An electrolytic cell is formed between the electrode and the grinding wheel. In this electrolytic cell, the grinding wheel works as an anode and the electrode works as a cathode. As a result of electrolysis, an oxide layer is formed on the grinding wheel, which partially covers the projected abrasive particles. This oxide layer, to some extent, reduces grinding efficiency. Also, due to the oxide layer formation, the bonding of the abrasive particles with the grinding wheel becomes weak, and in due course, the abrasive particles get detached from the grinding wheel. It permits the fresh/new particles to interact with the workpiece material and the grinding efficiency starts going up. Thus, both operations, the grinding and dressing of the grinding wheel, keep going on simultaneously. This process is gaining popularity on the shop floor in different types of industries.

Chemomechanical magnetorheological finishing (CMMRF): This process combines the essential features of the CMP and MRF processes. In CMP, the forces cannot be controlled on-line or externally, and in MRF, there is no chemical reaction that makes the parent material softer to increase both MRR and FR.

In this process, chemical reactions (used for creating of passivated superficial layer) associated with CMP are used to improve MRR and FR, as well as the final surface finish, whereas MR fluid and magnets of the MRF setup are used to control the magnitude of the abrading forces acting on the workpiece for nano-abrasion as well as to control the flexibility of the FMAB for finishing non-planer surfaces. The CMMRF process is capable of finishing a wide variety of materials, say, ductile or brittle, electrically conductive or electrically non-conductive up to a few nanometres and in some cases, even up to sub-nanometre range. This process was used to finish a single crystal silicon blank to the final Ra value as 0.468 nm (Jain et al., 2010). Figure 1.10 shows a single crystal silicon wafer mirror finished by this process (Ranjan, 2009).

Electric discharge diamond grinding (EDDG): Apart from the earlier stated problems of grinding, it cannot be used for finishing very hard materials because it imposes the constraint on the hardness of the abrasive particles used in the grinding wheel. To solve this problem, researchers have blended grinding with EDM and named it as electric discharge abrasive grinding (EDAG), or EDDG depending upon whether normal abrasive particles are used or diamond particles are used as abrasive particles in the grinding wheel. In this process, an electric discharge/spark removes a very small amount of material, but it softens the hard-to-machine material within and surrounding the spark area. Thus, the abrasive/diamond particles following the spark are able to easily finish comparatively softer surface (became softer due to heating). Thus, a spark serves two purposes simultaneously: In EDAG, the EDM is able to machine hard-to-machine material and make it softer, and comparatively, softer abrasive particles are able to finish the converted soft layer of workpiece material. However, the EDM is a thermal process and grinding is a mechanical process producing lot of heat. A probability of the presence of thermal defects in the finished part still persists to some extent if the parameters are not optimised to minimising the thermal defects.

Mirror image of IITK BARC DST
on the finished 'Si' substrate

FIGURE 1.10
Single crystal silicon wafer finished by the CMMRF process. (From Ranjan, P., Development of Nanofinishing Technology for Si Substrate. MTech Thesis, IIT Kanpur, India, 2009.)

1.6 Optimisation, Simulation, Measurement and Applications

Optimisation: It is a well-known fact that the performance of any machining process is controlled by its input process parameters. The optimum value of the input process parameters depends on the objective function in case of single objective optimisation and objective functions and weights to different objective functions in case of multi-objective optimisation. Researchers have recognised this fact, and many techniques have been developed for single-objective and multi-objective optimisation (Jain, 2014), for example, genetic algorithm, teaching-learning-based optimisation, goal programming, geometric programming, multivariable regression analysis, artificial neural networks, etc. For achieving the best results out of any machining and finishing process, parameter optimisation is highly essential. To study the effects of input process parameters of a finishing processes on its performance measures (say, surface roughness, MRR, cutting forces, etc.), various optimisation techniques have been proposed by different researchers working in the field of optimisation (Rao and Kalyankar, 2014). With the advancements in machining and finishing machine tools, machine tools are usually equipped with the 'adaptive controls' which help in running the machine tool for most of the times under the optimum machining/finishing conditions without any intervention of the operator. The machine tool controlled input parameters are automatically on-line modified periodically depending upon the newly calculated optimum input parameters. However, the cost of such CNC machines with adaptive control is definitely much higher than that of a normal CNC machines.

Simulation: Almost all theories available in the literature related to machining and finishing deal with material deformation or cutting in the bulk material depending on the mechanical and physical properties of those materials. But it is a well-known fact that as you go to the material removal at a smaller and smaller size, the material properties beyond a particular size keep changing. The properties become completely different when the material is machined/finished at nanometre, in general, and sub-nanometre, in particular. There are certain machining processes that practically remove material atom by atom or a small group of atoms (say, focused ion beam machining and elastic emission machining). The modelling and then simulation of the sub-nano-level machining process at this level of material removal (atom by atom) are not available in the existing metal cutting theories. About a couple of decades back, some researchers (Chandrasekaran et al., 1997) proposed the molecular dynamics (MD) simulation of a nano-cutting process. This process is able to capture very fine details of nano-machining that are not possible otherwise. In fact, MD simulation is virtual computer simulation of nano-cutting that is not possible to analyse theoretically as well as experimentally. The same technique has been applied for nanofinishing processes as well. The results are interesting. Different researchers have discussed nano-cutting of copper, polishing of silicon wafers at the atomic scale, etc.

Measurement systems: Metrology has very special place in any manufacturing scenario. It will not be exaggeration if I say, 'you cannot manufacture a part if you cannot measure it; does not matter whether it is dimension, geometry, or surface integrity'. Measurement system's design becomes more important when one talks about micromanufacturing/nanomanufacturing. The measurement systems become more complex when one wants to handle 3D and free-form surfaces in general and at the micro level in particular. Hence, advanced-level research in micro- and nano-metrology becomes more important because it will help in understanding the performance of micro/nanostructured surfaces.

There are two classes of measurement/characterisation techniques for microfeatures and nanofeatures. They are contact and non-contact measuring techniques. The real challenge arises when feature size comes down to nanoscale as the contact of measuring instrument will damage the feature. Characterisation of free-form surfaces is still more difficult because of variation of x, y and z coordinates from point to point on the surface. To resolve the problem, various techniques have been proposed by different researchers.

Applications: Nanofinishing techniques have applications in various fields of engineering, biomedical sciences and others. Some human implants (Figure 1.1) require nanometre-level surface finish. In some cases, a differential surface finish is required in different areas of the same part, depending upon the functional requirements of the part. One of the examples requiring a differential surface finish is a knee joint implant. Different devices in many other areas require a nanolevel surface finish, for example, aerospace, automobile, optics, etc.

1.7 Remarks

In view of the brief discussion here about the nanofinishing processes, modelling, simulation, applications and metrology, it can be concluded that this is an important area in manufacturing that should be studied at both the levels of technology and science of these processes, including optimisation of the process parameters. Further, it is not the optimisation of the process parameters alone; rather, their machine tool design optimisation is equally important to extract the best from the existing technologies. With this in view, this book emphasises the technology as well as scientific basis of the processes that are discussed herein. However, the design optimisation of their machine tools is out of the scope of this book; hence, they have not been touched upon.

Bibliography

Ahn, Y. and Park, S.-S. (1997). Surface roughness and material removal rate of lapping process on ceramics. *KSME International Journal* 11(5): 494–504.

Chandrasekaran, N., Noori-Khajavi, A., Raff, L.M., and Komanduri, R. (1997). A new method for MD simulation of nanometric cutting. *Philosophical Magazine* 77(1): 7–26.

Choudhary, S.K., Jain, V.K., and Gupta, M. (1999). Electrical discharge diamond grinding of high-speed steel. *Machining Science and Technology* 3(1): 91–105.

Das, M., Jain, V.K., and Ghoshdastidar, P.S. (2010). Nano-finishing of stainless-steel tubes using rotational magnetorheological abrasive flow finishing process. *Machining Science and Technology* 14(3): 365–389.

EI-Taweel, T.A. (2008). Modelling and analysis of hybrid electrochemical turning – Magnetic abrasive finishing of 6061 Al/Al_2O_3 composite. *International Journal of Advanced Manufacturing Technology* 37: 705–714.

Fox, M., Agrawal, K., Shinmura,T., and Komanduri, R. (1994). Magnetic abrasive finishing of rollers. *CIRP Annals – Manufacturing Technology* 43: 181–184.

Gorana, V.K., Jain, V.K., and Lal, G.K. (2004). Experimental investigation into cutting forces and active grain density during abrasive flow machining. *International Journal of Machine Tools and Manufacture* 44: 201–211.

Huang, J., Zhang, J.Q., and Liu, J.N. (2005). Effect of magnetic field on properties of MR fluids. *International Journal of Modern Physics B* 19(01.3): 597–601.

Jacobs, S.D., Arrasmith, S.A., Kozhinova, I.A., Gregg, L.L., Shorey, A.B., Ramanofsky, H.J., Golini, D., Kordonski, W.I., and Hogan, S. (2000). An overview of magneto-rheological finishing (MRF) for precision optics manufacturing (invited contribution). *Ceramic Transactions* 102: 185–200.

Google Scholar: elastic emission machining. https://scholar.google.co.in/scholar?q =elastic+emission+machining&btnG=&hl=en&as_sdt=0%2C5.

Google Scholar: ion beam figuring. https://scholar.google.co.in/scholar?hl=en&q =ion+beam+figuring&btnG=.

Jain, V.K. (2009a). *Introduction to Micromachining*, Narosa Publishers, India.

Jain, V.K. (2009b). Magnetic field assisted abrasive based micro-/nano-finishing. *Journal of Materials Processing Technology* 209: 6022–6038.

Jain, V.K. (2014). *Introduction to Micromachining* (Edited), (Second Edition), Narosa Publishing House, New Delhi.

Jain, V.K. and Jha, S. (2005). Nano-finishing techniques. In: *Micromanufacturing and Nano-Technology*, Editor: N.P. Mahalik, Springer Verlag, Berlin Heidelberg: 171–195.

Jain, V.K., Kumar, P., Behera, P.K., and Jayswal, S.C. (2001). Effect of working gap and circumferential speed on the performance of magnetic abrasive finishing process. *Wear* 250: 384–390.

Jain, N.K., Naik, L.R., Dubey, A.K., and Shan, H.S. (2009). State-of-art-review of electrochemical honing of internal cylinders and gears. *Proceedings of the Institution of Mechanical Engineers, Part B: Journal of Engineering Manufacture* 223(6): 665–681.

Jain, V.K., Ranjan, P., Suri, V.K., and Komanduri, R. (2010). Chemo-mechanical magneto-rheological finishing (CMMRF) of silicon for microelectronics applications. *CIRP Annals – Manufacturing Technology* 59: 323–328.

Jain, V.K., Sidpara, A., Sankar, M.R., and Das, M. (2012). Nano-finishing techniques: A review. *Proceedings of the Institution of Mechanical Engineers, Part C: Journal of Mechanical Engineering Science* 226(2): 327–346.

Jha, S. and Jain, V.K. (2004). Design and development of magnetorheological abrasive flow finishing (MRAFF) process. *International Journal of Machine Tools and Manufacture* 44(10): 1019–1029.

Judal, K.B. and Yadava, V. (2013). Electrochemical magnetic abrasive machining of aisi304 stainless steel tubes. *International Journal of Precision Engineering and Manufacturing* 14: 137–143.

Jung, B., Jang, K.I., Min, B.K., Lee, S.J., and Seok, J. (2009). Magnetorheological finishing process for hard materials using sintered iron–CNT compound abrasives. *International Journal of Machine Tools and Manufacture* 49(5): 407–418.

Kim, J.D., Xu, Y.M., and Kang, Y.H. (1998). Study on the characteristics of magneto-electrolytic abrasive polishing by using the newly developed nonwoven-abrasive pads. *International Journal of Machine Tools and Manufacture* 38: 1038–1043.

Komanduri, R. (1996). On material removal mechanism in finishing of advanced ceramics and glasses. *Annals of CIRP* 45(1): 509–514.

Komanduri, R., Lucca, D.A., and Tani, Y. (1997). Technological advances in fine abrasive processes. *CIRP Annals* 46(2): 545–597.

Koshy, P., Jain, V.K., and Lal, G.K. (1996). Mechanism of material removal in electrical discharge diamond grinding. *International Journal of Machine Tools and Manufacture* 36(10): 1173–1185.

Kozak, J. and Kazimierz, E.O. (2001). Selected problems of abrasive hybrid machining. *Journal of Materials Processing Technology* 109: 360–366.

Kumar, S. (2013). Nanofinishing of freeform surfaces by magnetorheological abrasive flow finishing process, MTech Thesis, IIT Kanpur.

Kumar, S., Jain, V.K., and Sidparac, A. (2015). Nanofinishing of freeform surfaces (knee joint implant) by rotational-magnetorheological abrasive flow finishing (R-MRAFF) process. *Precision Engineering* 42. doi: 10.1016/j.precisioneng.2015.04.014.

Mali, H.S. and Manna, A. (2009). Current status and application of abrasive flow finishing processes: A review. *Proceedings of the Institution of Mechanical Engineers, Part B: Journal of Engineering Manufacture* 223: 809–820.

Marinescu, I.D., Hitchiner, M.P., Uhlmann, E., Rowe, W.B., and Inasaki, I. (2016). *Handbook of Machining with Grinding Wheels.* CRC Press, Taylor & Francis Group, Boca Raton, FL.

Miao, C., Shafrir, S.N., Lambropoulos, J.C., Mici, J., and Jacobs, S.D. (2009). Shear stress in magnetorheological finishing for glasses. *Applied Optics* 48(13): 2585–2594.

Mohan, R. and Ramesh Babu, N. (2012). Ultrafine finishing of metallic surfaces with the ice bonded abrasive polishing processes. *Materials and Manufacturing Processes* 27: 412–419.

Mori, Y., Yamauchi, K., and Endo, K. (1987). Elastic emission machining. *Precision Engineering* 9(3): 123–128.

Mulik, R.S. and Pandey, P.M. (2010). Mechanism of surface finishing in ultrasonic-assisted magnetic abrasive finishing process. *Materials and Manufacturing Processes* 25: 1418–1427.

Nagdeve, L., Jain, V.K., and Ramkumar, J. (2016). Experimental Investigations into nano-finishing of freeform surfaces using negative replica of the knee joint. Proceedings of the 18th CIRP Conference on Electro Physical and Chemical Machining (ISEM XVIII).

Nagel. Honing technology. www.nagel.com.

Pathak, S., Jain, N.K., and Palani, I.A. (2015). On surface quality and wear resistance of straight bevel gears by pulsed electrochemical honing process. *International Journal of Electrochemical Sciences* 10(11): 1–18.

Ranjan, P. (2009). Development of nanofinishing technology for Si substrate. MTech Thesis, IIT Kanpur, India.

Rao, R.V. and Kalyankar, V.D. (2014). Optimization of modern machining processes using advanced optimization techniques: A review. *International Journal of Advanced Manufacturing Technology* 73: 1159–1188.

Rhoades, L. (1991). Abrasive flow machining: A case study. *Journal of Materials Processing Technology* 28: 107–116.

Sankar, M.R., Ramkumar, J., and Jain, V.K. (2009). Experimental investigation and mechanism of material removal in nano finishing of MMCs using abrasive flow finishing (AFF) process. *Wear* 266: 688–698.

Sarkar, M. (2013). Nanofinishing of freeform surfaces by abrasive flow finishing (AFF) process. MTech Thesis, Indian Institute of Technology Kanpur, India.

Sarkar, M. and Jain, V.K. (2015). Nanofinishing of freeform surfaces using abrasive flow finishing process. *Proceedings of the Institution of Mechanical Engineers, Part B: Journal of Engineering Manufacture* 1–15. doi: 10.1177/0954405415599913.

Seok, J., Lee, S.O., Jang, K.I., Min, B.K., and Lee, S.J. (2009). Tribological properties of a magnetorheological (MR) fluid in a finishing process. *Tribology Transactions* 52(4): 460–469.

Shinmura, T. and Aizawa, T. (1989). Study on internal finishing of non-ferromagnetic tubing by magnetic abrasive machining process. *Bulletin Japan Society of Precision Engineering* 23(1): 37–41.

Shinmura, T., Takazawa, K., Hatano, E., Matsunaga, M., and Matsuo, T. (1990). Study on magnetic abrasive finishing. *CIRP Annals–Manufacturing Technology* 39: 325–328.

Sidpara, A. (2013). Magnetorheological fluid based nanofinishing of flat and freeform surfaces, Ph.D. Thesis, Indian Institute of Technology Kanpur, India.

Sidpara, A. and Jain, V.K. (2012a). Magnetorheological and allied finishing processes. In: *Micromanufacturing Processes*, Editor: V.K. Jain, CRC Press (Taylor & Francis), Boca Raton, FL: 133–153.

Sidpara, A.M. and Jain, V.K. (2012b). Nanofinishing of freeform surfaces of prosthetic knee joint implant. *Proceedings of the Institution of Mechanical Engineers, Part B: Journal of Engineering Manufacture* 226(11): 1833–1846.

Sidpara, A. and Jain, V.K. (2012c). Nano-level finishing of single crystal silicon blank using magnetorheological finishing process. *Tribology International* 47: 159–166.

Sidpara, A. and Jain, V.K. (2014). Magnetic float polishing: An advanced finishing process for ceramic balls. In *Introduction to Micromachining*, Editor: V.K. Jain, Narosa Publishing House, New Delhi: 10.1–10.4.

Sidpara, A., Das, M., and Jain, V.K. (2009). Rheological characterization of magnetorheological finishing fluid. *Materials and Manufacturing Processes* 24(12): 1467–1478.

Singh, D.K., Jain, V.K., and Raghuram, V. (2006). Experimental investigations into forces acting during a magnetic abrasive finishing process. *The International Journal of Advanced Manufacturing Technology* 30: 652–662.

Singh, A.K., Jha, S., and Pandey, P.M. (2011). Design and development of nanofinishing process for 3D surfaces using ball end MR finishing tool. *International Journal of Machine Tools & Manufacture* 51: 142–151.

Singh, A.K., Jha, S., and Pandey, P.M. (2015). Performance analysis of ball end magne-
    torheological finishing process with MR polishing fluid. *Journal of Materials and
    Manufacturing Processes* 30(12): 1482–1489.
Smolkin, M.R. and Smolkin, R.D. (2006). Calculation and analysis of the magnetic
    force acting on a particle in the magnetic field of separator. Analysis of the equa-
    tion used in the magnetic methods of separation. *IEEE Transactions on Magnetics*
    42: 3682–3693.
Sooraj, V.S. and Radhakrishnan, V. (2014a). A study on fine finishing of hard work
    piece surfaces using fluidized elastic abrasives. *International Journal of Advanced
    Manufacturing Technology* 73(9): 1495–1509.
Sooraj, V.S. and Radhakrishnan, V. (2014b). Fine finishing of internal surfaces using
    elastic abrasives. *International Journal of Machine Tools and Manufacture* 78: 30–40.
Wang, A.C. and Lee, S.J. (2009). Study the characteristics of magnetic finishing with
    gel abrasive. *International Journal of Machine Tools and Manufacture* 49: 1063–1069.
Wei, G., Wang, Z., and Chen, C. (1987). Field controlled electrochemical honing of
    gears. *Precision Engineering* 9: 218–221.
Wikipedia. Honing (metal working). https://en.wikipedia.org/wiki/Honing
    _(metalworking).
Wikipedia. Lapping. https://en.wikipedia.org/wiki/Lapping.
Yadava, V., Jain, V.K., and Dixit, P.M. (2002). Temperature distribution during electro-
    discharge abrasive grinding. *Machining Science and Technology – An International
    Journal* 6(1): 97–127.
Yamaguchi, H. and Shinmura, T. (1999). Study of the surface modification resulting
    from an internal magnetic abrasive finishing process. *Wear* 225: 246–255.
Yamaguchi, H. and Shinmura, T. (2004). Internal finishing process for alumina ceramic
    components by a magnetic field assisted finishing process. *Precision Engineering*
    28: 135–142.
Yi, J., Zheng, J., and Yang, T. (2002). Solving the control problem for electrochemical
    gear tooth-profile modification using an artificial neural network. *International
    Journal of Advanced Manufacturing Technology* 19: 8–13.
Zeng, S., Blunt, L., and Racasan, R. (2014). An investigation of the viability of bon-
    net polishing as a possible method to manufacture hip prostheses with multi-
    radius femoral heads. *International Journal of Advanced Manufacturing Technology*
    70: 583–590.

Section II

Traditional Nanofinishing Processes

2

![line]

Honing

Pramod Kumar Jain and Harpreet Singh

Department of Mechanical and Industrial Engineering, Indian Institute of Technology Roorkee, Roorkee, India

CONTENTS

2.1 Introduction

Honing is an abrasive finishing process, most often used to improve the dimensional and geometrical accuracy of the functional surfaces of engineering parts, and it is characterised by (a) a large area of abrasive contact, (b) a low cutting pressure, (c) a low speed-sizing and (d) a floating part or tool. The surface finish so achievable influences functional characteristics like wear resistance, fatigue strength, corrosion resistance and power loss due to friction. Hence, modern machineries are required to be assembled with high dimensional and geometrical accuracy along with high surface finish parts. Normal machining methods like turning, milling and classical grinding cannot meet these stringent requirements mainly due to process limitations. Being an abrasive machining process, honing uses a bonded abrasive tool and is recommended to finish hard materials and hardened surfaces. Honing is applied post precision machining such as grinding to achieve the surface characteristics necessary. The surface finish achievable in honing may be in the range of 0.13–1.25 μm compared to 0.9–5 μm in grinding (Mahajan and Tajane, 2013). It is desired that the honing stone/stick, which acts as an abrasive tool, should not leave the work surface at any time during machining and the stroke length must cover the entire work length. Rotational speed, oscillation speed, length and position of stroke, honing stick pressure, etc., are several process parameters that are important and affect the honing process.

According to the literature, experiments with the honing process started around 1920 with the aim of achieving a very high finish along with a faster material removal rate (MRR) for internal cylinders. Initially, the cylinder finishing was carried out by rough and finish-boring with wide tools, followed by reaming and grinding. Later, cast iron and copper laps were used. These methods are very slow and also could not meet the high finish demanded. This led to the design of honing machines. Initially, drilling machines were converted to single-spindle honing machines and the first honing head was fabricated in 1923. Subsequently, continuous improvements in the honing machines established honing as a mass production super-finishing process, owing to the intelligent machine layout and other combinations with defined cutting geometries. As such, it is considered a technologically and economically serious production process (Flores et al., 2014; Flores, 2015). Honing is widely used to finish bores of cylinders of internal combustion engines, hydraulic cylinders, gas barrels, bearing races, valve and valve seat, gears, components for aerospace applications, etc.

2.2 Description of Process

Honing is a metal removal process with bonded abrasives and is applied after precision machining (such as grinding). The honing process is recommended

TABLE 2.1

Types of Honing Processes

Internal Honing	External Honing
Longitudinal stroke honing	Centreless plunge honing
Through feed honing	Short stroke honing
Profile honing	Surface honing
Gear honing	Gear honing

for the finishing of both internal and external features of engineering parts (Table 2.1). During the honing process, honing stones are pressured radially on the work surface. Although honing uses an abrasive tool (hone), it is different from grinding in two major ways: (a) honing is a slow process and the hone moves at a low speed relative to the work surface (i.e. 0.2 m/s to 2 m/s) and (b) the honing tool can align flexibly to the work surface. As bonded abrasives are used in the form of a hone, the honing process involves two-body abrasion in an abrasive tool and the work surface is considered to be in pure sliding motion during the process. Moreover, honing improves the surface quality of the machined surface as follows:

- It improves geometrical accuracy by correcting out-of-roundness, waviness, bell mouth, barrel, taper, rainbow, reamer chatter, etc.
- It improves dimensional accuracy.
- It also improves surface character by improving roughness, lay pattern and integrity.

Usually, when the honing stroke is in the longitudinal direction, it is termed *internal honing* mainly to finish internal cylindrical surfaces, such as automobile cylindrical liners, brake drums, guide holes, brake cylinder, etc. It is also known as conventional honing or simply honing. The primary aim of honing in such applications is to remove micro-geometrical errors originating from the previous machining processes, removing any damage on the work surface due to cutting forces and temperature during previous machining, reducing surface roughness and formation of micro-scratch system beneficial to lubrication. Various types of geometrical errors originating from machining processes that can be corrected by internal honing are shown in Figure 2.1.

During internal honing, a controlled light pressure is applied to the honing stones to press against the work surface. The honing head floats inside the hole and is guided by the work surface without any external guiding mechanism. As shown in Figure 2.2, there is simultaneous rotation and reciprocation motion given to the bonded abrasive honing stick, resulting in the generation of a cross-hatched lay pattern on the finished work surface due to the abrasive nature of the honing tool. The generated cross-hatched

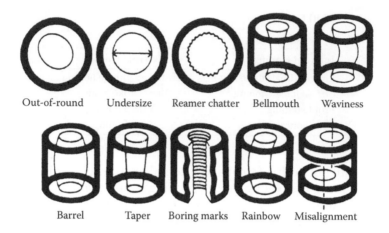

Out-of-round   Undersize   Reamer chatter   Bellmouth   Waviness

Barrel   Taper   Boring marks   Rainbow   Misalignment

FIGURE 2.1
Geometrical inaccuracies in cylindrical work piece.

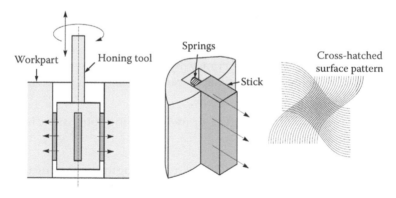

FIGURE 2.2
Schematic of honing process and cross hatch surface pattern.

lay pattern helps in improving the tribological properties of the work surface by way of improving its lubrication retention capacity during operation of the component, thus contributing to its improved function and service life.

For honing holes/small bores (diameter 2–12 mm) mostly single-stone tools equipped with two guide stones and one honing stone are used. The guide stones guide the tool inside the bore and the honing stone does the finishing through material abrasion on the work surface.

The axial feed given to the feeding cone is converted into a radial feed to the honing stone during finishing, as shown in Figure 2.3. This feeding system is highly important for the honing process as it can be controlled either by feed or by force. Bores of any size, long or short, blind, tandem, with keyways, etc., can be honed. The material can be ferrous or non-ferrous, hard or in soft condition.

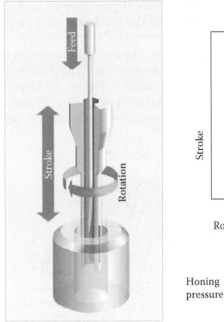 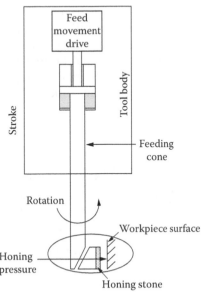

FIGURE 2.3
Schematic of an internal honing process with a single stone tool.

External Honing

Honing of external surfaces is performed to achieve the improved surface quality of the machined surface. It has generally been applied to remove the grinding feed marks on the functional surfaces such as crank shafts, gears, rotor shafts, complex rotating parts, cam shafts, piston pins, piston rods and piston valves for pneumatic and hydraulic drives, axles, pins, roller bearing rollers, tapered rollers, spherical rollers, bearing rings, bearing axles, rocker arm shafts, etc.

Most of the work on external honing is reported on the finishing of the gear tooth surface. Honing of gears is a hard-finishing method to eliminate the gear profile errors. This process was originally adopted and developed to remove nicks and burrs that are often unavoidably encountered during gear machining. It has also been seen that the honing process can correct minor tooth irregularities and improve tooth surface finish. The honing of gears facilitates ease of operation, reduced noise generation, higher input speed, increased surface durability and load bearing ability, improved gear efficiency and smoother performance in a power transmission. Honing also contributes to the reduced friction in the power train. Gear honing is performed at very low cutting speeds (0.5 and 10 m/s) in comparison to gear grinding and hence results in low thermal stress on the machined surface, thereby preventing burning of the ground surface.

In gear honing, an abrasive-impregnated helical-gear-shaped tool is used as a hone. This tool is run in a tight mesh with the hardened workpiece gear in a cross-axes relationship under low and controlled centre–distance pressure (Mehta and Rathi, 2013). Normally, the workpiece gear is driven by the honing tool in both directions during the honing cycle. This process is carried out with conventional honing oil as a coolant. Figure 2.4 depicts the gear honing and tooth flank topography of the workpiece gear. The material removal is a result of the kinematically imposed relative motion (sliding)

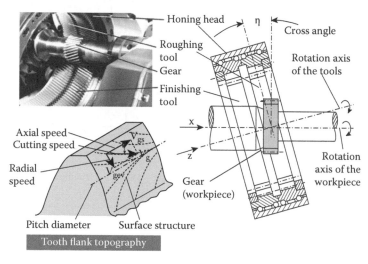

FIGURE 2.4
Gear honing and tooth flank topography.

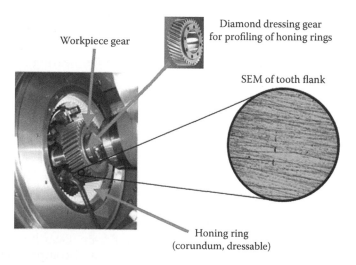

FIGURE 2.5
Layout pattern on tooth flank face.

between abrasive grits of the honing tool and the workpiece. The total sliding speed V_g is the resultant of the relative motion in direction of the involute V_{gev} through the rolling action between the gears and the relative motion in direction of the helix V_{gs} due to different helix angles of the tool and workpiece. The cutting angle α describes the geometrical deviation of the total sliding speed V_g from the direction of the helix. The relative movement is responsible for the typical pattern of scratches on the workpiece tooth flank topography and the resulting surface pattern with the corundum honing gear, as described in Figure 2.5.

2.2.1 Metal Removal Process

2.2.1.1 Kinematic Principle

The kinematic principle of the honing process is explained by considering the external flat surfaces and internal cylindrical surfaces. On a flat surface (Figure 2.6), it can be observed that there are three speed components due to the relative motion between the work surface and tool mating surface (i.e. axial feed rate v_{fa}, tangential feed rate v_{ft} and a perpendicular feed rate v_{fn}).

Cutting speed (V_c) can be calculated accordingly as

$$V_c = \sqrt{V_{fa}^2 + V_{ft}^2 + V_{fn}^2} \, . \tag{2.1}$$

As the MRR is low, v_{fn} is very small in comparison to v_{fa} and v_{ft}. Hence, cutting speed (V_c) can be calculated as

$$V_c = \sqrt{V_{fa}^2 + V_{ft}^2} \, . \tag{2.2}$$

Speed and direction depend on the kinematics of the tool drive. The tangential feed rate v_{ft} is generally constant. The axial feed rate v_{fa} is an

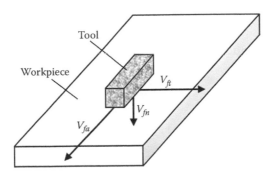

FIGURE 2.6
Speed components in short-stroke honing.

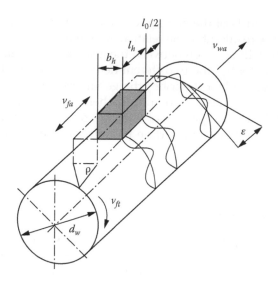

FIGURE 2.7
Movement in centreless short-stroke plunge and through feed honing.

oscillating movement describing a sine wave. Proceeding from the path function (Figure 2.7), we obtain the following formula:

$$y = \frac{l_0}{2}\sin(\omega t)$$

(2.3)

with y being the path of oscillation and l_0 being the oscillation amplitude. Following this, we obtain

$$V_{fa} = \dot{y} = \frac{l_0}{2}\omega\cos(\omega t).$$

(2.4)

Alternately, introducing the stroke frequency $f = \dfrac{\omega}{2\pi}$ yields

$$V_{fa} = l_0\pi f\cos(\omega t).$$

(2.5)

According to Equations 2.2 and 2.5, taking account of the superposed axial workpiece speed v_{wa}, the speed V_c is calculated as

$$v_c = \sqrt{v_{ft}^2 + (l_0\,\pi\,f\,\cos(\omega t) + v_{wa})^2}$$

(2.6)

and

$$v_c = \sqrt{v_{ft}^2 + (l_{0\max} \pi f + v_{wa})^2}.$$ (2.7)

The maximum possible value of speed can be

$$v_c = v_{ft}.$$ (2.8)

The cross-hatch angle α is calculated as

$$\tan \alpha = \frac{V_{fa} + V_{wa}}{V_{ft}}.$$ (2.9)

In the case of longitudinal stroke honing, the relative movement between the tool and workpiece can be due to the speed components like axial feed rate v_{fa}, tangential feed rate v_{ft} and perpendicular feed rate (feed rate v_{fn}). The axial speed remains constant over the entire stroke length except at the turning points. The cutting speed V_c is calculated as

$$\overrightarrow{V_c} = \overrightarrow{V_{fa}} + \overrightarrow{V_{ft}} \text{ or } V_c = \sqrt{V_{fa}^2 + V_{ft}^2}.$$ (2.10)

Usually, the cross-hatch angle or included angle is required in the range of 20° to 60°. A constant cross-hatch angle is obtained in case of constant axial and tangential speed, which yields smaller values at the ends of the workpiece, i.e. in the turning position of the tool. The highest MRR can be obtained at a cross-hatch angle between 40° and 60° with grit size 220G (Bai et al., 2007). Figure 2.8 shows the kinematics involved and the resultant surface pattern due to axial and rotational movement of the honing stone. The oscillatory movements of the honing stone are graphically presented in Figure 2.9a without stroke delay. Sometimes, a stroke delay can also be included for reasons of the shape accuracy in one or both turning positions as shown in Figure 2.9b.

2.2.1.2 Honing Stone Pressure

As discussed, a constant pressure is exerted by the honing stone on the work surface, which can be calculated as follows:

$$P_n = \frac{F_n}{A_h} = \frac{F_n}{b_h . l_h}.$$ (2.11)

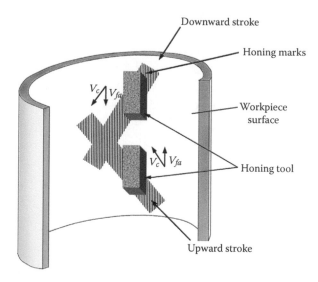

FIGURE 2.8
Kinematics of longitudinal stroke honing.

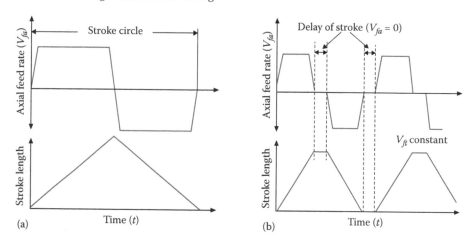

FIGURE 2.9
Path and speed profile during longitudinal stroke honing: (a) without delay and (b) with delay.

The principle of depth of cut is described in Figure 2.10a and b, by considering the in-line contact honing pressure between the honing tool and workpiece under two conditions, namely, (a) force bound (force-fit) and (b) track bound (form-fit) depth of cut (Klocke, 2009).

2.2.1.3 Honing Control Strategy

For better control of the honing process and its results, the interface between the honing tool and work surface and the relative movement

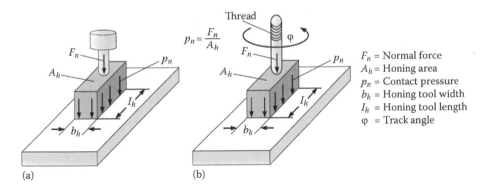

FIGURE 2.10
The principle of depth of cut: (a) force bound and (b) track bound.

between the two are worth considering. Mainly, two control strategies are employed during the honing process to give required feed motion to the honing tool, viz., feed controlled and force controlled. Force-controlled honing tries to maintain a constant pressure between the work surface and honing stick during the process, whereas feed-controlled honing tries to maintain the constant cycle time between honing steps (Bahre et al., 2012). Feed-controlled honing (Figure 2.3) is the traditional form of honing, which is an open-loop control. Thereby, the radial feed movement of the honing stone is performed by an axial movement of the feeding cone. Due to the overlaying movements, the honed surface shows a characteristic cross-hatch pattern. The force-controlled honing gives better results in terms of geometric and form accuracy, surface structure and tool wear. In this strategy, the forces generated during the process are measured and kept constant through a regulation of the feed movement (closed-loop control). Being the last finishing step, honed parts bear high costs, and hence, the requirements of process stability and repeatability for honing are very high (Schmitt and Bahre, 2013), which is high in case of force-controlled honing. The positions of the honing stone and process force during feed-controlled and force-controlled honing are graphically illustrated in Figure 2.11a and b, respectively.

2.2.2 Honing Methods

Honing methods can be divided into three categories:

- Conventional honing
- Single-pass honing
- Plateau honing

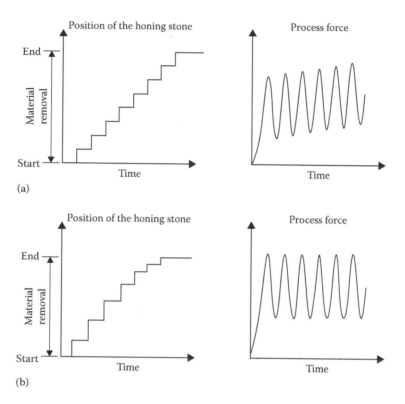

FIGURE 2.11
The position of honing stone and process force: (a) feed-controlled honing and (b) force-controlled honing.

2.2.2.1 Conventional Honing

In conventional honing, the tool rotates with an abrasive honing stick pressed against a bore surface to finish it. Generally, the bore is finished after repeated strokes. In this process, the work surface is abraded back and forth by the tool. This process is capable of finishing the bore to close diameter value and within tight tolerances. Accurate size control is achieved with an automatic size control mechanism. Process repeatability in terms of size control is very high and, in the case of bore diameter, ranges from 0.00254 mm to 0.00508 mm. In automatic machines, the operator only loads and unloads the part and fixture; everything else is automatic. Conventional honing offers several advantages: tightly controlled finished diameter size, cylindricity, and surface finish.

2.2.2.2 Single-Pass Honing

In this process, the bore is finished in a single-pass. The honing tool is in the form of a sleeve and is mounted on an expandable tapered arbor. The sleeve and tapered arbor mechanism help in pushing the honing tool against the work surface. Due to single-pass finishing, this process is a faster and more accurate method of honing a bore to its final size. The bore is finished in the forward stroke of the tool and backward stroke is used only to withdraw the tool out of the bore. Single-stroke honing offers better size control and repeatability and honed bores do not require gauging. This process offers higher production rates due to its single-pass nature.

2.2.2.3 Plateau Honing

Plateau honing is a two-step process. In the first step, rough honing with big size abrasive particles is carried out. In the second step, finish honing is done with very small size abrasive particles to remove only the peaks partially. The partial removal of peaks leads to improved tribological properties of the work surface and results in reduced wear of work surface during its useful life. The surface topographies of the honed surfaces with the above methods are shown in Figure 2.12.

2.2.3 Tooling

Honing machine tools can be of horizontal or vertical spindle types.

2.2.3.1 Horizontal Spindle Honing Machines

These machines are used for handheld work with bore sizes up to 6 in. and the machine rotates the hone in the range of 30–76 m/min. Work is moved

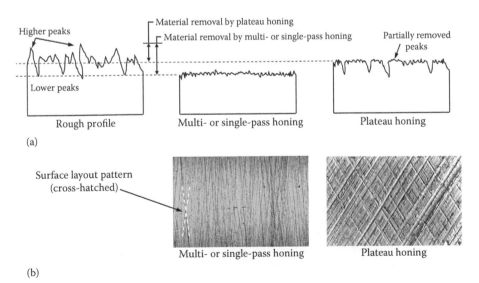

FIGURE 2.12
A comparison of surface topography: (a) roughness profiles and (b) micro-structure images.

back and forth over the rotating hone. During finishing, work is never pressured against the hone, but it is floated over the hone. Any hard pressing of the work will lead to ovality of the finished bore. Sometimes, the workpiece is rotated as well. If the machine is equipped with power stroking, then the work is held in a self-aligning fixture and the speed and length of the stroke are regulated by controls on the machine (Williams, 1928). Honing is expanded by hydraulic or mechanical means until the desired hole diameter is achieved. The rate of expansion (radial feed) can be controlled using various mechanical and electrical devices. These machines are among the most widely used ones.

2.2.3.2 Vertical Spindle Machines

These machines are recommended especially for larger and heavier work and are equipped with power stroking at a speed in the range of 6–35 m/min. The length of the stroke is controlled by the operator by setting the mechanical stops at appropriate positions. These machines are also equipped (in many cases) with multiple spindles to machine several hones in one go (as in automobile cylinders). The honing tool is made in several designs using a single stone for small holes and two to eight stones for larger holes. The stones are available in a variety of sizes and shapes. It is recommended to use a cutting fluid during honing with the objectives of removing tiny chips from the cutting zone, removing excessive heat generated during machining and lubricating of the hone–work interface. Cutting fluid is usually recycled through a fine mesh filtering system.

2.2.3.3 Honing Stone

The honing stones/sticks are made of hard materials (such as aluminium oxide, silicon carbide and diamond). The bonding material may be vitrified clay, resinoid, cork, carbon or metal. The bond should be of optimum strength to hold the grit properly and it should not rub the surface. The number and the distribution of stones depend on the extent to which the roundness error is to be corrected. The honing tool geometry influences the surface finish of the workpiece. Some frequently used honing tools are shown in Figure 2.13.

2.2.3.3.1 Length of Honing Stone

The length of the honing stone influences the shape correction required in the bore. Longer stone lengths correct the shape errors like cylindricity more effectively than do shorter lengths. Shorter-length tools are used for enlarging bores and also have an effect on enlarging the tilt moment in an upper reversal position. Figure 2.14 depicts the schematic views of tool paths of varying tool lengths and their effects on bore geometry.

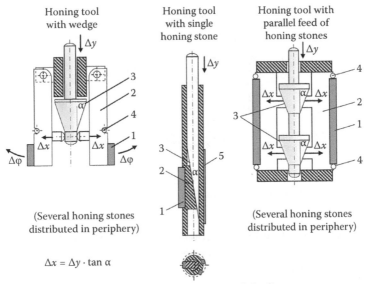

Honing tool
with wedge

Honing tool
with single
honing stone

Honing tool with
parallel feed of
honing stones

(Several honing stones
distributed in periphery)

(Several honing stones
distributed in periphery)

$$\Delta x = \Delta y \cdot \tan \alpha$$

1. Honing stone, 2. Honing stone carrier, 3. Feeding cone,
4. Return spring, 5. Guide rail

FIGURE 2.13
Types of honing tools.

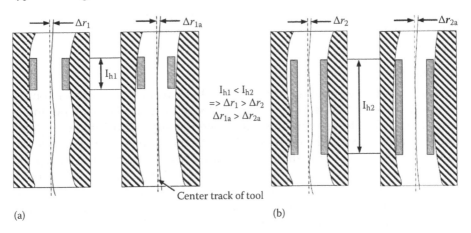

$$I_{h1} < I_{h2}$$
$$\Rightarrow \Delta r_1 > \Delta r_2$$
$$\Delta r_{1a} > \Delta r_{2a}$$

Center track of tool

(a)   (b)

FIGURE 2.14
Effect of honing stone length on shape irregularities: (a) tool follows shape irregularity and
(b) tool eliminates shape irregularity.

2.2.3.3.2 Width of Honing Stone

Honing stones of greater width can correct the out-of-roundness more effectively than narrower stones can. Also, increasing width decreases the contact
pressure between tool and work surface. Hence, the tool width should be
changed by considering its effect on contact pressure accordingly. The process

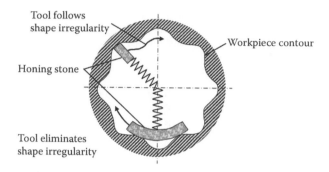

FIGURE 2.15
Roundness correction by wider honing stone.

of error correction with wider/narrower stones is shown in Figure 2.15. It has also been reported that narrower stones have more undesirable vibrations as they remove material.

2.2.3.4 Specifications of Honing Stone

The honing stones are specified by particle type, particle size, bond type, hardness and treatment, and accordingly, honing stones are selected for specific applications. The marking system of honing stones is similar to that of grinding wheels and specifies the type of abrasive, grit size, hardness and type of bond.

2.2.3.4.1 Particle Size and Abrasive Materials

Normally, grit sizes ranging from 36 to 600 are available for use in honing stones. The recommended range of the grit size to be used during honing is from 120 to 320. However, if one is interested in super-finishing, even finer than 600 grits also may be used in honing. As the particle size increases, the MRR also increases, but with poorer surface finish. However, increasing the particle size has a negative effect on the cylindricity of the finished bore. Summaries of the effects of particle size on surface roughness are shown in Tables 2.2 and 2.3 for various combinations of abrasive material and work material. The number of contact particles increases with increasing abrasive concentration, nominal contact area and static load. The number of contact particles decreases with an increase in abrasive particle size and yield strength of workpiece material. The maximum particle depth of cut increases with increasing abrasive particle size and static load (Feng et al., 2015).

Finer particles tool and higher honing speed are beneficial for achieving higher MRRs with a moderate increase in surface roughness, which sometimes may be implemented to increase the productivity of the process (Vrac et al., 2014). Sometimes, with a decrease in the hardness of abrasive material (only up to certain extent), the MRR increases provided that sharp particles

TABLE 2.2

Effect of Corundum Honing Stone Particle Size on Surface Roughness

Particle Size (Mesh)	Surface Roughness, R_t (μm)							
	Steel				Cast Iron			
	50 HRC		62 HRC		180 HB		250 HB	
	Vitrified	Bakelite	Vitrified	Bakelite	Vitrified	Bakelite	Vitrified	Bakelite
80	8–12	8–10	5–7	4–5	10	–	6–8	–
120	7–9	6–8	4–6	3–4	7–9	–	4–6	–
150	5–7	4–6	3–5	2–3	5–7	–	3–5	2–3
220	3–5	2–4	2–4	1.5–2.5	4–6	2–4	2–4	1–2
400	2–4	1–2	2–3	1–2	3–4	1–3	1–3	0.5–1.5
700	1–3	0.5–1	1–2	0.2–1	–	0.5–1	–	0.5–1
1000	0.5–1	0.2–0.5	0.2–1	–	–	0.5	–	0.3–0.8

Source: Klocke, F., *Manufacturing Processes 2 – Grinding, Honing, Lapping*, Springer-Verlag, Berlin Heidelberg, 2009.

TABLE 2.3

Effect of Diamond Honing Stone Particle Size on Surface Roughness

	Surface Roughness, R_t (µm)			
	Steel		Cast Iron	
Particle Size	50 HRC	62 HRC	180 HB	250 HB
D7	0.8	0.3	0.8	0.6
D15	1.8	0.6	1.8	1.2
D20	2.0	0.8	2.0	1.8
D30	2.5	1.2	2.5	2.0
D40	3.0	1.5	3.5	2.5
D50	3.5	2.0	4.0	3.5
D60	4.0	2.5	4.5	4.0
D70	4.5	3.0	5.5	4.5
D80	5.5	3.5	6.0	5.5
D100	6.0	4.0	6.5	6.0
D120	6.5	4.5	7.0	6.5
D150	7.0	5.0	8.0	7.0
D180	8.0	5.5	9.0	8.0
D200	9.0	6.0	10.0	9.0

Source: Klocke, F., *Manufacturing Processes 2 – Grinding, Honing, Lapping,* Springer-Verlag, Berlin Heidelberg, 2009.

are in continual engagement. Too less hardness of the abrasive material decreases the MRR due to low bond strength of particle engagement. A higher hardness of abrasive may lead to blunting of particles, sometimes, if bonding material is not selected properly.

The use of diamond or cubic boron nitride (CBN) abrasive materials helps to improve the process performance of the honing process. Also, the processing of the work surface can be completed in just one complete stroke. Microcrystalline CBN grit has enhanced the capability further. It can maintain sharp cutting conditions with consistent results over long durations. Super-abrasive honing sticks with a monolayer configuration, where a layer of CBN grits is attached to stick by a galvanically deposited metal layer, are typically found in single-stroke honing applications. Figure 2.16 depicts the super-abrasive honing tool. The most significant variables for MRRs are particle size and pressure, followed by tangential speed and density.

Nowadays, for super-finishing, different types of bonded diamond stones are used, namely,

a. Metallic bonded diamond (Dia-M): This stick contains 50% of diamond particles with a particle size of 91 µm embedded on a bronze matrix.

b. Vitrified bonded diamond (Dia-V): This is a new type of composite consisting of microsized diamonds of 107 µm and green silicon

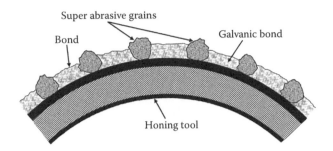

FIGURE 2.16
Super abrasive honing stick with single layer configuration.

carbide in the size of 60 μm which arebonded with vitreous C7 glass material. The composition is 64.6% SiO_2, 4.1% $Al_2O_3 \cdot Fe_2O_3$, 13.4% CaO, 3.3% MgO, 9.6% $Na_2O \cdot K_2O$, 4.7% B_2O_3 and 0.9% BaO. The symbol C was originally chosen for chemical resistance.

c. Resinoid bonded diamond (Dia-R): This stone consists of diamond particles of 91 μm bonded on a resinoid bond. Abrasive diamond particles were coated with nickel to offer good adhesion for high diamond retention in resin bond.

The performance and functionality of the work surface are affected mainly by three standard criteria, as follows:

i. Cr: Running-in criterion is applicable when the bearing ratio varies between 33% and 1%. It informs about the top portion of the surface that will be worn out during the run-in period.

ii. Cf: Operating criterion is applicable when the bearing ratio varies between 75% and 15%. It corresponds to the long-term running of the surface and the surface behaves as a stable surface, which ultimately influences the performance and life of the bearing surface.

iii. Cl: Lubrication criterion is applicable when the bearing ratio varies between 99% and 45%. It concerns the lubricant retention capacity of the work surface (i.e. at the lowest part of the surface and in the form of valleys).

A comparison of workpiece cylidricity between Dia-V, Dia-M and Dia-R stones is shown in Figure 2.17. The surface quality differentiation with different bonded stones is shown in Figure 2.18a–c.

The selection of a suitable abrasive material depends on the hardness and composition of the work material, the finish desired and cost (such as, for steel, aluminium oxide is used; for cast iron and non-ferrous material, silicon carbide is used). Super-abrasive honing stones (such as diamond) have a very high wear-resistant property than conventional stones do (made of

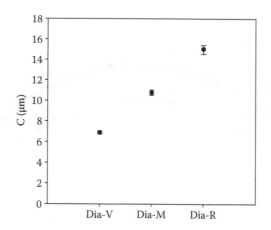

FIGURE 2.17
Comparison of cylindricity between Dia-V, Dia-M and Dia-R stones. (From Sabri, L., Mexghani, S., Mansori, M.E.I., *Surf. Coat. Technol.*, 205, 1515–1519, 2010.)

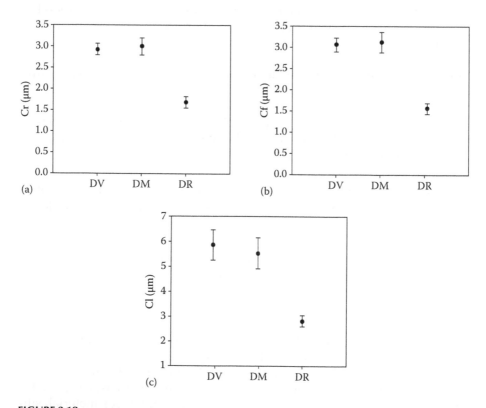

FIGURE 2.18
Surface roughness performances of the Dia-V, Dia-M and Dia-R: (a) running-in criteria (Cr), (b) functional criteria (Cf) and (c) lubrication criteria (Cl).

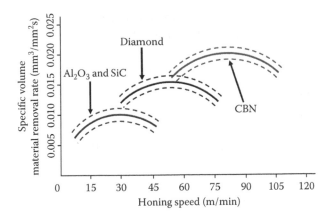

FIGURE 2.19
MRRs with different honing stones.

corundum or silicon carbide) and mainly hard and wear-resistant materials are machined with these stones. However, these stones are expensive also. The form (such as cylindricity) and dimensional accuracy achievable with these super abrasive stones are much higher than the conventional honing stones. Figure 2.19 compares different types of particle materials over MRR (Cabanettes et al., 2015).

2.2.3.5 Work Material Properties

The work material properties (mainly hardness) influence the surface quality. Harder materials can be machined to a better quality than ductile materials because tiny chips produced during machining of ductile materials clog the honing stone during operation and hence hampers good surface quality. For steel having hardness 50–65 RC, surface finishes in the range of 0.01 μm can be obtained with 500G Al_2O_3 stones.

2.2.3.6 Honing Oil

During the honing process, normally, the heat generation is less as compared to the grinding operation, due to a small contact area in honing. Therefore, the honing oil is primarily used abundantly with a provision for flushing, not for cooling. The viscosity of the oil should be low to encourage high MRR and productivity of the process.

The honing grooves, the volume and the direction of the valleys control the amount of oil retained and its uniform spread on the bore surface. Sometimes, during honing with a diamond tool, metal chips are found on the work surface, resulting in an interruption in oil flow in the grooves. This causes abrasive wear as well as axial scratches on the cylinder surface. The conventional abrasives are better than diamond in improving the spreading of the oil, self-sharpening properties and higher friability (Hurpekli et al., 2014).

2.3 Process Parameters

As with most of the machining processes, in honing also, process parameters are categorised into fixed and variable process parameters as shown in Table 2.4.

2.3.1 Surface Integrity

The surface integrity of the work surface is the most important aspect of the honing process. It can remove the top of the peaks with the intention of providing a good bearing surface. The work surface topography can be designed to give the best performance using the different variants of honing process. In this section, the effects of honing process parameters on the surface finish and shape deviation of the workpiece surface are described in detail.

2.3.1.1 Effect of Contact Pressure

The abrasive particles put more pressure on the high spots and the bore is straightened after the crests are removed. Uniform surface finish is obtained because a large number of particles is in contact. With an increase in tool contact pressure, the surface finish worsens as it leads to more particle imprints on the surface. Also, an increase in contact pressure leads to increased breaking away of the particles from the bond and, thus, greater tool wear and more material removal and higher surface roughness as shown in Figure 2.20. The fallen out particles have unpredictable trajectories over the workpiece surface, making irregular creases, and when these are pronounced, their impact on inconsistent lubricant flow is higher. The unit pressure should be selected so as to get minimum surface roughness with the highest possible MRR. The

TABLE 2.4

Process Parameters of Honing Processes

Fixed Process Parameters	Variable Process Parameters
Machine tool performance	Axial movement of both tool and workpiece
Workpiece property	Rotation of the workpiece or tool for cylindrical surfaces
Honing tool design	Length and position of the stroke
Cooling lubricant	Contact pressure between the workpiece and tool
	The pressure and the quantity of the cooling media

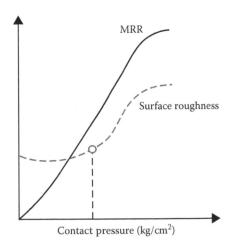

FIGURE 2.20
Effect of contact pressure.

range of pressure applied is around 1.0 to 3.2 MPa (Rao, 2013). When a coarser particle tool is used, the honing speed has a higher impact on maximum roughness than specific honing pressure does. Specific honing pressure is more significant than honing speed when a finer particles tool is applied (Vrac et al., 2014). The effect of contact pressure on cylindricity and roundness is also depicted in Figure 2.21. The cylindricity worsens with increasing

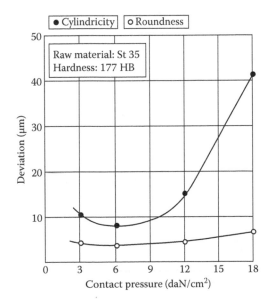

FIGURE 2.21
Effect of contact pressure on surface roughness and MRR on cylindricity and roundness.

contact pressure. The roundness is not much affected by the increase in contact pressure.

2.3.1.2 Effect of Peripheral Honing Speed

Any increase in the peripheral honing speed increases the MRR and also the surface quality, as depicted in Figure 2.22. Also, micro irregularities created from the previous operations are smoothed out due to the increased MRR with higher honing speed. Honing speed is more influential on achievable maximum peak height than feed and depth of cut are when a coarser particle tool is used. For finer particle tools, the maximum peak height depends mostly on feed, honing speed and depth of cut respectively in order of impact. Hard-to-machine materials are normally finished at lower cutting speeds. It has also been an observed that at higher cutting speeds, surface finish improves with a slight deterioration in dimensional accuracy due to overheating and dulling of the abrasives.

2.3.1.3 Effect of Stroke Length

Stroke length primarily affects bore roundness and cylindricity, as shown in Figure 2.23. An appropriate selection of stroke length will produce a straight bore. An excessive stroke length may result in an enlarged diameter at the opening. Shorter stroke length will result in internal ovality.

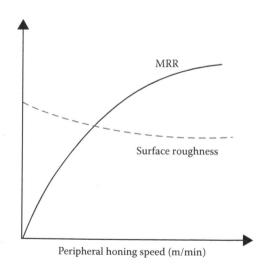

FIGURE 2.22
Effect of peripheral honing speed.

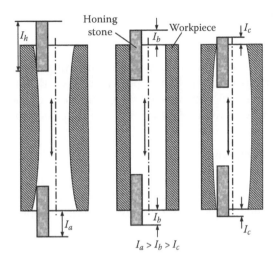

FIGURE 2.23
Shape errors due to variations in stroke length and over-run length.

2.3.1.4 Effect of Machining Time

Initially, the higher peaks are removed very quickly, resulting in higher MRRs and fast reduction in surface roughness. Then after getting a relatively smoother surface, the roughness reduction rate decreases and almost no further improvement is observed with increasing time. It is shown in Figure 2.24 that after an optimum machining time, surface roughness starts increasing with continuously decreasing MRR. The honing time is thus estimated on the basis of desired surface roughness and the type of geometrical errors to be removed during honing.

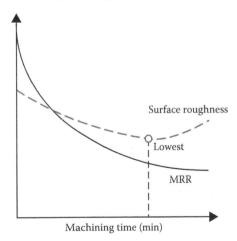

FIGURE 2.24
Effect of machining time during honing.

2.4 Applications

Honing has many applications in precision finishing of various machined components. Various machined parts honed with vertical honing machines are shown in Figure 2.25. Several specific applications of the honing process are discussed herewith with reference to various applications.

2.4.1 Cylinder Block Honing

Bores sometime require a preliminary rough honing operation to remove stock, followed by finish honing to get the desired surface finish. The cross-hatch pattern thus obtained helps in improved oil retention property (Akbari, 2002).

2.4.2 Position Honing

This is a tooling- or machine-driven system developed in Europe to replace the rough honing and finish honing with one process called position honing. It can remove as much as 1 mm of material in 25–30 sec, reducing abrasive cost by 30%, reducing tool rework costs and eliminating one spindle step and one inspection step in processing. Also, better quality and higher consistency are observed. This process is used mainly in small-engine applications. There are several variants of this process, such as the following:

a. *Life hone system:* This speeds up the MRR by moving the workpiece (such as bushes) up and down rather than the hone spindle. Due to small and lightweight parts, it can be performed at a higher speed

FIGURE 2.25
A series of parts hones on a vertical honing machine.

FIGURE 2.26
Cross-hatched finish in cylinder bore with laser structuring.

(i.e. 7000–8000 rpm). A high processing speed results in better straightness and roundness of the bore.

b. *Thermal spray honing:* This process is used for small, lightweight aluminium engines and for remanufacturing worn-out engine cylinders. This process replaces cast-iron cylinder sleeves with a 100–150 µm layer of low-carbon steel sprayed onto the cylinder inner surface, which is then finish honed.

c. *Laser structuring:* This uses a laser incorporated in a conventional honing spindle to burn microscopic pockets inside the engine cylinder. It is a technology that allows engine builders to further improve engine oil effectiveness. It creates a kind of enhanced cross-hatching and also reduces friction in the cylinder wall (Figure 2.26). This structure cuts emissions and enhances gas mileage by over 6% when the entire cylinder surface is treated. Application of the laser-honing process requires three steps. In the first step, rough honing is performed to macro-hone the bore surface. In the second step, precisely defined lubricant reservoirs (cross-hatch pattern) are produced with the laser. In the third step, finish honing is done to get an extremely fine surface finish. This process enhances engine life by reducing wear on the cylinder surface and on the piston rings.

2.5 Solved Examples

1. Calculate the honing pressure if a honing tool has six stones, and the stone measures are 30 mm × 200 mm and piston diameter: 150 mm (tg $^{\alpha/2}$ constant is 0.414).

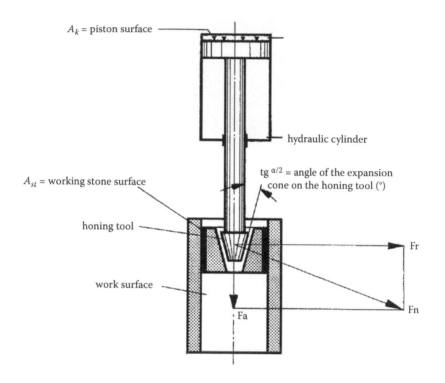

Total honing stone area, $A_{st} = 3$ cm * 20 cm * 6 = 360 cm²

Piston area, $A_k = 15$ cm² * 0.785 = 176.625 cm²

$$P = \frac{q * tg^{\alpha/2} * A_{st}}{A_k}$$

$$= \frac{20 * 0.414 * 360}{176.625}$$

$$= 16.8764 \text{ atü} = 165.43819 \text{ N/cm}^2$$

2. Calculate the honing cutting speed and honing stone pressure if the quantity of honing stones is 6, dimensions are 4 × 4 × 25 mm, load is 6 N, axial feed rate is 85 m/min and tangential feed rate is 50 m/min.

   Honing cutting speed:

$$V_c = \sqrt{V_{fa}^2 + V_{ft}^2}$$

$$V_c = \sqrt{85^2 + 50^2}$$

$$V_c = 98.60 \text{ m/min.}$$

Honing stone pressure:

$$P_n = \frac{F_n}{A_h}$$

$$P_n = \frac{F_n}{b_h . l_h}$$

$$P_n = \frac{6}{4 * 25}$$

$$P_n = 0.06 \text{ N/mm}^2 = 60 \text{ kPa.}$$

References

Akbari, J. (2002). An experimental study on bore honing operation, *Initiatives of Precision Engineering at the Beginning of a Millennium*, pp. 426–430. Available at http://link.springer.com/chapter/10.1007%2F0-306-47000-4_83. Accessed 29 April 2015.

Bahre, D., Schmitt, C., Moos, U. (2012). Analysis of the differences between force control and feed control strategies during the honing of bores, *Procedia CIRP*, Vol. 1, pp. 377–381.

Bai, Y.J., Zhang, L.H., Ren, C.G. (2007). Experimental investigation on honing of small holes, *Key Engineering Materials*, Vol. 329, pp. 303–308.

Cabanettes, F., Dimkovski, Z., Rosén, B.G. (2015). Roughness variations in cylinder liners induced by honing tools' wear. Available at http://www.sciencedirect.com/science/article/pii/S0141635915000057. Accessed 8 May 2015.

Feng, Q., Ren, C., Pei, Z. (2015). A physics-based predictive model for number of contact grains and grain depth of cut in honing, *Machining Science and Technology*, Vol. 19, pp. 50–70, doi:10.1080/10910344.2014.991024.

Flores, G.K. (2015). Graded freeform machining of cylinder bores using form honing. Available at http://papers.sae.org/2015-01-1725. Accessed 8 April 2015.

Flores, G., Wiens, A., Stammen, O. (2014). Honing of gears, gear technology. Available at www.thielenhaus.com/.../Honing_of_Gears_-_Gear_Technology. Accessed 15 June 2015.

Honing, Lapping, Superfinishing and Burnishing, Production Technology, HMT Banglore, Tata McGraw-Hill Publishing Company Limited, New Delhi, 1980.

Hurpekli, M., Yilmaz, R., Kondakci, E., Solak, N. (2014). Effects of Ceramic and Diamond Honing on Bore/Liner Surface in View of Oil Retention, SAE Technical Paper 2014-01-1660, 2014, doi:10.4271/2014-01-1660.

Klocke, F. (2009). *Manufacturing Processes 2 – Grinding, Honing, Lapping*, Springer-Verlag, Berlin, Heidelberg. doi:10.1007/978-3-540-92259-9.

Mahajan, D., Tajane, R. (2013). A review on ball burnishing processes, *International Journal of Scientific and Research Publications*, Vol. 3, Issue 4, pp. 1–8.

Mehta, D.T., Rathi, M.G. (2013). A review on internal gear honing, *International Journal of Engineering Research & Technology*, Vol. 2, Issue 5.

Rao, P.N. (2013). *Manufacturing Technology, Metal Cutting and Machine Tools*, Vol. 2, 3rd edition, pp. 236–238. Tata McGraw-Hill Publishing Company Limited, New Delhi.

Sabri, L., Mexghani, S., Mansori, M.E.I. (2010). A study on the influence of bond material on honing engine cylinder bores with coated diamond stones, *Surface and Coatings Technology*, Vol. 205, pp. 1515–1519.

Schmitt, C., Bähre, D. (2013). An approach to the calculation of process forces during the precision honing of small bores, *Procedia CIRP*, Vol. 7, pp. 282–287.

Vrac, D., Sidjanin, L., Balos, S. (2014). The effect of honing speed and grain size on surface roughness and material removal rate during honing, *Acta Polytechnica Hungarica*, Vol. 11, No. 10, pp. 163–175.

Williams, C.G. (1928). Progress in Honing-Machines and the Honing Process, SAE Technical Paper 280060, doi:10.4271/280060.

3

Lapping

Pramod Kumar Jain and Harpreet Singh

Department of Mechanical and Industrial Engineering, Indian Institute of Technology Roorkee, Roorkee, India

CONTENTS

3.1 Introduction

Lapping is defined as a chipping process involving a relatively low volume of material removal. In this process, material removal takes place due to the cutting action of loose abrasive particles distributed in a medium (fluid or paste or lapping slurry), which move in a undirected cutting pattern on the work surface with the help of a lapping tool (similar to a copying tool). During lapping, a group of geometrically dissimilar cutting edges remove unwanted material in a simultaneous and slow (rpm < 80) cutting action from the work surface and can finish almost any material. A suitable range of size of abrasive particles used is 5–20 μm. These abrasive particles are suspended in a viscous or liquid vehicle (such as soluble oil, mineral oil or grease) during cutting action. Lapped surfaces can be categorised as smooth and flat but lacking in mirror finish (i.e. unpolished). Lapping can be used to generate geometrically accurate functional surfaces with defined shape, surface quality and accurate dimensional tolerances (generally less than 2.5 µm). Lapping is also used to remove any subsurface damage caused by sawing or grinding to produce surface with geometrical precision.

The ultra-precision finishing of the work surface is achieved in a two-step lapping process (i.e. pre-lapping and polishing). The pre-lapping is characterised by higher material removal while achieving good surface quality with tight dimensional and form tolerances. During polishing, the highest order of surface qualities is achieved with a mirror-like finish. Lapped surfaces exhibit two major characteristics, such as undirected processing traces and a semi-gloss appearance. Also, when such surfaces are under strain, they can be distinguished by little wear. During super finishing of flat or cylindrical workpieces, the lapping process is preferred to honing, ultra-precision grinding and ultra-precision machining. Other geometrical shapes can also be finished using modified lapping method/tooling. Over time, the lapping process is found be an ideal option to achieve the highest quality requirements. However, economic considerations may not support the use of a lapping process in many applications.

Lapping has major applications in the finishing of many materials, such as metals, ceramics, glasses, natural materials (like marble, granite and basalt), gems, plastics, semiconductor materials, carbon, graphite and even diamond. Some of the important characteristics of the lapping process that vary in degree according to the particular system and equipment are as follows:

1. Material removal rate (MRR) is low due to the low cutting speed and low depth of cut due to superficial penetration of the abrasive particles into the work surface.

2. The lapping process is a low-temperature process and hence does not cause any thermal damage to the work surface.

3. During lapping, no or very low stresses are induced, and hence, virtually, there is no distortion to the workpiece.

4. If used appropriately, lapping can produce very accurate surfaces in terms of form accuracy. Some of the achievable surface characteristics for flat surfaces are the following:

   • Flatness to less than one light band 0.0000116″/0.3 µm
   • Roughness of less than 1 µin. R_a/0.025 µm
   • Size control to less than 0.001″/2.5 µm
   • Parallelism within 0.00005″/1.3 µm

5. Lapping is also used to finish brittle materials and fragile parts because a relatively uniform and low pressure is applied during the processing.

6. Lapping helps in maintaining very close dimensional control (within 0.5 µm).

7. Soft non-ferrous materials as well as hardened tool steel, carbide ceramics and even diamond can be lapped to surface roughness below 2 µin.

3.2 Description of Process

Lapping is the abrading of a work surface by a lap (which is made of a softer material than the work surface material to be lapped), which has been charged by the fine and hard abrasive particles. When the lap and work surface are rubbed together with the fine abrasive particles (slurry) between them, these particles become embedded in the softer lap and then it becomes a holder for the hard abrasive particles as described in Figure 3.1 [1]. The charged lap is thus rubbed against a hard work surface, and during this rubbing, the hard abrasive particles embedded in the lap surface act as cutting edges to remove small amount of the material from the harder work surface. During the lapping process, the lap is protected against any wear by the layer of embedded abrasive particles on it [2].

The material removal usually ranges from less than 0.025 mm (finish lapping) to up to 0.075 mm (rough lapping). If lapping is used to remove scratch marks left by grinding or honing, then it can produce parts to the size limits of 0.000625 mm. A sample of work surface finished by lapping is shown in Figure 3.2. It can be observed that no cutting or scratch marks are visible [3].

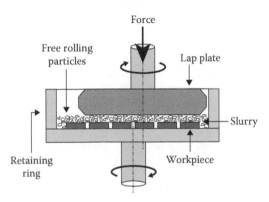

FIGURE 3.1
A schematic view of the lapping process.

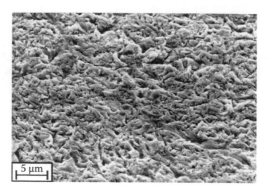

FIGURE 3.2
Lapped workpiece surface.

Laps are made of any soft material (such as soft cast iron, wood, leather, brass, copper, lead or soft steel) that can receive and retain the abrasive particles during rubbing against the work surface. However, fine-grained cast iron is recommended as a lap material in many applications. But Cu is the preferred choice for lapping of diamonds. Soft cloth is a good choice when hardened materials are lapped for metallographic examination. The selection of abrasive particles is based on the work material and its application. For steel surfaces, artificial corundum is recommended as an abrasive for pre-lapping and again for polishing. Silicon carbide is preferred for pre-lapping of cast iron and alumina for finest lapping. When small work surface is lapped, diamond dust or boron carbide in fine particle size gives good results. The recommended range of particle size is from 120 grit up to the finest available sizes. In nearly all cases, a paraffin lubricant is recommended as a viscous medium to be used during lapping, the exception being for soft materials when a soluble oil or water lubricant is used. A summary of advantages and disadvantages of the lapping process is presented in Table 3.1. Table 3.2

TABLE 3.1

Advantages and Disadvantages of the Lapping Process

Advantages	Disadvantages
• Can process almost every material and part size • Low cost of workpiece holders • Several workpieces can be processed in one operation cycle • No visible tool marks • No deformation or changes in structure of the processed workpieces • Extremely accurate shape control in terms of flatness and plane parallelism	• Disposal of the lapping slude/paste as special waste • Low MRR but high rate of wear of abrasive particles • Necessity of final cleaning of workpieces • Processability only of basic geometries of workpiece

TABLE 3.2

Evaluation of the Types of Precision Machining Processes

	Grinding	Honing	Lapping
Shape accuracy	+	0	++
Dimensional accuracy	+	++	++
Positional accuracy	+	−	−
Surface quality	0	+	++
Positive influence on surface integrity	0	+	+
MRR	++	0	−
Manufacturing cost	−	+	++

Note: −, minor influence; +, major influence; 0, moderate influence; ++, superior influence.

shows a comparative statement on the process capabilities of various precision machining processes in terms of operational results and efficiency.

3.2.1 Metal Removal Process

3.2.1.1 Kinematic Principle

The material removal rate (MRR) during lapping is calculated as follows:

$$MRR = \Delta V/\Delta t, \tag{3.1}$$

where MRR is material removal rate (mm³/min) and ΔV is the volume of material removed (mm³) during time interval Δt (min) [2].

It can also be calculated as

$$MRR = \Delta THK/\Delta t \; (\mu m/min), \tag{3.2}$$

where MRR is material removal rate (μm/min) and ΔTHK is the thickness removed (μm) during time interval Δt (min).

3.2.1.2 Lapping Methods

Lapping can be performed in two ways on the work surface depending on its size, surface requirements and work material [4]:

- Hand lapping
- Machine lapping

3.2.1.2.1 Hand Lapping

Hand lapping is the process to finish the flat surface by rubbing it over an accurately finished flat surface of a master lap (lapping tool). Usually, the

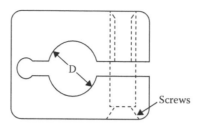

FIGURE 3.3
A ring-shaped lap.

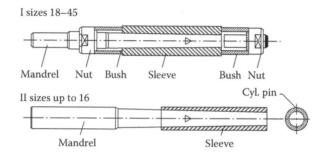

FIGURE 3.4
Manual lapping of internal cylindrical surfaces.

surface of the master lap is made of a thick close-grained cast iron block. During abrading action, very fine abrasive particles are placed between the master lap and the work surface. Lap movements are controlled manually by a human hand and it requires a highly skilled worker to apply the controlled lap pressure and speed during undirected lap movements. Flat surfaces are lapped with a lapping disc made of cast iron and external cylindrical surfaces are lapped by a ring-shaped lap. A ring-shaped lap is shown in Figure 3.3. The internal diameter of the ring is very close to the external diameter of the cylindrical workpiece, and it can be adjusted accurately with the help of a setscrew mechanism. Ring lapping has found applications in the precision finishing of machine spindles and plug gauges. This type of lapping can also be used for finishing external threads where the lap is in the form of a bush having internal threads. Large size laps are made of cast iron, while those of small size are made of steel or brass. A schematic diagram of internal cylindrical surface lapping using manual lapping is given in Figure 3.4. This figure shows drift elements of the lap and it is used for manual lapping.

3.2.1.2.2 Machine Lapping

Machine lapping is recommended to be used to finish parts in batches for economic reasons. In this process, metal laps are used and abrasive particles are held in suitable vehicles during finishing. In some cases, abrasive paper

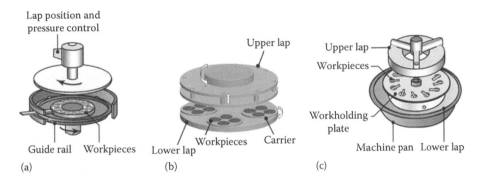

FIGURE 3.5
Production lapping on (a, b) flat surfaces and (c) cylindrical surfaces.

or abrasive cloth is also employed as a lapping medium. A lapping setup for flat surfaces on a single- and double-side lapping machine and cylindrical surfaces on double-side lapping machine is shown in Figure 3.5.

In centreless roll lapping, two cast iron rolls are used. One of them serves as the lapping roller and the second one as a regulating roller. During lapping, the abrasive compound is applied to the rolls rotating in the same direction while the workpiece is fed across the rolls. This process is suitable for lapping a single piece at a time and mostly used for lapping plug gauges, measuring wires and similar straight or tapered cylindrical parts. Centreless lapping is similar to centreless grinding in its working. This technique is used to produce high roundness accuracy and fine finish. In multi-pass lapping, the workpiece passes through a number of progressively finer lapping wheels. Centreless lapping offers high throughput and is suitable for workpieces requiring a small amount of shape rectification. Lapping of internal cylindrical surfaces is done on a machine similar to a honing machine with the help of a power stroke. These machines provide rotational and reciprocation movements simultaneously to the workpiece or to the lap. The lap is usually made of cast iron and is either solid or adjustable in nature as per the convenience on the setup.

Recently, researchers have tried to use ultrasonic vibrations during lapping. In this process, lapping is performed with loose abrasive particles, which are evenly distributed in a fluid or paste and are excited with ultrasonic impulses. These impulses improve the cutting action of the abrasive particles.

3.2.2 Tooling

3.2.2.1 Abrasive Material in Lapping

The most commonly used abrasive materials during the lapping process include silicon carbide (SiC), corundum (Al_2O_3), boron carbide (B_4C) and,

TABLE 3.3

List of Most Commonly Used Lapping Materials and Their Applications

Abrasive Material	Area of Applications in Lapping
Corundum	Soft steels, light and non-ferrous metals, carbon, semiconductor materials
Silicon carbide	Quenched and tempered steels, steel alloys, grey cast iron, glass, porcelain
Boron carbide	Carbides, ceramics
Diamond	Hard materials

increasingly, diamond. The most common applications of these abrasive materials are listed in Table 3.3 [5].

3.2.2.2 Lapping Disc

The lapping disc plays a critical role during the process as it carries lapping slurry and dressing rings and provides a rotational movement to the work surface. To avoid high centrifugal forces, proper selection of rotational speed is important. The lapping disc is made of a special fine-grained cast iron or a hardened steel alloy. Among the many physical and chemical properties of the material of lapping disc, the penetration depth of the abrasive particles in it is an important aspect. For this purpose, lapping discs are distinguished: soft (<140 HB), medium hard (140 to 220 HB) and hard (>220 HB). Low hardness of the lapping disc makes it easy to attach abrasive particles into the disc surface, and these abrasive particles act as cutting edges to initiate a chip formation on the work surface. In this, a sliding motion leads to the removal of material by a plowing action. However, hard-lapping discs tend to cause rolling of the abrasive particles in the active gap, and the stresses induced in the work lead to micro-fracture and thereby remove the material. This sometimes may become problematic, as the abrasives tend to get embedded in the workpiece and contaminate it [3]. A softer lapping disc gives a good surface finish, but at the expense of loss of planarity. In order to achieve both, modern lapping machines use a hard lapping disc and very fine abrasive grits. In addition to the particle's attachment behaviour, the disc hardness also determines disc wear and the possible amount of material removal. Whatever material is used, the lap disc should be softer than the workpiece; otherwise, the workpiece will become charged and cut the lap.

During rough lapping, a lap disc with fine grooves that are usually located about a half inch apart and extending both lengthwise and crosswise and forming a series of squares similar to those on a checkerboard is used. This scored surface helps to generate better and faster cutting in the roughing operation. For finishing operations, a planar lapping disc with no scores is used. To produce a perfectly smooth and flat surface that is free of scratches, the lapping disc needs to be charged with very fine abrasive particles. Abrasive slurry is gradually poured on the lapping disc as described in

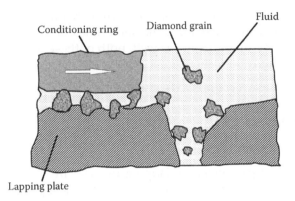

FIGURE 3.6
Charging of a lapping plate.

TABLE 3.4

A List of Lapping Plates Used for the Lapping Process

Lapping Disc Material	Description
Iron disc	• Used as primary/roughing lap disc • Has long service life • Produces good surface finish on most materials, especially metals and ceramics • Generally used with coarse to medium size diamond abrasives
Copper disc	• Mostly used when primary and final lapping are combined in one-step process • Used for any solid material such as metals, ceramics, plastics and glass • Used with medium to fine diamond particles sizes
Tin disc	• Used to avoid any lead-type contamination • Used with extra-fine abrasive particles
Ceramic disc	• Used to avoid any metallic-type contamination • Used with both coarse to fine diamond particles sizes
Composite disc	• Used to finish two or more dissimilar materials to uniform plane of flatness

Figure 3.6. A list of the various lapping disc materials and their applications are illustrated in Table 3.4.

3.2.2.3 Lapping Slurry

The role of lapping slurry is many fold during the lapping process. It helps in cooling down the work surface along with the lapping disc, moves away the debris from the lapping zone and lubricates in between the work surface and the lapping disc. The removal of debris further helps in preventing the lapping disc from overloading and thus extends lap life between dressings.

The control of temperature in lapping is extremely important, as otherwise, it may lead to poor surface quality, and sometimes, large temperature gradients may cause thermal warping as well. Both of these will lead to a serious loss of dimensional and form tolerances [6].

3.3 Process Parameters

The input process parameters of lapping process are broadly divided into five categories, as described in Table 3.5. The technical parameters affecting lapping process performance are unit pressure, particle size of abrasive, concentration of abrasives in the vehicle and lapping speed.

In the lapping process, mainly two input process parameters, namely, lapping pressure and lapping speed, are considered. The MRR is dependent

TABLE 3.5

List of Input Parameters of Lapping Process

Kinematics	Lapping Disc	Slurry	Machine	Workpiece
• Rotational speeds • Dimensions • Carrier configuration	• Material • Slot configuration and geometry • Evenness	• Particle material • Particle size distribution • Carrier fluid • Volume flow	• Normal force • Machine power • Oscillation behaviour • Thermal stability	• Material • Geometry • Thickness distributions

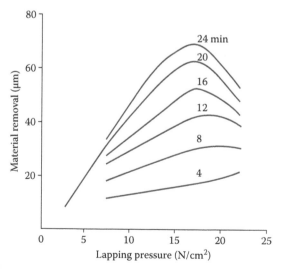

FIGURE 3.7

The influence of lapping pressure and duration on workpiece removal.

largely on the pressure of the lapping disc. It is clear from Figure 3.7 that the MRR rises constantly given identical lapping durations up to a load of approximately 16 N/cm^2; a further increase in lapping pressure, however, causes the MRR to sink. This course of events can be ascribed to the breaking of the lapping particles added at the beginning of operation. Also, as the lapping grain size increases, the MRR increases. In addition, the MRR decreases with an increasing rotational speed of disc up to a certain level. It has been observed that a large centrifugal force throws away the lapping slurry too quickly from the active cutting zone to the edges of the lapping disc.

3.3.1 Effect of Contact Pressure

Figure 3.8 shows the effect of unit pressure on the MRR and surface roughness of the finished surface during lapping with the variation in the contact pressure. It can be easily observed that the surface quality deteriorates with the increment in lapping pressure, mainly due to the impregnation of abrasive particles on the work surface. The optimum range of lapping pressure can be obtained with a trade-off between surface roughness and MRR [7].

3.3.2 Effect of Abrasive Particle Size and Grain Content

The variations in MRR and surface roughness with particle size of abrasives are shown in Figure 3.9. Again, there is an increasing trend of MRR and surface roughness with an increase in particle size. However, with lapping being a finishing process, achieving desired surface finish is more important. Hence, accordingly, a suitable abrasive particle size is selected to achieve the desired results [4]. Figure 3.10 depicts the variation in MRR with change in abrasive content of the lapping slurry at constant lapping pressures (p_1, p_2, p_3). It can be seen that at a given lapping pressure, MRR starts decreasing after a

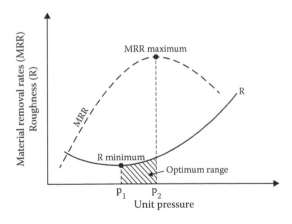

FIGURE 3.8
Effect of unit pressure on surface roughness and MRR.

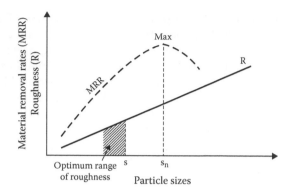

FIGURE 3.9
Effect of abrasive particle size on MRRs and surface roughness.

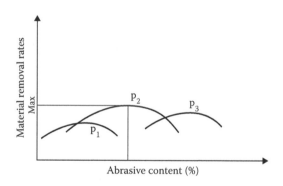

FIGURE 3.10
Effect of abrasive content on MRRs.

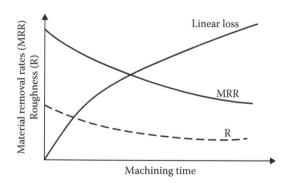

FIGURE 3.11
Effect of machining on MRRs and surface roughness. (From Chattopadhyay, A.K. et al., *Superfinishing Processes, Honing, Lapping and Superfinishing, Lesson 30, Version 2 ME*, IIT Kharagpur, Kharagpur, India.)

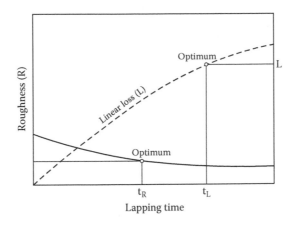

FIGURE 3.12
Criteria for choosing lapping time.

FIGURE 3.13
A series of typical parts finished by the lapping process.

certain percentage abrasive content due to wear of abrasive particles among themselves. This phenomenon may lead to the rubbing of the work surface with worn-out abrasive particles, resulting in a reduced MRR.

3.3.3 Effect of Machining Time

The dependence of MRR, surface roughness and linear loss (L) of workpiece dimension is shown in Figure 3.11. It can be observed that linear loss of workpiece dimensions increases with the increase in machining time, while the trend of MRR and surface roughness is reverse; that is, they decline with an increment in machining time. It may occur due to deterioration in the sharpness of lapping particles with the increase of machining time. Lapping conditions are so chosen that the designed surface finish is obtained with the permissible limit of linear loss of workpiece dimension, as shown in Figure 3.12.

3.4 Applications

There are various applications of lapping and are shown in Figure 3.13 [8].

References

1. Deaconescu, A., Deaconescu, T. (2014). Improving the quality of surfaces finished by lapping by robust parameter design, *Journal of Economics, Business and Management*, Vol. 2, No. 1, pp. 1–4.
2. Sharma, P.C. (2011). *A Textbook of Production Technology: Manufacturing Processes*, S. Chand Publishing, New Delhi, 727–739.
3. *Lapping and Polishing Basics*, Applications Laboratory Report 54, South Bay Technology, San Clemente, CA.
4. Chattopadhyay, A.K., Chattopadhyay, A.B., Paul, S. (2008). *Superfinishing Processes, Honing, Lapping and Superfinishing, Lesson 30, Version 2 ME*, IIT Kharagpur, Kharagpur, India.
5. Klocke, F. (2009). *Manufacturing Processes 2 – Grinding, Honing, Lapping*, Springer-Verlag, Berlin, Heidelberg. doi: 10.1007/978-3-540-92259-9.
6. Stahli, A.W. (2006). *The Technique of Lapping*, CH-2542, Maschinen Corporate Publications, Pieterien, Biel, Switzerland.
7. Marinescu, I.D., Uhlmann, E., Doi, T. (2006). *Handbook of Lapping and Polishing*, CRC Press, Taylor & Francis Group, Boca Raton, FL.
8. Schneider, G. (2011). *Cutting Tool Applications, Chapter 18: Lapping and Honing*, American Machinist, Cleveland, OH.

Section III

Advanced Nanofinishing Processes

4

Abrasive Flow Finishing Process and Modelling

Sachin Singh,[1] M. Ravi Sankar,[1] Vijay K. Jain[2] and J. Ramkumar[2]

[1]*Department of Mechanical Engineering, Indian Institute of Technology Guwahati, Guwahati, India*

[2]*Department of Mechanical Engineering, Indian Institute of Technology Kanpur, Kanpur, India*

CONTENTS

4.1 Introduction

Surface finish determines the wear resistance of the component and power loss due to friction. If the surface is not smooth, it can lead to component failure because of crack initiation from the surface. Advancing technology made the miniaturisation of components a necessity. Surface roughness becomes more important as there is miniaturisation of the component dimensions. Even a slight alteration in the dimension can altogether change component functionality. The traditional finishing processes, viz. grinding, honing and lapping, are incapable of meeting the increasing demand of achieving the end surface finish (nano and sub-nanometer) on the complicated geometries made up of difficult-to-finish materials. Hence, many advanced finishing processes came into existence. Abrasive flow finishing (AFF) is one such advanced finishing process developed to finish internal as well as external, simple to complex geometries.

In the 1960s, the Extrude Hone Corporation, USA, developed this technology for deburring hydraulic control blocks, which were deburred by hand at that time. Since deburring is a machining process, this process is called the abrasive flow machining (AFM) process. By 1968, this process was successfully developed by the Extrude Hone Corporation and later used in a vast number of fields, including nanofinishing applications. Thus, this process is referred in the present chapter as the AFF process. AFF is gaining importance because of its ability to give predictable, repeatable and consistent results [1]. The AFF operation is analogous to the grinding operation such that the grinding wheel acts as the 'self-deformable stone'. It takes the shape of the pathway that is needed to be finished. In the AFF process, the finish is obtained by using the viscoelastic medium, which acts as the backbone of the abrasive particles mixed in it for shearing the surface peaks. The major drawback of the conventional finishing technologies lies in their incapability of controlling abrading forces during finishing with close tolerances and without harming the surface topography, which in turn affects surface finish. Such difficulty is overcome in AFF as it has the ability to control and select the intensity and location of the abrading forces through proper fixture design, medium selection and process parameters.

Principal finishing parameters that determine the end finish of the component during the AFF operation are as follows:

1. Setting parameters: These include the input parameters, which are given before the start of the AFF process, e.g. extrusion pressure, medium flow velocity, number of AFF cycles and flow volume.

2. Tooling: The design of the fixture depends on the workpiece to be finished. Some of the basic functions of the fixture/tooling (Figure 4.1) are as follows:

   a. To position the workpiece at the desired location and hold it tightly.

   b. To have enough strength in order to withstand the forces acting during finishing operation.

   c. It should act as a restriction passage for guiding the abrasive medium across the workpiece properly without any leakage through them.

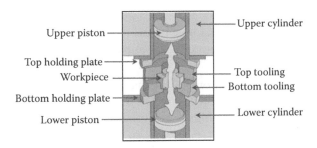

Upper piston

Top holding plate

Workpiece

Bottom holding plate

Lower piston

Upper cylinder

Top tooling

Bottom tooling

Lower cylinder

FIGURE 4.1
Tooling used to hold the workpiece during the AFF process.

3. Medium: The AFF medium plays a vital role in the finishing of workpiece. The medium mainly comprises of the base polymer, which is viscoelastic in nature; additives (plasticisers, softeners) and abrasive particles. A good AFF medium should be mechanically stable, chemically non-reactive and self-deformable and have good fluidity and good abrading ability. The composition and quantities of all these ingredients depend on the dimensions of the pathway through which the medium is to be extruded for finishing the component.

4. Workpiece: The various types of machining operations are performed before final finishing by the AFF process. Hardness of material plays a vital role in the improvement of surface finish.

The AFF process is capable of achieving surface finishes up to 50 nm, deburr the holes as small as 0.2 mm in diameter, radius the edge from 0.025 to 1.5 mm and give hole tolerance up to ±5 μm. Up to 90% time can be saved by using AFF as compared to hand finishing operations [1]. In the first few cycles, the major part of the total improvement in the surface finish occurs.

Some of the characteristics of AFF process are the following:

- Deburr, polish and radius difficult to reach surfaces.
- Produces radii even on complex edges.
- Surface roughness is reduced by 90% on cast, machined or electric discharge machined surfaces.
- AFF can simultaneously process multiple parts or many areas of a single workpiece.
- As there is no pre-defined relative motion of the tool relative to the workpiece, inaccessible areas and complex internal passages can be finished economically as well as effectively.

- It is easy to integrate AFF in any automatic manufacturing environment. Automatic AFF systems are capable of handling thousands of parts per day, greatly reducing labour costs by eliminating tedious handwork.
- By understanding and controlling the process parameters, AFF can be applied to an impressive range of finishing operations by changing tooling, process settings and abrasive medium.
- It is possible to control and select the intensity as well as the location of abrasion through fixture design, medium selection and process parameters.

4.2 Mechanism of Material Removal during the AFF Process

The medium used in finishing of workpieces during the AFF process is composed of a soft base polymer carrier, additives and fine abrasive particles. The base polymer is viscoelastic in nature; i.e. it contains both elastic as well as viscous properties when undergoing deformation. It acts as a binder for holding abrasive particles. As shown in Figure 4.2a and b, when extrusion pressure (P) is applied on the medium, the elastic component of the medium results in generation of radial force (F_R), while viscous component generates axial force (F_A). F_R is responsible for indentation of abrasive particles in the workpiece surface while F_A pushes the indented abrasive particles in the axial direction to remove the material by shearing action in the form of microchip.

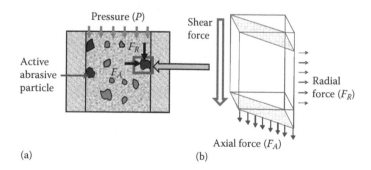

(a)   (b)

FIGURE 4.2
(a) Forces acting on abrasive particle in AFF process and (b) viscoelastic medium behaviour of exerting radial force normal to the applied shear force direction. (From Singh, S., Sankar, M.R., Jain, V.K., Ramkumar, J., *Modeling of Finishing Forces and Surface Roughness in Abrasive Flow Finishing Process Using Rheological Properties*, 5th International and 26th All India Manufacturing Technology, Design, and Research, 2016.)

It is assumed that the amount of F_R and F_A forces generated in the medium are fully transmitted to the abrasive particles.

4.3 Types of AFF Processes

Depending on the geometry of the component to be finished, the AFF process can be broadly classified in to three main types: one-way AFF, two-way AFF and orbital AFF.

4.3.1 One-Way AFF

As the name suggests, in one-way AFF, the viscoelastic medium is extruded unidirectionally. As shown in Figure 4.3, the one-way AFF setup consists of a hydraulically actuated reciprocating piston and an extrusion chamber whose role is to receive the medium flowing out of the workpiece after one completed cycle to extrude the same medium unidirectional through the workpiece for next cycle. It is used mainly to finish components for which it is difficult to make tooling for one of the two ends such as engine blocks and multiple holes (Figure 4.4).

The fixture directs the flow of medium from the extrusion chamber through the internal/external passages on the workpiece, while a collector is put to gravitationally collect the extruded medium flowing out of the workpiece (Figure 4.3). The medium then flows periodically from the collector to the extrusion chamber through the access port provided in the extrusion chamber. The access port is controlled by the hydraulically actuated piston, which opens to collect the medium in the extrusion chamber and the next cycle resumes as soon as the medium chamber is charged. The piston then advances into engagement with the chamber, thus sealing the medium for the next extrusion cycle. The low pressure created by the piston in the chamber during

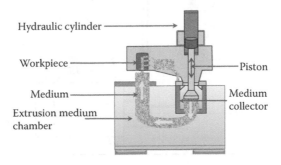

FIGURE 4.3
Unidirectional AFF process.

(a)   (b)

FIGURE 4.4
Applications of unidirectional AFF process: (a) multiple holes and (b) engine block. (Courtesy of Extrude Hone.)

withdrawal often assists the gravity flow of the medium and accelerates the filling process. One-way AFF offers the advantage of faster processing of parts, easy clean up and use of quick simple change tooling.

4.3.2 Two-Way AFF

As shown in Figure 4.5, a two-way AFF setup consists of upper and lower medium cylinders with pistons, a workpiece fixture, hydraulic drive and supporting frame. The workpiece to be finished is held in the fixture securely and placed in the space between the upper and lower medium cylinders. The primary function of the medium cylinder is to contain the abrasive and guide the reciprocating motion of the piston during extruding the medium. At the start of the finishing cycle, usually, the lower medium cylinder is filled with the

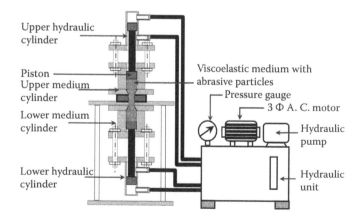

FIGURE 4.5
Schematic diagram of AFF process experimental setup. (Source: Ravi Sankar, M., Nanofinishing of Metal Matrix Composites Using Rotational Abrasive Flow Finishing (R-AFF) Process. PhD thesis. IIT Kanpur, India.)

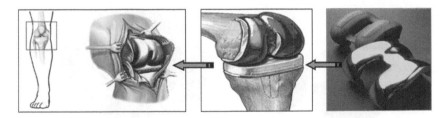

FIGURE 4.6
Knee joint implant. (Courtesy of Extrude Hone.)

medium while the upper cylinder is empty or nearly empty. Under the action of an external force, hydraulically or mechanically, the medium is extruded via the restricted passage through or past the workpiece surface to be finished.

When the upper medium cylinder discharges the whole medium to the lower medium cylinder, the lower medium cylinder's piston begins to push the medium having abrasive particles, back to the upper cylinder, completing one cycle. Thus, the flow of the medium is done back and forth through the workpiece until the desired results are obtained. An advantage of the two-way AFF process lies in its capability to finish simple to complex components such as knee joint implants (Figure 4.6) without much difficulty.

4.3.3 Orbital AFF

In orbital AFF, the finishing of the workpiece is achieved by providing rapid, low-amplitude oscillations in two or three dimensions within a slow-flowing 'pad' of elastic/plastic medium. As compared to two-way AFF, the medium here is more viscous and elastic in nature. The basic principle employed in the orbital AFF is to provide translational motion to the medium with respect to the workpiece. A small orbital oscillation (0.5 to 5 mm), i.e. circular eccentric planar oscillation, is applied to the workpiece relative to the self-deforming viscoelastic finishing medium, as shown in Figure 4.7. The pad is positioned on the surface of the displacer, which itself is a mirror image of the workpiece, plus or minus a gap accommodating the layer of medium and a clearance. Different portions of the workpiece are finished as the circular eccentric oscillation continues.

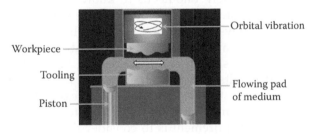

FIGURE 4.7
Orbital AFF process.

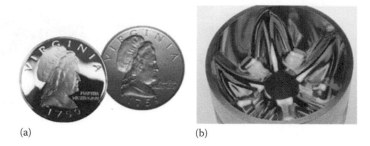

(a)   (b)

FIGURE 4.8
Applications of orbital AFF process: (a) coins and (b) plastic bottle bottom dies. (Courtesy of Extrude Hone.)

Since small orbital oscillation is provided to the workpiece, this helps the medium to flow into every internal complex feature, which is even smaller than the amplitude of the oscillation. Also, some beneficial compressive residual stresses are induced in the workpiece due to the repeated bumping of the self-shaped layer of viscoelastic abrasive medium on the workpiece. There is a remarkable improvement in the surface finish due to the small amplitude oscillatory motion to the workpiece across the compressed viscoelastic medium. It is used to finish components such as coins and plastic bottle bottom dies (Figure 4.8).

4.4 Modelling of AFF Process

In order to apply any process successfully, the foremost thing is to understand the process. With proper understanding of the process, further improvements in the same can be done to achieve optimum results with minimum efforts. AFF consists mainly of three components: setup, tooling and medium. These are the basic requirements that one needs to explore and understand in order to apply the process. Further, in order to understand the process completely, one needs to know the physics of the process. Starting from the late 1990s, researchers have been investigating the AFF process from the physics point of view. Modelling of the process can be done mainly in four ways: theoretical, numerical/simulation, empirical and soft computing. The success of any modelling lies in its capability to match its predicted results with the experimental results with least possible error. The more realistic the modelling is, the less is the error between the predicted and experimental results. In order to make the modelling realistic, the foremost requirement is to consider the number of assumptions as less as possible.

The following sections highlight the work done by various researchers in the field of modelling the AFF process. Several models to predict the finishing forces, surface roughness and material removal (MR) in terms of input variables of the AFF process have been reported. Their main objective is to derive the models that can predict the output results of the AFF process in terms of the input variables as accurately as possible with respect to the experimental results.

4.4.1 Numerical/Simulation Modelling

To have a better understanding of the AFF process, Jain et al. [2,3] developed models for MR and surface roughness during the AFF process. The authors reported the application of the finite element method (FEM) to determine the stresses and forces generated during the AFF process. Secondly, these forces were then used in the theoretical models to predict the surface finish and MR during the AFF process.

In order to make the AFF modelling less complex, the authors made several assumptions. Three governing equations, viz. continuity, momentum and constitutive equations of steady state, incompressible and axi-symmetric natures are used as base to build the FEM model. After applying the Glakerin weighted residual method and appropriate weighting functions, governing equations are changed into weak formulations. The final global FEM equation after application of the boundary conditions [2] is

$$\begin{bmatrix} [0] & [k_{pv}] \\ [k_{vp}] & [k_{vv}] \end{bmatrix} \begin{Bmatrix} p \\ v \end{Bmatrix} = \begin{Bmatrix} 0 \\ F \end{Bmatrix}, \tag{4.1}$$

where

$$[k_{pv}] = -\int_{A^e} \{N_p\}\{m\}^T [B] 2\Pi r \, dr \, dz, \tag{4.2}$$

$$\{m\}^T = \{1101\}, \tag{4.3}$$

$$[k_{vv}] = \int_{A^e} \eta [B]^T [B] 2\Pi r \, dr \, dz, \tag{4.4}$$

$$[k_{vp}] = -\int_{A^e} [B]^T \{m\}\{N_p\}^T 2\Pi r \, dr \, dz, \tag{4.5}$$

and

$$\{F\} = \int_{\Gamma^b} \{N_b\}\{N_b\}^T \{t^b\} 2\Pi r \, d\Gamma, \tag{4.6}$$

where $[B]$ is a matrix consisting of differential coefficients of shape functions for the velocities, $\{N_p\}$ is a column vector of bilinear shape functions for the approximation of pressure, $\{N_b\}$ is one-dimensional matrix of biquadratic shape functions and $\{t^b\}$ is traction vector at boundaries.

The stresses generated in the medium are determined from the finite element analysis. These stress values are used to theoretically calculate the MR rate (MRR) and surface roughness at different locations along the medium flow directions. As shown in Figure 4.9a, a simplified model of an abrasive particle indenting on the workpiece surface is assumed [2]. The MR from the surface is considered to be taking place in the form of microcutting action. They considered uniform roughness profile [2], which is triangular (Figure 4.9b) in shape, as initial surface roughness profile of the workpiece.

The total volumetric MR in ith stroke, V_i, is derived [2] and is given as

$$V_i = 2\pi N_a L_s \frac{R_m^2}{R_w} \left(\frac{d_p^2}{4} \sin^{-1}\left(\frac{2\sqrt{(t(d_p - t))}}{d_p} \right) \right.$$

$$\left. - \sqrt{(t(d_p - t))}\left(\frac{d_p}{2} - t \right) \right)\left[1 - \frac{R_a^i}{R_a^0} \right] L_w. \tag{4.7}$$

The surface roughness value after the ith stroke, R_a^i, during the AFF process [2] is given as

$$R_a^i = R_a^{i-1} - \frac{1}{7} N_a L_s \left(\frac{R_m^2}{R_w^2} \right)\left(\frac{d_p^2}{4} \sin^{-1}\left(\frac{2\sqrt{(t(d_p - t))}}{d_p} \right) - \sqrt{(t(d_p - t))}\left(\frac{d_p}{2} - t \right) \right), \tag{4.8}$$

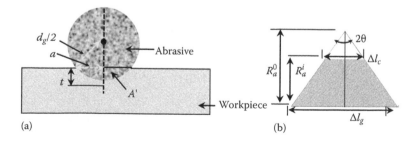

(a)   (b)

FIGURE 4.9
(a) Schematic diagram of the indentation of an abrasive particle on the workpiece surface.
(b) Simplified surface geometry.

where R_w is the radius of the workpiece, R_m is the radius of the medium cylinder, N_a is the number of active abrasive particles per unit workpiece surface area, L_s is the stroke length, d_p is the abrasive particle diameter and t is the depth of indentation of the abrasive particle in the workpiece surface.

This model is validated by comparing the theoretical results with the experimental results. It was concluded that the finite element model predicts the radial stresses at the workpiece surface with reasonable accuracy. The MR increases with an increase in extrusion pressure and abrasive concentration in the medium and decreases with an increase in abrasive particle mesh size. For a certain number of cycles, with an increase in piston velocity, piston pressure, percentage concentration of abrasives particles and abrasive particle mesh size, the surface roughness decreases. Increasing extrusion pressure and piston velocity beyond a threshold value increase the depth of indentation of the abrasive particle. This leads to an increase in surface roughness.

Later on, Jain and Jain [4] did a simulation of a final surface finish obtained after the AFF process. The forces are calculated as was explained previously. The simulated results are in good agreement with the experimental results. The small error between the experimental and simulated results is due to the fact that researchers have not considered the real shape of abrasive particles, the wear of abrasive particles and pile up of material and actual surface profile of the work. Jain and Jain [5] further used the same numerical approach to determine the effect of percentage abrasive concentration and mesh size on active abrasive particles density and also verified the same experimentally. The flowchart for the AFF simulation is shown in Figure 4.10 [5].

Jain et al. [6] extended the finite element analysis of the AFF process for complex workpieces. Instead of assuming medium as a Newtonian flow, it was modelled as a Bingham plastic fluid. The governing equations (conservation of mass, balance of momentum and conservation of energy) are modified accordingly. Later on, a theoretical model was developed for the volume of material removed during the AFF process. The authors [6] also conducted experiments to see the variation of surface roughness and MR by varying extrusion pressure and number of cycles. A regression model was derived using the experimental data and showed a similar trend.

To improve the efficiency of the AFF process, it has been modified in various ways. One of the modified versions of the AFF process is the centrifugal-force-assisted AFF (CFAAFM) process. In this process, a centrifugal force generating (CFG) rod is introduced in the tooling. This centrifugal force adds up with the finishing forces generated during the AFF process. These two finishing forces together enhance the efficiency of the AFF process. Walia et al. [7] developed a two-dimensional finite element model to simulate the flow of medium during the CFAAFM process. After applying assumptions and boundary conditions, the final finite element equation for steady flow of the medium during a CFAAFM process is given as

$$C(u)U + KU + R(u)\tau = F \tag{4.9}$$

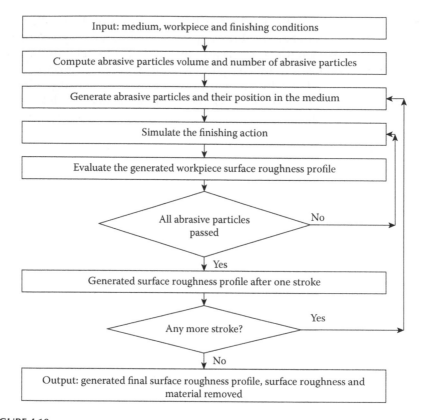

FIGURE 4.10
Flowchart for AFF simulation. (From Jain, R.K., Jain, V.K., *J. Mater. Process. Technol.*, 152, 17–22, 2004.)

$$N\tau - D(u)\tau + C'(u)\tau - LU = 0, \qquad (4.10)$$

where $C(u)$, K and $R(u)$ are the coefficient matrices that the authors [7] derived after changing momentum and continuity equations into weak form. Similarly, N, $D(u)$, $C'(u)$ and L are the coefficient matrices for the constitutive equations and U is the vector matrix for unknowns (i.e. velocity and pressure). Equations 4.9 and 4.10 are solved to evaluate the resultant pressure, velocity and radial stresses developed during the working of the CFAAFM process. It was concluded that modifying the AFM process by using CFG rod enhances the performance of the AFF process.

Finishing of complex workpieces by the AFF process leads to a non-uniform surface finish across the workpiece surface. Researchers tried to eliminate this shortcoming. Wang et al. [8] attempted to achieve uniform surface finish on a complex hole surface during finishing with the AFF process (Figure 4.11). As the complex holes have varying cross-sectional areas, there is a change in medium velocity flows, strain rates and shear forces

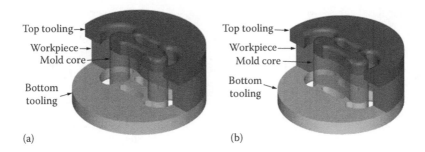

FIGURE 4.11
Two types of mold core in chain holes: (a) mold core with two cylinders and (b) mold core with chain shape. (From Wang, A.C., Tsai, L., Liang, K.Z., Liu, C.H., Weng, S.H., *Trans. Nonferrous Met. Soc. China*, 19, 250–257, 2009.)

throughout the workpiece. This results in a non-uniform surface finish. So researchers used computational fluid dynamics (CFD) to determine the velocity, strain rates and shear forces acting on the surface under constant pressure condition. The objective is to design an appropriate passageway in the core of the workpiece so as to have the same shear force that will result in uniform surface finish. If the gap between the workpiece surface and the core is kept constant throughout, it results in uniform velocities, strain rates and shear forces on the workpiece surface. This results in achieving uniform surface finish.

Cheng et al. [9] studied the rheological properties of the medium and also tried to achieve uniform surface finish of deep circular holes by developing various passageways that act as cores inside the workpiece. Researchers laid emphasis on the fact that by creating a multiple direction flow of the abrasive particles, uniform surface finish can be achieved. It is mainly due to an increase in abrasive surface area and radial shear forces. The motion of the abrasive particles was studied by using circular, hollow and helical passageway. Numerical simulations were studied using CFD-ACE software. Experimental results showed that the highest surface uniformity was achieved with a passageway with six helices. Extending the same concept, Chen and Cheng [10] attempted to achieve uniform radial distribution of surface finish in polygonal holes. As the radial distance of polygonal holes is not constant, it resulted in the generation of non-uniform radial strains and radial stress, which lead to the non-uniformity of surface finish. Researchers carried out the CFD simulation to study strain rates and velocity distribution during finishing of the polygonal holes. Holes of various passageways (square, polygon, hexagon and octagon) are finished by introducing helical cores in the workpieces. The authors showed that use of helical cores not only increases the efficiency of the AFF process by increasing the amount of finishing forces but also helps in achieving uniform surface finish on different passage ways. Researchers also concluded that the use of helical pathways narrowed down the range of strain rates

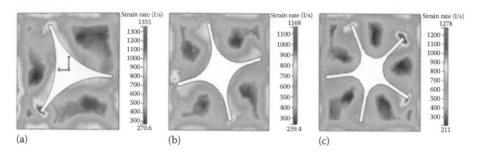

FIGURE 4.12
A cross-section showing radial strain rates in a square hole with different helical cores: (a) three-helix passageway, (b) four-helix passageway and (c) five-helix passageway. (From Chen, K.Y., Cheng, K.C., *Int. J. Adv. Manuf. Technol.*, 74, 781–790, 2014.)

produced, which resulted in uniform surface finish. Further, the higher the symmetry of the helix, the smaller will be the variation between the minimum and maximum values of the strain rates and the more is the uniformity of surface finish achieved (Figure 4.12).

To increase the interaction between the abrasive particles and the workpiece surface, another hybrid form of AFF process is spiral flow assisted AFM (SFAAFM). SFAAFM is similar to the AFF process except that in SFAAFM, a spiral rod is rotated inside the workpiece during finishing operation. Arora et al. [11] did a finite element analysis of the medium flow during the SFAAFM operation. It is reported that the extrusion pressure and velocity on the surface of the workpiece decrease along the length of the workpiece. Wan et al. [12] did CFD simulation of the AFF process while finishing straight tubes with ellipsoidal cross-sections. They coupled three models to predict R_a and MR, which can be given as follows:

1. Non-Newtonian model to predict the medium flow:

$$\tau_{ij} = \eta \dot{\gamma}_{ij} \tag{4.11}$$

$$\eta = K \,|\, \dot{\gamma}_{ij} \,|^{g-1}. \tag{4.12}$$

2. Wall slip model to calculate the relative motion between the medium and the bounding wall:

$$v_s = v_b \left(\frac{\tau_w}{\tau_b} \right)^m. \tag{4.13}$$

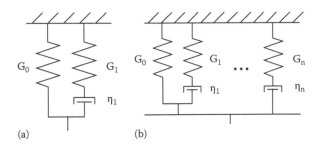

FIGURE 4.13
Maxwell model: (a) standard and (b) generalised. (From Uhlmann, E., Doits, M., Schmiedel, C., *Proc. CIRP*, 8, 351–356, 2013.)

3. Roughness, R_a, and h, model:

$$R_a = (R_i - R_\infty)e^{\frac{-k_T pv_s}{H}t} + R_\infty \tag{4.14}$$

$$h = a(R_i - R_\infty)\left(1 - e^{\frac{-k_T pv_s}{H}t}\right) + \frac{k_s pv_s}{H}t, \tag{4.15}$$

where τ_{ij} is shear stress, $\dot{\gamma}_{ij}$ is the shear rate, η is the non-Newton viscosity, v_s is the wall slip velocity, τ_w is the wall shear stress, R_i and R_∞ are the initial and limiting surface roughness, H is the workpiece Vickers hardness, p is the medium pressure, k_t and k_s are wear coefficients and g, K and a are constants. The authors [12] also designed an insert to get a more uniform surface finish on tubes with ellipsoidal cross-sections. By introducing an insert, the wall slip velocity distribution was uniform, which in turn would generate more uniform surface finish.

In order to predict accurately the finishing forces and other output parameters of the AFF process, it is necessary to model the physical behaviour of the abrasive medium as exact as possible. Uhlmann et al. [13] laid emphasis on the previously stated concept and found that the standard Maxwell model (Figure 4.13a), when extended to a generalised Maxwell model (Figure 4.13b), can more accurately model the AFF medium in terms of storage and loss modulus.

4.4.2 Theoretical/Mathematical Models

Gorana et al. [14] developed an experimental setup to measure radial and axial forces developed in the AFF process. The authors also calculated the active abrasive particle density and studied its effect on surface roughness produced. Later on, the theoretical model of radial and axial forces acting on a single abrasive particle was developed [15]. Scratching experiments were performed to see the mechanism of MR during the AFF process. The extrusion pressure, when applied to the medium, results in the generation of radial (normal) forces and axial forces. After applying assumptions, the final equations derived are as follows:

The radial force, F_r, acting on a single abrasive particle can be given as

$$F_r = \sigma_r \Pi \left(\frac{b}{2} \right)^2. \tag{4.16}$$

The axial force, F_a, acting on a single abrasive particle is

$$F_a = \left[\frac{(2Rd' - d'^2)^{3/2}}{2R} + 1.2566(2Rd' - d'^2) \right] \sigma_r, \tag{4.17}$$

where σ_r is flow stress of workpiece material, b is the diameter of projected area of indentation of the abrasive particle on the workpiece surface, d' is the depth of indentation of the abrasive particle into the workpiece surface and R is the radius of the abrasive particle. Experiments were conducted to verify the theoretical results. The authors [15] showed that the axial force, radial force, active abrasive particle density and abrasive particle depth of indentation all have a significant influence on the scale of material deformation. Gorana et al. [16] developed an analytical model to simulate surface roughness during AFF and verified the same experimentally. The authors [16] laid the emphasis on the fact that active abrasive particle density and forces generated in the medium play an important role in determining the end surface finish of the workpiece during the AFF operation. The effect of percentage abrasive concentration and extrusion pressure on the active abrasive particles density as well as surface roughness of the workpiece are also studied. The analytical equation developed by the authors to determine the total number of active abrasive particles in a cylindrical workpiece, n_a, is

$$n_a = \left[\frac{48 F_{am} G A_1}{\Pi^2 \sigma_r D_g^4} \right]^{1/2}, \tag{4.18}$$

where F_{am} is the axial force exerted on the workpiece by the abrasive particles, G is the volume ratio (ratio of the volume of the abrasive particles to the total volume of the cuboidal element) and A_1 is area of the internal surface of the cylindrical workpiece.

Assuming that the material is removed in the form of ploughing, the surface roughness value, R_a, after completing ith number of pass is given as

$$R_a = \frac{l_t \sqrt{3}}{8} - \frac{2(i \times d'')^2}{l_t \sqrt{3}},\tag{4.19}$$

where l_t is the base length of the equilateral triangle assumed as initial surface profile of the workpiece and d'' is the depth up to which the material is removed in one stroke over the triangular roughness peak by the abrasive particles.

Medium rheology plays a major role in deciding the surface finish of the component finished by the AFF process. Davies and Fletcher [17] used different viscosities of polyborosiloxane medium and silicon carbide abrasives. It was found that the temperature of the medium plays an important role during the AFF process due to its effect on the medium viscosity. The authors used the following expression to calculate viscosity, η_i, of the fluid flowing in a circular pipe:

$$\eta_i = \frac{\Pi r^4 \Delta P}{8 Q L_c}.\tag{4.20}$$

The shear rate, $\dot{\gamma}$, is calculated by the following relation:

$$\dot{\gamma} = \frac{4Q}{\Pi r_{ct}^3},\tag{4.21}$$

where r_{ct} is the radius of the capillary tube, L_c is the length of capillary, Q is volumetric flow rate and ΔP is axial pressure drop through the capillary. Fletcher and Fioravanti [18] reported about the thermal properties of the medium (polyborosiloxane and silicon carbide abrasive) used during the honing process. They developed theoretical models for thermal conductivity and specific heat capacity of the medium as a function of abrasive concentration and abrasive particle size. An empirical model for surface heat transfer coefficient was also proposed. Thermal conductivity, λ_0, is given by the following equation:

$$\lambda_0 = \frac{3.33 v_d + 1}{g} \left[\frac{\lambda_d \lambda_c f^2}{\lambda_c f + \lambda_d j} \right] + \lambda_c c \left(1 + \frac{f}{g} \right),\tag{4.22}$$

where v_d is the volume fraction of abrasive particles, λ_d is the thermal conductivity of the dispersed phase (abrasive particles), λ_c is the thermal conductivity of the continuous phase (base polymer), f is characteristic abrasive particle dimension, g is the length of the side of the control cube and j is the difference between f and g (i.e. f-g).

Based on the relative proportion of mass of each component in the mixture, a simple weighted mean model for specific heat capacity of the mixture, $C_{p,mix}$, is used:

$$C_{p,mix} = (1 - x)C_{p,c} + xC_{p,d}, \tag{4.23}$$

where x is the percentage concentration of the dispersed phase in the medium, $C_{p,c}$ is the specific heat capacity of the continuous phase and $C_{p,d}$ is the specific heat capacity of the dispersed phase.

The surface heat transfer coefficient, h_s, is calculated by the general relation

$$h_s = K_1\theta_{intf} + K_2, \tag{4.24}$$

where θ_{intf} is interface temperature, K_1 is a coefficient of θ_{intf} and K_2 is a constant in the heat transfer coefficient model. The model's output data are validated with the experimental data and found to be in good agreement. Theoretical models for determination of specific energy, heat transfer and temperature produced during the AFF process were developed by Jain and Jain [19] as follows:

1. Specific energy, u:

$$u = F_{tgt} \frac{v_m^2}{A'}. \tag{4.25}$$

2. The thermal flux generated over the medium–workpiece interface area, q:

$$q = unA'v_p \frac{r_c^2}{r_w^2}. \tag{4.26}$$

3. Average temperature on the workpiece surface, T_{sav}:

$$T_{sav} - T_i = \frac{4}{3}q_m \left(\frac{l_w}{\Pi(k_m\rho_m c_{pm}v_m)} \right)^{\frac{1}{2}}. \tag{4.27}$$

4. Fraction of total heat flux entering into the workpiece, S:

$$S = \frac{\dfrac{2}{3}\left(\dfrac{l_w}{k_m \rho_m c_{pm} v_m}\right)^{\frac{1}{2}}}{\left(\dfrac{\tau_{ft}}{\rho_w c_{pw} k_w}\right)^{\frac{1}{2}} + \dfrac{2}{3}\left(\dfrac{l_w}{k_m \rho_m c_{pm} v_m}\right)^{\frac{1}{2}}}. \tag{4.28}$$

Here, F_{tgt} is total tangential force per active abrasive particle; v_m is the velocity of flow of the medium across the workpiece surface; A' is area of groove generated by the abrasive particle in the workpiece surface; q_m is heat flux in the medium; l_w is workpiece length; ρ_m, k_m and c_{pm} are density, thermal conductivity and specific heat of medium, respectively; c_{pw} is the specific heat of the workpiece material and τ_{ft} is finishing time. The authors [19] claimed that theoretical results are in good agreement with the experimental results. Specific energy remains almost constant with respect to a change in abrasive mesh size. The heat transfer model is capable of predicting change in the workpiece temperature.

Walia et al. [20] conducted an experimental study of the CFAAFM process. The effects of input variables (shape and rotational speed of CFG rod, extrusion pressure, number of cycles and abrasive mesh size) on output parameters (MR and surface roughness) were studied. They also developed mathematical models to determine the resultant velocity, V_r, and angle of strike of the abrasive particle, θ_a, on the workpiece surface, which can be given as

$$V_c = \frac{F_c}{3\Pi\mu d_p}, \tag{4.29}$$

$$V_r = \sqrt{V_a^2 + V_c^2}, \tag{4.30}$$

and

$$\theta_a = \tan^{-1}\left(\frac{V_c}{V_a}\right), \tag{4.31}$$

where V_c and F_c are the centrifugal velocity and centrifugal force on the abrasive particle due to CFG rod, V_a is the axial velocity of the abrasive particle (calculated from extrusion pressure), μ is the viscosity of the medium and d_p

is the diameter of the abrasive particle. Through experiments, it is proved that changing the shape and velocity of the CFG rod influences the output parameters of the CFAAFM process. Later on, Walia et al. [21] further investigated the CFAAFM process. They developed mathematical models to determine the number of active abrasive particles in the AFF process, N_{AFF} and CFAAFM process, N_{CFAAFM}, which can be given as

$$N_{AFF} = \frac{4V_v \left(Dd_p - d_p^2\right)}{V_0 D^2} \tag{4.32}$$

and

$$N_{CFAAFM} = \frac{\Pi D_c^2 C_d}{4} L_w, \tag{4.33}$$

where V_v is the volume of the abrasive particle in full volume of medium; V_0 is the volume of one abrasive particle; D, d_p and D_c are the diameter of the workpiece, abrasive particle and medium cylinder, respectively; C_d is the number of dynamic active abrasive particles per unit cross sectional area of work passage and L_w is the length of the workpiece. The action of centrifugal force acts on the abrasive particles and throws them on the internal surface (workpiece finishing region) of the workpiece. The value of N_{CFAAFM} is more than that of N_{AFF}. Thus, more abrasive particles take part in the finishing operation during the CFAAFM process, making it a more effective process. The abrasive particle movement pattern during the AFF process is predicted by Fang et al. [22]. The moving pattern of the abrasive particle decides the mode of MR during the AFF process. There are generally two kinds of abrasive particle motions in the AFF process, viz. sliding–rubbing and rolling. The effects of normal load, abrasive particle size and hardness of machined part are taken as the input parameters for deciding the abrasive particle movement.

After analysing various forces acting on the abrasive particle during the AFF process, the following criteria are derived (Figure 4.14):

1. For the abrasive particle to groove on the workpiece surface:

$$\frac{e}{h} \geq \mu_s = \mu_k. \tag{4.34}$$

2. For the abrasive particle to roll over the workpiece surface:

$$\frac{e}{h} < \mu_r < \mu_k. \tag{4.35}$$

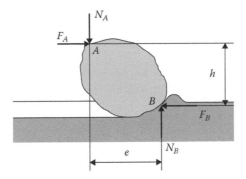

FIGURE 4.14
Schematic illustration of force analysis on an abrasive particle. (From Fang, L., Sun, K., Cen, Q., *Tribotest*, 13, 195–206, 2007.)

Here, e is the horizontal distance between the normal load acting on the abrasive particle, h is the normal distance between friction force and axial force, μ_s is abrasive particle grooving friction coefficient, μ_k is friction coefficient causing material yield during abrasive particle grooving and μ_r is abrasive particle rolling friction coefficient. They concluded that below a certain normal load, most of the abrasive particles tend to groove. Also, as the normal load is increased, the proportion of grooving particles in the case of small size abrasive particles increases, while in case of large-sized abrasive particles, as the normal load is increased, the proportion of rolling particles increases.

To enhance the performance of the AFF process, it is modified into rotational AFF (R-AFF) process. The R-AFF process is same as the AFF process except that in the R-AFF process, the tooling is externally rotated and the medium reciprocates inside it with the help of hydraulic actuators. During the R-AFF process, as the rotational speed of the workpiece increases, abrasive particles hit the workpiece surface with high kinetic energy. It results in generating μ-chip, and also, an extra component of force known as the tangential force adds up in the finishing force generated during AFF process (Figure 4.15). Also, as shown in Figure 4.16, the length of the finishing path increases. Thus, ΔR_a in the case of the R-AFF process increases as compared to the AFF process.

Sankar et al. [23–25] performed experiments to finish Al/SiC metal matrix composite (MMC) using the R-AFF process. The authors explained in detail the finishing mechanism of MMC when processed with the R-AFF process. Mathematical modelling to calculate the helix angle and helical path was done and their validation is done with the experimental results. Helix angle, φ, is given as

$$\varphi = \tan^{-1}\left[\frac{(30V_{pv})}{(\Pi R_w N_{wr})}\right], \tag{4.36}$$

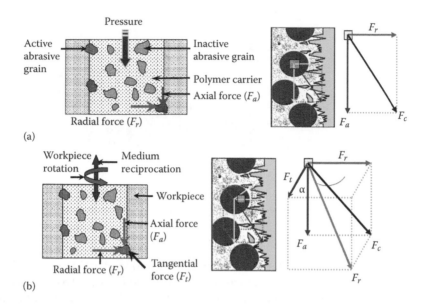

FIGURE 4.15
Forces in finishing region in the (a) AFF process and (b) rotational abrasive flow finishing process. (From Sankar, M.R., Jain, V.K., Ramkumar, J., *Int. J. Mach. Tools Manuf.*, 50, pp. 637–650, 2010.)

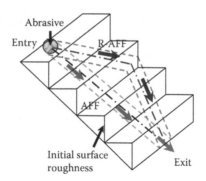

FIGURE 4.16
Abrasive finishing path in the AFF and R-AFF processes on initial surface roughness. (From Sankar, M.R., Jain, V.K., Ramkumar, J., *Int. J. Mach. Tools Manuf.*, 50, pp. 637–650, 2010.)

while helical path length, s, is given as

$$s = \left(\frac{V_{pv} t_s}{\sin \varphi} \right), \tag{4.37}$$

where V_{pv} is piston velocity, R_w is the radius of the workpiece, N_{wr} is the number of workpiece revolutions (rpm) and t_s is the time taken for one stroke.

The authors concluded that R-AFF can give 44% better ΔR_a and 81.8% more MR compared to the AFF process. Rheology of the medium is one of the important parameters that decides the amount of finishing forces that are generated in the medium. Sankar et al. [26–29] conducted a detailed static and dynamic rheological study of the medium and explained the role of its various properties in deciding the end surface finish of the component. The viscoelastic behaviour of the medium is also modelled by the same authors with the help of various viscoelastic material models. Forwarding the same concept ahead, Singh et al. [30] conducted a theoretical analysis to find out theoretically the finishing forces generated during the AFF process. Later on, by using the same forces, a theoretical modelling based on the guideline given by Jain et al. [2] is done. To determine elastic and viscous stresses, a frequency sweep test is performed on an MCR 101 parallel plate rheometer.

As shown in Figure 4.17, the sample is placed in between the two parallel plates. The top plate (tool master) and bottom stationary plate have a diamond cone pattern to hold the semisolid medium sample. During the test, the sample is given sinusoidal deformation, which results in the development of sinusoidal stresses that are of same frequency as the applied deformation but shifted by a phase angle δ [31]:

$$\gamma = \gamma_0 \sin \omega t \qquad (4.38)$$

and

$$\tau = \tau_0 \sin(\omega t + \delta), \qquad (4.39)$$

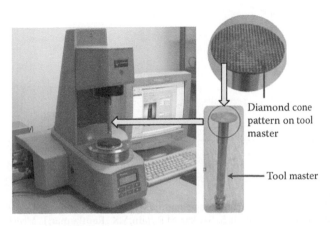

Diamond cone pattern on tool master

Tool master

FIGURE 4.17
Rheometer used for frequency sweep test (Anton par-MCR-301 series).

where γ_0 is the maximum amplitude of applied shear strain, γ is the amplitude of applied shear strain, τ_0 is the maximum amplitude of shear stress produced, τ is the amplitude of shear stress produced and ω is the applied angular frequency.

The stress generated is decomposed into two components with the same frequency. One component is in phase with the strain wave ($\sin \omega t$), giving storage modulus (Figure 4.18a) and is responsible for elastic stress, τ'. The second component is $90°$ out of phase ($\cos \omega t$), which gives loss modulus (Figure 4.18b) and is responsible for viscous stress, τ'', i.e.

$$\tau = \tau' + \tau'' = \tau_0' \sin \omega t + \tau_0'' \cos \omega t. \tag{4.40}$$

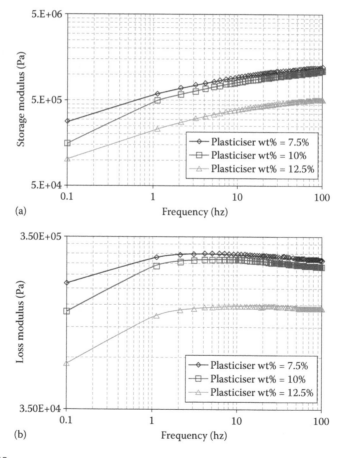

(a)

(b)

FIGURE 4.18
Variation with frequency (Hz) of (a) storage modulus (Pa) and (b) loss modulus (Pa) (logarithmic scale on X and Y axis). (From Singh, S., Sankar, M.R., Jain, V.K., Ramkumar, J., *Modeling of Finishing Forces and Surface Roughness in Abrasive Flow Finishing Process Using Rheological Properties*, 5th International and 26th All India Manufacturing Technology, Design, and Research, 2016.)

Now, the relation between in-phase or storage (elastic) modulus, G', and elastic stress, τ_0', is given by the following equation [31]:

$$G' = \frac{\tau_0'}{\gamma_0}. \tag{4.41}$$

Thus, the radial force (F_R) acting on a single abrasive particle can be calculated as

$$F_R = \tau_0' \pi \frac{D_g^2}{4}, \tag{4.42}$$

where D_g is the diameter of the abrasive particle.

On the other hand, the relation between an out-of-phase or loss (viscous) modulus, G'', and viscous stress, τ_0'', is given by the following equation [31]:

$$G'' = \frac{\tau_0''}{\gamma_0}. \tag{4.43}$$

Thus, the axial force (F_A) acting on an abrasive particle is given by the following relation [32]:

$$F_A = \tau_0''(A - A'), \tag{4.44}$$

where A is the cross-sectional area of the abrasive particle and A' is the cross-sectional area of the groove generated in the workpiece surface.

4.4.3 Empirical Modelling

The final surface finish of the component after the AFF process does not depend solely on the AFF input parameters but also on the prior machining operation used to prepare the workpiece. This point was proved by Loveless et al. [33]. The authors prepared the workpiece by turning, milling, grinding and wire electrical-discharge machining. Analysis of variance (ANOVA) was used to model the relationship between the input parameters and output responses of the AFF process. They concluded that the prior machining operation did influence the amount of MR during the AFF process. Also, machining operation prior to finishing and viscosity of the medium play a major role in determining the surface finish improvement. Further, a data-dependent system, a stochastic modelling and analysis technique, was used to study the surface roughness before and after the AFF operation.

Jain and Adsul [34] performed experiments to study the effects of the number of cycles, concentration of abrasives, abrasive mesh size and medium flow speed on MR and surface finish during the AFF process. They used a multivariable curve fitting technique to obtain equations for the MR and surface finish improvement. Jain et al. [35] conducted an experimental study to find the effect of temperature of medium, concentration and mesh size of abrasive particles on the medium viscosity. Later, by using experimental data, regression analysis was done to evaluate the viscosity equation in terms of input variables. They concluded that viscosity decreases with an increase in medium temperature and mesh size but increases with an increase in abrasive concentration. It was observed that as the viscosity of the medium increases, MR increases, but surface roughness shows the opposite trend.

Tzeng et al. [36,37] performed AFF on micro-slits and micro-holes and conducted ANOVA to find out the significant parameters during the finishing operation. They explored the effects of AFF parameters, such as motion mode (rotary or axial), extruding pressure, extruding period, abrasive concentration, abrasive particle size and machining time, on MRR and surface roughness of micro-hole. The same experiments were performed with micro-slits, except the parameters motion mode and extruding time. Optimal combination of finishing parameters is found by signal-to-noise ratio (S/N) response graphs. The three general categories of quality characteristics of S/N response graphs are higher is better (*HB*), lower is better (*LB*) and nominal is better (*NB*) and can be given as

$$HB : \varepsilon = -10 \log \left[\frac{1}{n} \sum_{i=1}^{n} y_i^{-2} \right], \tag{4.45}$$

$$LB : \varepsilon = -10 \log \left[\frac{1}{n} \sum_{i=1}^{n} y_i^{2} \right], \tag{4.46}$$

and

$$NB : \varepsilon = -10 \log \left[\frac{1}{n-1} \sum_{i=1}^{n} (y_i - \mu)^2 \right], \tag{4.47}$$

where ε is the S/N ratio calculated from the observed values, y_i is experimentally observed values (output) in the ith experiment and n is the number of repetitions in each experiment. The authors used the LB characteristic of the S/N response graph for surface roughness in case of micro-slit to find the optimal combination of parameters. On the other hand, the HB and LB characteristics of the S/N response graph were used for MRR and differences between the dimensions of the entrance and exit of micro-hole respectively. The authors reported that during finishing of the micro-slit, abrasive particle size is the dominant factor, while for micro-holes, finishing time is the dominant parameter for MRR. Walia et al. [38] optimised the CFAAFM process by using utility theory and Taguchi quality loss function. The developed approach is used for multi-objective optimisation of the CFAAFM process. They considered three outputs as percentage improvement in surface roughness and MR as the 'higher the better' type of quality characteristic and scatter of surface roughness (SSR) as the 'lower the better' type of quality characteristic.

To arrange the input variables in order of their importance in deciding the end surface finish, Walia et al. [39,40] conducted an ANOVA. The effect of rotation speed of the CFG rod, extrusion pressure and mesh size is studied on R_a, MRR and SSR. The authors reported that rotational speed of the CFG rod is the most important and mesh size is the least important parameter in deciding R_a and MR. In case of SSR, rotational speed of the CFG rod and extrusion pressure are the significant input parameters, while the effect of abrasive mesh size on SSR is insignificant. The processing time required to achieve the same R_a is reduced by 70%–80% in the case of the CFAAFM process as compared to the AFF process. Based on the experimental results, Reddy et al. [41] concluded that the combination of high speed of the CFG rod and high extrusion pressure are favourable conditions to obtain a high degree of surface finish. On the other hand, if high speed of the CFG rod is combined with a larger size of abrasive particles, the results is an enhanced MR.

Mali and Manna [42,43] studied the effect of AFF process parameters while finishing Al/SiC MMC. Experiments were designed using the Taguchi experimental design concept. The effect of input parameters (abrasive mesh size, number of cycles, extrusion pressure, percentage of abrasive concentration and medium viscosity) on output parameters (R_a, ΔR_a, R_t, ΔR_t and MR) were reported. While determining the optimal combination of parameters through the S/N response graph, quality characteristic used was 'the lower the better' for R_a and R_t and 'the higher the better' for the MR. They also established a polynomial regression equation using multivariable curve fitting technique and reported that abrasive mesh size is the most significant parameter for the MR and R_t, while percentage of abrasive concentration in the medium is the most significant parameter for R_a. By using different sets of input parameters, Mittal et al. [44] also finished Al/SiC MMC. The variation of input parameters (extrusion pressure, percentage of oil in medium, mesh number, concentration of abrasives, workpiece material and number of cycles) was studied on output parameters (R_a and MRR) by using the Taguchi design of experiments. With the help of ANOVA, they have concluded that the extrusion pressure is the most significant factor in deciding the magnitude of output parameters followed by workpiece material and number of cycles. Other input parameters have little or no effect on output responses.

4.4.4 Soft Computing

Parameters that affect AFF process performance (surface roughness and MRR) typically fall into five categories: characteristics of the workpiece, characteristics of the medium, machining parameters, technical tolerances and process objectives. To capture knowledge of the process and to reduce the cycle time on new application development, Petri et al. [45] developed a neural network model that predicts surface finish and dimensional changes. The training of neural network is done by using the back-propagation (BP) algorithm, which adjusts weights according to the gradient descent method. After the successful development of the neural network models, the research-ers integrated those models with a heuristic optimisation routine to select the best sets of machine setup parameters for the AFF process and also to decide on the set of parameters for new applications. The authors concluded that the process model can be used to reduce the development time for new applications.

Jain et al. [46] also used the neural network model to predict the process parameters' effect on the output of AFF process. The inputs were medium flow rate, percentage concentration of abrasives, abrasive mesh size and number of cycles; the outputs measured were MRR and surface roughness. BP neural network, usually referred to as feed-forward multi-layered net-work with hidden layers trained with a gradient descent technique, was used in the study. Simulation results showed a good agreement with the experimental results for a wide range of finishing conditions. In another study, Jain et al. [47] developed a neural network system to optimise it using the augmented Lagrange multiplier (ALM) and genetic algorithm (GA). The results closely matched with the experimental results. The trained network successfully synthesised the optimal input conditions for the AFF process. The optimal conditions maximise MRR subjected to appropriate process constraints.

The pseudo-objective function used in the ALM is

$$A(x, \lambda, r_p) = L(x) + \sum_j \left[\lambda_j \left\{ g_j(x) + s_j^2 \right\} + r_p \left\{ g_j(x) + s_j^2 \right\}^2 \right], \qquad (4.48)$$

subjected to

$$x_i(\text{min}) \le x_i \le x_i(\text{max}) \qquad (4.49)$$

$$g_j(x) = (d_j - y_j) \le 0, \qquad (4.50)$$

where x_i(min), x_i(max) are the minimum and maximum values of the input variables, x_i and s_j are slack variables added to inequality constraints, λ_j is the Lagrange multiplier and r_p is the penalty parameter.

The constrained optimisation problem used for GA is

$$\max MRR \tag{4.51}$$

subjected to

$$R_a \leq R_{a\,max} \tag{4.52}$$

$$x_i^l \leq x_i \leq x_i^u, \tag{4.53}$$

where MRR is material removal rate and x_i^l and x_i^u are the lower and upper bounds on process variables x_i.

Mollah and Pratihar [48] modelled the AFF process using a radial basis function network (RBFN). RBFN may be computationally faster than the conventional feed-forward neural network as it has a simpler structure and faster learning speed. Similar to the previous study, here also, medium flow rate, percentage concentration of the abrasives, abrasive mesh size and number of cycles were taken as input parameters and the output responses are MRR as well as surface roughness. Batch mode of training is used and the network is tuned in two ways, viz., using the BP algorithm and a GA separately. The GA optimised network produced good results as compared to the BP tuned RBFN because there is a chance of the solution getting stuck in the local minima while using the BP algorithm, as it works on the principle of steepest descent. Ali Tavoli et al. [49] used group method of data handling–type neural networks to model the AFF process. The authors modelled the effect of number of cycles and abrasive concentration on surface finish and MR. In the second step, the polynomial models obtained from the neural network modelling were used for the optimisation of the AFF process. The Pareto multi-objective optimisation method was used by the authors [49], considering the MR and surface finish as conflicting objectives. Multi-objective optimisation is mathematically defined as finding the vector X^* to optimise $F(X)$ subjected to m inequality constraints and p_c equality constraints such that

$$X^* = \left[x_1^*, x_2^*, \dots x_n^* \right]^T, \tag{4.54}$$

$$F(X) = [f_1(X), f_2(X), \dots f_k(X)]^T, \tag{4.55}$$

$$g_i(X) \leq 0 \quad i = 1 \text{ to } m, \tag{4.56}$$

and

$$h_j(X) = 0 \quad j = 1 \text{ to } p_c. \tag{4.57}$$

4.5 Comments

The AFF process has good finishing capabilities for easy-to-finish, simple to complex geometries made up of difficult to finish materials. The AFF process is gaining importance in today's machining industry as this process can simultaneously finish multiple parts or many areas of a single workpiece. In order to employ a process effectively and efficiently, it is necessary to have in-depth knowledge of the physics of the process so that it can be modified accordingly to achieve maximum gain. Various researchers since 1990s developed models for the AFF process by theoretical analysis, simulation, empirical modelling and soft computing techniques to predict the output responses of the AFF process in terms of AFF input parameters. Some of the shortcomings of the AFF process are generation of non-uniform surface finish on complex workpieces and incapability to control online the finishing forces in the AFF medium. Researchers tried to overcome these shortcomings in different ways. To further maximise the outcome of the AFF process, many versions of the AFF process have been proposed, such as CFAAFM, R-AFF and SFAAFM. Many assumptions are made during the modelling of the AFF process, such as the shape of the abrasive particle, forces acting on abrasive particle, cutting edge of the abrasive particle, etc. If these assumptions can be relaxed, it can further help in producing more realistic responses.

Nomenclature

A'	Area of groove generated by abrasive particle in the workpiece surface
A_1	Area of the internal surface of the cylindrical workpiece
b	Diameter of projected area of indentation of the abrasive particle on the workpiece surface
$[B]$	Matrix consisting of differential coefficients of shape functions for the velocities
$C(u), K, R(u)$	Coefficient matrices after changing momentum and continuity equations into weak form

C_d	Number of dynamic active abrasive particles per unit cross sectional area of work passage
$C_{p,c}$	Specific heat capacity of continuous phase
$C_{p,d}$	Specific heat capacity of dispersed phase
c_{pm}	Specific heat of medium
$C_{p,mix}$	Specific heat capacity of the mixture
c_{pw}	Specific heat of workpiece material
D	Diameter of work piece
d'	Depth of indentation of the abrasive particle into the workpiece surface
d''	Depth up to which the material is removed in one stroke over the triangular roughness peak by the abrasive particles
D_c	Diameter of medium cylinder
d_p	Diameter of abrasive particle
e	Horizontal distance between two normal load acting on the abrasive particle
f	Characteristic abrasive particle dimension
F_A	Axial force acting on the abrasive particle
F_{am}	Axial force exerted on the workpiece by the abrasive particles
F_c	Centrifugal force on the abrasive particle due to CFG rod
F_R	Radial force acting on the abrasive particle
F_{tgt}	Total tangential force per active abrasive particle
$F(X)$	Function
g	Length of side of control cube
G	Volume ratio (ratio of the volume of the abrasive particles to the total volume of the cuboidal element)
G'	Storage (elastic) modulus
G''	Loss (viscous) modulus
h	Normal distance between friction force and axial force
H	Workpiece Vickers hardness
h_s	Surface heat transfer coefficient
j	f–g
K_1	Coefficient of θ_{intf}
K_2	Constant in heat transfer coefficient model
k_m	Thermal conductivity of medium
k_t, k_s	Wear coefficients
L_c	Length of capillary
L_s	Stroke length
l_t	Base length of the equilateral triangle assumed as initial surface profile of the workpiece
l_w	Workpiece length
m	Inequality constraints
m, K, a	Constants

n	Number of repetitions in each experiment
N_a	Number of active abrasive particles per unit workpiece surface area
$\{N_b\}$	One-dimensional matrix of biquadratic shape functions
$N, D(u), C'(u), L$	Coefficient matrices for the constitutive equation
$\{N_p\}$	Column vector of bilinear shape functions for the approximation of pressure
N_{wr}	Number of workpiece revolutions
p	Medium pressure
P	Extrusion pressure
p_c	Equality constraints
ΔP	Axial pressure drop through the capillary
q	Thermal flux generated over the medium–workpiece interface area
Q	Volumetric flow rate
q_m	Heat flux in the medium
R	Radius of the abrasive particle
r_{ct}	Radius of the capillary tube
R_i	Initial surface roughness
R_m	Radius of the medium cylinder
R_w	Radius of the workpiece
R_∞	Limiting surface roughness
S	Fraction of total heat flux entering into the workpiece
t	Depth of indentation of the abrasive particle in the workpiece surface
$\{t^b\}$	Traction vector at boundaries
t_s	Time taken for one stroke
T_{sav}	Average temperature on the workpiece surface
u	Specific energy
U	Vector matrix for unknowns (i.e. velocity and pressure)
V_0	Volume of one abrasive particle
V_a	Axial velocity of the abrasive particle
v_b	Wall shear stress, when $\tau_w = \tau_b$
V_c	Centrifugal velocity on the abrasive particle due to CFG rod
v_d	Volume fraction of abrasive particles
v_m	Velocity of flow of medium across workpiece surface
V_{pv}	Piston velocity
v_s	Wall slip velocity
V_v	Volume of abrasive particle in full volume of medium
x	Percentage concentration of dispersed phase in the medium
X^*	Vector
x_i^u	Upper bound on process variables x_i
$x_i(\text{max})$	Maximum value of the input variable

$x_i(\mathbf{min})$	Minimum value of the input variable
x_i, s_j	Slack variables added to inequality constraints
y_i	Experimentally observed values (output) in the *i*th experiment
σ_r	Flow stress of workpiece material
τ	Amplitude of shear stress produced
τ'	Elastic stress
τ''	Viscous stress
τ_0	Maximum amplitude of shear stress produced
τ_{ij}	Shear stress
τ_{ft}	Finishing time
τ_w	Wall shear stress
λ_0	Thermal conductivity
λ_c	Thermal conductivity of continuous phase
λ_d	Thermal conductivity of dispersed phase
λ_j	Lagrange multiplier
η	Non-Newton viscosity
η_i	Viscosity of the fluid
μ	Viscosity of medium
μ_k	Friction coefficient causing material yield during abrasive particle grooving
μ_r	Abrasive particle rolling friction coefficient
μ_s	Abrasive particle grooving friction coefficient
γ	Amplitude of applied shear strain
γ_0	Maximum amplitude of applied shear strain
$\dot{\gamma}_{ij}$	Shear rate
θ_{intf}	Interface temperature
ε	S/N ratio calculated from the observed values
ω	Applied angular frequency
ρ_m	Density of medium

References

1. Rhoades, L., 1991, 'Abrasive flow machining: A case study', *Journal of Material Processing Technology*, 28, pp. 107–116.
2. Jain, R.K., Jain, V.K., Dixit, P.M., 1999, 'Modelling of material removal and surface roughness in abrasive flow machining process', *International Journal of Machine Tools and Manufacture*, 39, pp. 1903–1923.
3. Jain, R.K., Jain, V.K., 2003, 'Finite element simulation of abrasive flow machining', *IMechE Part B: J. Engineering Manufacture*, 217, pp. 1723–1736.
4. Jain, R.K., Jain, V.K., 1999, 'Simulation of surface generated in abrasive flow machining process', *Robotics and Computer Integrated Manufacturing*, 15, pp. 403–412.

5. Jain, R.K., Jain, V.K., 2004, 'Stochastic simulation of active grain density in abrasive flow machining', *Journal of Materials Processing Technology*, 152, pp. 17–22.

6. Jain, V.K., Kumar, R., Dixit, P.M., Sidpara, A., 2009, 'Investigations into abrasive flow finishing of complex workpieces using FEM', *Wear*, 267, pp. 71–80.

7. Walia, R.S., Shan, H.S., and Kumar, P.K., 2006, 'Finite element analysis of medium used in the centrifugal force assisted abrasive flow machining process', *Proceedings of IMechE Part B: Journal of Engineering Manufacture*, 220, pp. 1775–1785.

8. Wang, A.C., Tsai, L., Liang, K.Z., Liu, C.H., Weng, S.H., 2009, 'Uniform surface polished method of complex holes in abrasive flow machining', *Transactions of Nonferrous Metals Society of China*, 19, pp. 250–257.

9. Cheng, K.C., Chen, K.Y., Wang, A.C., Lin, Y.C., 2010, 'Study the rheological properties of abrasive gel with various passageways in abrasive flow machining', *Advanced Materials Research*, 126–128, pp. 447–456.

10. Chen, K.Y., Cheng, K.C., 2014, 'A study of helical passageways applied to polygon holes in abrasive flow machining', *International Journal of Advanced Manufacturing Technology*, 74, pp. 781–790.

11. Arora, G., Sharma, A.K., Bhowmik, P., Priyadarshini, S., 2013, 'An FEM approach to analysis of medium used in the spiral flow assisted abrasive flow machining', *International Journal of Mechanical Engineering and Research*, 3, pp. 635–640.

12. Wan, S., Ang, Y.J., Sat, T., Lim, G.C., 2014, 'Process modeling and CFD simulation of two-way abrasive flow machining', *International Journal of Advanced Manufacturing Technology*, 71, pp. 1077–1086.

13. Uhlmann, E., Doits, M., Schmiedel, C., 2013, 'Development of a material model for visco-elastic abrasive medium in abrasive flow machining', *Procedia CIRP*, 8, pp. 351–356.

14. Gorana, V.K., Jain, V.K., Lal, G.K., 2004, 'Experimental investigation into cutting forces and active grain density during abrasive flow machining', *International Journal of Machine Tools and Manufacture*, 44, pp. 201–211.

15. Gorana, V.K., Jain, V.K., Lal, G.K., 2006, 'Forces prediction during material deformation in abrasive flow machining', *Wear*, 260, pp. 128–139.

16. Gorana, V.K., Jain, V.K., Lal, G.K., 2006, 'Prediction of surface roughness during abrasive flow machining', *International Journal of Advanced Manufacturing Technology*, 31, pp. 258–267.

17. Davies, P.J., Fletcher, A.J., 1995, 'The assessment of the rheological characteristics of various polyborosilixane/grit mixtures as utilized in the abrasive flow machining', *Proceedings of the Institute of Mechanical Engineers*, 209, pp. 409–418.

18. Fletcher, A.J., Fioravanti, A., 1996, 'Polishing and honing processes: An investigation of the thermal properties of mixtures of polyborosiloxane and silicon carbide abrasive', *Proceedings of the Institute of Mechanical Engineers*, 210, pp. 255–266.

19. Jain, R.K., Jain, V.K., 2001, 'Specific energy and temperature determination in abrasive flow machining process', *International Journal of Machine Tools and Manufacture*, 41, pp. 1689–1704.

20. Walia, R.S., Shan, H.S., Kumar, P.K., 2006, 'Abrasive flow machining with additional centrifugal force applied to the medium', *Machining Science and Technology*, 10, pp. 341–354.

21. Walia, R.S., Shan, H.S., Kumar, P.K., 2008, 'Determining dynamically active abrasive particles in the medium used in centrifugal force assisted abrasive flow machining process', *International Journal of Advanced Manufacturing Technology*, 38, pp. 1157–1164.
22. Fang, L., Sun, K., Cen, Q., 2007, 'Particle movement patterns and their prediction in abrasive flow machining', *Tribotest*, 13, pp. 195–206.
23. Sankar, M.R., Jain, V.K., Ramkumar, J., 2010, 'Rotational abrasive flow finishing (R-AFF) process and its effects on finished surface topography', *International Journal of Machine Tools and Manufacture*, 50, pp. 637–650.
24. Sankar, M.R., Jain, V.K., Ramkumar, J., 2009, 'Experimental investigations into rotating workpiece Abrasive flow finishing', *Wear*, 267(1–4), pp. 43–51.
25. Sankar, M.R., Jain, V.K., Ramkumar, J., 2012, *Nano Finishing of Hard Cylindrical Steel Tubes Using Rotational Abrasive Flow Finishing (R-AFF) Process*, 4th International and 25th All India Manufacturing Technology, Design and Research (AIMTDR-2012).
26. Sankar, M.R., Jain, V.K., Ramkumar, J., Joshi, Y.M., 2011, 'Rheological characterization of styrene-butadiene based medium and its finishing performance using rotational abrasive flow finishing process', *International Journal of Machine Tools and Manufacture*, 51, pp. 947–957.
27. Sankar, M.R., Jain, V.K., Ramkumar, J., Kar, K.K., 2010, 'Rheological characterization and performance evaluation of a new medium developed for abrasive flow finishing', *International Journal of Precision Technology*, 1, pp. 302–313.
28. Sankar, M.R., Jain, V.K., Ramkumar, J., Joshi, Y.M., 2012, 'Dependence of R-AFF process on rheological characteristics of soft styrene based organic polymer abrasive medium', *International Journal of Manufacturing Technology Research*, 2012, 4, pp. 89–104.
29. Sankar, M.R., Jain, V.K., Ramkumar, J., 2012, 'Effect of Abrasive medium ingredients on finishing of Al alloy and Al alloy/SiC metal matrix composites using rotational abrasive flow finishing', *Applied Mechanics and Materials*, 110–116, pp. 1328–1335.
30. Singh, S., Sankar, M.R., Jain, V.K., Ramkumar, J., 2016, *Modeling of Finishing Forces and Surface Roughness in Abrasive Flow Finishing Process Using Rheological Properties*, 5th International and 26th All India Manufacturing Technology, Design and Research (AIMTDR-2014).
31. Macosko, C.W., 1994, *Rheology Principles, Measurements and Applications*, VCH Publishers, Inc., USA.
32. Das, M., Jain, V.K., Ghoshdastidar, P.S., 2008, 'Fluid flow analysis of magneto-rheological abrasive flow finishing (MRAFF) process', *International Journal of Machine Tools and Manufacture*, 48, pp. 415–426.
33. Loveless, T.R., Willams R.E., Rajurkar K.P., 1994, 'A study of the effects of abrasive flow finishing on various machined surfaces', *Journal of Material Processing Technology*, 47, pp. 133–151.
34. Jain, V.K., Adsul, S.G., 2000, 'Experimental investigations into abrasive flow machining', *International Journal of Machine Tools and Manufacture*, 40, pp. 201–211.
35. Jain, V.K., Ranganatha, C., Muralidhar, K., 2001, 'Evaluation of rheological properties of medium for AFM process', *Machining Science and Technology*, 5(2), pp. 151–170.

36. Tzeng, H.J., Yan, B.H., Hsu, R.T., Chow, H.M., 2007, 'Finishing effect of abrasive flow machining on micro slit fabricated by wire-EDM', *International Journal of Advanced Manufacturing Technology*, 34, pp. 649–656.

37. Lin, Y.C., Chow, H.M., Yan, B.H., Tzeng, H.J., 2007, 'Effects of finishing in abrasive fluid machining on microholes fabricated by EDM', *International Journal of Advanced Manufacturing Technology*, 33, pp. 489–497.

38. Walia, R.S., Shan, H.S., Kumar, P.K., 2006, 'Multi-response optimization of CFAAFM process through Taguchi method and utility concept', *Materials and Manufacturing Processes*, 21, pp. 907–914.

39. Walia, R.S., Shan, H.S., and Kumar, P.K., 2006, 'Parametric optimization of centrifugal force-assisted abrasive flow machining (CFAAFM) by the Taguchi method', *Materials and Manufacturing Processes*, 21, pp. 375–382.

40. Walia, R.S., Shan, H.S., and Kumar, P.K., 2009, 'Enhancing AFM process productivity through improved fixturing', *International Journal of Advanced Manufacturing Technology*, 44, pp. 700–709.

41. Reddy, K.M., Sharma, A.K., Kumar, P., 2008, 'Some aspects of centrifugal force assisted abrasive flow machining of 2014 Al alloy', *Proceedings of IMechE Part B: Journal of Engineering Manufacture*, 222, pp. 773–783.

42. Mali, H.S., Manna, A., 2009, 'An experimental investigation on abrasive flow finishing of Al-6063 alloy cylindrical surface', *IE (I) Journal–PR*, 90, pp. 3–7.

43. Mali, H.S., Manna, A., 2010, 'Optimum selection of abrasive flow machining conditions during fine finishing of Al/15 wt% SiC-MMC using Taguchi method', *International Journal of Advanced Manufacturing Technology*, 50, pp. 1013–1024.

44. Mittal, S., Kumar, V., Kumar, H., 2015, 'Experimental investigation and optimization of process parameters of Al/SiC MMCs finished by abrasive flow machining', *Materials and Manufacturing Processes*, 30(7), pp. 902–911.

45. Petri, K.L., Billo R.E., Bidanda, B., 1998, 'A neural network process model for abrasive flow machining operations', *Journal of Manufacturing Systems*, 17(1), pp. 52–64.

46. Jain, R.K., Jain, V.K., Kalra, P.K., 1999, 'Modelling of abrasive flow machining process: A neural network approach', *Wear*, 231, pp. 242–248.

47. Jain, R.K., Jain, V.K., 2000, 'Optimum selection of machining conditions in abrasive flow machining using neural network', *Journal of Material Processing Technology*, 108, pp. 62–67.

48. Mollah, A.A., Pratihar, D.K., 2007, 'Modelling of TIG welding and abrasive flow machining processes using radial basis function networks', *International Journal of Advanced Manufacturing Technology*, 37, pp. 937–952.

49. Ali Tavoli, M., Nariman-Zadeh, N., Khakhali, A., Mehran, M., 2006, 'Multi-objective optimization of abrasive flow machining processes using polynomial neural networks and genetic algorithms', *Machining Science and Technology*, 10, pp. 491–510.

5

Elastic Emission Machining

Ajay Sidpara

Mechanical Engineering Department, Indian Institute of Technology Kharagpur, Kharagpur, India

CONTENTS

5.1 Introduction

Requirements of ultra-smooth surfaces have been increasing continuously in the field of optics. Optical materials have been used in many technologies such as telescopes, cameras, astronomical mirrors, metrology, navigation, vision systems, modulation of energy beams (laser, electron, x-rays etc.) and many more. These applications of optics require extreme control over figure (profile) and surface roughness. Apart from the figure and surface roughness control, fabrication of stress-free surfaces is equally important to reduce the probability of unexpected failure of optical materials. Very few processes are available to process optical materials when the requirement is of figure accuracy in the sub-micron level and surface roughness in the angstrom level. One such process is 'elastic emission machining (EEM)'.

5.2 Polishing and Machining Processes

Figure 5.1 shows different types of errors due to the machining of a curved component. Roughness is due to the irregularities that are inherent in the production process (e.g. cutting tool and feed rates). The roughness also depends on the material composition and heat treatment. Figure error or waviness may result from vibration, chatter or workpiece deflection and strain in the material. Form error is the general shape deviation of the surface from the intended shape, neglecting variations due to roughness and figure error.

The polishing process should remove the form error without introducing any defect. Hence, a high-precision optical surface can be fabricated by accurately compensating or removing form errors. It can be performed by processes such as diamond turning, high-precision grinding, magnetorheological finishing, chemo-mechanical polishing (CMP), EEM and a few more. Diamond turning (Figure 5.2a) and high-precision grinding are considered as trajectory decided machining methods where a cutting tool or abrasive particles are fixed or embedded and they follow a certain path to achieve the desired surface profile (Su et al., 1996a). In such a method, the tool trajectory should accurately follow the programmed or desired surface profile. To achieve this, the deviation between the actual tool trajectory and the desired surface profile should be less than the error profile. When the error profile is very small (<0.1 μm), the rigidity of the machine tool, the position accuracy and precision of the servo motion controller, temperature variation,

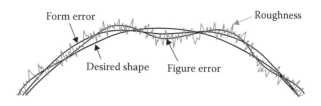

FIGURE 5.1
Types of surface errors on the curved component. (From Khan, G.S., Characterization of surface roughness and shape deviations of aspheric surfaces, PhD Thesis, University of Erlangen-Nuremberg, Germany, 2008.)

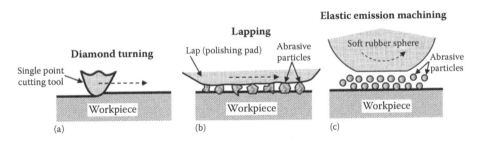

FIGURE 5.2
Interaction of cutting tool in different processes: (a) diamond turning, (b) lapping and (c) EEM.

vibration from the surroundings and machining environment significantly affect the performance of the process (Su et al., 1996b). Such machines (or processes) are very costly due to the high level of control requirements, and many times, they cannot be used when precision and accuracy requirement are in the order of nanometre level.

A polishing process is considered as deterministic if the polishing behaviour is always consistent whenever the process parameters are set to particular values. There are a few advantages of using a deterministic polishing process compared with diamond turning or grinding for removal of profile error at the sub-micron or nanometre level.

1. A polishing machine does not need a very high precision position movement. Hence, a high-precision servo controller is not required, which is necessary for diamond turning and grinding. This requirement is fulfilled by an accurate and controlled material removal rate (MRR) of the polishing process. When it is possible to achieve a controlled MRR, the machining precision of the work surface is higher than the positioning precision of a machine tool (Su et al., 1996b).

2. Structural deflection and vibration have very little effect on the machining or finishing performance of a polishing process (Su et al., 1996c). As a result, the cost of error compensation can be significantly reduced due to relief of a high-precision controller, machine rigidity and controlled machining environment, which are the prime requirements of diamond turning and grinding.

Based on these two advantages, the polishing process is a better choice for profile error compensation.

In polishing processes such as lapping (Figure 5.2b) and CMP, abrasive particles are directly pressed by the tool (lap) on the workpiece surface. Hence, MRR in such polishing processes is dictated by the effectiveness of abrasive particles and the number of abrasive particles participating in material removal (active abrasive particles). Selection of abrasive particles depends on the workpiece material being processed. Hence, polishing efficiency depends on abrasive particle selection, their velocity, as well as the distribution and normal load acting on them. These parameters must be controlled stringently to achieve deterministic polishing. However, abrasive particle shapes and surface irregularity of the workpiece are random. It makes these processes difficult to adopt as an ultra-precision polishing method (Su et al., 1996c).

There is another polishing process where abrasive particles are not pressed directly by the tool on the workpiece but they are dragged by the flow of liquid over the workpiece surface, as shown in Figure 5.2c. When two solid materials of different properties (mechanical and chemical) interact with each other, many reactions (mechanical and chemical) occur at the interface. Bonding between these two materials occurs due to the release of their surface energy before they contact each other. When these solid materials are separated by some external mechanical force or fluid flow, there is a high probability that the moving solid material detaches some atoms from the stationary solid material. If it continues for a long time, atom-by-atom material removal is possible without much mechanical force (Mori et al., 1987). The workpiece material is one solid body and the loose abrasive particle is another solid body in the EEM process. This is how, in brief, the EEM process works. The details of the process are discussed in the following sections.

5.3 Elastic Emission Machining (EEM)

Figure 5.3a shows a schematic diagram of the EEM process. A polyurethane sphere is attached to a variable speed motor to set a different rotational speed of the sphere to maintain a uniform working gap. The sphere is made of polyurethane because of its elasticity and stability to water. It is manufactured to a surface roughness of about 0.1 μm or less by diamond turning, grinding or similar high-precision cutting process.

The size of the abrasive particles is smaller than 1 μm, and they are mixed with ultra-pure (deionised) water. A temperature controller is used to maintain constant temperature during finishing and to reduce the effect of variable temperatures on the abrasive added liquid medium. The workpiece is fixed on a numerical controlled table so that a large area can be finished by providing raster or circular motion to the workpiece. Figure 5.3b shows a magnified view of the finishing zone where abrasive particles are dragged over the workpiece surface by rotation of the polyurethane sphere with little normal load and without physically pressing abrasive particles on the workpiece surface. Figure 5.3c shows a schematic of molecular dynamic simulation of the atomic interaction of abrasive particles and the workpiece surface.

Abrasive particles of 1 μm or less in size are uniformly mixed with ultra-pure water. They are accelerated by the flow of ultra-pure water and transported to the workpiece surface. A little normal load is applied to adjust or

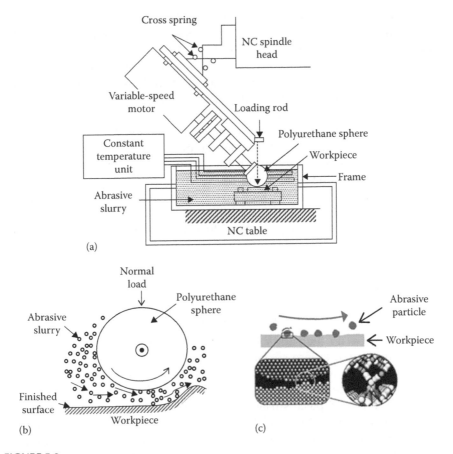

FIGURE 5.3
(a) Schematic diagram of EEM machine. (From Mori, Y., Yamauchi, K., Endo, K., *Prec. Eng.*, 9, 123–128, 1987.) (b) Closer view of material removal zone. (c) Atomic interaction between abrasive particle and workpiece surface. (From Yamauchi, K., Kikuji, H., Hidekazu, G., Kazuhisa, S., Koji, I., Kazuya, Y., Yasuhisa, S., Yuzo, M., *Comput. Mater. Sci.*, 14, 232–235, 1999.)

maintain thin film lubrication condition. When they contact the workpiece surface, atom-by-atom material removal takes place, as explained in Figure 5.3.

It is possible to remove material only when the atoms from both the surfaces create bonding. Furthermore, the binding energy of the topmost layer should be smaller than the binding energy of the layer below the topmost layer of the workpiece. The workpiece surface and abrasive particles are composed of different types of atoms. So, their interface may be electrically polarised and local charge-up may be induced due to various states of electrons. As a result, atoms of the workpiece located on the top surface are in a different condition than those of bulk material and it may help in releasing the top atoms during finishing (Yamauchi et al., 1999).

In EEM, the MRR strongly depends on the properties of the workpiece and abrasive particles. It is similar to the relationship between the workpiece material and the etchant in chemical etching. Therefore, selection of abrasive particles that are chemically active with workpiece material is one of the most important ways to enhance the MRR. The binding energy of zirconia (ZrO_2) particles is smaller than that of silica (SiO_2) in the case of silicon as a workpiece material. Hence, removal of atom from the silicon surface is easier in the case of ZrO_2 and the MRR is higher with ZrO_2 than SiO_2 (Inagaki et al., 2001).

The objective of EEM is to remove material atom by atom without inducing any plastic deformation, cracks, deep indentations, etc. In EEM, abrasive particles have kinetic energy which they transfer to the workpiece surface and they may cause damage to the workpiece surface due to their momentum. But due to very low mass, they do not aggressively strike the workpiece surface even though they may have high velocity. Therefore, it is necessary to very small sized abrasive particles (Mori et al., 1988).

5.4 Mechanism of Material Removal

Figure 5.4 shows a schematic explanation of probable steps of removal of atoms from the workpiece surface (Inagaki et al., 2001). It is believed that atoms of both the abrasive particle and the workpiece surface are terminated by hydroxide (OH) species (Figure 5.4a). When an abrasive particle approaches the workpiece surface, hydroxide species from both the solid surfaces contact each other and form a bonding (Figure 5.4b). After bonding, H_2O is separated by dehydration (Figure 5.4c). Finally, the flow of liquid medium drags the abrasive particle, and it may separate an atom from the workpiece (Figure 5.4d). This cycle continues, and atom-by-atom removal occurs from the workpiece surface.

Material removal in EEM depends on the chemical reaction, but it is different from conventional chemical etching process. In chemical etching (Figure 5.5), angstrom-sized reactive species (molecules and ions) attack the

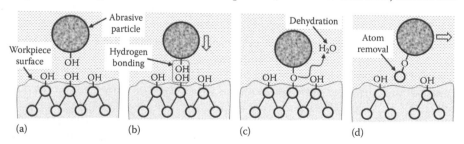

(a)   (b)   (c)   (d)

FIGURE 5.4

Schematic explanation of the interaction between abrasive particles and workpiece surface: (a) before interaction, (b) hydrogen bonding, (c) dehydration and (d) atom removal.

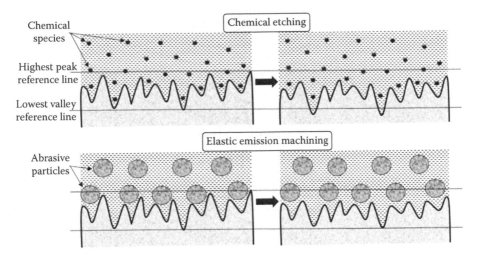

FIGURE 5.5
Surface chemical etching vs. EEM.

top surface of the workpiece. Due to the very small size of these reactive species, they enter micro pits of the surface and increase their depth. They do not flatten the surface, and a high-precision surface is difficult to achieve, while in the case of EEM (Figure 5.5), reactive species are abrasive particles, which are around 1 μm in size, and they do not enter in the micro pits of the workpiece surface. As a result, they selectively interact with the irregularities of the topmost layer of the workpiece only, and the probability of entering the abrasive particles inside the micro pits is much less (Sano et al., 2008). Hence, it is possible to achieve a high degree of surface planarisation.

5.5 Force Analysis

A study of magnitude and direction of forces acting on an abrasive particle provides insight of the abrasive particle behaviour during finishing. It helps in understanding the mechanism of material removal at a micro/nano scale. Figure 5.6 shows the forces acting on a particle passing through a lubrication film (Kanaoka et al., 2008). The behaviour of a particle can be traced from the equations of motion involved in these forces. It was found that without considering F_L and F_F forces, a theoretically calculated MRR from these forces has a positive correlation with the rotational speed of the tool and concentration of abrasive particles. However, the trend does not agree well with the experimental MRR. When these forces (F_L and F_F) are included, the theoretical MRR is in good agreement with the experimental MRR. It shows a strong effect of these forces on material removal.

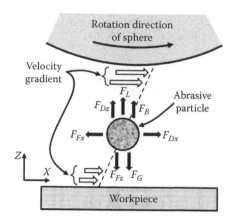

FIGURE 5.6
Forces acting in the EEM process.

Stokes drag forces (F_{Dx} and F_{Dz}) are given by

$$F_D = 6\pi\eta RV,$$ (5.1)

where η is fluid viscosity, R is the radius of a particle and V is the velocity of the particle.

Gravity force (F_G) is given by

$$F_G = mg,$$ (5.2)

where m is the mass of the object and g is acceleration due to gravity.

Buoyancy force (F_B) is given by

$$F_B = (\rho_{object} - \rho_{fluid})Vg,$$ (5.3)

where ρ is density.

Lift force or Suffman force (F_L) is generated in fluid flow with a velocity gradient.

F_{Fx} and F_{Fz} are the forces generated by a mutual interaction of particles in the x and z directions, acting on a particle to hinder it from moving forward.

5.6 EEM and Hydrodynamic Lubrication

The working principle of EEM is synonymous to hydrodynamic lubrication theory. The polishing spherical tool transmits energy to the slurry by high-speed rotation. It is believed that the abrasives and the liquid move together and have the same velocity. When abrasive-particle-laden liquid is dragged into the converging working gap, it gets compressed and its viscosity increases sharply to resist the compression. As a result, pressure rises sharply due to the viscosity-pressure effect (Hamrock et al., 2004). When sufficient pressure is developed, the spherical tool floats above the workpiece surface. At some particular combination of rotating speed and normal load on the sphere, a thin fluid film of a few microns is formed. As the viscosity of the slurry is low, elasto-hydrodynamic lubrication happens at high rotational speed and light load of the tool (Zhang et al., 2012).

5.7 Machining Strategies

Material removal depth in the polishing process depends on the dwell time distribution of the tool. Depending on the profile error of the workpiece surface, the motion of x–y–z axes controllers is programmed to set a particular dwell time. High dwell time increases the finishing spot area and material removal depth. The size of the finishing spot and MRR are directly related to the spatial resolution and machining time in EEM (Takei and Mimura, 2013).

When the variation in profile to be machined is high, a differential dwell time for machining is necessary at different locations to achieve the desired surface profile. Let the machining depths of points 1 to 4 be denoted as d_1 to d_4, respectively, and let the required machining times for points 1 to 4 be denoted as t_1 to t_4, respectively, as shown in Figure 5.7.

The MRR may change with location and time due to variation in the curvature of the workpiece surface even when all the process parameters are kept constant. That is because the surface profile curvature influences the

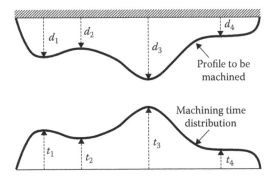

FIGURE 5.7
Relationship between machining depth (d) and machining time (t).

shear stress of slurry. If the curvature effects of points 1 and 2 on MRR are negligible and the machining spot is infinitely small, the relation between machining depth (d) and machining time (t) can be written as

$$\frac{d_1}{d_2} = \frac{t_1}{t_2}.$$

(5.4)

In the same way, it can be written for points 2, 3 and 4.

This relation states that the machining time distribution of tool motion is a linear function of the profile to be machined when the curvature effect of a desired profile is negligible and the machining spot area is infinitely small (Su et al., 1996c). However, in actual practice, the machining spot has a finite area because of that machining outside the boundary of machining area takes place when the tool is moving over the boundary of the machining area, as shown in Figure 5.8. This results in over-cutting (shaded area in Figure 5.8) of the desired dimension of the machining zone. Over-cutting can be reduced either by reducing the machining spot (small diameter or high elastic modulus of the tool or low normal load) or reducing the machining depth by proper selection of process parameters.

When the machining area is finite and the shape (area and depth) of the spot is non-uniform, it is very difficult to program the motion planning of the tool. In EEM, the MRR depends on the shear stress or fluid film thickness distribution between the tool and the workpiece surface. The film thickness is not uniform and it is also a function of lubricating conditions. Due to that, it results in different MRRs at different locations and create surface ripples similar to turning process. When the big area is machined/ finished by a small tool, raster motion is given to the tool or workpiece to cover the whole area of the surface to be machined. When the tool is given offset in the y axis and the material removal spot is located side by side (spots are connected continuously), the final surface generation has a high variation in surface roughness on the top surface of the workpiece, as shown in Figure 5.9a. This variation can be reduced when the machining spots are overlapped with each other, as shown in Figure 5.9b. However,

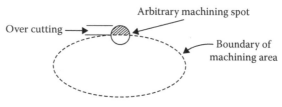

FIGURE 5.8
Schematic showing overcutting on the top surface of the workpiece.

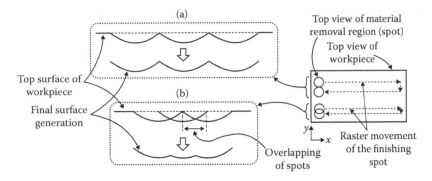

FIGURE 5.9
Generation of surface ripples when material removal spots are (a) side by side and (b) overlapping each other.

the overlapping length should be chosen properly and it should be decided by profiling the machining spot and the other process parameters.

5.8 Shear Stress in EEM

Sufficient velocity is required to move the abrasive particles, break the bonding with the workpiece atom and take away the atom from the workpiece. This is the primary condition to remove material in EEM. Drag force (velocity of the fluid) ensures that the abrasive particles are not firmly bonded with the workpiece atom, flush away the removed atoms from the working zone and allow the subsequent abrasive particles to interact with the workpiece surface. Drag force results from the shear flow of the fluid and is a function of the lubrication condition of the abrasive-particle-laden liquid at the region between the tool and the work surface (Su et al., 1995).

Consider a sphere rotating in a fluid relative to a fixed plate without touching it. If the flow of the fluid between the sphere and the plate is Newtonian, then the shear stress τ at any point of the flow is given by

$$\tau = \mu \frac{du}{dz}, \tag{5.5}$$

where μ is fluid viscosity, u is the velocity of the point and z is the height perpendicular to the flow direction.

If a small particle exists in the flow, the periphery of the particle will be subjected to shear stress due to velocity gradient (du/dz), as shown in Figure 5.6.

In the EEM process, the lubrication condition at the fluid film generated between the rotating tool and workpiece surface can be approximated by

$$\tau \approx \mu \frac{r\omega}{h}, \tag{5.6}$$

where r is the radius of the tool, ω is the rotational speed of the tool and h is the film thickness.

For a fixed diameter of the sphere, shear stress can be determined by the abrasive slurry viscosity, tool speed and film thickness. The slurry viscosity depends mainly on the concentration of abrasive particles and the properties of liquid and particle. However, under high pressure, the slurry viscosity can be evaluated by the following relationship (Hamrock et al., 2004):

$$\mu = \mu_0 e^{\zeta P}, \tag{5.7}$$

where μ_0 is slurry viscosity at the atmosphere, ζ is pressure-viscosity coefficient and P is pressure on the film.

When the normal load is large and the elastic modulus of the tool is high, viscosity can be increased significantly under a fully flooded lubricating condition. However, if the tool is made of soft material as in the case of EEM, the viscosity of slurry can be assumed to be independent of the pressure at the fluid film unless ζ is high, which is a slurry property. Shear stress in EEM is affected by the fluid film thickness, which should be small enough to remove abrasive particles effectively from the workpiece surface (Su et al., 1995). Abrasive particles have variation in their size, and many particles are bigger than the average particle size. When the fluid film thickness is small, there is a probability of scratching the workpiece surface by these big size abrasive particles. Therefore, proper control of the fluid film thickness is important to create a scratch-free surface. The fluid film thickness depends on a specific region of lubrication in case of fully flooded lubricating condition.

Depending on the magnitude of the elastic deformation of the solid tool under normal load and increase in fluid viscosity with pressure, fluid film lubrication is divided into four different regimes (Hamrock et al., 2004).

Isoviscous-rigid (IR) regime: The amount of elastic deformation of the solids is insignificant relative to the thickness of the fluid film separating them and the maximum contact pressure is too low to increase the fluid viscosity significantly.

Viscous-rigid regime: If the contact pressure is high enough to increase the fluid viscosity significantly within the conjunction, it may be necessary to consider the pressure-viscosity characteristics of the lubricant while assuming that the solids remain rigid (deformation of the solids is insignificant relative to the fluid film thickness).

Isoviscous-elastic (IE) regime: Elastic deformation of the solids is significant relative to the thickness of the fluid film separating them but the contact pressure is quite low and insufficient to increase the viscosity significantly.

Viscous-elastic regime: The elastic deformation of the solids is significant relative to the thickness of the fluid film separating them in fully developed elasto-hydrodynamic lubrication and the contact pressure is high enough to increase the viscosity of lubricant significantly within the conjunction.

It is observed in the simulation that if the tool is made of soft material and the pressure-viscosity coefficient (ζ) has an order of 10^{-8} m^2.N, the lubrication of the EEM process is most likely in the IE and IR regions (Su et al., 1995). Hence, minimum fluid film thickness in these two regions can be written as

$$h_{IR} = C_{IR}\omega^2\mu^2 w^{-2} \tag{5.8}$$

and

$$h_{IE} = C_{IE}\omega^{0.66}\mu^{0.66}w^{-0.2111}E'^{-0.4422}, \tag{5.9}$$

where w is applied normal load, E' is effective elastic modulus of the workpiece and the tool and C_{IR} and C_{IE} are geometric parameters, which are functions of the geometries of the tool and workpiece.

Putting Equations 5.8 and 5.9 into Equation 5.6, shear stress (τ) in different regions can be written as

$$\tau_{IR} = r\omega^{-1}\mu^{-1}w^2 C_{IR}^{-1} \tag{5.10}$$

and

$$\tau_{IE} = r\omega^{0.34}\mu^{0.34}w^{0.2111}E'^{0.4422}C_{IE}^{-1}. \tag{5.11}$$

Table 5.1 shows the relationship of the process parameters of Equations 5.10 and 5.11 with the shear stress of different regions. It is observed that

TABLE 5.1

Relationship of Process Parameters with Shear Stress in Different Regions

Parameters	τ_{IR}	τ_{IE}
Tool rotation (ω)	Negative	Positive
Viscosity (μ)	Negative	Positive
Applied normal load (w)	Positive	Positive
Effective elastic modulus (E')	–	Positive

when the normal load is low and tool speed is high, lubrication is likely in the IR regime, and when the speed is low, lubrication tends to be in the IE region (Su et al., 1995).

5.9 Effect of Tool Roughness

The tool is assumed as a perfectly smooth sphere. The tool is made of soft rubber-like material and it is difficult to fabricate the tool surface having the surface roughness less than the thickness of the lubrication film formed during machining/finishing (Su et al., 1996b). Therefore, assuming a perfectly smooth tool surface is not correct as it has surface micro-irregularities, as shown in Figure 5.10.

The surface roughness of the tool and work surface affects the formation and thickness of the fluid film between the tool and the workpiece surface (Hamrock et al., 2004). Due to the random distribution of tool surface roughness, the fluid film thickness also varies at different locations, as shown in Figure 5.10. As a result, shear stress (τ) also varies (Equation 5.6) and subsequently MRR also. This roughness effect may be significant when the film thickness has the same order as the dimension of surface roughness.

Due to irregularities (surface roughness and waviness) of the tool and workpiece surface, a certain portion of the high asperities of the tool directly presses the abrasive particles to contact the work surface. This particular

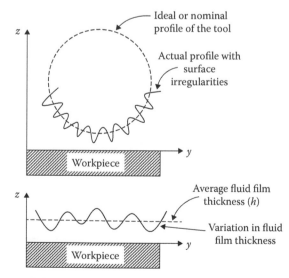

FIGURE 5.10
Effect of tool surface irregularities on the variation of the fluid film thickness.

situation is similar to mixed lubrication (Hamrock et al., 2004). Hence, the assumption of non-contact between tool and workpiece may not be valid.

When abrasive particles directly press the workpiece surface due to the tool asperities, they exert high normal load and shear stress on the workpiece surface. These normal and tangential forces are much higher than those forces that act when the abrasive particles do not directly press the workpiece surface. Hence, the MRR is quite high under such circumstances. The effect of tool surface irregularities on the MRR can be minimised by either reducing the surface roughness of tool or increasing the fluid film thickness (Su et al., 1996c).

5.10 Types of Tool, Workpiece and Abrasive Interactions

The MRR is inversely proportional to the fluid film thickness distribution. The maximum MRR is achieved at the thinnest fluid film. Hence, the working gap between the tool and workpiece surface affects the lubrication state and MRR (Zhang et al., 2012). Depending on the process parameters, the fluid film thickness and the interaction of tool-workpiece-abrasive, the lubrication in the elasto-hydrodynamic zone may be a semi-contact type (mixed lubrication) or a non-contact type (Hamrock et al., 2004). Abrasive particles are dragged in the converging working gap in both cases.

5.10.1 Semi-Contact Case

In this case, the asperities of the tool may press the abrasive particles against the workpiece surface. Due to the direct interaction between tool-abrasive-workpiece, mechanical abrasion or adhesion may occur. When a larger number of asperities are in contact with the workpiece surface, more abrasive particles are directly pressed on the workpiece surface and it results in a higher MRR. The area of contact depends on the distribution of surface roughness (peak to valley) and relative magnitude between the fluid film thickness and surface roughness of the tool and workpiece surface. When the fluid film thickness is small, the probability of direct contact of tool-abrasive-workpiece is higher, and subsequently, the MRR increases. Similarly, if the tool speed is high, the frequency of direct contact of the tool and the abrasive particle and the pressing abrasive particles on the workpiece surface increases and it increases the MRR.

The tool rotational speed has two contradictory effects on MRR. Based on lubrication theory, the fluid film thickness increases with increasing tool rotational speed. It reduces the probability of contact of asperities and, hence, the MRR also. On the other end, the high rotational speed of the tool increases the capability of the abrasive particles to overcome the resistance (interatomic bond of workpiece) of the workpiece atoms and it increases the MRR. Apart from the tool speed, viscosity also plays an important role in controlling fluid film thickness. However, an increase in MRR is much higher with an increase in tool speed as compared to an increase in viscosity. A high normal load increases the probability of a semi-contact mode as a high normal load reduces the film thickness and increases the proportion of contact asperities and finally the MRR increases.

5.10.2 Non-Contact Case

When the fluid film thickness is high, the tool does not press the abrasive particles directly on the workpiece surface and the tool also does not come in contact with the workpiece surface. Under such a condition, the abrasive particles may be subjected to three kinds of forces when they interact with the workpiece surface (Su et al., 1995):

1. *Eccentric force driven by the tool*

   The trajectory of the abrasive particle is different before and after entering the finishing or machining zone. So, the eccentric force near the entry of the machining zone pushes the abrasive particle to move toward the workpiece surface.

2. *Van der Waals force*

   When the abrasive particles are dragged to the workpiece surface with atomic size distance, Van der Waals force acts between the abrasive particle and workpiece surface atoms (Gane et al., 1974). As a result, the abrasive particle probably bonds with the workpiece surface atoms (Cook, 1990).

3. *Shear force*

   The film thickness is of micrometer order; slurry flow in the machining zone generates high shear stress due to the large velocity gradient. Hence, shear force keeps the abrasive particle in motion and does not allow the abrasive particle to bind with the workpiece surface atoms.

5.11 Factors Affecting Shear Stress

Material removal in EEM significantly depends on the shear stress acting on the abrasive particles. Some of the parameters affecting shear stress are shown in Figure 5.11 (Zhang et al., 2012). In the laminar flow condition, shear stress can be approximately calculated by tool rotational speed, viscosity of slurry and fluid film thickness. Out of these three parameters, the first two can be directly measured and the last parameter is a function of applied normal load, tool rotational speed, elastic moduli, as well as the geometric properties of the tool and workpiece surface (Su et al., 1996c).

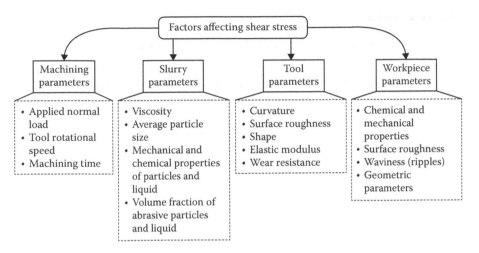

FIGURE 5.11
Factors affecting shear stress in EEM.

Apart from shear stress, the adhesive force between the abrasive particle and workpiece surface atoms also affects the MRR in EEM. The adhesive force depends on the topography of abrasive particle's atoms (Gane et al., 1974), chemical properties of the abrasive particles, liquid medium and workpiece (Mori et al., 1987).

5.12 Process Environment

Since material removal is at the atomic scale and the required surface finish to be achieved is at the angstrom level, the process environment plays an important role in achieving the desired results. Material removal in EEM is based on a chemical reaction between an abrasive-based slurry and the workpiece surface. Hence, introduction of chemical and other impurities should be avoided to achieve the desired precision. Generally, two types of impurities create more problems during finishing (Yamauchi et al., 2002). The first one is organic contamination on the processed surface. It acts as a blocking mask and hinders chemical reaction during EEM. It results in a low MRR and more processing time to reduce micro roughness. The second one is oxygen dissolved in ultra-pure water. It oxides the processed surface particularly during silicon finishing as the formation of the oxide layer is high (~1 nm/min) in some cases. Hence, the introduction of oxygen gas and organic materials such as oil vapour should be avoided/minimised. Therefore, a hydrostatic supporting system having ultra-pure water is used for guides, bearings and other moving stages.

5.13 Agglomerated and Separated Abrasive Particles

The MRR is also affected by the dispersion of abrasive particles. A single sphere has a smooth surface, while agglomerated particles have significant surface irregularities due to the size and the shape of the primary particles. Hence, agglomerated particles have a high probability of increasing the contact area with the workpiece surface and it results in a high MRR. It is found that an MRR with agglomerated particles is 100 times more than that achieved with free (non-agglomerated) abrasive particles (Kubota et al., 2006).

5.14 Applications

EEM is a widely used process for fabrication of atomically smooth and stress-free surfaces of optical materials, such as 4H-SiC(0001) (Kubota et al., 2005), ULE and Zerodur (Kanaoka et al., 2007), silicon carbide (Kubota et al., 2007), adaptive bimorph mirror (Sawhney et al., 2013) and many more. A few more specific applications are briefly described in the following.

5.14.1 Optics for Extreme Ultraviolet Lithography

Extreme ultraviolet lithography is a fabrication technique suitable for printing features smaller than 50 nm. It can be achieved by using light at a wavelength of 13.5 nm, which is much shorter than the wavelength of 193 generally used today (Kanaoka et al., 2007). An optical lens system operating in such a short wavelength range demands extremely high accuracy as a small figure error (even in sub-micron level) or surface roughness may distort the wave front of the reflected light (Takei and Mimura, 2013). Hence, such an application requires high-precision optics with absolute figure accuracy and a root mean square (rms) roughness of 0.1 nm or so. Such optics are made of materials having a low coefficient of thermal expansion such as Zerodur, ULE, etc., because around 30% of the light illuminating on optical surface is not reflected back and is converted to heat.

5.14.2 Ellipsoidal Mirrors in X-Ray Microscopy

An ellipsoidal mirror is a rotationally symmetric ring-shaped reflective focusing device that has many advantages such as a large aperture, long focal length, extremely low chromatic aberration and nanometre focal size (Motoyama et al., 2014). Ellipsoidal mirrors are utilised in microscopes using next-generation synchrotron radiation (SR) and laboratory X-ray sources such as X-ray microscopy. These mirrors have been developed using epoxy resin (Hasegawa et al., 1994), but they are not precise enough to condense coherent soft X-rays to the diffraction limit. Therefore, EEM is used for the fabrication of mandrels of ellipsoidal focusing mirrors (Takei et al., 2013).

5.14.3 Focusing Mirrors in SR Beamlines

Focusing mirrors are an inevitable component of SR beamlines. They are used for steering and modulating the hard X-ray to the focal point. These reflective mirrors have many advantages such as achromaticity, nearly 100% efficiency and a range of working distances (Sawhney et al., 2013). This application requires an atomically flat and/or crystallographically perfect surface of mirror and EEM is one of the processes used for fabrication of these mirrors.

5.15 Remarks

EEM is a very promising process for atomic-scale finishing of optical materials. It has proven its capability of producing surface roughness at the sub-nanometre level and figure accuracy at the sub-micron level. Shear stress and the formation of fluid film play vital roles in the controlled MRR and surface finish. Apart from that, the environmental condition also affects the final results.

Questions

1. Explain the terms 'surface roughness' and 'figure error'.
2. Why is diamond turning a costly process for removing form error?
3. What are the advantages of using a deterministic polishing process in place of diamond turning for form error removal?
4. Explain the working principle of the EEM process.
5. How is EEM better than the lapping process for achieving sub-nanometre surface roughness?
6. State the differences between EEM and the conventional etching process.
7. Draw a diagram of forces acting in the EEM process.
8. How does an overlapping finishing spot reduce the height of surface ripples in the EEM process?
9. What are the different regimes of fluid film lubrication?
10. How do the irregularities on tool surfaces affect the formation of fluid film in the EEM process?

11. What are semi-contact and non-contact modes in the EEM process?
12. What are the process parameters affecting shear stress in the EEM process?
13. State some applications of the EEM process.

References

Cook, L. M. (1990). Chemical processes in glass polishing. *Journal of Non-crystalline Solids*, 120(1), 152–171.

Gane, N., Pfaelzer, P. F., & Tabor, D. (1974, October). Adhesion between clean surfaces at light loads. *Proceedings of the Royal Society of London A: Mathematical, Physical and Engineering Sciences*, 340(1623), 495–517.

Hamrock, B. J., Schmid, S. R., & Jacobson, B. O. (2004). *Fundamentals of Fluid Film Lubrication*, Marcel Dekker, Inc., New York.

Hasegawa, M., Taira, H., Harada, T., Aoki, S., & Ninomiya, K. (1994). Fabrication of Wolter-type x-ray focusing mirror using epoxy resin. *Review of Scientific Instruments*, 65(8), 2568–2573.

Inagaki, K., Yamauchi, K., Mimura, H., Sugiyama, K., Hirose, K., & Mori, Y. (2001). First-principles evaluations of machinability dependency on powder material in elastic emission machining. *Materials Transactions*, 42(11), 2290–2294.

Kanaoka, M., Liu, C., Nomura, K., Ando, M., Takino, H., Fukuda, Y., Mimura, H., Yamauchi, K., & Mori, Y. (2007). Figuring and smoothing capabilities of elastic emission machining for low-thermal-expansion glass optics. *Journal of Vacuum Science & Technology B: Microelectronics and Nanometer Structures*, 25(6), 2110–2113.

Kanaoka, M., Takino, H., Nomura, K., Mimura, H., Yamauchi, K., & Mori, Y. (2008, October 19–24). Factors affecting changes in removal rate of elastic emission machining, Proceedings of ASPE 2008 Annual Meeting and the Twelfth ICPE; Portland, Oregon: 615–618.

Khan, G. S. (2008). Characterization of surface roughness and shape deviations of aspheric surfaces, PhD Thesis, University of Erlangen-Nuremberg, Germany.

Kubota, A., Mimura, H., Inagaki, K., Arima, K., Mori, Y., & Yamauchi, K. (2005). Preparation of ultrasmooth and defect-free 4H-SiC (0001) surfaces by elastic emission machining. *Journal of Electronic Materials*, 34(4), 439–443.

Kubota, A., Mimura, H., Inagaki, K., Mori, Y., & Yamauchi, K. (2006). Effect of particle morphology on removal rate and surface topography in elastic emission machining. *Journal of the Electrochemical Society*, 153(9), G874–G878.

Kubota, A., Shinbayashi, Y., Mimura, H., Sano, Y., Inagaki, K., Mori, Y., & Yamauchi, K. (2007). Investigation of the surface removal process of silicon carbide in elastic emission machining. *Journal of Electronic Materials*, 36(1), 92–97.

Mori, Y., Yamauchi, K., & Endo, K. (1987). Elastic emission machining. *Precision Engineering*, 9(3), 123–128.

Mori, Y., Yamauchi, K., & Endo, K. (1988). Mechanism of atomic removal in elastic emission machining. *Precision Engineering*, 10(1), 24–28.

Motoyama, H., Saito, T., & Mimura, H. (2014). Error analysis of ellipsoidal mirrors for soft X-ray focusing by wave-optical simulation. *Japanese Journal of Applied Physics*, 53(2), 022503.

Sano, Y., Yamamura, K., Mimura, H., Yamauchi, K., & Mori, Y. (2008). Plasma chemical vaporization machining and elastic emission machining, in *Crystal Growth Technology: From Fundamentals and Simulation to Large-scale Production* (eds H. J. Scheel and P. Capper), Wiley-VCH Verlag GmbH & Co. KGaA, Weinheim, Germany. doi: 10.1002/9783527623440.ch19.

Sawhney, K., Alcock, S., Sutter, J., Berujon, S., Wang, H., & Signorato, R. (2013, March). Characterisation of a novel super-polished bimorph mirror. *Journal of Physics: Conference Series*, 425(5): 052026.

Su, Y. T., Wang, S. Y., Chao, P. Y., Hwang, Y. D., & Hsiau, J. S. (1995). Investigation of elastic emission machining process: Lubrication effects. *Precision Engineering*, 17(3), 164–172.

Su, Y. T., Horng, C. C., Sheen, J. Y., & Jar-Sian, H. (1996a). A process planning strategy for removing an arbitrary profile by hydrodynamic polishing process. *International Journal of Machine Tools and Manufacture*, 36(11), 1227–1245.

Su, Y. T., Horng, C. C., Hwang, Y. D., & Guo, W. K. (1996b). Effects of tool surface irregularities on machining rate of a hydrodynamic polishing process. *Wear*, 199(1), 89–99.

Su, Y. T., Horng, C. C., Wang, S. Y., & Jang, S. H. (1996c). Ultra-precision machining by the hydrodynamic polishing process. *International Journal of Machine Tools and Manufacture*, 36(2), 275–291.

Takei, Y., & Mimura, H. (2013). Effect of focusing flow on stationary spot machining properties in elastic emission machining. *Nanoscale Research Letters*, 8(1), 1–6.

Takei, Y., Kume, T., Motoyama, H., Hiraguri, K., Hashizume, H., & Mimura, H. (2013, September). Development of a numerically controlled elastic emission machining system for fabricating mandrels of ellipsoidal focusing mirrors used in soft x-ray microscopy. In SPIE Optical Engineering + Applications (pp. 88480C–88480C). International Society for Optics and Photonics.

Yamauchi, K., Kikuji, H., Hidekazu, G., Kazuhisa, S., Koji, I., Kazuya, Y., Yasuhisa, S., & Yuzo, M. (1999). First-principles simulations of removal process in EEM (elastic emission machining). *Computational Materials Science*, 14(1), 232–235.

Yamauchi, K., Mimura, H., Inagaki, K., & Mori, Y. (2002). Figuring with subnanometer-level accuracy by numerically controlled elastic emission machining. *Review of Scientific Instruments*, 73(11), 4028–4033.

Zhang, L., Wang, J., & Zhang, J. (2012). Super-smooth surface fabrication technique and experimental research. *Applied Optics*, 51(27), 6612–6617.

6

Elasto-Abrasive Finishing

V.S. Sooraj and V. Radhakrishnan

Indian Institute of Space Science and Technology, Kerala, India

CONTENTS

6.1 Introduction

Elasto-abrasive finishing is a newly developed procedure for ultra-fine finishing of engineering surfaces in the micro/nano range, without altering the basic surface form. As the name indicates, it is an abrasive-based methodology using specially developed elasto-abrasive balls or spheres. The elasto-abrasive balls discussed in this chapter refer to *abrasive embedded elastomeric beads* in the form of spherical balls of meso-/micro-scale dimensions. Elasto-abrasives in the form of balls (diameter in the range of 3 to 4 mm) are easy to handle and have the advantages of both loose as well as bonded abrasive particles. This allows them to be used for multiple applications based on *abrasion* as well as *erosion* approaches.

The presence of an elastomeric medium that bonds the abrasives makes these balls capable of deforming in conformity to the work surface and thus reduces the effective contact stress in its interaction with the work surface. There is a significant reduction in the contact force, as it is transferred to the work surface through an elastomeric medium (which acts like a spring), facilitating very fine material removal. Due to this special feature, the technique can be employed to yield micro/nano finishing without altering the surface form.

This chapter discusses the configuration and characteristics of elasto-abrasives, their multi-application capabilities, types of applications and details of the finishing systems.

6.2 Elasto-Abrasive Finishing: Concept and Scope of the New Technique

Fine refinement of surface irregularities without severe sub-surface damage is a critical requirement in many of the abrasive-based finishing operations. One of the significant strategies for this is to reduce the magnitude of the abrading forces. Use of rubber bonded wheels in grinding, soft-flexible lap or pad in lapping and polishing are typical methodologies practiced for the same [1–4]. When using a hard polishing pad, the abrasive grains will be pushed deeper into the surface, leading to poor surface quality. It is also reported that a soft pad may deteriorate the flatness of the part, as the stock removal progresses. However, mechano-chemical polishing is a viable method to sustain the flatness, although the chemical reaction by the slurry medium is a serious concern [4]. When the discussion is about achieving an optical quality surface on intricate components, lapping and polishing have major limitations. The major bottleneck lies in the control of the abrading forces and the accessibility of intricate features. Even though polishing operations can be fine-tuned to address some of these limitations, the need for expensive equipment and the long processing time make them economically non-viable [5].

A number of advanced fine-finishing techniques have been reported in the recent past [5], with and without a deterministic control over in-process finishing forces. Micron/sub-micron-sized abrasives are preferred in many of these micro/nanofinishing operations because of the following characteristics:

- Low magnitude of finishing forces
- Miniaturised cutting edges
- Reduced depth of penetration
- Nano-scale material removal
- Possibility of more number of active grains per contact area

Fine abrasives are capable of yielding ultra-fine finishes as well as dimensional tolerance, even in hard and difficult-to-cut materials, because of these characteristics [1,5,6].

However, handling of micron/sub-micron-sized abrasive particles in their loose state and feeding them in the working zone are relatively cumbersome. Unbounded magnetic abrasive finishing is an example of such a situation. During magnetic abrasive finishing, it is found that the abrasive grains in their loose form may easily flow away from the working zone and the abrasives cannot be recycled after the finishing operation. These observations motivated researchers like Wang and Lee [7] to think about a flexible bonding of abrasives through silicone gel in magnetic abrasive polishing. But it will be of interest to have a promising methodology that can replace the usage of these gels or slurry medium.

Even though loose abrasives are effective in many of the advanced methodologies, a number of experiments have demonstrated the merits of bonded abrasives. Usage of bonded magnetic abrasive particles, with abrasives held in a ferromagnetic matrix formed by sintering, showed improved results in finishing compared with an unbounded homogeneous mixture of ferromagnetic powders and abrasives [5,8]. An alternative mode of utilising abrasive grains bonded in an ice matrix, which can expose fresh grains periodically without any dressing, is reported [9–12]. But the preparation of bonded particles and the requirement of special operating conditions (such as cryogenic environment in an ice-bonded technique) are typical difficulties associated with these bonded techniques. Accessibility restrictions for complex-profiled surfaces are also of major concern.

Alternatively, erosion of surfaces using free abrasives has been attempted for fine-finishing applications [13–16]. Fluidisation of loose abrasive grains and impinging them against the target surface (at relatively low velocities of the order of 5 m/s or less) by revolving the workpiece is the strategy proposed in this technique. Unlike abrasive jet machining having jet velocities of the order 300 m/s, this procedure results in fine surface erosion, giving excellent surface finish. But this technique has limitations in achieving nano-scale finishing, as fluidisation of micron- and sub-micron-sized abrasives (Geldart group C particles) is impossible [17].

From an application point of view, the major bottlenecks observed in the existing techniques (in addition to the difficulties associated with the handling of fine grains) are as follows:

- Preparation, selection and feeding of abrasive carrier media
- Variation in the rheological behaviour of the abrasive carrier medium, according to the method opted
- Limitations in the use of the same medium for abrasion as well as erosion-based techniques

To illustrate these points in detail, consider the polymer-abrasive medium used in abrasive flow finishing (AFF). Detailed literature review confirms that there is significant variation in the viscosity requirement of this medium, even to make it effective for the hybrid variants of AFF. More specifically, a low-viscosity medium allowing the migration of ferromagnetic particles towards the magnet is ideally suited for magnetic AFF. But for spiral polishing, the results were excellent with a high-viscosity medium. A low-viscosity medium cannot be scooped up and down as a lump. Then again, a low-viscosity medium is preferred for centrifugal-force-assisted AFF for facilitating easy flow of the medium [5]. In addition, it is very evident that the AFF medium is in no way suited for erosion-based techniques.

Elasto-abrasive finishing is a new technique developed to address some of these challenges. Elasto-abrasives (or elastic abrasives) are near spherical elastomeric beads embedded with abrasive grains. The term 'grain' is used in this chapter as a synonym for particle. In the form of balls, they can be moved freely like loose particles; at the same time, the fine abrasive grains are bonded together in an elastomeric medium. This approach does not belong to the category of rigid bonded finishing or to the fully loose abrasives in a viscous media. Thus, elasto-abrasive finishing is capable of bridging the gap between loose abrasive finishing and rigid bonded abrasive finishing, combining the practical advantages of both. Elasto-abrasive finishing is a flexible, multi-application-oriented approach, which can be customised according to the processing requirements. Applications of this technique to finish internal, external, flat, non-circular and grooved surfaces, without altering the surface form and without severe sub-surface damage, are clearly illustrated in this chapter. It is also easy to fluidise elasto-abrasive balls, which makes them suitable for erosion-based fluidised finishing systems as well.

6.3 Elasto-Abrasives

6.3.1 Configuration and Characteristics

The major constituents of an elasto-abrasive ball are (i) elastomeric polymer beads, (ii) fine abrasive grains and (iii) an appropriate organic solvent. The role of the organic solvent is to soften the polymer bead and make it compliant to embed the abrasive grains over it, through a proper heating cycle (Figure 6.1). The desirable characteristics for the selection of polymer bead are the following:

- High resilient elastomer with relatively low flexural modulus and shore hardness, combined with reasonable shear/tear strength
- Surface softening and swelling of polymer achieved by using an appropriate organic solvent to facilitate easy embedding of abrasives

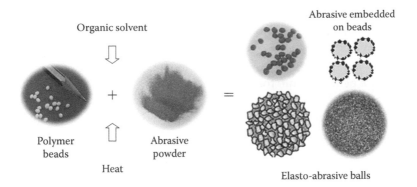

FIGURE 6.1
Concept and configuration of elasto-abrasive balls (exact composition and preparation are covered under patent application).

- Reasonably high temperature stability to avoid fusing of elastic abrasives during the process
- Good resistance to environmental factors, high resistance to corrosion/abrasion, as well as good oil resistance

The abrasive grains for the preparation of these balls can be selected from the list of commonly used natural or synthetic abrasives. However, it is ideal to use common abrasives that are cost effective, to make the process economical. Silicon carbide (SiC) and aluminium oxide (Al_2O_3) grains of size range 10 to 250 μm were used for the studies presented in this chapter.

In a simple way, the preparation of elasto-abrasive balls can be done by embedding the abrasives over the surface of polymer beads, following a direct and simple chemical procedure. After softening the elastomeric beads using the organic solvent, abrasive grains can be embedded directly over the beads using a controlled heating cycle. These balls can also be prepared in micro-scale dimensions using a suspension polymerisation procedure. Some of the unique features of elasto-abrasive balls are as follows:

- Elastic abrasives in the form of spherical ball are suitable for erosion as well as for abrasive action. Using custom-made application setups, elastic abrasive balls can be applied over a variety of surfaces, including flat surfaces, internal surfaces, external surfaces, grooves, free-form surfaces, etc. Hence, it can be categorised as a multi-mode, multi-application oriented approach.
- Elastic abrasives in the form of small balls are convenient to handle and transport and allow easy cleaning of the part after finishing. These balls can be applied without any slurry medium, making them environment friendly.

- The presence of an elastomeric medium makes these balls self-deformable in conformity to the work surface upon loading, absorb energy and reduce the contact stress.
- While the elastic ball deforms, a number of fine abrasive grains on its surface penetrate into the work surface. These are referred to as active abrasive grains, which are directly engaged in micro-cutting action.
- It is easy to control the overall size and characteristics of the elasto-abrasive balls, as per application needs. This can be achieved by the proper selection of polymer beads and additives.
- The type and size of embedded abrasive grains can be varied according to application needs.
- Reusable to a great extent under normal processing conditions. In case of severe abrasive wear, re-embedding of abrasive grains on elastomeric polymer bead can also be done.
- Reasonably good oil and water resistance under normal conditions of operation.

6.3.2 Types of Elasto-Abrasive Balls

Elasto-abrasive balls can be prepared with magnetic or non-magnetic characteristics. While preparing magnetic balls, fine ferromagnetic powder is used as an additional ingredient along with abrasive grains during the process of embedding. The applications of these two types are discussed later.

By changing the volume fraction of elastomer and abrasives, the elastic characteristic of the ball could be altered for different applications.

Table 6.1 shows the general characteristics of elasto-abrasive balls.

Figure 6.2 shows macroscopic images of elasto-abrasives embedded with various sizes of silicon carbide grains.

6.3.3 Effect of Elastomeric Medium

The effect of an elastomeric medium when the elasto-abrasives are pressed against a surface is shown in Figure 6.3. Due to the presence of a resilient polymer medium, the elastic abrasive balls have a relatively lower modulus of

TABLE 6.1

Characteristics of Elasto-Abrasive Balls

Average mass per ball	0.03 g
Average diameter	3.5 mm
Density	1.8 g/cc
Size of embedded abrasive grains	Flexible range (selected range: 10 μm to 250 μm)
Volume fraction (elastomer: abrasive)	Flexible (typical ratio: 70:30)

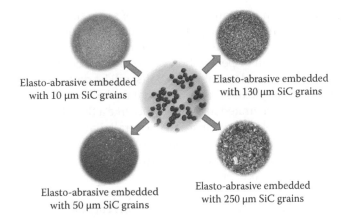

FIGURE 6.2
Macroscopic images of elasto-abrasive grains (×33).

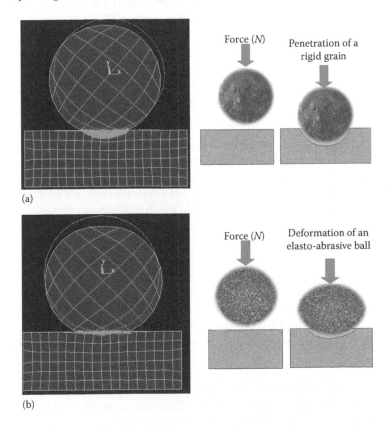

FIGURE 6.3
Effect of elastomeric medium during contact. (a) Behaviour of a rigid SiC grain during its loading on to the surface, similar to a Brinell hardness indenter. (b) Elastic deformation and low depth of penetration of an elasto-abrasive ball loaded into work surface.

elasticity, compared to a standard abrasive grit of same size. As a result, the use of elastic abrasives will result in significant reduction in equivalent elastic modulus at the contact interface. This in turn will reduce the contact pressure and thereby reduce the depth of penetration of the embedded abrasive grains into the surface. Moreover, the elasto-abrasive balls will be deformed while loading, unlike a hard abrasive sphere. As an example, if the ball is of silicon carbide with diameter D, it will be penetrated into the surface like a Brinell hardness indenter. Alternatively, if it is an elasto-abrasive ball of overall diameter D (with SiC grains of diameter d embedded over the periphery of elastomeric bead), the deformation of the ball will facilitate a number of fine SiC grains on the surface to get penetrated in to the surface. Since the pressure is transferred into the surface through the elastomeric medium, the depth of penetration will be much lower in this case. This unique feature makes these balls capable of yielding an ultra-fine finish without altering the nominal surface form.

The previous point can be explained using Hertzian contact theory [18], according to which the maximum contact pressure (P), developed during the interaction of elasto-abrasive sphere with a surface (at ideal conditions), can be written as

$$P = \frac{1}{\pi}\left[\frac{6 F_r E_e^2}{R_b^2}\right]^{\frac{1}{3}}.$$

(6.1)

In this equation, F_r, R_b and E_e denote the normal (radial) contact force, the radius of the elastic abrasive ball and the equivalent elastic modulus, respectively. Depth of penetration of the embedded abrasives depends mainly on the size of grains embedded as well the contact pressure mentioned previously.

More specifically, the equivalent elastic modulus is defined as

$$\frac{1}{E_e} = \frac{1-v_s^2}{E_s} + \frac{1-v_b^2}{E_b}$$

(6.2)

and is a critical parameter influencing the depth of penetration of abrasive grains. Here, E and v are the elastic modulus and Poisson's ratio, with the suffixes s and b representing the work surface and elasto-abrasive ball, respectively.

The same effect is also valid for the impingent of elasto-abrasives. Here again, the depth of penetration will be relatively low compared to a hard abrasive sphere of the same dimension. This can be attributed to the elastic rebound during impact as well as a reduction in the equivalent elastic modulus due to the presence of a resilient polymer medium. The theory of this elastic impact as well as an experimental illustration of this hypothesis are well described in the article by Sooraj and Radhakrishnan [19].

6.4 Types of Elasto-Abrasive Finishing Processes

As mentioned earlier, elasto-abrasives can be used for various types of finishing applications, using custom-made finishing setups. In this section, three major types of finishing processes developed using elasto-abrasives are discussed.

6.4.1 Elasto-Abrasive Squeeze Finishing

This is a technique developed to finish internal surfaces, using the basic principle of 'abrasion'. The process can be applied effectively to finish inner surfaces of tubes, sleeves, bushes etc., which can be customised easily for geometrically symmetric, circular as well as non-circular workpieces. A simple finishing machine developed for the purpose is shown in Figure 6.4. The system mainly consists of a pneumatic piston-cylinder arrangement controlled by a specially developed electronic circuit. In the machine, two double acting piston-cylinders (2 and 3) are located on both sides of a base plate (1). The workpiece (4) to be finished is located at the centre using two guide bushes (5) on either side of it, allowing the extension rods of pistons to reciprocate *to and fro* inside them in tandem. The extension rods are connected to the pistons using flexible couplings to compensate for any misalignment during workpiece clamping. The tolerances on the reciprocating parts are maintained precisely to eliminate any chances of rubbing with the work surface. The pistons are controlled using a solenoid actuated directional control valve (two-position five-port four-way spring return type), to move them in the same direction. The triggering of solenoid and velocities of pistons are controlled using a microprocessor circuit.

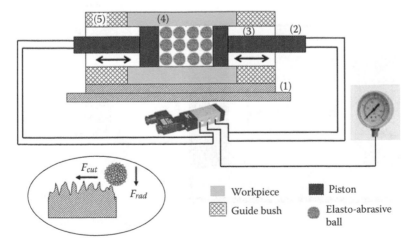

FIGURE 6.4
Details of the elasto-abrasive squeeze finishing setup.

To start the finishing operation, elasto-abrasive balls are filled inside the work-piece and the pistons are brought in position to squeeze these balls axially. The micro-cutting operation will be initiated when the pistons are reciprocated linearly. The action of embedded abrasive grains can be of three modes: (i) rubbing, (ii) ploughing or (iii) micro-cutting. The size and shape of the embedded grains and the cutting forces are the major factors that decide the mode of abrasive action. Since the abrasive grains are of random shape, it is very difficult to predict this nature accurately. Therefore, it can be considered as a combination of three modes, leading to the removal of material in the form of micro/nano-chips.

6.4.1.1 Mechanism of Finishing and the Effect of Process Variables

Elastic abrasive balls undergo radial deformation while they are squeezed axially. The radial deformation will pressurise the embedded abrasive grains into the work surface. This penetration is analogous to the depth of cut in a cutting operation. As the piston reciprocates, the penetrated grains will induce a shear force on the work material, resulting in micro-cutting action. Since the depth of penetration is controlled by an elastomeric medium, the chips produced will be of micro/nano size and the quantity of material removed will be relatively lower compared to existing finishing techniques.

Figure 6.5 illustrates the mechanism of material removal in elasto-abrasive squeeze finishing. The condition for initiating material removal can be expressed as

$$F_{axial} \geq \tau_w A_{proj},$$

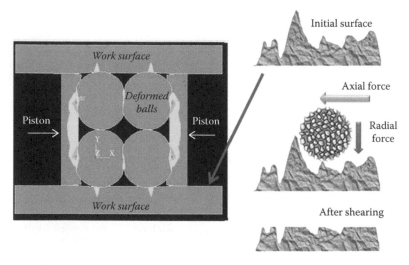

FIGURE 6.5
Mechanism of material removal in elasto-abrasive squeeze finishing.

where τ_w represents the shear strength of work material and A_{proj} is the projected area of grain penetration (which is a function of embedded abrasive grain size and its depth of penetration). A detailed analysis of material removal in elasto-abrasive squeeze finishing is reported by Sooraj and Radhakrishnan [20].

As per the configuration of the finishing setup discussed earlier, the major process variables involved in elasto-abrasive squeeze finishing are the following:

- Axial pressure
- Linear (reciprocating) velocity (cutting speed)
- Size of embedded abrasive grains
- Processing time
- Initial roughness and properties of the work material

As the axial pressure increases, the radial force transmitted to the embedded grains through the elastomeric medium also increases. This will enhance the removal of surface peaks, leading to an improvement in finish. But the increase in axial pressure after a certain limit may lead to loading and jamming of balls, deteriorating the finish (Figure 6.6a). The roughness value decreases significantly with an increase in cutting velocity. This is attributed to the increase in the frequency of cutting (the number of times the surface

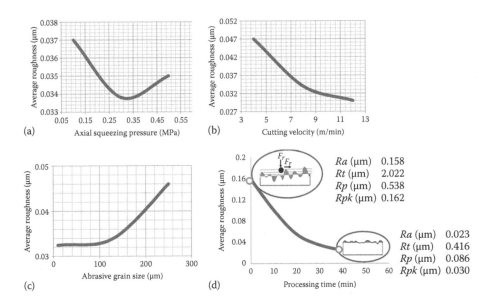

FIGURE 6.6
Influence of process variables in elasto-abrasive squeeze finishing: Effect of (a) axial squeezing pressure, (b) cutting velocity, (c) abrasive grain size and (d) processing time, on average roughness.

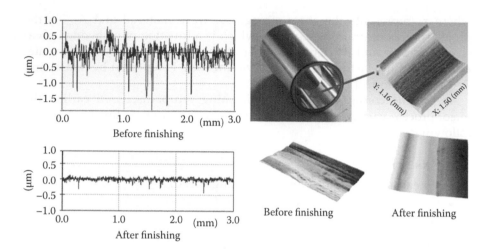

FIGURE 6.7
Surface profile before and after elasto-abrasive squeeze finishing.

peaks come across the active grains) per stroke (Figure 6.6b). As observed in many other techniques, the finish was considerably improved by the use of fine abrasive grains (Figure 6.6c) due to low depth penetration and shear area. Finally, the variation in roughness with processing time (Figure 6.6d) and the modification in surface profile (Figure 6.7) clearly indicate the effectiveness of elasto-abrasive balls in fine finishing. It should be noted that all these results are obtained on hardened steel tubes, using the setup shown previously.

6.4.2 Rotary Elasto-Abrasive Squeeze Finishing

The squeeze finishing machine described previously was improvised further by incorporating workpiece rotation as shown in Figure 6.8. The basic working of the machine remains the same, with a pneumatic piston-cylinder arrangement as discussed in the previous section. But the new machine is attached with a bearing-collect assembly (which holds the tubular work piece inside it), rotated by a motor through a Vee belt. Thus, an additional rotational movement of the workpiece is introduced in the finishing cycle.

The rotation of the workpiece was found to be effective in yielding a better finish compared to the one without rotation. In this methodology, there is an extra component of force in the tangential direction (due to the rotation, as shown in Figure 6.9) in addition to the axial and radial (normal) forces. By this approach, roughness was reduced further down to 0.019 μm (from an initial roughness of 0.158 μm), after a processing time of 40 minutes.

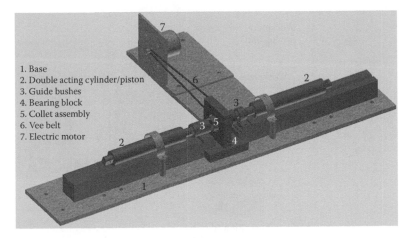

FIGURE 6.8
Rotary elasto-abrasive squeeze finishing setup.

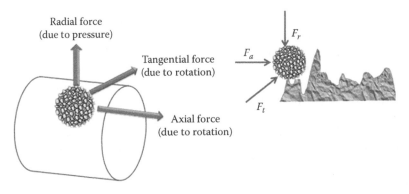

FIGURE 6.9
Component of forces in rotary elasto-abrasive squeeze finishing.

6.4.3 Fluidised Elasto-Abrasive Finishing

The fundamental principle behind this technique is the *impingement and erosion of surfaces* by elasto-abrasive balls. The schematic diagram of a finishing machine developed for the same is shown in Figure 6.10. The system mainly comprises a cylindrical fluidisation chamber with a detachable porous base plate (mesh plate). Elasto-abrasive balls placed on the base plate are fluidised when air is supplied from a centrifugal blower, through an air-distributing window. A specially designed work clamping and driving unit (using a planetary gear system) is incorporated in the machine to facilitate revolution as well as low-speed spinning of the workpiece. It is also provided with a choice to tilt the workpiece, for making the impingement of balls at different angles. Unlike the direct fluidisation of fine abrasive powders, elasto-abrasives in

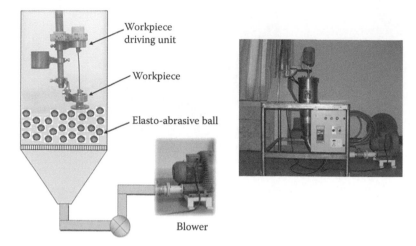

FIGURE 6.10
Details of a fluidised elasto-abrasive finishing setup.

the form of meso-scale balls are easy to fluidise. The elastomeric medium will act like a carrier that facilitates easy fluidisation of the fine abrasive powders embedded over it. It is also noted that the hitting of balls (with meso-scale dimension) will provide adequate energy for effective erosion, at the same time allowing only the fine (micron/sub-micron) grains to interact with the surface. However, the presence of a resilient elastomeric medium will have a significant role in energy transfer, ensuring controlled erosion without damaging the sub-surface. In addition, the airflow and work rotation are also regulated in such a way that the velocity of impingement is of the order 5 m/s or less. Compared to an abrasive jet machining, this velocity is very low, and thus, the erosion is also 'very gentle' in nature.

As the workpiece is rotated inside a homogeneously fluidised chamber, elasto-abrasive balls will impinge against the work surface randomly. When the balls are assumed to be in minimum fluidisation condition (condition at which weight of the balls is counterbalanced by drag force), they will be simply suspended in air. At this stage, the rotation of the workpiece will initiate the hitting of balls against the surface. Here, the angle of impingement can be varied by tilting the work surface at different angles. But in actual practice, this hypothesis may not be exactly valid. There are always chances of random hitting, rotation of balls, as well as turbulence.

To understand the mechanism of material removal, consider the simple situation of minimum fluidisation. For a particular ball hitting the work surface with an impingement velocity V, at an angle α, there will be mainly two components of velocities. The normal component of velocity, $V \cos\alpha$, will be responsible for an erosion effect, whereas the tangential component, $V \sin\alpha$, will create a sliding/scribing effect on the surface. Even though the interaction of balls with work surfaces is quite complex in nature, the end results were quite convincing. After a reasonable processing time, the multiple erosions were producing a significant improvement in surface roughness. This is attributed to the fine erosion caused at low velocity of impingement, interaction of fine grains, as well as the effect of an elastomeric medium.

Thus, the mechanism of material removal in this operation can be summarised as the overlapping of finely eroded craters on multiple impacts of elastic abrasives, leading to a gradual refinement of the surface profile, as indicated schematically in Figure 6.11. Theoretical analyses of low-velocity impact, rebound characteristics of elasto-abrasive balls, energy transferred during elastic impact, effect of fluidisation and the modelling of multiple impacts etc., are reported in detail by Sooraj and Radhakrishnan [19,21].

The major process variables involved in this operation are as follows:

- Degree of fluidisation (ratio of fluidised bed height to static bed height)
- Rotational speed of workpiece
- Angle of impingement
- Size and characteristics of elasto-abrasive balls
- Work material properties
- Initial roughness of work surface
- Size of embedded abrasive grains

Here, the static bed height is defined as the height to which elasto-abrasive balls are filled inside the static chamber. As the air flows in, fluidised balls will lift up and the bed will expand to a height referred as the 'fluidised bed height'. A typical variation of average roughness with bed height ratio (fluidised to static) is shown in Figure 6.12. The roughness values can be significantly reduced with an increasing bed height ratio to an optimal level. This can be attributed to lower bed density and lesser number of active balls impinging against the work surface at higher bed expansion. Similar to this,

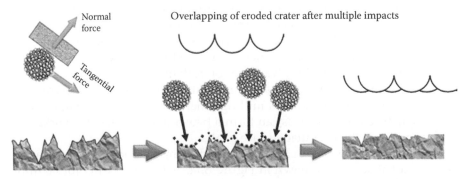

FIGURE 6.11
Mechanism of material removal in fluidised elasto-abrasive finishing.

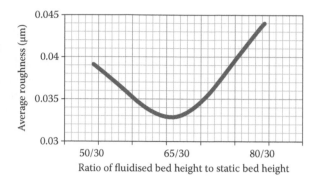

FIGURE 6.12
Variation of average roughness with fluidised bed height.

lower ranges of rotational speeds may not create sufficient number of hitting, whereas too high a rotational speed may create unnecessary turbulence that may lead to uneven scratch marks on the surface. While conducting experiments in a cylindrical chamber of diameter 300 mm, at a bed height ratio of 2.2 (=65/30), the optimum speed range is observed to be 350 to 450 rpm. This process can be customised to finish flat as well as free-form surfaces. The method can be fine-tuned for selective polishing too. For selective polishing, the surface is masked by a thin layer of polymer film and the area to be selectively polished is exposed by removing the mask.

A hardened steel specimen (flat end of cylindrical workpiece with an initial roughness $Ra = 0.2$ μm) finished using the aforementioned fluidisation chamber (for a processing time of 60 minutes) showed an improved Ra value of 0.034 μm. The two-dimensional parameters like peak height (Rp), theoretical peak to valley roughness (Rt), etc., were also reduced significantly, without changing the flatness of the specimen [21].

6.4.4 Magnetic Elasto-Abrasive Finishing

In this operation, the second kind of elasto-abrasive, the magnetic type, is used for finishing applications. Magnetic-type elasto-abrasives can be used very easily to form flexible magnetic brushes (as seen in magnetic abrasive finishing). They are more convenient compared to fine-loose abrasive particles, and they combine the advantages of both fixed and loose abrasives.

The setup mainly consists of a magnet (either permanent or electromagnet) attached to the spindle of a milling machine, as shown in Figure 6.13. The magnetic elasto-abrasives can form a brush that abrades the workpiece surface and cuts the irregular surface peaks.

A typical improvement in finish obtained using magnetic elasto-abrasive balls on a flat surface is shown in Figure 6.14. After processing for 60 minutes, significant reduction in Ra, Rt and Rp values were observed and the flatness of the surface remained unaltered.

FIGURE 6.13
Details of magnetic elasto-abrasive finishing.

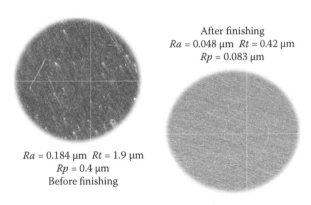

After finishing
$Ra = 0.048$ μm $Rt = 0.42$ μm
$Rp = 0.083$ μm

$Ra = 0.184$ μm $Rt = 1.9$ μm
$Rp = 0.4$ μm
Before finishing

FIGURE 6.14
Images of surface before and after magnetic elasto-abrasive finishing.

6.5 Potential Applications of Elasto-Abrasive Finishing

6.5.1 Fine Finishing of Internal Circumferential Grooves

Finishing of an internal circumferential groove, without altering its basic form, is an application of industrial relevance. But it is not so easy to use the conventional methods of finishing. The experimental setup shown in Figure 6.8 can be used to finish internal circumferential grooves. The elasto-abrasive balls can be filled inside the specimen and squeezed using the pistons. Then, the machine can be operated without reciprocation of pistons, and only rotating the workpiece (with internal groove) located inside the collect. This can

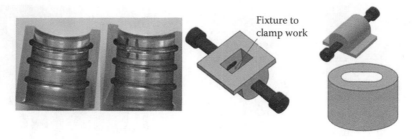

FIGURE 6.15
Details of internal circumferential grooves and non-circular specimens.

also be done using a general purpose lathe with some simple fixtures [22]. A typical grooved specimen after finishing is shown in Figure 6.15.

6.5.2 Fine Finishing of Non-Circular Internal Surfaces

The elasto-abrasive squeeze finishing setup (Figure 6.4) can be used for fine finishing of geometrically symmetric non-circular internal surfaces such as oval bores, D slots, etc. It can be achieved by clamping the workpiece (single or multiple numbers) in the setup, using simple fixtures that can be prepared using rapid prototyping (as shown in Figure 6.15). In this case, the cutting motions are different from the groove finishing discussed previously. Only reciprocation of squeezed balls is provided without rotation of the workpiece.

6.6 Summary

It should be noted that elasto-abrasives in the form of meso-scale balls can be used for both erosion- as well as abrasion-based finishing approaches. The application of this approach spans a wide range of surfaces, as indicated in Table 6.2. The scope of using elastic abrasives for multiple applications is a significant advantage, as many of the existing procedures demand radical changes in the abrasive media for different applications.

The finish achieved in terms of average roughness (Ra) is of the order of 0.035 to 0.02 µm (measured at a cutoff value 0.25), from an initial value of 0.2 µm. Since elasto-abrasive finishing is developed as an ultra-fine finishing approach, capable of removing material in very low quantity, the initial roughness of the workpieces should be selected appropriately. The parameters like Rt, Rp and Rpk also showed significant improvement, without affecting the sub-surface and the nominal form. The performances can be enhanced further by the usage of finer abrasive grains and improved hardware/control

TABLE 6.2

Multiple Applications of Elasto-Abrasive Finishing: Summary of Results

Sl. No.	Type of Surface	Material	Initial *Ra* (μm)	Final *Ra* (μm)	Principle	Time
1	Internal–tubular	Hardened steel	0.158	0.024		
2	Internal-oval	Hardened steel	0.18	0.026	Abrasion	
3	Internal circumferential groove	Hardened Steel	0.24	0.02		Processing time is of the order of 40 to 60 minutes
4	Cylindrical flat (disc shaped)	Hardened steel	0.2	0.034	Erosion	
5	Cylindrical flat (disc shaped)	Stainless steel	0.2	0.048	Magnetic assisted abrasion	

systems. But an important observation is that the approaches and outcomes reported in this chapter are of great industrial relevance.

Questions and Discussions

Q1. Design a simple experimental procedure to illustrate the effect of elastomeric medium during impingement.

Hint: A simple dropping mass experiment can be devised to illustrate the effect of elastic impact. Figure 6.16a shows the image of a crater formed by impinging a small metal block attached with a silicon carbide grit of diameter 3 mm directly on a surface, dropped from a height of 0.15 m. On the other hand, Figure 6.16b shows one formed by dropping the same metal block glued with an elasto-abrasive ball of diameter 3 mm (embedded with 250 μm grains). Comparison of these two figures clearly shows the effect of an elastomeric medium and the reduction in depth of penetration. A similar experimental investigation can be referred from the publication by Sooraj and Radhakrishnan [19]. In this, the result obtained by dropping an abrasive grit with and without rubber pad is described to demonstrate the effect of elastomeric support.

Q2. Suggest a typical *value addition* in elasto-abrasive finishing

Hint: Elasto-abrasives can be made in the form of wires, which can be used to finish high aspect ratio holes. A finishing system similar to the one discussed in Section 6.4.1 can be used for the same. Typical experimental investigation on the application of these wires is reported by Sooraj et al. [23].

Q3. What are the difficulties associated with the fluidisation of fine abrasive particles? What are the options to improvise fluidisation?

Hint: Fine abrasive particles (typically in Geldart group C category) are difficult to fluidise because of agglomeration and cohesiveness.

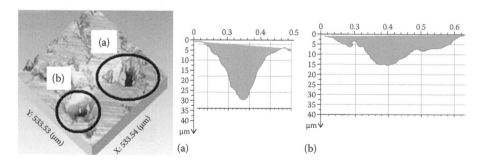

FIGURE 6.16
Effect of elastomeric medium during impact. Impact of (a) standard abrasive grit and (b) elasto-abrasive ball.

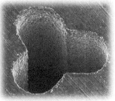

FIGURE 6.17
Parts to be finished using the elasto-abrasive technique.

Vibration-assisted fluidisation is a method to enhance fluidisation. But some difficulties are observed while using it for finishing applications. At this point, carrier type fluidisation using elasto-abrasive balls is a feasible approach [24].

Questions for Open Thoughts

Q1. In Section 6.5.1, it is mentioned that the finishing of an internal circumferential groove using elasto-abrasive balls can be performed using a general purpose lathe. Can you design a setup for the same?

Q2. Can you modify the design of rotary-elasto abrasive finishing machine discussed in Section 6.4.2 to finish the external surface of a shaft?

Q3. Devise the strategies and experimental systems to finish the intricate profiled components in Figure 6.17 using elasto-abrasive balls.

Q4. In the case of brittle and ductile materials, compare the mechanics of low-velocity impact and quasi-static indentation.

References

1. Komanduri, R., Lucca, D. A., Tani, Y. (1997). Technological advances in fine abrasive processes. *CIRP Annals*, 46(2): 545–597.
2. Marinescu, L. D., Hitchiner, M., Uhlmann, E., Rowe, B. W., Inasaki, I. (2007). *Handbook of Machining with Grinding Wheels*. CRC Press (Taylor & Francis): New York.
3. Marinescu, L. D., Rowe, B. W., Dimitrov, B., Inasaki, I. (2004). *Tribology of Abrasive Machining Processes*. William Andrew Publishing Inc., New York.
4. Marinescu, L. D., Uhlmann, E., Doi, T. K. (2007). *Handbook of Lapping and Polishing*. CRC Press (Taylor & Francis): New York.
5. Jain, V. K. (2013). *Micromanufacturing Processes*. CRC Press (Taylor & Francis): New York.
6. Inasaki, I., Tonsoff, H. K., Howes, T. D. (1993). Abrasive machining in future. *Annals of CIRP*, 42: 723–732.

7. Wang, A. C., Lee, S. J. (2009). Study the characteristics of magnetic finishing with gel abrasive. *International Journal of Machine Tools and Manufacture*, 49: 1063–1069.

8. Fox, M., Agrawal, K., Shinmura, T., Komanduri, R. (1994). Magnetic abrasive finishing of rollers. *Annals of CIRP*, 43(1): 181–184.

9. Belyshkin, D. V. (1966). Using ice for polishing glass and crystals. *Glass and Ceramics*, 23(10): 523–525.

10. Mohan, R., Ramesh Babu, N. (2011). Design, development and characterisation of ice bonded abrasive polishing process. *International Journal of Abrasive Technology*, 4(1): 57–76.

11. Mohan, R., Ramesh Babu, N. (2012). Ultrafine finishing of metallic surfaces with the ice bonded abrasive polishing processes. *Materials and Manufacturing Processes*, 27: 412–419.

12. Zhang, F., Han, R., Liu, Y., Pei, S. (2001). Cryogenic polishing method of optical materials. In: Ichiro Inasaki (Ed.), *Initiatives of Precision Engineering at the Beginning of a Millennium*. In International Conference on Precision Engineering (ICPE), Yokohama, Japan, July 18–20, 2001, pp. 396–400.

13. Jaganathan, R., Radhakrishnan, V. (1997). A preliminary study on fluidized bed abrasive polishing. *Transactions of NAMRI/SME*, 25: 189–194.

14. Massarsky, M. L., Davidson, D. A. (1997). Turbo-abrasive machining and finishing. *Metal Finishing*, 95(7): 29–31.

15. Massarsky, M. L., David, D. A. (2007). Turbo abrasive machining for edge and surface finishing. *Gear Solutions*, 5(53): 23–58.

16. Massarsky, M. L., Davidson, D. A. (1995). Turbo-abrasive machining: Theory and application, SME technical paper MR 95-271, *Proc. of the first International Machining and Grinding Conference*, 1995, SME: Dearborn, MI.

17. Geldart, D. (2005). The Characterization of bulk powders. In: Don McGlinchey (Ed.), *Characterization of Bulk Solids* (pp. 132–149). CRC Press, Blackwell Publishing Ltd: Oxford, UK.

18. Johnson, K. L. (1985). *Contact Mechanics*. Cambridge University Press: Cambridge, UK.

19. Sooraj, V. S., Radhakrishnan, V. (2013). Elastic impact of abrasives for controlled erosion in fine finishing of surfaces. *Manufacturing Science and Engineering*, 135: 051019.

20. Sooraj, V. S., Radhakrishnan, V. (2014). Fine finishing of internal surfaces using elastic abrasives. *International Journal of Machine Tools and Manufacture*, 78: 30–40.

21. Sooraj, V. S., Radhakrishnan, V. (2014). A study on fine finishing of hard work piece surfaces using fluidized elastic abrasives. *International Journal of Advanced Manufacturing Technology*, 73(9): 1495–1509.

22. Sooraj, V. S., Radhakrishnan, V. (2013). Feasibility study on fine finishing of internal grooves using elastic abrasives. *Materials and Manufacturing Processes*, 28: 1110–1116.

23. Sooraj, V. S., Alisha, S., Manohar, P., Radhakrishnan, V. (2014). *Finishing of Small Diameter High Aspect Ratio Holes Using Elastic Abrasive Wires*. Technical Paper MQ 20, Proc. of International Conference on Precision Meso Micro Nano Engineering (COPEN), December 2013, India.

24. Sooraj, V. S., Radhakrishnan, V. (2014). Prospective methodologies to use impact wear for micro/nano finishing of surfaces. *International Journal of Manufacturing Technology and Management*, 28: 94–113.

7

Focused Ion Beam (FIB) Nanofinishing for Ultra-Thin TEM Sample Preparation

Rakesh G. Mote[1] and Li Xiaomin[2]

[1]*Department of Mechanical Engineering, IIT Bombay, Mumbai, India*

[2]*WinTech Nano-Technology Services Pte Ltd, Singapore, Singapore*

CONTENTS

7.1 Introduction

The focused ion beam (FIB) has been a well-established tool in the semiconductor industry mainly for applications like integrated circuit repair and debugging, device modification and failure analysis. Advances in FIB technology in achieving a tighter focus and integrated analysis tools (scanning electron microscope [SEM], X-ray diffraction, etc.) have made micromachining by FIB more attractive for researchers. This is mainly due to its high-resolution, one-step maskless fabrication and possibility to work with a variety of materials and geometries (two- and three-dimensional). As is evident from Figure 7.1a, micro-sized logos of IIT Bombay and WinTech

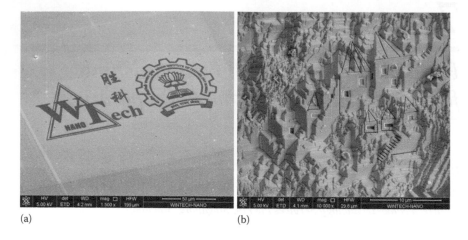

(a)   (b)

FIGURE 7.1
FIB nanopatterning capabilities: (a) IIT Bombay and WinTech logos with nanometer features
on silicon and (b) nano-patterning and fabrication of a castle on a nickel material.

Nano, with nanometer features, are directly patterned, without using
any mask, on silicon using FIB and imaged using SEM in the same setup.
Figure 7.1b demonstrates patterning of an artistic castle on a Ni material.
In addition to the nano-fabrication capability, the FIB can be further used
to polish surfaces to attain low loss, highly reflective mirrors, typically
required in micro-devices/components such as planar photodiodes, surface emitting lasers, optical interconnects, etc. [1]. The flexibility of the
direct milling and the ion beam manipulation enables polishing along
various orientations.

The use of FIB has been established as a major tool for the preparation
of specimens for analysis in transmission electron microscopy (TEM).
Miniaturisation of opto-electronic devices, developments in material sciences and new materials demand advanced characterisation techniques for
material properties investigations, failure analysis, etc. State-of-the-art SEM
or FIB imaging is not adequate to detect and analyse sub-nanometric device
features and structural layers. TEM analysis offers a much higher image resolution in the sub-nanometre regime (<0.1 nm). TEM analysis has also been
applied to certain unique applications like Li et al. [2] have investigated the
NASA Mars meteorite from which an extremely small sample (~20–30 nm)
was picked out. However, in order to carry out TEM analysis, the samples are
required to be meticulously prepared with minimum damage. The sample
has to be polished to a uniform thickness to acquire enough transparency
to the electron beam of the TEM equipment. The typical sample thickness is
about 100 nm or less and the sample is to be mounted on TEM grids of millimetre order.

This chapter presents a review of the dual-beam FIB/SEM instrumentation and its capabilities and various conventional polishing methods for TEM sample preparation and discusses in detail the use of FIB nano-structuring and polishing for site-specific TEM sample preparations.

7.2 FIB System

In the early 1970s, Levi-Setti [3] and Orloff and Swanson [4] demonstrated the application of a field emission ion source to produce FIB. The FIB instrument operates similar to SEM as it makes use of focused charged particles. The possibility of applying FIB to micromachining was first reported by Kubena et al. [5] of Hughes Research Laboratory in 1981. They discovered that FIB could make a V-shaped trench in a single scan of the beam. Since then, FIB has evolved to a sophisticated platform to cater to needs in material characterisation and nanotechnology. Its application ranges from geological to biological domains.

The FIB system in simple terms comprises an ion beam formed by extraction of ions from an ion source that is then tightly focused through a chamber of lenses onto a sample while operating in a high vacuum (Figure 7.2a). The three main components of the system are the ion source, the ion column and the beam writing mechanism. The system can be used for imaging, milling, deposition and implantation on a sample. The typical dual-beam or cross-beam system comprises an ion beam column and SEM column in the same chamber, as shown in Figure 7.2b. The addition of an SEM column enhances the imaging capability and makes it possible to have a real-time observation of the FIB processing.

Gallium ions are the most widely used ions in FIB systems. Gallium ions are extracted from ion sources in the form of liquid metal. They are then accelerated with typical accelerating voltages in the range of 5–30 keV. Figure 7.2a shows the components of an FIB column. The ion column has two lenses, a condenser lens and an objective lens. The condenser lens collimates ions into a parallel beam forming a probe. A set of apertures of various diameters define the probe size and ion current.

The stage is motorised, and software provides control of the lateral plane (x, y axis) and rotation, and it can be manually tilted in the xz plane. Two different types of gases can be released above the specimen surface at a distance of approximately 100 mm; one is used for gas enhanced etch, and the other, for material deposition. The whole setup is usually placed in an ultra-low-pressure chamber of the order of 10^{-7} Torr. The high vacuum avoids any interference of air particles with the beam. Hence, the mean free paths of the ions are increased and the beam strength is uniformly maintained.

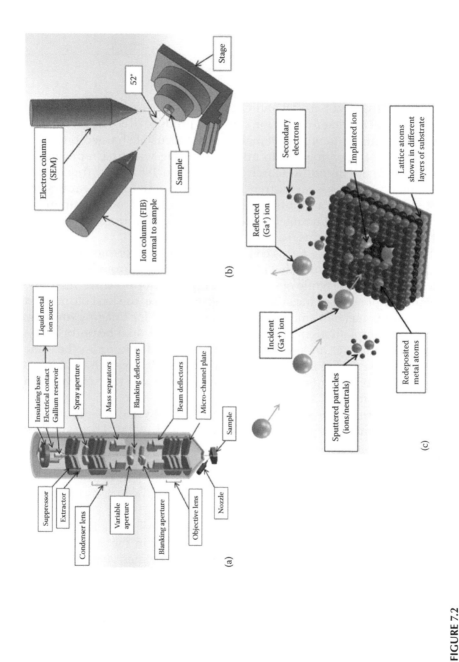

FIGURE 7.2
(a) FIB column. (b) Dual-beam FIB and SEM configuration. (c) Ion beam interaction with material.

7.2.1 Ion Source

Focusing of an ion beam into a submicron diameter requires a source emitting limited solid angle. Liquid metal ion sources (LMISs) or gaseous field ionisation sources satisfy these conditions. LMIS gained wide acceptance as an ion source after the successful demonstration of high-performance FIB system by Seliger et al. [6].

This source type usually consists of an emitter needle with a tip radius of the order 10 mm, which is coated with a pure metal (e.g. Ga, In) or an alloy (e.g. $Au_{82}Si_{18}$, $Co_{64}Nd_{36}$) with a provision of heating. The wetted tip end of the emitter is subjected to an electric field ($\sim 10^8$ V/cm), which causes the liquid to form a point source (~ 2–5 nm diameter) in the shape of a cone. The cone is termed as a *Taylor cone*. Such conical shape is the result of dynamic equilibrium among electrostatic forces, surface tension and flow of the liquid metal. A field evaporation process leads to ion emission from a liquid jet based on this cone. The current density of ions may be of the order of $\sim 10^8$ A/cm^2.

Many metallic elements or alloys can serve as LMIS. Commercial FIB system uses gallium as LMIS. Gallium as an ion source has following advantages:

1. Gallium is a heavy enough ion (atomic number 31) enabling the sputtering mechanism.
2. Gallium has a low melting point ($\sim 30°C$) and therefore exists in the liquid state around room temperature. In addition, the lower temperatures minimise the risk of any reaction or inter diffusion between the liquid and the tungsten needle substrate.
3. A large proportion of the charged gallium of the beam exhibits narrow energy distribution ($[Ga^{2+}]/[Ga^+] \sim 10^{-4}$ at 10 μA), which eliminates the need for a mass filter, leading to a simpler ion column design.
4. Gallium can be focused to a beam size as small as 5 nm diameter.

New ion species are now gaining popularity due to the innovative applications of FIB. Recently, FIB applications in biological sciences, magnetic-film fabrication and secondary ion mass spectroscopy (SIMS) demanded beams of noble-gas ions (e.g. Ar^+), metallic ions (e.g. Mn^+) and molecular ions (e.g. C_{60}^+). Also, rare earth elements like Er^+, Pr^+ and Nd^+ are used for optoelectronic applications or metals, like Au^+ and Co^+ for plasmonic structures. The Plasma and Ion Source Technology group at Lawrence Berkeley National Laboratory has developed FIB systems employing multi-cusp ion sources, which can generate various ion beams, such as O^+, B^+, BF_2^+, P^+ and Ar^+ [7,8]. These ion sources make maskless and resistless lithography possible.

Recently, Hanssen et al. [9] proposed laser-cooled neutral atoms in a magneto-optical trap as an ion source. The source exhibits a narrower energy width than other ion sources do. This feature is useful in achieving smaller chromatic aberrations and nano-scale resolution at much lower energies. An ion beam can be created from a larger selection of atomic species than is available with liquid metal, gas phase or plasma sources, due to the range of atoms amenable to being trapped in a magneto-optical trap.

The nature of ion–solid interactions defines the milling/depositing and imaging capability of a FIB instrument (Figure 7.2c). The study of ion–solid energy transfer mechanism, which is a function of a material undergoing ion bombardment, is vital to understand phenomena like sputtering, the emission of X-rays, Auger electrons and secondary electrons (SEs), phonons, as well as the generation of lattice defects. Such fundamental information may be used to predict sputtering behaviour, imaging, as well as damage to the material.

7.2.2 Imaging

FIB imaging is similar to SEM in their principle of operation except that the probing beam consists of ions instead of electrons. FIB imaging can visualise different contrasts from SEM and can provide significant information about material structure. Due to the much larger size of the ions, the penetration depth is much smaller in FIB. This makes FIB imaging much more sensitive to the surface.

During FIB imaging, the fine FIB is raster scanned over a substrate, and secondary particles (neutral atoms, ions and electrons) are generated in the sample (Figure 7.2c). As they leave the sample, the electrons or ions are collected via a biased detector. The detector bias is a positive or a negative voltage depending on the nature of collecting species, viz., the SEs or secondary ions.

Two imaging modes are possible with an FIB system:

1. Scanning ion microscopy (SIM): This is similar to SEM, where charged particles (ions) are focused onto the surface of a sample and scanned across it. The interaction of impinging ions gives rise to ion-induced SEs (ISEs), ion-induced secondary ions, ion-induced secondary atoms and phonons. SIM has higher material contrast than SEM imaging does owing to the higher ion mass and less penetration into the target compared to the electron beam. For crystalline materials, at certain grain orientations, ions penetrate to deeper regions, in other words, ion 'channels' better at certain grain orientations. The main consequence is the lower ISE yield at such orientations. Hence, the SIM also offers better channelling contrast, enabling the orientation details for polycrystals, and using a different angle of incidence, accurate grain size can be estimated.

2. SIMS: As the FIB scans the surface, the emitted secondary ions are collected and mass analysed. The analysis yields the atomic composition of the target area being scanned with the ion beam. One complete scan of the area of interest is required for atomic mass mapping. The spatial resolution of the atomic map is governed by the ion beam

diameter; hence, excellent precision is expected. SIMS has lateral (20 nm) and depth (1 nm) resolution. The minimum detectable concentration of the material is about few hundred ppm. This is excellent when compared with that of energy dispersive X-ray analysis [10].

7.2.3 Sputtering

Simple milling in FIB refers to the sputtering phenomenon due to energetic impingement of FIB on the target material. The ion milling process is a combination of physical sputtering, back sputtering and material redeposition. As shown Figure 7.2c, the material removal occurs when a surface atom receives kinetic energy from the incident ion high enough to cross the surface binding energy of the target material. Through a series of collision cascades, it is possible to process several tens on a nanometre-level area without using a mask. This is the reason it is also termed as 'maskless etching'.

In addition, a provision to observe the processing condition under a microscope makes very accurate and highly precise results possible. It must be noted here that the sputtering process is a physical process depending on momentum transfer. This enables any material to be sputtered using the FIB milling process. Sputtered particles generally have energy about 2–5 eV. The sputtering yield is defined as the number of ejected particles per incident ion. If the target is made of several elements, there is a separate sputtering yield for each element.

Figure 7.3 shows patterning capabilities of FIB systems. The FIB can be used to create nano or micro patterns and SEM is used to visualse the structures. Additional details can be obtained via making a cross-section using FIB milling. As shown in Figure 7.3b, a cross-section nano hole reveals tapered hole walls with diameters varying from 52 nm at the top to 21 nm at the base over a depth of 74 nm. Figure 7.3d shows an individual zone depth profile of a micro zone plate lens on glass [11]. Usually, platinum is deposited using FIB-induced deposition (as will be discussed in Section 7.2.4) before making a cross-section in order to render a better contrast to reveal the topography details.

Ion–solid interactions in FIB involve ion implantation into the target surface, redeposition or selective removal of target atoms. This process is inherently destructive. The damage can be impact damage due to the heavier mass of ions or surface/subsurface charging effects. Atoms that are displaced from their equilibrium position by the interaction with energetic bombarding ions generate a collision cascade within the target material. During interaction with the target atoms, the ions, particularly metal ion (like gallium), may alter the electronic property of the substrate material locally. Interaction with higher ion doses may cause the destruction of the substrate crystal lattice, which eventually leads to amorphisation of the target material. Although the amorphous phase is metastable, restoration of the disorder depends upon a coordinated motion of the alloying atoms. Usually, the more complex the unit cell, the larger the amorphous layer. For example, under FIB milling

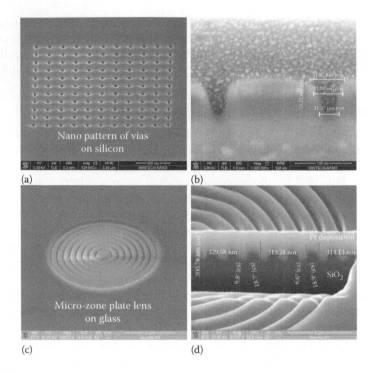

FIGURE 7.3
FIB nanopatterning and cross-sectioning to reveal fabricated structure details: (a) nano-scale pattern of vias on the silicon fabricated using 30 keV Ga + ion with ion beam current 24 pA. (b) FIB sectioned view of vias shows non-vertical side walls. (c) Phase micro-zone plate lens on glass used for near-field focusing (focal length, 0.5 μm). (d) FIB sectioned micro lens reveals patterned zones. ([c, d] From Mote, R.G., Yu, S.F., Kumar, A., Li, X.F., *Appl. Phys. B (Lasers Opt.)*, 102, 95–100, 2011.)

using gallium ions, Si undergoes amorphisation but Cu does not due to the formation of Cu_3Ga phase.

The low throughput of FIB sputtering limits its applications to prototype or nanostructure formation. For instance, milling of silica is challenging owing to its poor conductivity when compared with semiconductor materials. Introducing reactive gases like Cl_2, XeF_2, etc., results in an improved etch rate compared to the sputtering rate of many materials like Si and Al. The process is termed as gas assisted etching (GAE) or chemical assisted FIB machining. Although high milling rates are achieved through GAE, the milled surfaces needs further cleaning/polishing, which are then carried out at slower milling rates without gas enhancement.

7.2.4 FIB-Induced Chemical Vapour Deposition

Small, high-resolution structures without the need of conventional lithographic techniques are possible to fabricate using the deposition technique possible with FIB. This is also termed as maskless deposition. Gamo et al. [12] were the first to investigate FIB-induced deposition. They reported the deposition of Al from Al $(CH)_3$ and W from WF_6.

Chemical-assisted deposition involves organometallic precursor gases. Precursor gases are locally sprayed on the surface, where structure has to be built, by a fine nozzle/needle of a gas injection system. Usually, the needle is controlled at a height of a sub-millimetre order above the target surface at an angle of 30°–60°, with the precursor gas being evaporated from a heated container. This leaves the gas to be locally adsorbed onto the surface. Then, the ion beam is raster scanned in a customisable pattern and it decomposes the adsorbed precursor gases. Usually, the products are metal, insulator or carbon. The volatile reaction products moving away from the surface are removed through the vacuum system of FIB chamber, while the desired reaction products remain adhered on the surface as a thin film or as a desired pattern scanned by ion or electron beam. Thus, it is important that the precursor gas possesses sufficient sticking ability and a rapid decomposition rate than the sputtering rate from the surface (Table 7.1).

Direct deposition involves ionisation of the material to be deposited and separating it from undesired ion species. Then, these ions of the material are focused, deflected and decelerated to the optimum deposition energy. The deposition is carried out by using a very-low-energy FIB. The major advantage of direct deposition over a chemical-assisted deposition is higher purity of the deposited film due to high vacuum conditions maintained during the deposition process. The absence of any gas during the deposition process prevents the deposit from being contaminated with absorbed gas molecules or irradiated ions. Also, the application of a very-low-energy beam (30–200 eV)

TABLE 7.1

Precursor Gases and Typical Deposit Composition

Element	Precursor Gas	Deposit
W	Tungsten hexacarbonyl [$W(CO)_6$], WF_6	W:C:Ga 60:20:20
Pt	Methylcyclopentadienyl (trimethyl) platinum [$C_9H_{16}Pt$], $C_7H_{17}Pt$, Trimethylcyclopentadienyl-platinum [$(CH_3)_3CH_3C_5H_5Pt$]	Pt:C:Ga 30:60:10
Au	$C_7H_7F_6O_2Au$	Au:C:Ga 80:10;10
Al	$Al(CH)_3$, $(CH_3)_3NAlH_3$	Al:O:Ga:C 37:27:26:10
Si, SiO_2	Siloxane compound TEOS: $Si(C_2H_5O)_4$, Silane TMOS: $Si(CH_3O)_4$, O_2	Si:O:C:Ga 11:64:11:13
C	Phenanthrene $C_{14}H_{10}$	–

Source: Jin, I., Chen, C.H., Pai, S.P., Ming, B., Kang, D.J., Venkatesan, T., Machalett, F., Edinger, K., Orloff, J., Melngailis, J., *IEEE Trans. Appl. Supercond.*, 9, 2894–2897, 1999; Giannuzzi, L.A., Stevie, F.A., *Introduction to Focused Ion Beams: Instrumentation, Theory, Techniques and Practice.* New York: Springer, 2005; Li, P.G., Jin, A.Z., Tang, W.H., *Phys. Status Solidi A*, 203, 282–286, 2006.

in direct deposition prevents a substrate and a deposit from inducing irradiation damage.

Nagamachi et al. designed and constructed a low-energy FIB apparatus and developed liquid–alloy ion sources for direct deposition of conductive [16], superconductive [17] and magnetic [18] materials. The FIB direct deposition method can be applied to IC modification, surface acoustic wave devices, superconducting quantum interference devices (SQUIDs), multi-layers and probing on small crystals [19].

7.2.5 Material Self-Organisation

The bombardment of solid surfaces by energetic particles may result in substantial morphological changes. Such structures can have highly periodic features as 'ripple'-like contours [20] or as 'nanodots' or 'nanoholes' [21], with feature sizes in the nanometre range as shown in Figure 7.4. The driving forces behind these self-organised structures have been the interactions between smoothening or roughening mechanisms. Ripple patterns with periodicity in the nanometre range have been observed under oblique ion incidence. However, for normal ion incidence, the absence of a driving direction leads to the formation of hillocks or pits or typically hexagonal dot patterns on compound materials.

During nanofinishing of surfaces with FIB, such self-organised periodic structure formation needs to be minimised. The self-organised ripples are usually a cause of concern while preparing mirror polished surfaces. The optimised beam energy and angle of incidence are the key parameters governing the self-organised patterns.

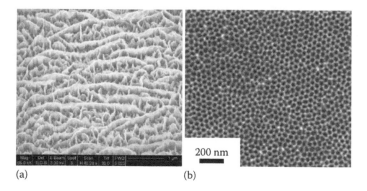

(a)   (b)

FIGURE 7.4
Self-organised periodic pattern formation due to FIB irradiations. (a) Ridge-shaped ripple formation with a spacing of 100–400 nm on Ge (001) and (b) a nanohole pattern on Ge (100) surface after Ga + ion beam irradiation under normal incidence. ([a] From Zhou, W., Cuenat, A., Aziz, M.J., *Microscopy of Semiconducting Materials 2003: Proceedings of the 13th International Conference on Microscopy of Semiconducting Materials*, 2003. [b] From Fritzsche, M., Muecklich, A., Facsko, S., *Appl. Phys. Lett.*, 100, 223108, 2012.)

7.3 TEM Sample Preparation

TEM is an electron microscopy technique whereby a beam of electrons is transmitted through a critically thin, polished sample. The electrons interact with the specimen as it passes through it and an image is formed. Usually, an objective lens is used to form the diffraction in the back focal plane and the image of the sample in the image plane. The images or the diffraction patterns are captured on an imaging screen or detected by a sensor such as a charge-coupled device (CCD) camera.

Ultrathin sample preparation is a critical step in TEM analysis. Almost all TEM studies involve a starting material to be much larger and a thin slice is required to be cut and further thinned to a required thickness. Samples are prepared in the form of thin and flat foils or lamellae. The thinness of such lamellae is of the order of 10–200 nm so as to ensure transparency towards the electron beam. The lamellae should exhibit minimum thickness variation. The sample preparation must ensure clean and damage-free lamellae. It must be noted here that the thickness of the sample is a function of electron scattering, which in turn depends on the atomic number of the elements. For higher atomic number materials, the sample thickness should be thinner. For crystalline materials, a tilted stage is used to bring in certain orientation to match Bragg's diffraction conditions in order to enhance the transparency of the sample to the electrons. Any surface variation or roughness causes an unwanted diffraction of electrons and thus deteriorates TEM images. Thus, the success of TEM analysis depends upon achieving required critical thickness of samples ensuring electron transparency and the high surface quality of the sample to minimise electron scattering or unwanted diffractions.

FIB can be used to make site-specific TEM specimens with sub-micron positional accuracy. Samples prepared can be thin enough to be used for both techniques that can make TEM membranes thin enough for energy-dispersive X-ray spectroscopy (EDX) analysis, electron energy loss spectroscopy (EELS) analysis, diffraction patterns and lattice imaging. FIB milling was found to be more popular than other techniques like chemical thinning, ion beam milling and wedge polishing with a tripod tool (Table 7.2).

The major benefits of using an FIB for TEM specimen preparation can be listed as follows:

- *Site-specific processing*: With the aid of a high-resolution dual beam, an ion and an electron beam, it is possible to precisely choose the site of interest and cut out the sample with a spatial accuracy of about 20 nm.
- *Surface finish*: A high-resolution ion beam, accurate beam positioning/orientation and ease of choosing beam currents enable polishing the TEM samples to the required accuracy. They also make it possible to repair damages during FIB processing.

TABLE 7.2

Comparison of Various Conventional TEM Sample Preparation Techniques

Methods	Pros	Cons
Mechanical polishing (dimpling, grinding, polishing)	Large areas	Slow and requires intensive, skilled labour, prone to mechanical damages, not site specific
Electrochemical polishing	Large areas, quick	Limited to electrically conductive materials, preferential etching, not site specific
Chemical thinning	Quick	Suited for standard materials, not site specific
Ultra microtomy	Large areas, soft biological samples	Slow, not suited for hard materials, not site specific
Ion polishing (low energy inert gas, Ar, ion beam)	Large areas, not labour intensive	Differential milling, slow, not site specific

- *Orientation details*: An FIB/SEM dual-beam configuration offers better contrast via ion channelling effects to reveal local crystallographic orientations. It is also possible to make use of SEM-based electron backscatter diffraction (EBSD) for identifying grain orientations.

- *Fast and repeatable processing*: FIB sample processing is faster than other TEM samples preparation methods, typically ranging from 20 minutes to a couple of hours.

- *Universal material range*: FIB sample preparation techniques are virtually independent of the nature of the material.

- *In situ and real-time observation*: With the aid of a dual-beam technology, the sample preparation can be observed in situ via ion- or electron-beam-induced SE imaging capability.

7.3.1 FIB for TEM Sample Preparation

The evolution of FIB systems over the late 1980s and 1990s, especially the development of the dual-beam configuration, has enabled TEM sample preparation with an edge over traditional techniques. The techniques can be seen along two categories, the conventional 'H-bar techniques' and 'lift-out techniques.'

7.3.1.1 The H-Bar Technique

The H-bar technique is the initial developed technique during the mid-1990s [22]. The procedure is not a stand-alone FIB sectioning. It requires a sample to be prepared using conventional methods, followed by FIB thinning.

A volume to be investigated is first located and a thin slab is cut from the bulk specimen with diamond wheels or the wire sawing method. The slab length and the thickness are of the order of 2–3 mm and 1 mm, respectively. As FIB milling is a slow process, the sample must be further thinned before being put in the FIB chamber to reduce the FIB milling time. The vertical walls of the cut-out sample are subjected to mechanical polishing. The tripod polisher [23] is best suited for such initial thinning of the sample. At this stage, the sample thickness is expected to about 50–100 μm, as shown in Figure 7.5b. The samples are now ready to be placed onto a TEM half-grid. The TEM half-grid, along with the glued sample, is then placed into the FIB chamber. The sample is then subjected to an ion beam sputtering along vertical directions to obtain further thinning of the area of interest (Figure 7.5c). A typical TEM sample prepared by the H-bar technique [24] is shown in Figure 7.5d. Before FIB sputtering action, the area of interest is protected through a deposition of a W/Pt thin line on the top of the specimen using an FIB-induced chemical vapour deposition process. During the FIB sputtering, the beam current is gradually reduced so as to remove the re-deposited sputtered material and to minimise the ion-induced damage to the specimen. FIB sputtering is continued until the required thickness (of the order of 100 nm or less) for electron transparency is attained. Thus, the prepared lamella if seen from the top, which resembles the alphabet 'H', and hence, the techniques is referred to as the 'H-bar technique'.

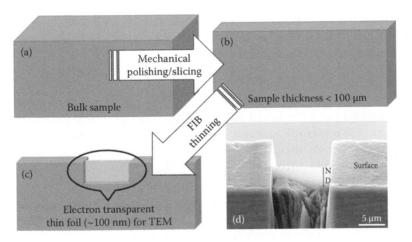

FIGURE 7.5
FIB H-bar technique. (a) Bulk sample. (b) Slicing using diamond wheel and thinning by mechanical polishing to thickness less than 100 μm. The sample is then attached to a TEM grid and transferred to the FIB chamber. (c) Region of interest is thinned down to the electron transparency using FIB milling on both sides. (d) TEM specimen prepared using the FIB H-bar technique. The sample was mechanically thinned to about 40 μm before transferring to the FIB chamber for further thinning via FIB milling. ([d] From Mayer, J., Giannuzzi, L.A., Kamino, T., Michael, J., *MRS Bull.*, 32, 400–407, 2011.)

It must be emphasised that the accuracy of locating the region of the interest is limited by the accuracy of a mechanical cut and polish technique prior to the FIB thinning. Further, the H-bar technique is most suitable to non-site-specific TEM studies. It has very limited applications in site-specific sample preparations. As the technique involves a mechanical way of cutting and polishing before putting in the FIB chamber, the sample is subjected to mechanical forces compromising the properties of the region of the interest. Such destructive sample preparation steps often limit the application of the method to certain studies like stress corrosion crack analysis [25].

7.3.1.2 The Lift-Out Technique

As described in the previous section, the H-bar technique is inadequate for site-specific TEM analysis owing to mechanical cutting and polishing. The sophistication of FIB instrumentation and the developments in dual-beam technology made it possible to make TEM samples directly from the bulk materials. The techniques are termed as 'lift-out techniques' [26,27] as the FIB prepared thin lamella is lifted or picked from the bulk specimen. The TEM samples are milled out of a very small region, typically volume of 50 µm × 50 µm × 50 µm, thus making the technique site-specific and thus leaving the bulk specimen unaffected.

Based on the lamella removal procedure, the lift-out technique can be ex situ, where the lamella is removed outside the FIB chamber after the sample preparation via FIB milling. The in situ lift-out involves lamella preparation, removal from the bulk sample and placement on the TEM grid all inside the FIB chamber.

The process of in situ lift-out comprises the following steps:

1. Identification of the site of interest for TEM analysis. The navigation is carried out with the aid of an electron beam available with a dual-beam FIB system. The choice of electron gun over ion-beam-induced imaging avoids the potential damage to the region due to ion bombardment. It is also possible to mark the required crystallographic grain orientation using EBSD.

2. The deposition of the protective cap (~0.5 µm thickness of Pt or W material) onto the surface in the selected area of the specimen (Figure 7.6a). This is done in order to protect the area from the redeposition of the sputtered material and to minimise the damage arisen owing to the ion beam milling process.

3. Now, using the protective cap as a reference, FIB milling is carried out on both sides of the cap, leaving a thin lamella at the centre via the formation of a trench, as shown in Figure 7.6a. The trench resembles a typical staircase pattern. The stair-step trenches make

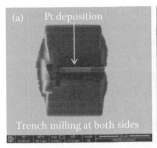
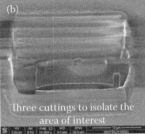
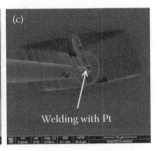
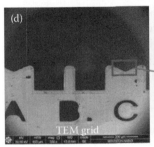
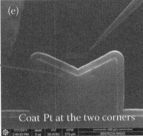
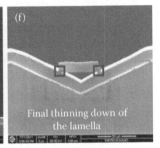

FIGURE 7.6

The FIB in situ lift-out technique. (a) Pt coating to protect the place of interest, (b) in situ lift-out, (c) in situ lift-out: probe attachment, (d) the location of the lamella place on the grid, (e) insert nano-probe with the attached lamella and (f) location of the Pt-coated lamella.

it possible to mill the lamella further. The lamella obtained until this stage is of the order of couple of microns and the side walls are not perfectly vertical or the sidewalls are of 'V' shape. The 'V' shape is typically observed in ion beam milling vertical walls as the ion beam profile bears a Gaussian distribution with long tails [28].

4. The FIB parameters, mainly the ion beam current, are slowly controlled to get the required thickness of the lamella. Generally, a lamella thickness of 1–2 μm is obtained by rough or bulk ion beam milling with current in the range of 10–15 nA at 30 keV. The medium thinning (~500 nm) is carried out with a current of about 1 nA alternatively on both sides of the lamella with a sample tilt of 1°–2°; the current is to be set to 30–100 pA for fine milling to obtain a thickness of about 100 nm. The sample tilt is required in order to offset the non-vertical or tapered side walls ('V' shape) of the lamella obtained as explained in Step 3. A fine cleaning or polishing is then carried out using a low-energy beam 2–5 keV. It is also possible to polish ultra-thin sections, lesser than 50 nm, with a very low energy, down to 500 eV ion beam.

5. Inside the FIB chamber, a nano-probe along with a micromanipulator can be placed. The micromanipulator can be navigated and the nano-probe movements are precisely attained via a control panel outside the FIB chamber. The lamella is welded to the nano-probe tip point, and with the aid of three cuts, the lamella is isolated from the bulk sample as shown in Figure 7.6b and c. The sample is tilted by about a 30°–40° angle to enable one cut to be made along the bottom edge and the two vertical cuts along sides of the lamella. Typically, the FIB-assisted Pt deposition is carried out to weld the lamella (Figure 7.6c). The lamella attached to the nano-probe is then moved to a TEM sample grid/holder.

6. The lamella is then welded to the TEM sample grid (Figure 7.6d and e) at the bottom corners in the similar procedure followed to weld the lamella to the nano-probe. Once the lamella is secured to the TEM grid, the nano-probe is detached from the lamella via FIB milling. These processes are as depicted in Figure 7.6d through f.

7. Finally, the lamella sitting on the TEM grid can be further thinned using the FIB milling at low beam energy. This final step ensures that the lamella thinness is right enough for an electron-transparency required for TEM analysis.

The ex situ lift-off technique involves similar steps of the thin lamella preparation for the required TEM transparency inside the FIB chamber via ion beam milling. The prepared lamella, along with the bulk sample, is then taken out of the FIB vacuum chamber and put under a light optical microscope. Under the optical microscope, the lamella is taken out of the trench using a micromanipulator carrying a glass probe. In this case, the lamella adheres to the probe via electrostatic attraction forces, unlike the welded joint via FIB-based deposition as required in the course of in situ method. There exists a risk to lose the sample during this stage. The electrostatic charges may sometimes repel the thin lamella away. The lamella is then mounted on the TEM grid. A large amount electrostatic charges may also cause difficulty in transferring the lamella to the TEM grid from the probe. The ex situ method is significantly quick. However, once the sample is mounted on the TEM grid, it becomes extremely difficult to further thin it down.

7.3.2 FIB-Induced Damage

Bombardment of energetic ions on the target materials have many consequences apart from sputtering of the atoms/molecules. As discussed in the previous section, the beam energy manipulation renders an effective control over the milling rates and hence on the thickness and the uniformity of the lamella. In addition to sputtering, there are adverse effects due to the ion beam interaction with the sample material. The most serious consequence of the ion beam interaction is the ion implantation into the target material, which may reach critical composition for second phase formation. For instance, while processing semiconductors, such ion implantation leads to amorphisation in the interaction zone. For metals, various defects or intermetallic phases are observed due to ion beam processing.

The thickness of the damaged layer for a given material depends on the ion type, ion beam incidence angle to the sample surface, as well as the ion beam energy. The degree of amorphisation also depends upon the degree of the complexity of the material unit cell. For example, silicon exhibits amorphisation due to FIB milling, but copper does not; however, FIB interaction with copper along certain orientations results in Cu_3Ga phase.

Apart from considerations like thickness uniformity and electron transparency, it is important to pay attention to the FIB-induced damage to the TEM sample during the sample preparation procedure. The damage layer puts a limit on the final thickness of the lamella useful for TEM analysis, and it also compromises the high-resolution TEM image quality. It has been observed that a normal ion beam of 30 keV energy may lead to a complete amorphisation of the 50 nm Si sample, owing to 20 nm damage per side of the sample. Thermal annealing or chemical polishing methods may be useful in removing the amorphous damage layer. However, such post-processing treatments are not possible in the FIB vacuum chamber and it is almost impossible to apply them to the site-specific thin sample.

The reduction of beam energy and using low-incidence angle has been found to be an effective way to reduce the amorphisation of the materials due to ion beam interactions. Amorphisation damage due to FIB can be minimised or even removed with a high-resolution beam with low energies. It must be noted that the FIB resolution deteriorates with the reduction in beam energies. However, recent advances in FIB instrumentation have made it possible to maintain high resolution at beam energies as low as 500 eV. For example, with the FIB used to clean the sample at 5 keV, the damage layer was reduced to 5 nm, and using 2 keV beam for fine cleaning, this is further reduced to less than 2 nm [29]. Figure 7.7 shows high-resolution TEM (HRTEM) images of amorphisation damage layer reduction as ion beam energy with the reduction in ion beam energy from 30 keV to 2 keV.

The amorphisation layer can also be removed by using broad ion beam (BIB) milling after taking the lamella out after the FIB processing. In this method, the lamella is placed in a BIB equipment and a low-energy broad argon ion beam is used at low incident angles. By using 250 eV argon ion beam for processing of silicon, it is possible to reduce the amorphous layer to thickness as small as 1 nm [30].

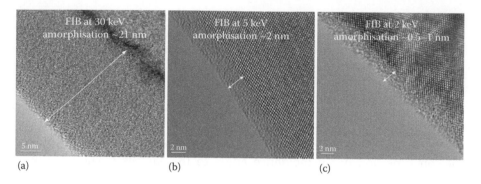

FIGURE 7.7
FIB-induced damage on the sidewalls of silicon. The TEM image shows an amorphous layer in the sidewalls of silicon after FIB milling. With the reduction of ion beam energy, the amorphous layer thickness is significantly reduced. The amorphous layer thickness observed is (a) ~21 nm at 30 keV, (b) ~2.5 nm at 5 keV and (c) ~0.5–1.5 nm at 2 keV. (From Giannuzzi, L.A., Geurts, R., Ringnalda, J., *Microsc. Microanal.*, 11, S02, 828–829, 2005.)

7.4 Conclusions

FIB has evolved into a powerful fabrication tool in achieving one-step maskless fabrication of special prototype structures. Prototype fabrication is facilitated with the use of different FIB modes like material removal through direct milling or material addition via chemical deposition. In addition, FIB is useful for site-specific repair/modification and as polishing tool for existing structures. The dual-beam, i.e. FIB and SEM, configuration allows unique capabilities like imaging and real-time observation.

The developments in material science and shrinking feature sizes demand advanced characterisation methods of which TEM analysis is an important tool. The TEM analysis requires preparation of site-specific, high-quality TEM samples with minimum damage to the bulk specimen. The dual-beam FIB with high-resolution nanofabrication capability enables site-specific TEM sample preparation with positioning accuracy of nanometer level. With the aid of channelling contrasts, a better visualisation is offered to locate accurate grain boundaries. Depending on the application, TEM sample can be prepared via in situ or ex situ lift-out techniques almost from any material. The FIB-based TEM sample preparation turns out to be much quicker than any other conventional mechanical or electrolytic polishing methods.

Acknowledgements

The authors acknowledge the financial support by IRCC, IIT Bombay, via seed grant Spons/ME/I14079-1/2014 and BRNS, Department of Atomic Energy (DAE), India, via project no. 34/14/06/2015/BRNS. Special thanks to Mr. Khoo Bing Sheng, WinTech Nano-Technology Services Pte. Ltd., for the careful FIB work and Abhishek, Bhavesh and Vivek of IIT Bombay for the schematics.

Example 7.1

1. For Ga^+ ion bombardment on an area of 20×20 µm², the ion dose and milling time are 10^{15} ions/cm² and 10 seconds, respectively. Calculate the ion beam current.

   Ion dose is calculated as

$$Dose = \frac{Ion\, number\, (ions)}{Area\,(cm^2)} = \frac{Q/q}{A} = \frac{I_B\,(Amp) \times T_M\,(sec)}{q \times A\,(cm^2)},$$

   where I_B is beam current, T_M is total milling time, A is milled area and q is electron charge, i.e. 1.6e–19 Coulomb/ion.

   Now, ion beam current can be calculated as

$$I_B = \frac{q \times A \times Dose}{T_M}$$

$$= \frac{1.6e-19 \times 400e-6 \times 1e15}{10} = 6.4\,nA$$

   Thus, the ion beam current is 6.4 nA.

2. A rectangular pocket of volume 100 µm³ is milled by Ga^+ FIB milling of energy 30 keV. The milling was carried out for 10.24 seconds. If the number of pixels in a row are 512 and the number of rows is 100, calculate the beam exposure time per pixel.

Milling time (T_M) = exposure time × numbers of pixels in area

$\qquad\qquad$ = exposure time × (# of pixels in row × # of rows)

$$\therefore \text{Exposure time} = \frac{Milling\ time\ (T_M)}{(\#\ \text{of pixels in row} \times \#\ \text{of rows})}$$

$$= \frac{10.24}{512 \times 100} = 200\ \mu s$$

Thus, the beam exposure time per pixel is 200 μs.

References

1. K. Watanabe, J. Schrauwen, A. Leinse, D. V. Thourhout, R. Heideman, and R. Baets, 'Total reflection mirrors fabricated on silica waveguides with focused ion beam', *Electron. Lett.*, vol. 45, no. 17, p. 883, 2009.

2. J. Li, V. Y. Gertsman, and J. Lo, 'Preparation of transmission electron microscope specimens from ultra-fine fibers by a FIB technique', *Microsc. Microanal.*, vol. 9, no. S02, pp. 888–889, 2003.

3. R. Levi-Setti, 'Proton scanning microscopy: Feasibility and promise', in *Seventh Annual Scanning Electron Microscope Symposium*, 1974, pp. 125–134.

4. J. H. Orloff and L. W. Swanson, 'Study of a field-ionization source for microprobe applications', *J. Vac. Sci. Technol.*, vol. 12, pp. 1209–1213, 1976.

5. R. L. Kubena, R. L. Seliger, and E. H. Stevens, 'High resolution sputtering using a focused ion beam', *Thin Solid Films*, vol. 92, no. 1/2, pp. 165–169, 1981.

6. R. L. Seliger, R. L. Kubena, R. D. Olney, J. W. Ward, and V. Wang, 'High-resolution, ion-beam processes for microstructure fabrication', *J. Vac. Sci. Technol.*, vol. 16, no. 6, pp. 1610–1612, 1979.

7. X. Jiang, Q. Ji, A. Chang, and K. N. Leung, 'Mini RF-driven ion sources for focused ion beam systems', *Rev. Sci. Instrum.*, vol. 74, no. 4, pp. 2288–2292, 2003.

8. J. Qing, L. Ka-Ngo, K. Tsu-Jae, J. Ximan, and B. R. Appleton, 'Development of focused ion beam systems with various ion species', *Nucl. Instruments Methods Phys. Res. Sect. B (Beam Interact. with Mater. Atoms)*, vol. 241, no. 1–4, pp. 335–340, 2005.

9. J. L. Hanssen, E. A. Dakin, J. J. McClelland, and M. Jacka, 'Using laser-cooled atoms as a focused ion beam source', *J. Vac. Sci. Technol. B (Microelectronics Nanom. Struct.*, vol. 24, no. 6, pp. 2907–2910, 2006.

10. J. Orloff, M. Utlaut, and L. Swanson, *High Resolution Focused Ion Beams: FIB and Its Applications*. New York: Kluwer Academic/Plenum Publishers, 2003.

11. R. G. Mote, S. F. Yu, A. Kumar, and X. F. Li, 'Experimental demonstration of near-field focusing of a phase micro-Fresnel zone plate (FZP) under linearly polarized illumination', *Appl. Phys. B (Lasers Opt.)*, vol. 102, no. 1, pp. 95–100, 2011.

12. K. Gamo, N. Takakura, N. Samoto, R. Shimizu, and S. Namba, 'Ion beam assisted deposition of metal organic films using focused ion beams', *Japanese J. Appl. Physics, Part 2*, vol. 23, no. 5, pp. 293–295, 1984.

13. I. Jin, C. H. Chen, S. P. Pai, B. Ming, D. J. Kang, T. Venkatesan, F. Machalett, K. Edinger, J. Orloff, and J. Melngailis, 'Fabrication of HTS Josephson junctions on substrates prepared by focused ion beam system', *IEEE Trans. Appl. Supercond.*, vol. 9, no. 2, pp. 2894–2897, 1999.
14. L. A. Giannuzzi and F. A. Stevie, *Introduction to Focused Ion Beams: Instrumentation, Theory, Techniques and Practice.* New York: Springer, 2005.
15. P. G. Li, A. Z. Jin, and W. H. Tang, 'Pt/Ga/C and Pt/C composite nanowires fabricated by focused ion and electron beam induced deposition', *Phys. Status Solidi A*, vol. 203, no. 2, pp. 282–286, 2006.
16. S. Nagamachi, Y. Yamakage, H. Maruno, M. Ueda, S. Sugimoto, and M. Asari, 'Focused ion beam direct deposition of gold', *Appl. Phys. Lett.*, vol. 62, no. 17, pp. 2143–2145, 1993.
17. S. Nagamachi, Y. Yamakage, M. Ueda, H. Maruno, K. Shinada, Y. Fujiyama, M. Asari, and J. Ishikawa, 'Focused ion beam direct deposition of superconductive thin film', *Appl. Phys. Lett.*, vol. 65, no. 25, p. 3278, 1994.
18. S. Nagamachi, M. Ueda, and J. Ishikawa, 'Focused ion beam direct deposition and its applications', *J. Vac. Sci. Technol. B (Microelectron. Process. Phenom.)*, vol. 16, no. 4, pp. 2515–2521, 1998.
19. D. C. Shaver, L. A. Stern, and N. P. Economou, 'Mask and circuit repair applications of focused ion beam deposition', *Microelectron. Eng.*, vol. 5, no. 1–4, p. 191, 1986.
20. W. Zhou, A. Cuenat, and M. J. Aziz, 'Formation of self-organized nanostructures on Ge during focused ion beam sputtering', in *Microscopy of Semiconducting Materials 2003: Proceedings of the 13th International Conference on Microscopy of Semiconducting Materials.* IOP Publishing Ltd., London, 2003, pp. 625–628.
21. M. Fritzsche, A. Muecklich, and S. Facsko, 'Nanohole pattern formation on germanium induced by focused ion beam and broad beam Ga+ irradiation', *Appl. Phys. Lett.*, vol. 100, no. 22, p. 223108, 2012.
22. D. M. Longo, J. M. Howe, and W. C. Johnson, 'Experimental method for determining Cliff–Lorimer factors in transmission electron microscopy (TEM) utilizing stepped wedge-shaped specimens prepared by focused ion beam (FIB) thinning', *Ultramicroscopy*, vol. 80, no. 2, pp. 85–97, 1999.
23. R. Anderson and S. J. Klepeis, 'Combined tripod polishing and FIB method for preparing semiconductor plan view specimens', *MRS Proc.*, vol. 480, p. 187, 1997.
24. J. Mayer, L. A. Giannuzzi, T. Kamino, and J. Michael, 'TEM sample preparation and FIB-induced damage', *MRS Bull.*, vol. 32, no. 05, pp. 400–407, 2011.
25. J. Li, M. Elboujdaini, V. Y. Gertsman, M. Gao, and D. C. Katz, 'Characterization of a stress corrosion crack with FIB and TEM', *Can. Metall. Q.*, vol. 44, no. 3, pp. 331–338, 2005.
26. L. A. Giannuzzi, J. L. Drown, S. R. Brown, R. B. Irwin, and F. A. Stevie, 'Focused ion beam milling and micromanipulation lift-out for site specific cross-section TEM specimen preparation', *MRS Proc.*, vol. 480, pp. 19–27, 1997.
27. B. I. Prenitzer, L. A. Giannuzzi, K. Newman, S. R. Brown, R. B. Irwin, T. L. Shofner, and F. A. Stevie, 'Transmission electron microscope specimen preparation of Zn powders using the focused ion beam lift-out technique', *Metall. Mater. Trans. A Phys. Metall. Mater. Sci.*, vol. 29A, no. 9, pp. 2399–2406, 1998.
28. A. A. Tseng, 'Recent developments in nanofabrication using ion projection lithography', *Small*, vol. 1, no. 6, pp. 594–608, 2005.

29. L. A. Giannuzzi, R. Geurts, and J. Ringnalda, '2 keV Ga+ FIB milling for reducing amorphous damage in silicon', *Microsc. Microanal.*, vol. 11, no. S02, pp. 828–829, 2005.
30. R. M. Langford and A. K. Petford-Long, 'Broad ion beam milling of focused ion beam prepared transmission electron microscopy cross sections for high resolution electron microscopy', *J. Vac. Sci. Technol. A (Vacuum, Surfaces, Film)*, vol. 19, no. 3, p. 982, 2001.

Section IV

Magnetic Field Assisted Nanofinishing Processes

8

Magnetic Abrasive Finishing Process and Modelling

Sachin Singh,[1] Deepu Kumar,[1] and M. Ravi Sankar[1] and Vijay K. Jain[2]

[1]Department of Mechanical Engineering, Indian Institute of Technology Guwahati, Guwahati, India

[2]Department of Mechanical Engineering, Indian Institute of Technology Kanpur, Kanpur, India

CONTENTS

8.1 Introduction

Finishing is the final machining operation performed to obtain a good quality surface integrity. Various finishing processes can be classified into two classes: (a) traditional finishing processes (viz., grinding, honing and lapping) and (b) advanced finishing processes (viz., abrasive flow finishing [AFF], rotational AFF, centrifugal-force-assisted abrasive flow machining process, magnetorheological AFF, etc.). A process is selected based on the level of accuracy required and the conditions to which the workpiece is subjected during finishing. The magnetic abrasive finishing (MAF) process is one of the advanced finishing processes used for finishing internal and external surfaces ranging from simple to complex shaped workpiece surfaces.

The MAF was developed to finish especially the flat large workpieces made up of hard-to-machine materials. This process employs ferromagnetic particles mixed with fine abrasive particles (SiC, Al_2O_3, CBN or diamond) called ferromagnetic abrasive particles (magnetic abrasive particles [MAPs]). The MAPs can be of two types: bonded and unbounded. Bonded MAPs are prepared by sintering ferromagnetic particles and abrasive particles at a very high pressure and temperature in inert gas atmosphere, while unbounded MAPs are just a mixture of ferromagnetic particles and abrasive particles mixed with a small amount of lubricant as an additive. The finishing action is controlled by controlling the applied magnetic field across the gap between the top surface of workpiece and bottom surface of magnet. On applying the magnetic field across the gap, a flexible magnetic abrasive brush (FMAB) is formed which acts as a multi-point cutting tool during finishing of the workpiece surface.

8.1.1 Working Principle of the MAF Process

In the MAF process, usually an electromagnet is used to produce a magnetic field. However, the electromagnet can also be replaced by a permanent magnet for the generation of a magnetic field. The only disadvantage in the latter case is that the magnetic flux, hence, the magnetic force, cannot be controlled effectively. As shown in Figure 8.1a–b, the MAPs align along the magnetic lines of force, thus forming a flexible magnetic abrasive brush. Depending upon the magnetic properties and the shape and size of the workpiece, there can be many alternate ways to arrange the magnetic poles. One of such arrangements is shown in Figure 8.1. The magnetic pole applies downward pressure through the FMAB, to press/indent the abrasive particles into the workpiece. The magnetic force also acts downward through the FMAB, on the workpiece. As the magnetic pole rotates, so does the FMAB brush, which

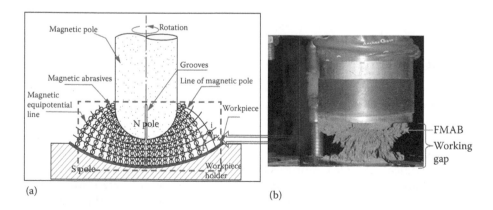

(a)   (b)

FIGURE 8.1
(a) Schematic diagram of the MAF process. (b) Photograph of FMAB at large working gap during filling of unbounded MAPs.

in turn rotates the abrasive particles and presses them against the workpiece surface. These result in the normal and tangential forces acting on the workpiece surface through the abrasive particles and remove the material in the form of micro/nano chips depending on the indention force. The pressure, P, due to magnetic force acting on the workpiece surface during the MAF process is given as [1]

$$P = \frac{1}{2}\left[\mu_0 H^2\left(1 - \frac{1}{\mu_m}\right)\right],$$ (8.1)

where μ_0 is permeability of vacuum, μ_m is the relative magnetic permeability of the MAP and H is the magnetic field strength.

Mainly two types of forces are generated by the magnetic field during the MAF process (Figure 8.2):

1. *Normal magnetic force, F_{mn}:* It helps in packing or concentrating MAPs into the working gap. Also, this is the force responsible for abrasive particles micro indenting the workpiece surface.

2. *Tangential magnetic force, F_{mi}:* There is a component of the magnetic force that acts in the tangential direction. Depending upon its direction, it may help/oppose the cutting action.

3. *Tangential (cutting) force, F_c:* The FMAB rotates the indented abrasive particles across the workpiece surface in a circular path. The tangential force helps these indented abrasive particles in shearing the roughness peaks in the form of micro/nano chips.

The MAPs align themselves along the magnetic lines of force in a non-uniform magnetic field. Apart from the above forces, a force is generated as

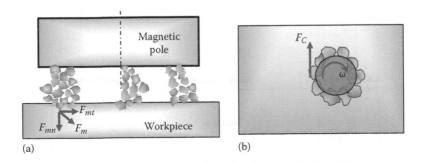

(a)

(b)

FIGURE 8.2
MAF process and modelling. (a) Normal and tangential magnetic force generated during MAF process. (b) Top view of figures showing tangential cutting force.

resistance to shear between the abrasive particles and the workpiece. Further, a centrifugal force due to the rotating magnetic pole also acts on the abrasive particles. These two forces try to splash out the ferromagnetic abrasive particles during the MAF operation. However, the strong magnetic field applied across the gap between the bottom surface of the magnet and the top surface of the workpiece prevents the splashing out of ferromagnetic abrasive particles. The intensity of magnetic field in the finishing zone is the most crucial factor during the MAF process. Too high intensity of magnetic field can cause formation of rigid FMAB. This may lead to the formation of deep scratches on the workpiece surface by the abrasive particles. Also, a rigid FMAB can cause the generation of a high amount of heat energy due to friction between the workpiece surface and the abrasive particles. On the other hand, too weak magnetic field intensity leads to the generation of a weak magnetic force, which may not be enough to form FMAB, and hence, there will be no or little surface finish improvement.

The formation of a magnetic abrasive brush requires energy. As shown in Figure 8.3, three types of energies are responsible for the formation of FMAB brush:

1. *Magnetisation energy, E_m:* This energy is required for the magnetisation of the MAPs to form bundles.
2. *Repulsion energy, E_r:* the bundles of the MAPs repel each other because of Faraday's law. Thus, E_r comes into existence.
3. *Tension energy, E_t:* It is required to counteract the repelling effect of the curved bundles of the ferromagnetic abrasive particles.

Therefore, the sum of these energies is responsible to form a FMAB as follows:

$$E = E_m + E_r + E_t. \tag{8.2}$$

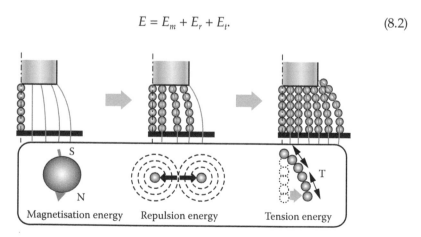

FIGURE 8.3
Configuration of magnetic abrasive brushes. (From Mori, T., Hirota, K., Kawashima, Y., *J. Mater. Proc. Technol.*, 143, 682–686, 2003.)

In order to form a stable FMAB, total energy, E, should be minimum, i.e.

$$dE = 0. \qquad (8.3)$$

8.1.2 Advantages of the MAF Process

a. In the MAF process, the finishing forces can be online controlled by changing the amount of current flowing in the circuit, depending upon the surface roughness value.

b. Due to the low magnitude of finishing forces acting on the abrasive particles, the MAF process reduces the chances of micro-crack occurrence, especially on the brittle materials during the finishing process.

c. MAF can produce surface roughness of the order of nanometer on flat surfaces as well as on cylindrical surfaces (internal and external both).

d. FMAB requires neither dressing nor compensation.

8.2 Mathematical Modelling of the MAF Process

Shinmura et al. [2,3] described the working principle and characteristic of the MAF process. They studied the effects of MAP size on surface finish and material removal of a cylindrical workpiece (SS41 steel) during finishing with the MAF process. The forces acting on the MAPs far from the working zone are shown in Figure 8.4. The force acting in the X direction, F_x, and the force acting in the Y direction, F_y, are given by

$$F_x = \chi_m vH\left(\frac{dH}{dx}\right) \qquad (8.4)$$

and

$$F_y = \chi_m vH\left(\frac{dH}{dy}\right), \qquad (8.5)$$

where v is volume of ferromagnetic particles, χ_m is susceptibility of MAPs, H is magnetic field strength at point A, x is the direction of line of magnetic force and y is the direction of magnetic equipotential line.

Considering the spherical shape of the MAP, Shinmura et al. [2,3] theoretically derived a formula for magnetic pressure between the workpiece and abrasive particles. Magnetic pressure, P, can be calculated by:

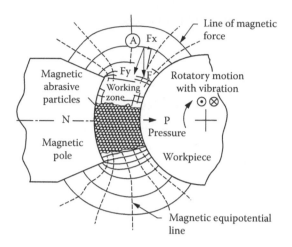

FIGURE 8.4
Magnetic forces acting on an abrasive particle in two-dimensional magnetic field. (From Shinmura, T., Takazawa, K., Hatano, E., Matsunaga, M., Matsuo, T., *CIRP Ann. Manuf. Technol.*, 39, 325–328, 1990.)

$$P = \frac{3\pi B^2 W}{4\mu_0}\left(\frac{(\mu_r - 1)}{3(\mu_r + 2) + \pi(\mu_r - 1)W}\right),$$ (8.6)

where B is magnetic flux density in the working zone, μ_r is the relative magnetic permeability of pure iron, μ_0 is magnetic permeability in vacuum and W is volumetric ratio of iron in MAPs. From Equation 8.6, it is evident that the working pressure is independent of MAP diameter in the working zone. From the experimental results, an expression was proposed that relates magnetic force acting (f) on one MAP and its diameter (D), given by

$$f = K_1 D^\alpha,$$ (8.7)

where K_1 is proportionality constant, D is the diameter of the MAP and the value of α lies between 2 and 3.

The force (Δf) acting on an abrasive particle is given by

$$\Delta f = \frac{f}{n} = \frac{K_1 D^\alpha}{n},$$ (8.8)

where n is the total number of abrasive particles in one MAP, which are acting on the workpiece surface at the same time.

Shinmura et al. [2,3] derived the following relationship for material removal by one MAP, M:

$$M = K_2 D^{(\alpha\beta-2)} n^{(1-\beta)},$$ (8.9)

where K_2 is a constant of proportionality and the value of β depends upon the shape of the abrasive particle cutting edge ($\beta = 1$ or 1.5 if the edge shape is a cone or sphere).

Shinmura et al. [2,3] concluded that the material removal increases with an increase in the diameter of the MAP (D) and the diameter of the abrasive particle (d). In order to produce a surface roughness of low value, the diameter of the abrasive particle should be smaller. With the help of the MAF process, the burrs produced in the grinding process can be removed and also with precision edge radii of about 0.01 mm can be obtained.

Later, Khairy et al. [4] proposed an analytical model to understand the kinematics of the MAF process. They [4] used the same analytical model to study working pressure (P) as outlined in Shinmura et al. [2,3]. Also, the maximum undeformed chip thickness, h_{max} is given by [4]

$$h_{max} = \sqrt{\left(\frac{\sqrt{3}}{CR}\right)\left(\frac{v_f}{v_S}\right)},\tag{8.10}$$

where C is the number of dynamic cutting points per unit surface area, R is the ratio of width to the thickness of the chip along the cutting path, v_S is workpiece surface speed and v_f is infeed rate (Figure 8.5), and it is given as

$$v_f = \left(\frac{\Delta d}{t}\right),\tag{8.11}$$

where Δd is the change in the diameter of the workpiece and t is the machining time of the MAF process.

To have a better understanding of the mechanism of material removal in the MAF process, Mori et al. [5] studied the magnetic field and types of forces

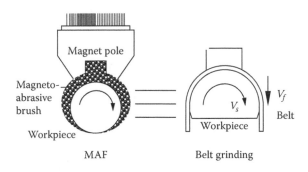

FIGURE 8.5
Correspondence of MAF and belt grinding. (From Khairy, A.B., *J. Mater. Process. Technol.*, 116, 77–83, 2001.)

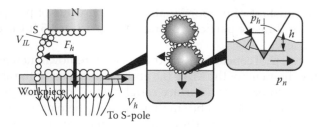

FIGURE 8.6
Configuration of magnetic abrasive polishing in the case of non-magnetic material. (From Mori, T., Hirota, K., Kawashima, Y., *J. Mater. Process. Technol.*, 143, 682–686, 2003.)

acting on the abrasive particles during the process. The tangential (f_t) and normal (f_n) forces acting on the abrasive particles during MAF (Figure 8.6)

$$f_n = nk\pi(1 + \theta) \tan^2 \theta h^2 \tag{8.12}$$

and

$$f_t = nk\pi(1 + \theta)\tan \theta h^2, \tag{8.13}$$

where n is the average number of active abrasive particles indented per MAP, h is depth of indentation and θ is included angle.

Raghuram and Joshi [6] developed an analytical model to predict output responses (R_a and MR) during the MAF process. The responses were found to be the function of various input parameters as given in

$$MR = f(H_w, N_s, P, d_m, d, n_a, t) \tag{8.14}$$

and

$$R_a = f\left(H_w, N_s, P, d_m, d, n_a, R_a^0, t\right), \tag{8.15}$$

where H_w is the workpiece hardness, N_s is the rotational speed of the magnetic pole, P is the abrasive pressure, d_m is the magnetic particle diameter, d_{ap} is the abrasive particle diameter, n_a is the number of active abrasive cutting edges, R_a^0 is the initial surface roughness value and t is the polishing time. In this analysis, several assumptions have been made to simplify the model:

$$MR = \frac{\Pi D_p N_s t N n_a}{4}\left[\frac{d^2}{4}\sin^{-1}\frac{2\sqrt{h_d(d - h_d)}}{d} - \sqrt{h_d(d - h_d)}\left(\frac{d}{2} - h_d\right)\right] \tag{8.16}$$

and

$$R_a = R_a^0 - \sqrt{\frac{\Pi D_p N_s t N n_a R_a^0}{2 l_l l_w} \left[\frac{d^2}{4} \sin^{-1} \frac{2\sqrt{h_d(d-h_d)}}{d} - \sqrt{h_d(d-h_d)} \left(\frac{d}{2} - h_d \right) \right]},$$

(8.17)

where D_p is the magnetic pole diameter, h_d is the indentation diameter of the abrasive particle into the workpiece surface and l_l and l_w are the length and width of the finished area, respectively. The analytical results were compared with the experimental results and they were found to be in good agreement.

8.3 Simulation Modelling

Kim and Choi [7] proposed a simulation model to predict the final surface roughness of a workpiece. This simulation model is used to automate the MAF process. The proposed model predicts the surface roughness as a function of time. Depending on the desired final surface roughness value and the present surface roughness value of the workpiece, the magnetic forces are changed online through the input current. The magnetic pressure between the workpiece and abrasives during MAF is given by [7]

$$P = \mu_0 \frac{H_a^2}{4} \left[\frac{3\pi(\mu_r - 1)W}{3(2+\mu_r) + \pi(\mu_r - 1)W} \right],$$

(8.18)

where μ_r is relative permeability of pure iron, μ_0 is magnetic permeability in vacuum and W is volumetric ratio of iron in a magnetic particle.

The strength of the magnetic field in the air gap, H_a, and relative permeability of pure ferromagnetic particle, μ_r, in Equation 8.18 are evaluated by using Equations 8.19 and 8.21 as follows:

$$H_a = \frac{N_c I}{A_a(L_a/A_a + L_m/\mu_{re} A_m)},$$

(8.19)

where N_c is the number of turns in the coil; I is current; A_a and A_m are cross-sectional areas of air gap and magnet, respectively; L_a and L_m are the length of the air gap and the magnet respectively and μ_{re} is the relative permeability of the electromagnet.

With the previous magnetic field strength, the MAPs are magnetised, which can be evaluated from the following expression:

$$H = \left(\frac{3}{\mu_r + 2} \right) H_a.$$

(8.20)

The relative permeability of a pure ferromagnetic particle (iron particle) (μ_r) is the slope of the magnetisation curve, and it is given as

$$\mu_r = \frac{1}{\mu_0} \frac{dB}{dH} = \frac{B_s}{\mu_0 H_c} \sec h^2 \left(\frac{H \mp H_c}{H_c} \right),$$

(8.21)

where B_s is the saturated magnetic flux density and H_c is the coercive force.

Kim and Choi [7] assumed the initial surface as a uniform surface roughness profile, which is triangular in nature without any statistical distribution (Figure 8.7).

Also, the abrasive particles move in the length direction of the scratches. The total material removal, M, by the FMAB during the MAF process is given by

$$M = \left(C \frac{nN\Delta f v_{ma} t}{H_w \pi \tan \theta L_m} \right)^{0.5} \left(R_a^0 \right)^{-0.25},$$

(8.22)

where n is the number of cutting edges of a MAP taking part in finishing, N is the number of MAPs acting in the finishing region simultaneously, C is constant of proportionality, θ is half of the mean angle of asperity of abrasive cutting edges, t is the machining time, v_{ma} is the speed of MAPs, H_w is the hardness of the workpiece material and R_a^0 is the initial surface roughness of the workpiece.

In Equation 8.22, the force, Δf, acting on an edge of an abrasive particle (assuming that the cross-section of the air gap is same as that of the magnetic brush) is given by

$$\Delta f = \frac{P d_a^2}{4n}.$$

(8.23)

where, d_a is the mean diameter of an abrasive particle

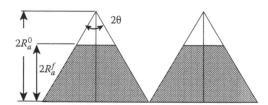

FIGURE 8.7
Simplified surface geometry. (From Kim, J.D., Choi, M.S., *J. Mater. Process. Technol.*, 53, 630–642, 1995.)

The final surface roughness value of the workpiece surface is given by

$$R_a^f = R_a^0 - C'\left(R_a^0\right)^{-1/8} (L_w)^{-5/4} \left(\frac{nN \; f v_{ma} t}{H_w \pi L_m \tan \theta}\right)^{0.25}, \tag{8.24}$$

where $C' = C^{1/4}$ is constant of proportionality.

These models can be used for the evaluation of material removal and surface roughness value, but they are based on the assumption of a homogeneous mixture having uniform properties (magnetic) throughout. Such non-homogeneity of the FMAB can be taken care of by numerical methods, say finite element, finite volume, finite difference, etc. Jayswal et al. [8–10] developed the finite element method (FEM) to numerically evaluate the distribution of magnetic forces generated during the MAF process. Later, these generated forces were used in the theoretical models to predict material removal and surface roughness during the MAF process. The final results were compared with the experimental results, and they were found to be in good agreement. Jayswal et al. [8–10] made several assumptions to simplify the analysis of the MAF process. Based on the Maxwell Equation, the governing equation in the axisymmetric form is given as [10]:

$$\frac{1}{r}\frac{\partial}{\partial r}\left[r\mu_r\frac{\partial\varphi}{\partial r}\right] + \frac{\partial}{\partial z}\left[\mu_r\frac{\partial\varphi}{\partial z}\right] = 0, \tag{8.25}$$

where φ is a magnetic scalar potential and μ_r is the relative permeability of MAPs.

After the implication of Galerkin's FEM with appropriate weight functions, the governing differential equations, with appropriate boundary conditions, are converted into integral form, i.e. weak formulation over a typical element. After assembling the coefficient matrices of all the elements, the global set of linear algebraic equation is given by

$$[K]\{\phi\} = 0, \tag{8.26}$$

where $\{\phi\}$ is global magnetic potential vector and $[K]$ is global coefficient matrix, which is obtained by assembling the elemental coefficient matrix ($[K]^e$) of all the elements as follows:

$$[K]^e = \int_{\Omega^e} \mu_r [A]^{eT}[A]^e \, 2\pi r \, dr \, dz, \tag{8.27}$$

where the matrix $[A]$ possesses the shape function derivatives for the approximation of scalar magnetic potential and Ω^e is the area of the elemental domain.

The normal, F_N and radial F_R, components of a magnetic force, F, acting on an abrasive particle are given by

$$F_N = \frac{\mu_0}{2} v \frac{\partial}{\partial z} (\chi_r H \cdot H) \tag{8.28}$$

and

$$F_R = \frac{\mu_0}{2} v \frac{\partial}{\partial r} (\chi_r H \cdot H), \tag{8.29}$$

where μ_0 is the permeability of free space, v is the volume of the abrasive particle and χ_r is its susceptibility.

Jayswal et al. [8–10] assumed that MAPs are tightly packed, and during the finishing operation, they move in a track (Figure 8.8a–b). The whole workpiece surface is divided into cells. The volume of material removal in a cell (i, j) in the nth revolution is given as

$$\Delta V_{(i,j)}^N = \frac{2 A_{p(tr)} h s_{(tr)}^n I_c R_{(tr)} n_a}{R_a^0 D_{map}}, \tag{8.30}$$

where
$A_{p(tr)}$ = sheared area in track tr

$$= \frac{(d)^2}{4} \left[\cos^{-1} \left(1 - \frac{2h_s}{D_s} \right) \right] - \left(\left(\frac{d}{2} - h_s \right) \sqrt{h_s(d - h_s)} \right) \tag{8.31}$$

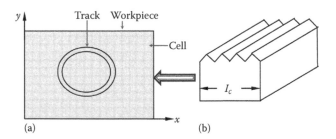

FIGURE 8.8
(a) Schematic diagram of workpiece with cells. (b) Exaggerated view of a typical cell. (From Jayswal, S.C., Jain, V.K., Dixit, P.M., *Int. J. Adv. Manuf. Technol.*, 26, 477–490, 2005.)

h_s = indentation depth into the surface profile

$$= \frac{d}{2} - \sqrt{\left(\frac{d}{2}\right)^2 - \frac{F_z}{H_w \pi}} \tag{8.32}$$

R_{tr} = the trth track radius

$$= (tr - 1)D_{map} + \frac{D_{map}}{2}. \tag{8.33}$$

H_w is the hardness of the workpiece, I_c is the length of a cell, d is the diameter of the abrasive particle, D_{map} is the diameter of MAP, n_a is the number of active cutting edges of MAP and R_a^0 is the initial surface roughness of the workpiece.

Therefore, in the nth rotation, the total material removal is evaluated by summing up the total material removal in each cell and is given as

$$\Delta V^n = \sum \Delta V_{(i,j)}^n. \tag{8.34}$$

Surface roughness after the nth revolution in a cell (i, j), $R_{\max(i,j)}^n$, during the MAF process is given by

$$R_{\max(i,j)}^n = R_a^0 - \left(\left(R_a^0 - R_{a(i,j)}^{(n-1)}\right)^2 + \left(\frac{\Delta V_{(i,j)}^n R_a^0 D_{map}}{\pi r_{tr} I_c R_{tr} n_a}\right)\right)^{1/2}, \tag{8.35}$$

where r_{tr} is the average width of cut produced by a cutting edge in track tr, $R_{\max(i,j)}^{(n-1)}$ is the surface roughness obtained after $(n-1)$th revolution of the N-pole.

Later on, extending the same concept of modelling of Jayswal et al. [8–10], in order to make simulation more realistic, Jain et al. [12] reported the modelling and simulation of surface roughness in the MAF process by considering a non-uniform surface roughness profile of the workpiece. The surface profile is assumed to be Gaussian in nature. Jayswal et al. [8–10] studied the effect of various MAF input parameters (flux density, height of working gap, size of MAPs and rotational speed of magnetic pole) on the output responses (surface roughness). They computed the magnetic forces acting on the abrasive particles with the previous model (FEM model) and did a simulation to find out the final surface roughness profile of the workpiece surface. Jain et al. [12] applied the probabilistic approach; i.e. the value of micro-indentation

The heat flux entering into the workpiece is

$$q_w = R_w \times P \times \mu_f \times v_{ap}, \tag{8.40}$$

where P is magnetic pressure supplied by the magnet, μ_f is the coefficient of workpiece, R_w is the energy partition and v_{ap} is the velocity of abrasive particles.

Judal and Yadava [12] used the Glakerin finite element (FE) formulation and temperature model over an element as follows:

$$[K_T]^e \{T\}^e = \{f\}^e, \tag{8.41}$$

where $[K_T]^e$ is the elemental coefficient matrix, which can be written as

$$[K_T]^e = k_w \int_{D^e} [B]^{et} [B]^e r \, dr \, dz. \tag{8.42}$$

Similarly, the $\{f\}^e$ force vector can be written as

$$\{f\}^e = \int_B \{N\}^b q_w r \, dB, \tag{8.43}$$

where $[B]^e$ is the derivative of the shape function matrix, $\{N\}$ is the column vector for shape function and B is magnetic flux density. Judal and Yadava [13] reported that the developed model for magnetic potential and heat flux predict the above phenomena reasonably accurately.

8.3.1 Hybrid Process Simulation Modelling

Electrochemical magnetic abrasive machining (EMAM) is a hybrid form of the MAF process developed to improve the efficiency of the MAF process. It can also be said that in this process, two operations (machining and finishing) are performed simultaneously to get rid of the post-machining operation (i.e. finishing). During the EMAM process, material is removed from the workpiece due to the mechanical abrasion by the MAF process and through electrochemical (EC) reaction due to the EC dissolution process. Judal and Yadava [13,14] used the C-EMAM process for finishing the cylindrical workpiece and named it the cylindrical EMAM (C-EMAM) process. They [13,14] proposed various models to calculate the material removal and surface roughness during the hybrid process. The model developed for material removal and surface roughness is the combination of three processes: abrasion, EC action and abrasion-assisted dissolution. In this model, the magnetic field distribution (scalar magnetic potential) on a cylindrical workpiece, which is placed between the two opposite magnetic poles, is computed in a similar manner as discussed in the literature [8–10]. After solving the FE formulation, final cutting forces acting on a ferromagnetic abrasive particle in the x-direction, F_x, and the force acting in the y-direction, F_y, are given as

$$F_x = \chi_m \mu_0 v H \left(\frac{dH}{dx} \right)$$ (8.44)

and

$$F_y = \chi_m \mu_0 v H \left(\frac{dH}{dy} \right),$$ (8.45)

where v is the volume of ferromagnetic abrasive particle, χ_m is the susceptibility of MAP, H is magnetic field strength and μ_0 is permeability in air.

As shown in Figure 8.11a–c, the normal force, F_N, along the N–N direction and the tangential force, F_T, along the T–T direction can be given as

$$F_N = F_x \cos \theta + F_y \sin \theta$$ (8.46)

and

$$F_T = -F_x \sin \theta + F_y \cos \theta.$$ (8.47)

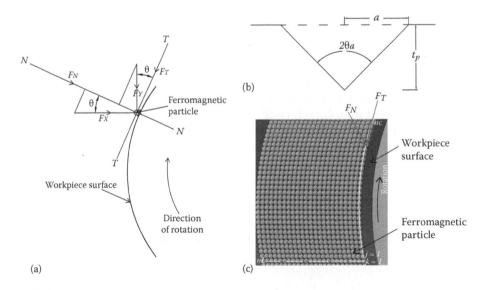

(a)   (b)   (c)

FIGURE 8.11
(a) Forces acting on a ferromagnetic particle during magnetic abrasive machining (MAM). (b) Indentation by a wedge-shaped abrasive particle. (c) Grid of ferromagnetic particles at contact surface. (From Judal, K.B., Yadava, V., *Int. J. Mach. Sci. Technol.*, 18, 221–250, 2014.)

Considering the roughness profile as triangular in nature without any statistical distribution, the total amount of volume of material removed by only the EC dissolution process, V_{ECD}, during C-EMAM is given as

$$
V_{MAF} = K_{MAF} \, \pi d_w s \, t \left(1 - \frac{R_a}{R_a^0} \right) \sum_{i=1}^{N_L} \sum_{j=1}^{N_C} \left(\frac{N_a f_n(j)}{H_m \pi \tan \theta_A} \right),
\tag{8.48}
$$

where K_{MAF} is a constant for the MAF process; s is the rotational speed of the workpiece; t is the finishing time of the MAF process; N_a is the number of abrasive particles below the ferromagnetic particle; N_C and N_L are number of ferromagnetic particles in contact with the workpiece surface along the circumference and length directions, respectively; d_w is the diameter of the cylindrical workpiece; f_n is the normal force acting on a single abrasive particle responsible for indentation and $2\theta_A$ is the included angle of wedge-shaped indentation.

The total amount of volume of material removed by only the EC dissolution process, V_{ECD}, during C-EMAM is given as

$$
V_{ECD} = \frac{\eta_c \, I_{ele} \, T_M}{\rho F \left(\displaystyle\sum_{j=1}^{n} X_j n_j / M_j \right)},
\tag{8.49}
$$

where I_{ele} is the electrolytic current, T_M is the machining time, F is Faraday's constant, X_j is the weight fraction of the jth species, n_j is the oxidation valence of the jth species, M_j is the atomic weight of the jth species, ρ is the density of the workpiece material and η_c is the current efficiency.

There is some extra material removed due to the abrasion-assisted dissolution, V_{Extra}. The total amount of material removal due to EC dissolution and abrasion-assisted dissolution, i.e. $V_{ECD} + V_{Extra}$, can be given by Equation 8.49 by replacing current efficiency, η_c, with current efficiency considering EC action and abrasion-assisted dissolution, η_{ecap}, which is given as

$$
\eta_{ecap} = 5.3523 f_{Nave}^{0.5} f^{-0.5} I_{ele} e^{(f/100)},
\tag{8.50}
$$

where f_{Nave} is the mean normal force acting on an abrasive particle [13].

Thus, the total amount of material removal during the C-EMAM, V_w, process is given as

$$
V_w = V_{MAF} + V_{ECD} + V_{Extra}.
\tag{8.51}
$$

Judal and Yadava [13] derived the surface roughness expression by using the total amount of material removal, V_w, during the C-EMAM process and considering the surface profile triangular in nature having arithmetic mean surface roughness, R_a^0. The finial surface roughness of cylindrical workpiece, R_a, is given as

$$R_a = R_a^0 - \sqrt{\frac{k_I R_a^0 V_W}{4\pi d_W l_W}},$$ (8.52)

where k_I is a constant, l_W is length and d_w is the diameter of the cylindrical workpiece.

Yamaguchi et al. [15] modified the MAF process to achieve internal finishing on stainless steel capillary tubes having an internal diameter of 800 μm or less. They suggested that the amount of MAPs and the magnetic force acting on them are the key components during finishing of such complicated workpieces. To know the amount of magnetic finishing forces generated during the MAF process, two-dimensional magnetic field analysis was done using FEM, before developing the actual experimental setup. Through the experiments performed, it was concluded that the number and size of the depressions on the workpiece surface diminished after the MAF process for 16 minutes. Kwak [16] customised the MAF process to increase its finishing efficiency during the finishing of non-ferrous materials. Kumar and Yadav [17] simulated the magnetic flux density produced during the MAF process and proposed the installation of a permanent magnet at the opposite side of the workpiece for achieving better finishing results (Figure 8.12). Installation of a permanent magnet increases the magnetic flux density (Figure 8.13), which in turn increases the magnitude of the magnetic finishing forces acting on the abrasive particles. Kumar and Yadav [17] verified the improved results through simulation and experimentally.

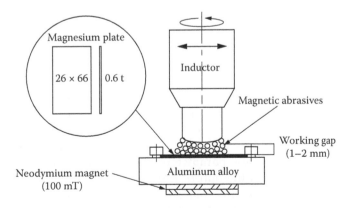

FIGURE 8.12
Schematic view of experimental setup for MAP of magnesium with permanent magnet. (From Kumar, G., Yadav, V., *Int. J. Adv. Manuf. Technol.*, 41, 1051–1058, 2009.)

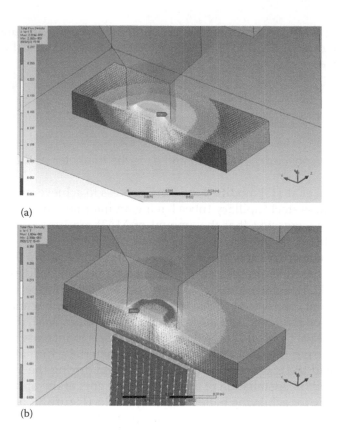

(a)

(b)

FIGURE 8.13
Magnetic flux density compared with and without permanent magnet in non-ferrous materials. Current: 2 A, max flux density: (a) 22.24 mT (without magnet) and (b) 30.04 mT (with magnet). (From Kumar, G., Yadav, V., *Int. J. Adv. Manuf. Technol.*, 41, 1051–1058, 2009.)

8.4 Empirical Modelling

Jain et al. [18] experimentally studied the effect of working gap and circumferential speed on the output responses (change in surface roughness, ΔR_a, and material removal) during the MAF process. They found that material removal decreases as the circumferential speed decreases, while the working gap increases. ΔR_a increases with an increase in circumferential speed. Later on, Singh et al. [19] used the design of experiments to study in detail the important MAF process parameters and their optimum combination, which significantly affect the output responses. The input parameters that are used during the study are voltage (direct current [DC]) applied to the electromagnet, working gap, rotational speed of the magnet and abrasive particle size. The change in surface roughness, ΔR_a, normal magnetic force, F_{mn}, and tangential cutting force, F_c, are the output responses. The regression models of the output responses in terms of input parameters derived by author's [18] are as follows:

$$\Delta R_a = 0.137 + 0.0317X_1 - 0.0367X_2 + 0.0100X_3 + 0.0167X_4, \qquad (8.53)$$

$$F_{mn} = 34.1 + 36.3X_1 - 32.3X_2 - 7.83X_3 + 25.2X_4, \qquad (8.54)$$

and

$$F_c = 29.3 + 8.68X_1 - 6.50X_2 - 2.85X_3 + 2.64X_4, \qquad (8.55)$$

where X_1, X_2, X_3 and X_4 are the coded level values of voltage, working gap, rotational speed of the magnet and abrasive mesh size, respectively. Singh et al. [19] concluded that the voltage and working gap are the most significant parameters affecting the output responses, while the rotational speed of the magnet and abrasive mesh size had negligible effect.

To gain a detailed insight of the MAF process, Singh et al. [20,21] performed a microscopic study of the MAF process. They found out that the MAF process during finishing creates micro scratches of about 0.5 μm size. Also, there were circular lays on the finished workpiece due to the rotation of the FMAB. Singh et al. [22] further studied the MAF process using the analysis of variance (ANOVA) approach. The effect of input parameters (current to the electromagnet, machining gap, abrasive size [mesh number] and number of cycles) on the output responses (change in surface roughness, ΔR_a, and material removal, MR) have been reported. Based on the findings, the regression equations for the output responses in terms of input parameters can be given as

$$R_a = 0.213 + 0.267Y_1 - 0.0208Y_2 + 0.0192Y_3 + 0.158Y_4 - 0.00717Y_1^2$$
$$- 0.00217Y_2^2 + 0.00283Y_3^2 - 0.00342Y_4^2 - 0.00875Y_1Y_2 + 0.00125Y_1Y_3$$
$$- 0.00750Y_1Y_4 + 0.00250Y_2Y_3 + 0.00875Y_2Y_4 + 0.00375Y_3Y_4$$
$$(8.56)$$

and

$$MR = 79.1 + 12.3Y_1 - 6.79Y_2 + 7.46Y_3 + 8.37Y_4 - 0.50Y_1^2$$
$$- 0.38Y_2^2 + 0.50Y_3^2 - 0.75Y_4^2 - 1.81Y_1Y_2 + 1.81Y_1Y_3 \qquad (8.57)$$
$$- 0.31Y_1Y_4 + 2.31Y_2Y_3 + 6.94Y_2Y_4 + 1.69Y_3Y_4,$$

where Y_1, Y_2, Y_3 and Y_4 are the coded level values of current, machining gap, grain mesh number and number of cycles, respectively. It was concluded that the current is the most influential parameter, followed by the machining gap, the abrasive particles' mesh number and the number of cycles. Based on the experimental results, a correlation between ΔR_a and MR is reported in the literature [22] for the given finishing conditions and workpiece material, as follows

$$\Delta R_a = 0.2549 - 0.0074 \times MR + 0.0001 \times MR^2 - 6 \times 10^{-04} \times MR^3. \quad (8.58)$$

It is the finishing force that decides the final surface roughness and material removal during the finishing of a workpiece with the MAF process. To understand the mechanism of the MAF process, it is essential to have knowledge of the forces acting during the finishing process. Singh et al. [23] experimentally studied the amount of finishing forces generated during the MAF process and also provided correlation between the finishing forces and the final surface finish obtained on the workpiece. ANOVA was used to find the regression model between the input parameters (current to the coil of electromagnet, working gap, percentage of lubricant, rotational speed of magnet and finishing time) and the output responses (F_{mn}, F_c and ΔRa). The regression models after neglecting the insignificant terms are given as

$$F_{mn} = 48 + 12Z_1 - 7.22Z_2 + 6.38Z_3 + 3.48Z_1^2 + 2.51Z_3^2 + 4.03Z_3Z_4, \quad (8.59)$$

$$F_c = 35.1 + 6.27Z_1 - 6.05Z_2 - 2.45Z_4 - 1.04Z_4^2, \quad (8.60)$$

and

$$\Delta R_a = 0.22 + 0.025Z_1 - 0.392Z_2 + 0.0175Z_4 - 0.025Z_1Z_2 + 0.0212Z_2Z_4, \quad (8.61)$$

where Z_1, Z_2, Z_3 and Z_4 are the coded level values for current to the coil of electromagnet, working gap, percentage of lubricant and rotational speed of magnet, respectively. As in any finishing operation, it is the force that plays the dominant role, so Singh et al. [23] also established the relationships between the finishing forces and ΔRa, which are given as

$$\Delta R_a = (0.0200566)(F_{mn})^{0.583}, \quad (8.62)$$

$$\Delta R_a = (0.008742)(F_c)^{0.885}, \quad (8.63)$$

and

$$\Delta R_a = 0.01324 F_{mn}^{0.293} F_c^{0.443}. \quad (8.64)$$

The use of a DC power supply forms a static-flexible magnetic abrasive brush (S-FMAB). The abrasive particles in the S-FMAB become dull as the finishing progresses due to the absence of a stirring effect that results in a low finishing rate. This shortcoming is eliminated by Jain et al. [24] by replacing the DC power supply by the pulsating DC power supply. They developed regression models correlating pulse on-time and duty cycle with percent decrease in Ra ($\%\Delta R_a$), normal magnetic force, F_{amn}, and tangential cutting force, F_{ac}, and average force ratio, F_{ar} (ratio of F_{amn} and F_{ac}) as follows:

$$\%\Delta R_a = 80.46 - 208.46\,x_1 - 3.64 \times 10^{-3} x_2 + 255.61 x_1^2 + 1.5 \times 10^{-6} x_2^2 + 2.58 \times 10^{-2} x_1 x_2,$$

(8.65)

$$F_{amn} = 1231.62 - 8843.72\,x_1 + 507.89 x_2 + 16488.64 x_1^2 + 246.91 x_2^2 - 2287.50 x_1 x_2,$$

(8.66)

$$F_{ac} = 138.08 - 211.19\,x_1 - 46.44 x_2 + 392.24 x_1^2 + 17.17 x_2^2 + 62.50 x_1 x_2, \quad (8.67)$$

and

$$F_{ar} = 115.61 - 106.52\,x_1 - 4.15 x_2 + 185.85 x_1^2 + 2 x_2^2 - 17.22 x_1 x_2, \quad (8.68)$$

where x_1 and x_2 are the coded level values for duty cycle and pulse on-time (ms). It was concluded that in a pulsed current MAF (P-MAF), the rate of surface finish is more as compared to the DC-MAF (D-MAF). Also, a more uniform surface texture was produced during a P-MAF as compared to a D-MAF.

Girma et al. [25] studied the effect of grain size, relative size of abrasive particles, feed rate and current on R_a and MR by response surface methodology (RSM) using central composite design. The models for R_a and MR evolved by RSM are given as

$$R_a = 0.305 + 0.191A - 0.060B + 0.106C + 0.056D + 0.009A^2 - 0.059AB + 0.045AC$$

$$+ 0.040AD - 0.008B^2 + 0.029BC - 0.076BD - 0.007C^2 + 0.011CD - 0.031D^2$$

(8.69)

and

$$MR = 11.407 + 1.501A - 1.376B + 0.246C + 1.35D - 1.091A^2 - 0.531AB + 0.406AC$$

$$- 0.094AD - 0.453B^2 + 0.219BC - 0.531BD - 0.071C^2 + 0.281CD - 0.326D^2,$$

(8.70)

where A, B, C and D are the values of MAP size, size ratio, feed rate and current, respectively. Jayswal et al. [25] concluded that all the considered input factors affect surface finish and material removal of the workpiece.

El-Taweel [26] suggested another way to increase the material removal rate and to further reduce the magnitude of the final surface finish, R_a, achieved during MAF process. El-Taweel [26] developed an experimental setup (Figure 8.14) to combine EC turning (ECT) process with MAF.

The basic principle of the hybrid ECT-MAF process is the production of an oxide layer on the workpiece surface during electrolysis. This layer is removed by mechanical abrasion during the MAF process. The new surface obtained after the MAF process again goes through oxidation and the same process repeats. ANOVA is done to investigate the influence of the hybrid ECT-MAF input parameters (magnetic flux density, applied voltage, tool feed rate and workpiece rotational speed) on the output responses (MRR and R_a). The regression models derived for the output responses in terms of input variables are given as follows:

$$MRR = -0.219 + 0.747K_1 + 0.034K_2 - 3.251K_3 + 0.001K_4 - 6.557K_1^2$$

$$- 0.001K_2^2 - 9.016K_3^2 + 0.026K_1K_2 + 22.917K_1K_3 \tag{8.71}$$

$$+ 0.125K_2K_3 + 0.004K_3K_4$$

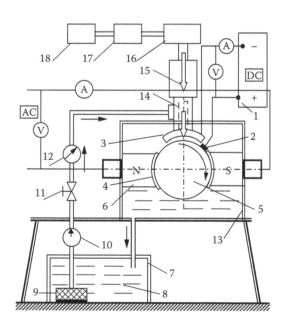

FIGURE 8.14
Schematic of ECT-MAF system 1 – DC power supply, 2 – carbon brush, 3 – tool, 4 – magnetic abrasives, 5 – workpiece, 6 – magnetic poles, 7 – electrolyte tank, 8 – electrolyte, 9 – filter, 10 – pump, 11 – control valve, 12 – flow meter, 13 – return tank, 14 – tool holder, 15 – feed direction, 16 – feed system, 17 – stepper motor, 18 – control unit. (From El-Taweel, T.A., *Int. J. Adv. Manuf. Technol.*, 37, 705–714, 2008.)

and

$$R_a = 0.6 - 1.724K_1 - 0.023K_2 - 0.791K_3 + 5.304K_1^2 + 0.001K_2^2$$
$$+ 28.984K_3^2 - 0.031K_2K_3 - 0.008K_3K_4, \qquad (8.72)$$

where K_1, K_2, K_3 and K_4 are the coded level values for the magnetic flux density, applied voltage, tool feed rate and workpiece rotational speed, respectively. It is concluded that an increase in the applied voltage and tool feed rate helps in achieving good surface finish.

Lin et al. [1] finished the free-form surfaces using MAF process. They [1] studied the optimal combination of finishing parameters by signal-to-noise ratio (S/N) response graphs. They [1] used ANOVA to find the significance of the input parameters (magnetic field, spindle revolution, feed rate, working gap, abrasive and lubricant) on the output response (R_a). Feed rate and working gap are found to be significant parameters. Wang et al. [27] finished the cylindrical rod through the MAF process. They followed the same thought as that of Lin et al. [1] to evaluate surface roughness through the S/N ratio 'smaller the better' characteristic. They concluded that the concentration of steel grit, machining time and the kind of abrasive particles dominate the behaviour of the MAF process. Naif [28] used ANOVA to study the effects of input MAF process parameters (applied current to the inductor, working gap between the workpiece and inductor, rotational speed and volume of abrasive powder) on output responses (ΔR_a and change in hardness, ΔH_a) while finishing a brass plate. The developed regression models are as follows:

$$\Delta R_a = 0.753 + 0.00029P_1 - 0.00283P_2 + 0.004P_3 + 0.00267P_4 \qquad (8.73)$$

and

$$\Delta H_a = 5.99 - 0.00257P_1 + 0.70P_2 + 3.85P_3 - 0.433P_4, \qquad (8.74)$$

where P_1, P_2, P_3 and P_4 are the values for rotational speed, coil current, volume of powder and working gap, respectively. Naif [28] concluded that rotational speed is the most significant parameter affecting the surface roughness and volume of powder is the most significant parameter affecting the surface hardness. To study the effectiveness of the MAF process during the finishing of hardened material, Mulik and Pandey [29] performed MAF experiments on hardened AISI 52100 steel and established the regression models. The regression model for ΔR_a, after neglecting the insignificant terms, can be given as

$$R_a = -674 + 5.48B_1 + 0.0628B_2 + 115B_3 + 15.8B_4 - 0.000018B_2^2 - 0.976B_1B_3$$

$$- 0.00203B_2B_4 - 2.41B_3B_4. \qquad (8.75)$$

where B_1, B_2, B_3 and B_4 are the values for voltage, mesh number, natural logarithm of rpm of electromagnet and percentage weight of SiC abrasives in MAPs, respectively. Mulik and Pandey [29] concluded that the finishing of the workpiece is a combination of shearing of the roughness peaks in the form of micro/nano-chips and brittle fracture of the roughness peaks in case of brittle materials. Later on, Mulik et al. [30] enhanced the efficiency of the MAF process by integrating MAF with ultrasonic vibrations. The resulting hybrid process, i.e. ultrasonic-assisted MAF process (UAMAF), uses ultrasonic vibrations combined with the MAF process to finish the workpiece to nanometre level within a relatively small time span as compared to the time span taken by the MAF process to finish the workpiece. They [30] used the Taguchi design of experiments to study the effect of the input parameters (percentage weight of abrasive, mesh number, rotation speed of the electromagnet, pulse-on time of ultrasonic vibrations and voltage) on the output responses (ΔR_a and MRR). The regression models for ΔR_a and MRR after neglecting insignificant terms can be given as

$$\Delta R_a = -3731 + 6.76c_1 - 0.097c_2 + 1232c_3 + 8.98c_4 + 4.10c_5 - 104c_2^2$$
$$- 0.0829c_3^2 + 0.00117c_1c_2 - 1.02c_1c_3 - 0.0642c_1c_4 \tag{8.76}$$

and

$$\Delta MRR = 13.0 + 0.0167c_1 - 0.362c_3 + 0.0292c_4 + 0.125c_5, \tag{8.77}$$

where c_1, c_2, c_3, c_4 and c_5 are the values for voltage, mesh number, rotational speed of magnet, percentage weight of abrasive particles and pulse-on time of ultrasonic vibrations, respectively. It is concluded that UAMAF is a more efficient process as compared to the MAF process. This is mainly due to the enhanced chances of interaction of the fresh abrasive particles with the workpiece roughness peaks.

8.5 Soft Computing

Hung et al. [31] processed the stainless steel SUS304 cylindrical tubes with the help of the MAF process. They experimentally studied the effect of input parameters (spindle speed, vibration frequency, discharge current and abrasive weight ratio) on the output response (R_a). ANOVA was used to find the relationship between the input process parameters and the output responses. After finding the significant input parameters, a surface prediction system was developed.

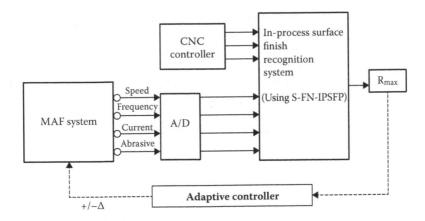

FIGURE 8.15
Experimental setup of statistical-assisted fuzzy nets in-process surface finish prediction system. (From Hung, C.L., Ku, W.L., Yang, L.D., *Mater. Manuf. Processes*, 25, 1404–1412, 2010.)

As shown in Figure 8.15, they developed a fuzzy-net-based MAF prediction system. The inputs for the system are vibration frequency (F), discharge current (C), spindle speed (S) and abrasive weight ratio (A). The predicted variable of the system was the surface finish (R_{max}). The network is developed in five steps:

1. Divide the input space into fuzzy regions.
2. Generate fuzzy rules for the given data pairs.
3. Avoid conflicting rules.
4. Statistical assisted fuzzy-nets rule base.
5. Determine a mapping based on the fuzzy rule base.

The predicted value of the output, i.e. R_{max}, can be given as

$$R_{max} = \frac{\sum_{i=1}^{m} \mu_{R_{max}^t}\left(R_{max^i}\right) \times c\left(R_{max^i}\right)}{\sum_{j=1}^{m} \mu_{R_{max}^t}\left(R_{max^i}\right)},$$ (8.78)

where $\mu_{R_{max}^t}$ is the degree of the output control responding to the input, $c\left(R_{max^i}\right)$ denotes the centre value of region R_{max^i}, m is the number of fuzzy rules in the combined fuzzy rule base and t denotes fuzzy rule number. Hung et al. [31] tested the developed statistical-assisted fuzzy-nets in-process surface finish prediction (S-FN-IPSFP) system with the experimental data and concluded that the system predicts the output response with 97% accuracy. Later, Oh et al. [32] also developed a system based on artificial neural networks (ANNs) to predict output during the MAF process. They investigated the MAF process by using the Taguchi design of experiments. They modified the experimental setup to measure acoustic emission (AE) signals and force signals (Figure 8.16). These samples were analysed using a relevant signal processing technique for giving input to the artificial neural networks to predict the generated surface roughness. The transient elastic stress waves generated by the rapid release of strain energy are captured by the AE signals. The AE energy (*AE r.m.s.*) released to the plastic work absorbed by the material can be expressed as

$$AE\,r.m.s. \propto \{\sigma \times \dot{\varepsilon} \times V_w\}^{1/2}, \tag{8.79}$$

where σ is effective stress, $\dot{\varepsilon}$ is effective strain and V_w is volume of material under deformation. The AE signals can be processed to find material removal rate, which in turn governs surface roughness, *Ra*. Similarly, force signal processing not only gives information about various forces involved during MAF operation but also helps in predicting surface roughness, *Ra*.

ANNs were used to predict the output, viz., sensor information-based ANN (AE and force signals) and data fusion-based ANN (AE + force signals). It was concluded that all the developed systems predict the results within ±10% error. But the fusion-based ANN performed better than each sensor-based ANN.

To study the performance of the ANN in predicting MAF process performance, Teimouri and Baseri [33] developed two networks, viz., a forward back-propagation neural network (FFBP-NN) and an adaptive neuro-fuzzy inference system (ANFIS) for modelling the MAF process, and conducted a

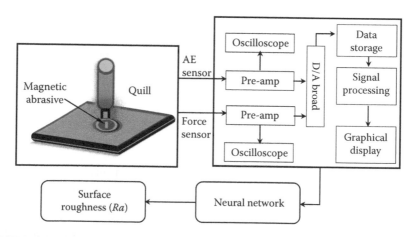

FIGURE 8.16
MAF monitoring setup. (From Oh, J.H., Lee, S.H., *Proc. Inst. Mech. Eng. Part B: J. Eng. Manuf.*, 225, 853–865, 2011.)

comparative study between the two. The input parameters of the MAF process were the electromagnet's voltage, mesh number of the abrasive particles, pole's rotational speed and weight percent of the abrasive particles, while percentage change in surface roughness (% ΔRa) was the output response. The two developed networks were compared on the basis of their mean absolute error and root mean-square error. Djavanroodi [34] studied the MAF input parameters (intensity of the magnetic field, work-piece velocity and finishing time) that affect the surface roughness of a brass shaft of CuZn37. Later, Djavanroodi [34] used ANN for simulating the MAF process to predict final surface roughness. The tangent sigmoid (*Tansig*) transfer function was used as the activation function for the hidden layer, which can be given as

$$Tansig(z) = 2(1 + \exp(-2z)) - 1, \tag{8.80}$$

where z is the weighted sum of the input. Djavanroodi [34] used mean square error, root mean-square error, absolute fraction of variance (R^2) and coefficient of variation in percent (cov) values to determine the performance of the trained network, which can be given as

$$RMS = \left(\frac{1}{T_s} \sum_j |t_j - a_j|^2 \right)^{1/2}, \tag{8.81}$$

$$R^2 = 1 - \left(\frac{\sum_j (t_j - a_j)^2}{\sum_j (a_j)^2} \right), \tag{8.82}$$

$$\mathrm{cov} = \frac{RMS}{O_{mean}} \times 100, \tag{8.83}$$

and

$$MSE = \frac{1}{T_s} \sum_j |t_j - a_j|^2, \tag{8.84}$$

where t_j is the target value, a_j is the output value and T_s is the number of test data.

Moosa [35] conducted an experimental study of the MAF process and developed the adaptive neuro fuzzy inference system to simulate the same. ANFIS is a hybrid network, formed by the combination of neural network and the fuzzy network that generates the mapping scheme between the input parameters and the output responses. Moosa [35] concluded that the error between the predicted and actual experimental results is below 5%.

8.6 Comments

The MAF process is an advanced finishing process used to achieve a mirror finish on flat and cylindrical components in the nanometer range. In order to understand the physics of the MAF process, several researchers modelled the MAF process. They carried out different approaches, viz., theoretical, numerical, empirical and soft computing, to model the MAF process. However, during their analysis, several assumptions were made that created a gap between the actual values and the modelled values. To enhance the efficiency of the MAF process, a few hybrid processes were developed, such as the EMAM and UAMAF processes. For the automation of the MAF process, researchers integrated the MAF process with artificial neural networks. Still, an adaptive control system needs to be developed that can map the relation between the input MAF parameters and output responses for optimising the same.

Nomenclature

a_i	FFBP-NN predicted value
$[A]$	Shape function derivatives matrix for the approximation of scalar magnetic potential
B	Magnetic flux density in the working zone
B_s	Saturated magnetic flux density
$[B]^e$	Derivative of shape function matrix
C	Number of dynamic cutting points per unit surface area
d	Diameter of the abrasive particle
d_a	Mean diameter of an abrasive particle
d_w	Diameter of the cylindrical workpiece
D_{map}	Diameter of the MAP
D_p	Magnetic pole diameter
f	Magnetic force acting on one MAP
f_{Nave}	Mean normal force acting on abrasive particle
f_n, f_t	Normal force and tangential forces acting during MAF process
F	Faraday's constant
F_{mn}, F_c	Normal magnetic force and tangential cutting force

F_N, F_R	Normal and radial components of magnetic force acting on an abrasive particle
F_x, F_y	Normal and tangential forces acting far from the working zone
h	Depth of indentation
h_d	Indentation diameter of the abrasive particle into the workpiece surface
h_i	Height of the ith profile
h_{max}	Height of the highest profile in the cell under consideration
H	Magnetic field strength
H_a	Strength of magnetic field in the air gap
H_c	Coercive force
H_{max}	Maximum undeformed chip thickness
H_w	Hardness of the workpiece
I	Input current
I_{ele}	Electrolytic current
$[K]$	Global coefficient matrix
$[K]^e$	Elemental coefficient matrix
k_w	Thermal conductivity of workpiece
l_l, l_w	Length and width of the finished area on workpiece surface
l_W	Length of machining portion on the workpiece
L_a, L_m	Length of the air gap and magnet
M	Stock removal of workpiece by one abrasive particle
M_j	Atomic weight of the jth species
n	Total number of abrasive particles in one MAP
n_a	Number of active cutting edges of MAP
n_e	Number of measurements made on the workpiece
n_j	Oxidation valence of the jth species
N	Number of MAPs acting on the finishing region simultaneously
N_c	Number of coil turns
N_C, N_L	Number of ferromagnetic particles in contact with workpiece surface along the circumferential and length directions
N_s	Rotational speed of magnetic electrode
$\{N\}$	Column vector for shape function
p_i	Probability of reducing the height of the ith profile
P	Magnetic pressure between workpiece and abrasives
q_t	Total heat flux generated due to abrasive particles
r_{tr}	Average width of cut produced by a cutting edge in track
r, z	Cylindrical coordinates
R	Ratio of width to the thickness of the chip along the cutting path
R_a	Final surface roughness of cylindrical workpiece
R_{tr}	trth track radius
R_w	Energy partition
R_a^0	Initial surface roughness of the workpiece
$R_{(i,j)}^n$	Surface roughness after nth rotation in a cell (i, j)
s	Rotational speed of workpiece

t	Machining time of MAF process
t_i	Target values
T	Temperature of the workpiece
T_M	Machining time of EC dissolution process
T_s	Number of test data
v	Volume of the ferromagnetic abrasive particle
v_{ap}	Speed of abrasive particles
v_f	Speed of in feed
v_{ma}	Speed of MAP
v_S	Workpiece surface speed
v_w	Volume of material under deformation
V_w	Total amount of material removal during C-EMAM process
W	Volumetric ratio of iron in a magnetic particle
x, y	Directions of line of magnetic force and magnetic equipotential line
X_j	Weight fraction of jth species
Y_i	ANFIS output value in training
Z	Weighted sum of the input
Δd	Change in diameter of the workpiece
ΔMR	Change in material removal
ΔH_a	Change in hardness
ΔR_a	Change in surface roughness
σ	Effective stress
$\dot{\varepsilon}$	Effective strain
θ	Included angle
θ_A	Included angle of wedge shaped indentation
φ	Magnetic scalar potential
ρ	Density of workpiece material
χ_m	Susceptibility of MAP
μ_0	Magnetic permeability in vacuum
μ_f	Coefficient of workpiece
μ_r	Relative magnetic permeability of pure iron
η_c	Current efficiency
η_{ecap}	Current efficiency considering EC action and abrasion-assisted dissolution
$\{\phi\}$	Global magnetic potential vector

References

1. Lin, C.T., Yang, L.D., and Chow, H.M., 2007, 'Study of magnetic abrasive finishing in free-form surface operations using the Taguchi method', *International Journal of Advanced Manufacturing Technology*, 34(1–2), pp. 122–130.

2. Shinmura, T., Takazawa, K., Hatano, E., Matsunaga, M., and Matsuo, T., 1990, 'Study on magnetic abrasive finishing', *CIRP Annals–Manufacturing Technology*, 39(1), pp. 325–328.

3. Shinmura, T., Takazawa, K., and Hatano, E., 1987, 'Study on magnetic abrasive finishing – Effects of various types of magnetic abrasives on finishing characteristics', *Bulletin of the Japan Society of Precision Engineering*, 21(2), pp. 139–141.

4. Khairy, A.B., 2001, 'Aspects of surface and edge finish by magneto abrasive particles', *Journal of Materials Processing Technology*, 116(1), pp. 77–83.

5. Mori, T., Hirota, K., and Kawashima, Y., 2003, 'Clarification of magnetic abrasive finishing mechanism', *Journal of Materials Processing Technology*, 143, pp. 682–686.

6. Raghuram, M.G.V.S. and Joshi, S.S., 2008, 'Modeling of polishing mechanism in magnetic abrasive polishing', 12th International Conference of International Association for Computer Methods and Advances in Geomechanics (IACMAG), Goa, India, pp. 344–352.

7. Kim, J.D. and Choi, M.S., 1995, 'Simulation for the prediction of surface-accuracy in magnetic abrasive machining', *Journal of Materials Processing Technology*, 53, pp. 630–642.

8. Jayswal, S.C., Jain, V.K., and Dixit, P.M., 2005, 'Modeling and simulation of magnetic abrasive finishing process', *International Journal of Advanced Manufacturing Technology*, 26, pp. 477–490.

9. Jayswal, S.C., Jain, V.K., and Dixit, P.M., 2004, 'Analysis of magnetic abrasive finishing with slotted magnetic pole', The 8th International conference on Numerical Methods in Industrial Forming Process, Columbus, Ohio, USA, American Institute of Physics conference Proceedings, 712, pp. 1435–1440.

10. Jayswal, S.C., Jain, V.K., and Dixit, P.M., 2005, 'Magnetic abrasive finishing process – A parametric analysis', *Journal of Advanced Manufacturing System*, 4(2), pp. 131–150.

11. Jefimenko, O.D., 1966, *Electricity and Magnetism*, Meredith, New York.

12. Jain, V.K., Jayswal, S.C., and Dixit, P.M., 2007, 'Modeling and simulation of surface roughness in magnetic abrasive finishing using non-uniform surface profiles', *International Journal of Materials and Manufacturing Processes*, 22, pp. 256–270.

13. Judal, K.B. and Yadava, V., 2013, 'Modeling and simulation of cylindrical electrochemical magnetic abrasive machining of AISI-420 magnetic steel', *International Journal of Material Processing Technology*, 213, pp. 2089–2100.

14. Judal, K.B. and Yadava, V., 2014, 'Modeling and simulation of cylindrical electrochemical magnetic abrasive machining process', *International Journal of Machining Science and Technology*, 18(2), pp. 221–250.

15. Yamaguchi, H., Shinmura, T., and Ikeda, R., 2007, 'Study of internal finishing of austenitic stainless steel capillary tubes by magnetic abrasive finishing', *Journal of Manufacturing Science and Engineering*, 129(5), pp. 885–892.

16. Kwak, J.S., 2009, 'Enhanced magnetic abrasive polishing of non-ferrous metals utilizing a permanent magnet', *International Journal of Machine Tools and Manufacture*, 49(7), pp. 613–618.

17. Kumar, G. and Yadav, V., 2009, 'Temperature distribution in the workpiece due to plane magnetic abrasive finishing using FEM', *International Journal of Advanced Manufacturing Technology*, 41(11–12), pp. 1051–1058.

18. Jain, V.K., Kumar, P., Behera, P.K., and Jayswal, S.C., 2001, 'Effect of working gap and circumferential speed on the performance of magnetic abrasive finishing process', *Wear*, 250(1), pp. 384–390.

19. Singh, D.K., Jain, V.K., and Raghuram, V., 2004, 'Parametric study of magnetic abrasive finishing process', *Journal of Materials Processing Technology*, 149(1), pp. 22–29.

20. Singh, D.K., Jain, V.K., Raghuram, V., and Komanduri, R., 2005, 'Analysis of surface texture generated by a flexible magnetic abrasive brush', *Wear*, 259(7), pp. 1254–1261.

21. Singh, D.K., Jain, V.K., Raghuram, V., and Komanduri, R., 2005, 'Analysis of surface roughness and surface texture generated by Pulsating flexible magnetic abrasive brush (P-FMAB)', Proceedings of WTC2005: World Tribology Congress III, Washington, DC, USA, September 12–16, 2005, pp. 1–2.

22. Singh, D.K., Jain, V.K., and Raghuram, V., 2005, 'On the performance analysis of flexible magnetic abrasive brush', *Machining Science and Technology*, 9(4), pp. 601–619.

23. Singh, D.K., Jain, V.K., and Raghuram, V., 2006, 'Experimental investigations into forces acting during a magnetic abrasive finishing process', *The International Journal of Advanced Manufacturing Technology*, 30(7–8), pp. 652–662.

24. Jain, V.K., Singh, D.K., and Raghuram, V., 2008, 'Analysis of performance of pulsating flexible magnetic abrasive brush (P-FMAB)', *Machining Science and Technology*, 12(1), pp. 53–76.

25. Girma, B., Joshi, S.S., Raghuram, M.V.G.S., and Balasubramaniam, R., 2006, 'An experimental analysis of magnetic abrasives finishing of plane surfaces', *Machining Science and Technology*, 10(3), pp. 323–340.

26. El-Taweel, T.A., 2008, 'Modeling and analysis of hybrid electrochemical turning-magnetic abrasive finishing of 6061 Al/Al2O3 composite', *International Journal of Advanced Manufacturing Technology*, 37(7–8), pp. 705–714.

27. Wang, A.C., Tsai, L., Liu, C.H., Liang, K.Z., and Lee, S.J., 2011, 'Elucidating the optimal parameters in magnetic finishing with gel abrasive', *Materials and Manufacturing Processes*, 26(5), pp. 786–791.

28. Naif, N.K.M., 2012, 'Study on the parameter optimization in magnetic abrasive polishing for brass CUZN33 plate using Taguchi method', *The Iraqi Journal for Mechanical and Material Engineering*, 12(3), pp. 596–615.

29. Mulik, R.S. and Pandey, P.M., 2011, 'Magnetic abrasive finishing of hardened AISI 52100 steel', *The International Journal of Advanced Manufacturing Technology*, 55(5–8), pp. 501–515.

30. Mulik, R.S. and Pandey, P.M., 2011, 'Experimental investigations and optimization of ultrasonic assisted magnetic abrasive finishing process', *Proceedings of the Institution of Mechanical Engineers, Part B: Journal of Engineering Manufacture*, 225(8), pp. 1347–1362.

31. Hung, C.L., Ku, W.L., and Yang, L.D., 2010, 'Prediction system of magnetic abrasive finishing (MAF) on the internal surface of a cylindrical tube', *Materials and Manufacturing Processes*, 25(12), pp. 1404–1412.

32. Oh, J.H. and Lee, S.H., 2011, 'Prediction of surface roughness in magnetic abrasive finishing using acoustic emission and force sensor data fusion', *Proceedings of the Institution of Mechanical Engineers, Part B: Journal of Engineering Manufacture*, 225(6), pp. 853–865.

33. Teimouri, R. and Baseri, H., 2013, 'Artificial evolutionary approaches to produce smoother surface in magnetic abrasive finishing of hardened AISI 52100 steel', *Journal of Mechanical Science and Technology*, 27(2), pp. 533–539.
34. Djavanroodi, F., 2013, 'Artificial neural network modeling of surface roughness in magnetic abrasive finishing process', *Research Journal of Applied Sciences, Engineering and Technology*, 6(11), pp. 1976–1983.
35. Moosa, A.A., 2013, 'Utilizing a Magnetic Abrasive Finishing Technique (MAF) Via Adaptive Nero Fuzzy (ANFIS)', *American Journal of Materials Engineering and Technology*, 1(3), pp. 49–53.

9

Magnetorheological Finishing

Ajay Sidpara[1] and Vijay K. Jain[2]

[1]*Mechanical Engineering Department, Indian Institute of Technology Kharagpur, Kharagpur, India*

[2]*Mechanical Engineering Department, Indian Institute of Technology Kanpur, Kanpur, India*

CONTENTS

9.1 Introduction

Surface finishing is an integral step in the fabrication of all components that require high precision and close tolerances for performing their tasks as an individual component or as a part of an assembly. These components have wide varieties in terms of size, shape, material properties, etc. Furthermore, new materials having superior properties are being developed continuously to meet the requirements of different fields, such as medical, aerospace, optics, automobile, electronics, etc. As a result, continuous up-gradation or development of new machining and finishing processes is necessary to process or fabricate such materials. Many processes are widely used for finishing, such as grinding, lapping, honing, etc., and they are well established. However, they are not very effective when it comes to finishing very soft materials that are used in optics, to achieve nanometre-level (or, in some cases, Angstrom) surface roughness value, to finish complex freeform surfaces, etc. Merely upgrading these processes will not work due to the previously mentioned requirements, which are not easy to fulfil. Hence, development of new finishing processes is the only solution to the present requirements. Many new and advanced finishing processes have been developed for different types of workpiece materials. These processes are chemical mechanical polishing (CMP), abrasive flow machining, magnetic abrasive finishing, elastic emission machining (EEM), electrolytic in-process dressing, bonnet polishing, magnetorheological (MR)-fluid-based finishing, etc. Each process has its own advantages and disadvantages. Some of these processes are discussed in detail in other chapters of this book.

One of the important characteristics of a nanofinishing process is to finish a surface gently without changing the figure or shape of the component. Hence, nanofinishing processes do not remove material at large scale as in the case of grinding. Therefore, nanofinishing processes cannot be used as material removal processes where the initial surface roughness is very high (say, more than 1 μm).

It is difficult to achieve nanofinishing by fixed or bonded abrasive particles due to the wear-out of abrasive particles, deep indentation and scratches, high friction, heat generation, etc. On the other side, loose abrasive-particles-based finishing generally does not have the previously stated problems. Loose abrasive particles are randomly oriented and distributed. Furthermore, they roll/tumble during finishing, which results in effective use of abrasive particles due to the interaction of different sharp edges of abrasive particles with the workpiece. This is not the case in bonded abrasive particles, where only one orientation or one edge of an abrasive particle interacts with the workpiece and the rest of the sharp edges located inside the grinding wheel (not projected on the wheel surface) have no role in finishing. In general, loose abrasive particles (three-body abrasion) provide good surface finish, while bonded abrasive particles (two-body abrasion) are good for comparatively high material removal.

In the case of loose abrasives finishing, control of forces is one of the most important parameters for achieving a nanometre-level surface finish. The magnitude of the forces dictates surface damage due to deep indentation and scratches. Therefore, a process having better control over forces acting on the surface will provide more satisfactory results. Many processes are included in the category of 'loose abrasives finishing', such as chemo-mechanical polishing, abrasive flow finishing, lapping, EEM, magnetic abrasive finishing, MR-fluid-based finishing, etc. Among these processes, the MR finishing (MRF) process has a few advantages, which makes it more suitable for nanofinishing requirements in general and nanofinishing of freeform surfaces in particular. Some advantages are real-time control over forces by changing current (in case of an electromagnet) or working gap (in case of a permanent magnet), lower rise in temperature due to the presence of water or other liquid in the MR fluid itself, etc.

9.2 MR Finishing

MR-fluid-based finishing (MRFF) relies on MR fluid composition, which consists of magnetic particles and abrasive particles mixed in a liquid carrier. Some chemicals and surfactants (or additives) are also added to increase the effective life of MR fluid and to increase the material removal rate (MRR). Figure 9.1 shows the different constituents of MR fluid. The primary process is called MRF, which was conceived at Minsk, Belarus, and further developed at the Center for Optics Manufacturing at University of Rochester. As an outcome of this effort, a commercial MRF machine was developed by QED Technologies in 1998 (Jacobs et al., 2000).

A schematic diagram of the MRF process is shown in Figure 9.2. An MR fluid circulatory system provides a continuous supply of MR fluid on the carrier wheel. The circulatory system also ensures constant or uniform liquid content by supplying additional water or other carrier fluid to maintain the viscosity of the MR fluid during finishing. Electromagnets are located below the carrier wheel, which helps in making the MR fluid rigid when it is supplied over the carrier wheel. Workpieces of different shape are placed over the carrier wheel and kept in contact with the MR fluid ribbon (Figure 9.2). The relative motion between the MR fluid ribbon and the workpiece results in material removal. Swinging and rotating the workpiece help in finishing the whole area of the workpiece.

Many different configurations of MRFs have evolved in the last few years as allied MRF processes with different names such as the MRFF process

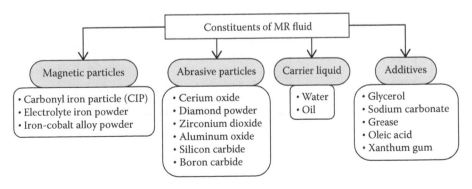

FIGURE 9.1
Constituents of MR fluid.

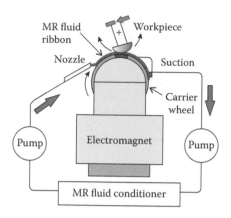

FIGURE 9.2
Schematic diagram of the MRF process. (From Jacobs, S.D., Arrasmith, S.A., Kozhinova, I.A., Gregg, L.L., Shorey, A.B., Ramanofsky, H.J. Romanofsky, Golini, D., Kordonski, W.I., Dumas, P., Hogan, S., *Ceram. Trans.*, 102, 185–200, 2000.)

(Sidpara and Jain, 2012a), ball end MRF (Singh et al., 2011), rotational-MR abrasive flow finishing (R-MRAFF) (Das et al., 2010), etc. They all use MR fluid as a finishing medium and work on the same principle as MRF. Some of these allied MRF processes are shown schematically in Figure 9.3 and discussed in detail in other chapters.

Nanofinishing processes such as CMP, EEM, MRF, etc., are the final finishing processes, and after these, only cleaning of the sample or component should be done before using them in specific applications. Very less amount of material should be removed without damaging the surface, which results in a very high processing time. Therefore, a chemical

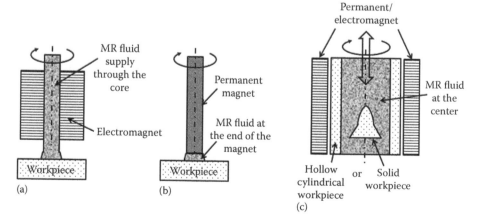

FIGURE 9.3
Allied MRF processes: (a) ball end MRF, (b) MRFF and (c) R-MRAFF.

synergy between the abrasive particles, liquid medium and workpiece material should be established in such a way that material removal takes place due to chemical reaction and mechanical abrasion. Chemical-mechanical action ensures a defect-free surface, relatively high MRR and low processing time.

9.3 Process Parameters

Since MRF is a fluid-based finishing process, most of the process parameters are related to MR fluid constituents and their concentration. Figure 9.4 shows some of the important parameters of the MRF process. A stringent control over these process parameters is required to get the maximum performance out of the MRF process, which is a very challenging task. The selection of these process parameters and their values depend on the properties of the workpiece to be finished, such as its composition, hardness, initial surface roughness, etc.

The MRF process significantly depends on the magnetic field and carbonyl iron particle (CIP) (size and concentration) as compared to other process parameters because MRF does not work if either of these two is not present. Hence, selecting the optimum level of these parameters is necessary to achieve the desired surface roughness value and MRR.

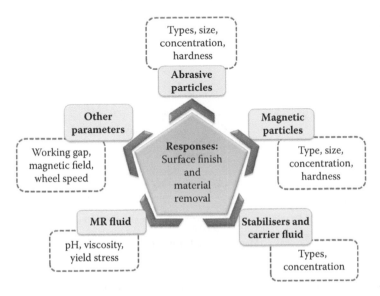

FIGURE 9.4
Process parameters of MRF. (From Sidpara, A., *Opt. Eng.*, 53, 092002–092002, 2014.)

9.3.1 Magnetic Flux Density

The magnetic field is the keystone parameter of the MR effect. The magnetic flux density controls the MR fluid's stiffness. Magnetic flux density is applied by an electromagnet or permanent magnet in the MRF process. An electromagnet is bulky, but it can regulate magnetic flux density by the varying electric current in real time during the finishing operation, while a permanent magnet is small in size and can be easily accommodated in a small space. However, magnetic flux density is fixed in a permanent magnet so variation in the magnetic field is not possible. One way to vary the magnetic field distribution on the workpiece surface is to adjust the working gap between the permanent magnet and the workpiece. However, it may change other conditions such as the immersion depth of the workpiece surface in an MR fluid ribbon, rheological parameters such as yield stress and viscosity, etc.

Figure 9.5 shows the effect of magnetic flux density on the final Ra (surface roughness) achieved on the workpiece, MRR, yield stress and viscosity (Sidpara and Jain, 2014). The model (Equation 9.1) derived by Ginder et al. (1995) shows that the yield stress is proportional to the square of the magnetic field strength (H^2):

$$\tau_y \propto C\mu_0 H^2, \tag{9.1}$$

where τ_y is yield stress, μ_0 is permeability of free space, H is the applied magnetic field strength and C is volume concentration of magnetic particles.

By increasing the magnetic field strength (H) or magnetic flux density (B), the dipole moment between CIPs increases and they are firmly aligned

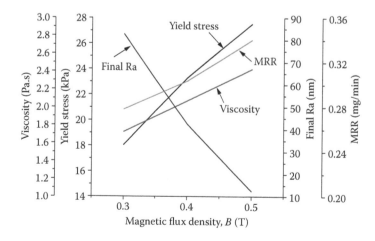

FIGURE 9.5
Effect of magnetic field (C = 40%, A = 5%, G = 8%). (From Sidpara, A., Jain, V.K. *Mach. Sci. Technol.*, 18, 367–385, 2014.)

along the magnetic lines of force. As a result, the yield stress of the MR fluid increases. The higher the magnetic flux density, the higher the stiffness of the MR fluid. Highly stiffened MR fluid exhibits high yield stress and results in high finishing forces (normal and tangential) when it interacts with the workpiece surface. Hence, MRR increases, but at the same time, the probability of scratches and deep indentation also increases.

When the magnetic flux density is high, the MR fluid exhibits high resistance before yielding. As the magnetic flux density (B) increases, the magnetic interaction force (F in Equation 9.2) between the two CIPs increases significantly (Huang et al., 2005). Hence, the strength of chains of CIPs increases, which in turn increases the stiffness of the MR fluid, i.e. yield stress. Furthermore, movement of the CIPs during the flow of MR fluid decreases due to dense and strong MR fluid when the magnetic flux density is high. It increases the viscosity of MR fluid.

$$F = \frac{\mu_0 \pi}{9} \left(\frac{r'^2 KM}{P} \right)^2,$$

(9.2)

where M is the intensity of magnetisation of magnetic fluid, r' is the radius of a CIP, P is the distance between two CIPs and K is a coefficient.

Due to high yield stress and viscosity, the MR fluid aggressively interacts without yielding with the workpiece, which results in a low final Ra value and a high MRR. This behaviour is true until it achieves the critical surface roughness value for the given finishing conditions.

9.3.2 Carbonyl Iron Particle Concentration

CIPs create chain-like structures under the influence of a magnetic field by aligning themselves along the line of magnetic flux (from the N-pole to the S-pole of a magnet). Figure 9.6 shows an increase in yield stress, viscosity and MRR and a decrease in final Ra with an increase in CIP concentration. The higher the concentration of CIPs, the thicker and denser the chain structure that will be formed. As a result, more force (yield stress) will be needed to break it.

Viscosity also increases with increasing CIP concentration (Figure 9.6) due to the high concentration of solid particles as well as the high MR effect due to the dense agglomeration of CIPs. The MRR and final Ra depend on the strength at which the abrasive particles are held/supported by the CIP chains. When the yield stress is high, abrasive particles are firmly gripped in the CIP chains' structure, and they efficiently interact with the workpiece without rolling or sliding over the workpiece surface. As a result, MRR increases and the final Ra decreases with an increase in CIP concentration.

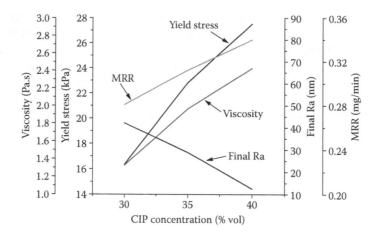

FIGURE 9.6
Effect of CIP concentration (M = 0.5 T, A = 5%, G = 8%). (From Sidpara, A., Jain, V.K. *Mach. Sci. Technol.*, 18, 367–385, 2014.)

9.3.3 Abrasive Particle Concentration

Abrasives particles are added in MR fluid to increase MRR (Shorey et al., 2001). Figure 9.7 shows a decrease in yield stress and viscosity with an increase in abrasive particle concentration, while the final Ra and MRR have optimum levels. Abrasive particles stay away from magnets and float over the chains of CIPs due to their non-magnetic nature and magnetic levitation force as discussed later by Equation 9.7. However, a few abrasive particles

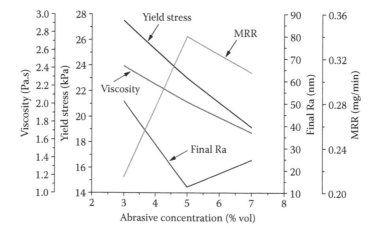

FIGURE 9.7
Effect of abrasive concentration (M = 0.5 T, C = 40%, G = 8%). (From Sidpara, A., Jain, V.K. *Mach. Sci. Technol.*, 18, 367–385, 2014.)

entangle in between CIP chains and hinder the formation of continuous and dense CIP chains. It is believed that continuous CIP chains have more resistance to shear as compared to short and discontinuous CIP chains. So, at a high shear rate, discontinuous chains are easily destroyed. When an abrasive particle comes in between two CIPs, it increases the distance between CIPs (P in Equation 9.2). As a result, the magnetic interaction force (F) between CIPs decreases, which results in low yield stress and viscosity (Huang et al., 2005). Hence, yield stress and viscosity decrease with increasing in abrasive particle concentration (Figure 9.7).

Final Ra and MRR have optimum levels with increasing abrasive particle concentration. Initially, abrasive particles increase the MRR, and they also result in a low final Ra. However, a further increase in abrasive particle concentration reduces the yield stress of the MR fluid. Hence, after a certain concentration of abrasive particles, CIP chain structures become very weak and do not provide enough support for the abrasive particles during finishing. As a result, it is difficult to reduce the final Ra and increase the MRR further. Furthermore, it is also presumed that once the contact zone between the MR fluid ribbon and the workpiece surface is saturated with abrasive particles, further addition of abrasive particles does not improve the process performance.

9.3.4 Carrier Wheel Speed

Carrier wheel speed is defined as the speed (velocity) at which the MR fluid interacts with the workpiece surface. Due to the high magnetic flux density, the MR fluid ribbon sticks on the surface of the carrier wheel. It is assumed that the MR fluid ribbon moves with the same speed as that of the carrier wheel (no-slip condition at the MR fluid and carrier wheel interface). Therefore, the velocity of the MR fluid also increases with the increase in wheel speed. Hence, the shear force applied by the abrasive particles on the workpiece surface (roughness peaks) also increases. Furthermore, as the carrier wheel speed increases, the frequency of interaction of the abrasive particles with the workpiece surface also increases, which accelerates the finishing action. Based on these facts, the final Ra decreases and the MRR increases with increasing carrier wheel speed, as shown in Figure 9.8.

9.3.5 Effect of pH

The pH value of slurry is an important parameter of a finishing process. It is little difficult to suggest/recommend the pH value for finishing different materials such as optics and metals as the slurry contains different types of solid particles (magnetic and non-magnetic) and other additives. In the MRF process, the pH of the MR fluid is generally kept on the higher side (>10 pH) to slow down the corrosion of magnetic particles (CIPs). The pH of the MR fluid can be increased by the addition of sodium carbonate (Na_2CO_3)

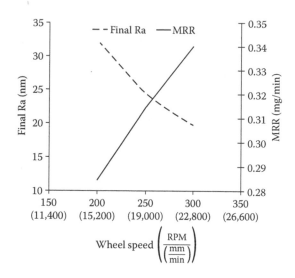

FIGURE 9.8
Effect of wheel speed on final Ra and MRR (CIPs = 40%, A = 5%, Ra_i = 1250 nm). (From Sidpara, A., Jain, V.K. *Tribol. Int.*, 47, 159–166, 2012.)

(Jacobs et al., 1998). It also acts as a buffer to maintain high pH over time, slowing changes due to dissolved oxygen and carbon dioxide from atmospheric exposure. Due to high pH, polishing of few metallic and rough (very high surface roughness value) materials becomes difficult or time-consuming as they may be efficiently finished in an acidic slurry (low pH). Kozhinova et al. (2005) used MR fluid having a pH around 4 for finishing of chemical vapor deposition (CVD) polycrystalline zinc sulfide (ZnS). Several microns of material were removed by acidic MR fluid along with mechanically soft CIPs and alumina abrasive particles. A surface roughness (rms) value below 2 nm was achieved on the surface pre-processed by microgrinding, diamond turning and pitch polishing. However, the corrosion of CIPs severely limited the stability and useful lifetime of the MR fluid.

Jacobs (2011) developed zirconia-coated CIPs to reduce corrosion of the CIPs even if they are used in low-pH (~4.4) MR fluid. Ceramic coating of zirconia over CIPs prevents direct contact of water with CIPs and reduces corrosion. It is observed that the coating is continuous and also resists the abrasion during finishing. As a result, no change in coating morphology is observed after 22 days of use of MR fluid (Shafrir et al., 2009). Furthermore, zirconia also works as abrasive particles in MR fluid.

9.3.6 Workpiece Parameters

A very high yield stress of MR fluid is also not desirable for finishing soft glass or silicon-based material because MR fluid rigorously interacts with the workpiece and generates deep micro/nano-scratches. Therefore, moderate yield stress is preferable for achieving a low final Ra as well as high MRR.

The selection of a process parameter mostly depends on the type of workpiece material (metal, optics such as glass or lens, ceramic, etc.) and initial surface roughness. If the material is optics, which is soft and in which surface roughness is low, MR fluid composition should be such that it should not aggressively abrade the workpiece surface. A low concentration of abrasive particles and low magnetic flux density ensure that the MR fluid is not very aggressive and its stiffness is low. However, the setting can be opposite of this if the workpiece material is metal or the initial surface roughness of the component is very high. It needs very stiff MR fluid and little more abrasive particles to reduce surface roughness in a short time.

9.4 Theoretical Analysis

The theoretical study of MRF process provides in-depth information on the interaction of MR fluid–abrasive particles with the workpiece. It provides models for theoretical calculations for estimation of different output parameters (responses) such as shear stress, magnetic force between magnetic particles, magnetic levitation force, centrifugal force, MRR, etc. Some of these mathematical models are discussed in the following sections.

9.4.1 Shear Stress

MRF utilises the state of flow of the magnetically stiffened MR abrasive fluid throughout a preset converging gap formed by the workpiece surface and a moving rigid wall, as shown in Figure 9.9. This type of flow is analogous to the classical problem commonly encountered in the theory and practice of hydrodynamic lubrication. Under conditions of routine lubrication theory, the basic equation of the flow is

$$\frac{\partial P}{\partial x} = \frac{\partial \tau_{xy}}{\partial y},$$
(9.3)

where $\dfrac{\partial P}{\partial x}$ is pressure gradient and τ_{xy} is shear stress.

By simplifying Equation 9.3, shear stress (τ_{xy}) can be written as (Kordonski and Golini, 1999)

$$\tau_{xy} = \frac{dP}{dx}\left(y - \frac{h(x)}{2}\right) - \frac{\eta_0 U}{h} - \tau_0(H),$$
(9.4)

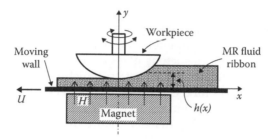

FIGURE 9.9
Schematic diagram of the MR finishing zone.

where H is magnetic field strength, h is working gap as a function of x, U is the linear velocity of the moving wall, η_0 is plastic viscosity, $\tau_0(H)$ is yield stress (function of magnetic field) and x and y are coordinates as shown in Figure 9.9.

9.4.2 Centrifugal Force

The theoretical model of centrifugal force acting on abrasive particles was proposed by Jung et al. (2009). It is concluded that this force, along with magnetic force, is the primary force governing the quasi-static state of an MR fluid. However, other force components, like the capillary viscous force due to the presence of water, squeezing force, gravitational force, etc., are ignored. A schematic diagram of the forces acting in the MRF process is shown in Figure 9.10.

The centrifugal force $F_{cen}(r)$ acting on a particle in a chain and located at a distance r from the centre of the carrier wheel is calculated by

$$F_{cen}(r) = \left(\rho_p \frac{4\pi r_p^3}{3} \right) \left(\frac{r_h - r}{2r_p} \right) \left(\frac{r_h + r}{2} \right) \omega^2, \tag{9.5}$$

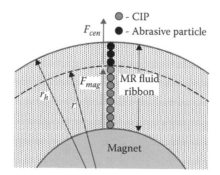

FIGURE 9.10
Schematic diagram of centrifugal force and magnetic force acting in an MRF process.

where ρ_p and r_p are the mass density and mean radius of CIPs, respectively; ω is the rotational speed of the wheel and r_h is the radius of MR fluid ribbon from the centre of the carrier wheel.

9.4.3 Magnetic Force

A simplified equation of the magnetic force $[F_{mag}(r)]$ of attraction between any two neighbouring CIPs in the radial direction is given by (Jung et al. 2009)

$$F_{mag}(r) = \frac{3m\left(\left\|\vec{H}(r)\right\|\right)^2}{4\pi\mu_0\mu_1(2r_p)^4},$$ (9.6)

where μ_0, μ_1, $\vec{H}(r)$ and m denote the permeability of the free space, the relative permeability, magnetic field intensity and the magnetic dipole moment, respectively.

The magnetic interaction force (F) is the attracting force between two magnetic particles as given by Equation 9.2 (Huang et al., 2005). This magnetic interaction force increases with increasing magnetic particle size. It means that the bigger the size of the magnetic particle, the higher the attracting force is between them. That results in more stiffness of the MR fluid and, subsequently, a higher MRR.

The magnetic field gradient causes the CIPs to be attracted towards the magnet (towards the carrier wheel surface), while the abrasive particles move towards the workpiece surface (away from the magnets) due to magnetic levitation force. Magnetic levitation force is defined as the force exerted on non-magnetic bodies by a magnetic fluid. The magnetic levitation force keeps abrasive particles on the outer periphery of the MR fluid ribbon and helps in indenting the abrasive particles into the workpiece surface when they are dragged in the converging gap, which results in material removal. The magnitude of the magnetic levitation force on an abrasive particle is determined by the magnetic field strength and the magnetic properties of the MR fluid. The levitation force (F_m) (Rosenweig, 1985) is given

$$F_m = -V\mu_0 M\nabla H,$$ (9.7)

where V is the volume of non-magnetic body, M is the intensity of magnetisation of magnetic fluid, μ_0 is the permeability of free space and ∇H is the gradient of a magnetic field.

9.4.4 Material Removal Rate

Seok et al. (2009) proposed a semi-empirical MRR model for explaining the tribological behaviour of MR fluid during the finishing process by considering both the solid- and fluid-like characteristics of the MR fluid under the magnetic field. Additionally, Archard's theory and Amonton's law of friction are applied to the model. The MRR in the MRF process is calculated by

$$\frac{dz}{dt} = \kappa \left(PV_p + \frac{\mu_l}{\mu_s} V_p^2 \right), \tag{9.8}$$

where dz/dt is the material removed (dz in μm) in a specific time (dt), P is the pressure applied on the workpiece, κ ($= \alpha\mu_s$) and μ_l/μ_s are treated as constant parameters to be determined from experiments, μ_s is the kinetic friction coefficient, V_p is actual particle velocity at the interface between the workpiece and the contacting MR fluid and α is the proportionality constant between MRR and shear work done at the workpiece surface.

Cheng et al. (2009) derived a mathematical model for material removal by taking into account the pressure (P) and the carrier wheel velocity (V). It is found that the magnetic pressure is about six orders of magnitude smaller than the hydrodynamic pressure. The material removal in MRF process is given by the Preston hypothesis as

$$\text{Material removal} = K \int PV \, dt, \tag{9.9}$$

where K is a constant that depends on the workpiece material and the properties of MR fluid, etc.

However, validation of the derived model is not supported by experiments. Furthermore, the Preston equation is used for mathematical derivation, which does not seem accurate enough. The chemistry of the MRFF process is difficult to explain by the Preston equation due to the consideration of only basic parameters, namely, pressure, velocity and time.

Miao et al. (2009) modified Preston's coefficient to calculate MRR in MRF in terms of hydrodynamic pressure, shear stress and a combination of the material figure of merits (FOMs) and shear stress. These calculated coefficients are in good agreement when both material FOMs and shear stress are considered for a range of optical glasses, which proves the predictive capabilities of the model. However, this model did not take into account the role of the MRF process parameters in material removal. The predictive MRR model in the MRF process is given by

$$MRR_{MRF} = C'_{p,MRF(\tau,FOM)} \frac{E}{K_c H_V^2} \tau V, \tag{9.10}$$

where $C'_{p,MRF(\tau,FOM)}$ is a modification of Preston's coefficient in terms of shear stress (τ) and the material FOM, V is the velocity of the wheel and H_v, K_c and E are Vicker's hardness, fracture toughness and Young's modulus of the workpiece material, respectively.

Miao et al. (2010) further modified the MRR model to incorporate the effect of the MRF process parameters (size as well as concentration of the abrasive particles and CIPs and penetration depth of the MR fluid ribbon in the workpiece surface). Nano-diamond particles are used as abrasive particles. The modified Equation 9.11 measures volumetric removal rate (VRR) instead of MRR. The VRR model improves understanding of the material removal mechanism in the MRF process and offers a direct estimation of MRR for glasses under the given conditions.

$$VRR \propto \left[B_{nd} \cdot \phi_{nd}^{-1/3} C_{nd}^{1/3} + B_{CI} \cdot \phi_{CI}^{-4/3} C_{CI} \right] \frac{E}{K_c H_V^2} \tau DV \qquad (9.11)$$

where B_{nd} and B_{CI} are coefficients (both empirically equal to 1), ϕ_{nd} is nanodiamond particle size, C_{nd} is nanodiamond concentration, ϕ_{CI} is CIP size, C is CIP concentration, and D is penetration depth of the MR fluid ribbon on the workpiece surface.

9.5 Influence Function (Finishing Spot)

The region of material removal in a specified time duration in a stationary finishing operation (without giving movement to x–y–z axes) is called the finishing spot or influence function, as shown in Figure 9.11. The profile of a finishing spot has different shapes depending on the configuration of the machine setup. Figure 9.11b shows a general shape of a finishing spot when the MRF setup has a wheel-type configuration. Other setups as shown in Figure 9.3 may have finishing spots with different profiles/shapes. The maximum depth of material removal is located at the minimum working gap between the carrier wheel and the workpiece surface (Figure 9.11c) because the squeezing of MR fluid is the highest at that location (or minimum working gap between the carrier wheel and workpiece surface). It contains the information of the characteristics of the region of material removal, such as the deepest penetration depth, finishing area, volume, etc., and it is further used to precisely remove the profile-error by adjusting x–y–z motion controllers and rotational speed of the finishing tool. The higher the accuracy of the influence function, the lower the remaining surface profile error and thus the higher the quality of the final polished surface. Hence, the polishing quality (MRR and surface finish) significantly depends on the finishing spot.

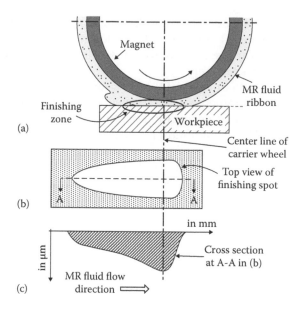

FIGURE 9.11
(a) Front view of wheel base in an MRF setup, (b) exaggerated top view of workpiece surface interacting with MR fluid ribbon and (c) depth profile of finishing spot in cross-sectional view of (b).

The finishing spot of a wheel-based MRF configuration depends on many parameters such as magnetic field strength, MR fluid volume on the wheel, wheel speed, workpiece penetration depth in MR fluid, duration of contact, MR fluid properties (viscosity, age, constituents, etc.), workpiece material (hardness, curvature, initial surface roughness, etc.), etc. (Schinhaerl et al., 2007).

9.6 Applications

MRF has been used for finishing almost all types of glasses and ceramic materials. Tricard et al. (2004) have reported many industrial applications where the MRF process has been successfully used. Table 9.1 shows a few very specific applications of MRF where extreme control over form and surface finish is the prime requirement.

It is also expected that the application of the MRF process will be extended to finishing materials other than optics such as metals, composites, etc. Metallic and ceramic components are used in aerospace, medical, nuclear reactors, automobiles, etc., and they require stringent control over surface roughness. Some of the components are prosthesis implants, aircraft bearings, dies or moulds, etc. MRF and allied processes have been efficiently used for finishing prosthesis implants such as knee joints (Sidpara and Jain, 2012a) and hip joints

TABLE 9.1

A Few Specific Applications of MRF

Authors	Applications
Arrasmith et al. (2001)	Relieving residual stress and subsurface damage on lapped semiconductor silicon wafers
Dumas et al. (2005)	Improving figure and finish of diamond turned surfaces by removing residual turning marks
Geiss et al. (2008)	Silicon nitride moulds for precision glass moulding
Tricard et al. (2008)	Polishing of continuous phase plates for manipulating and controlling laser beam-shapes, energy distributions and wave front profiles
Beier et al. (2013)	Figuring of aluminum lightweight mirrors with electroless nickel plating
Salzman et al. (2013)	Finishing of single crystal ZnS with chemically modified MR fluid

Source: Sidpara, A. *Opt. Eng.*, 53, 092002–092002, 2014.

(Sutton, 2011). MRF is very efficient for finishing of non-magnetic materials, but it is also necessary to explore the applications of MRF for magnetic materials. Only a few reports are available where MRF and allied processes are used for finishing magnetic materials (Shafrir et al., 2007; Singh et al., 2011). Further developments in this direction will provide an alternate and efficient way of nanofinishing the components made of magnetic materials. Finishing metals or hard materials by standard MR fluid is not appropriate because of the low volumetric MRR, which may result in high processing time. Therefore, it is necessary to increase the aggressiveness of the MR fluid by adding suitable chemicals that soften the metallic surface and assist the abrasive particles in mechanical abrasion (Sidpara and Jain, 2012a).

Questions

1. Why is it difficult to achieve nanofinishing by fixed or bonded abrasive particles?
2. MR fluid consists of
   a. Magnetic particles
   b. Abrasive particles
   c. Liquid medium
   d. All of them
3. What do we mean by the final finishing process?
4. What are the important process parameters of the MRF process?
5. What are the *most* important process parameters of the MRF process?
   a. Magnetic particles and magnetic field
   b. Magnetic particles and abrasive particles

  c. Abrasive particles and additives

  d. Additives and magnetic field

6. How does the MRR increase with increasing magnetic flux density?

7. What is critical surface roughness?

8. Explain the trend of the following parameters with reference to concentration of magnetic particles:

  a. Surface finish

  b. MRR

  c. Yield stress

  d. Viscosity

9. Optimum concentration of abrasive particles is required to achieve low surface roughness and high MRR. Elaborate on the statement.

10. How do you select a process parameter for the following workpieces?

  a. Metal with high initial surface roughness

  b. Optical glass with low initial surface roughness

11. State some of the applications of the MRF process.

References

Arrasmith, S. R., Jacobs, S. D., Lambropoulos, J. C., Maltsev, A., Golini, D., & Kordonski, W. I. (2001, December). Use of magnetorheological finishing (MRF) to relieve residual stress and subsurface damage on lapped semiconductor silicon wafers. In International Symposium on Optical Science and Technology (pp. 286–294). International Society for Optics and Photonics.

Beier, M., Scheiding, S., Gebhardt, A., Loose, R., Risse, S., Eberhardt, R., & Tünnermann, A. (2013, September). Fabrication of high precision metallic freeform mirrors with magnetorheological finishing (MRF). In SPIE Optifab (p. 88840S). International Society for Optics and Photonics.

Cheng, H. B., Yam, Y., & Wang, Y. T. (2009). Experimentation on MR fluid using a 2-axis wheel tool. *Journal of Materials Processing Technology*, 209(12), 5254–5261.

Das, M., Jain, V. K., & Ghoshdastidar, P. S. (2010). Nano-finishing of stainless-steel tubes using rotational magnetorheological abrasive flow finishing process. *Machining Science and Technology*, 14(3), 365–389.

Dumas, P., Golini, D., & Tricard, M. (2005, May). Improvement of figure and finish of diamond turned surfaces with magneto-rheological finishing (MRF). In Defense and Security (pp. 296–304). International Society for Optics and Photonics.

Geiss, A., Rascher, R., Slabeycius, J., Schinhaerl, M., Sperber, P., & Patham, F. (2008, August). Material removal study at silicon nitride molds for the precision glass molding using MRF process. In Optical Engineering + Applications (p. 706007). International Society for Optics and Photonics.

Ginder, J. M., Davis, L. C., & Elie, L. D. (1995). Rheology of magneto-rheological fluids: Models and measurements. In 5th International Conference on ER Fluids and MR Suspensions (pp. 504–514). Singapore: World Scientific.

Huang, J., Zhang, J. Q., & Liu, J. N. (2005). Effect of magnetic field on properties of MR fluids. *International Journal of Modern Physics B*, 19(01n03), 597–601.

Jacobs, S. D. (2011, September). MRF with adjustable pH. In SPIE Optical Systems Design (p. 816902). International Society for Optics and Photonics.

Jacobs, S. D., Kordonski, W., Prokhorov, I. V., Golini, D., Gorodkin, G. R., & Strafford, D. T., (1998), Magnetorheological fluid composition. U.S. Patent 5,804,095.

Jung, B., Jang, K. I., Min, B. K., Lee, S. J., & Seok, J. (2009). Magnetorheological finishing process for hard materials using sintered iron-CNT compound abrasives. *International Journal of Machine Tools and Manufacture*, 49(5), 407–418.

Kordonski, W., & Golini, D. (1999). Progress update in magnetorheological finishing. *International Journal of Modern Physics B*, 13(14n16), 2205–2212.

Kozhinova, I. A., Romanofsky, H. J., Maltsev, A., Jacobs, S. D., Kordonski, W. I., & Gorodkin, S. R. (2005). Minimizing artifact formation in metnetorheological finishing of chemical vapor deposition ZnS flats. *Applied Optics* 44, 4671–4677.

Miao, C., Shafrir, S. N., Lambropoulos, J. C., Mici, J., & Jacobs, S. D. (2009). Shear stress in magnetorheological finishing for glasses. *Applied Optics*, 48(13), 2585–2594.

Miao, C., Lambropoulos, J. C., & Jacobs, S. D. (2010). Process parameter effects on material removal in magnetorheological finishing of borosilicate glass. *Applied Optics*, 49(10), 1951–1963.

Rosenweig, R. E. *Ferrohydrodynamics*, Dover, New York, 1985.

Schinhaerl, M., Smith, G., Geiss, A., Smith, L., Rascher, R., Sperber, P., Pitschke E., & Stamp, R. (2007, September). Calculation of MRF influence functions. In Optical Engineering + Applications (p. 66710Y). International Society for Optics and Photonics.

Salzman, S., Romanofsky, H. J., Clara, Y. I., Giannechini, L. J., West, G., Lambropoulos, J. C., & Jacobs, S. D. (2013, September). Magnetorheological finishing with chemically modified fluids for studying material removal of single-crystal ZnS. In SPIE Optifab (p. 888407). International Society for Optics and Photonics.

Seok, J., Lee, S. O., Jang, K. I., Min, B. K., & Lee, S. J. (2009). Tribological properties of a magnetorheological (MR) fluid in a finishing process. *Tribology Transactions*, 52(4), 460–469.

Shafrir, S. N., Lambropoulos, J. C., & Jacobs, S. D. (2007). Toward magnetorheological finishing of magnetic materials. *Journal of Manufacturing Science and Engineering*, 129(5), 961–964.

Shafrir, S. N., Romanofsky, H. J., Skarlinski, M., Wang, M., Miao, C., Salzman, S., Chartier, T., Mici, J., Lambropoulos, J. C., Shen, R., Yang, H., & Jacobs, S. D. (2009). Zirconia-coated carbonyl-iron-particle-based magnetorheological fluid for polishing optical glasses and ceramics. *Applied Optics*, 48, 6797–6810.

Shorey, A. B., Jacobs, S. D., Kordonski, W. I., & Gans, R. F. (2001). Experiments and observations regarding the mechanisms of glass removal in magnetorheological finishing. *Applied Optics*, 40(1), 20–33.

Sidpara, A. (2014). Magnetorheological finishing: A perfect solution to nanofinishing requirements. *Optical Engineering*, 53(9), 092002.

Sidpara, A. M. & Jain, V. K. (2012a). Nanofinishing of freeform surfaces of prosthetic knee joint implant. *Proceedings of the Institution of Mechanical Engineers, Part B: Journal of Engineering Manufacture*, 226(11), 1833–1846.

Sidpara, A., & Jain, V. K. (2012b). Nano-level finishing of single crystal silicon blank using magnetorheological finishing process. *Tribology International*, 47, 159–166.

Sidpara, A., & Jain, V. K. (2014). Rheological properties and their correlation with surface finish quality in MR fluid-based finishing process. *Machining Science and Technology*, 18(3), 367–385.

Singh, A. K., Jha, S., & Pandey, P. M. (2011). Design and development of nanofinishing process for 3D surfaces using ball end MR finishing tool. *International Journal of Machine Tools and Manufacture*, 51(2), 142–151.

Sutton, J. K. (2011). Orthopedic component manufacturing method and equipment. US Patent 7959490B2.

Tricard, M., Dumas, P. R., & Golini, D. (2004, October). New industrial applications of magnetorheological finishing (MRF). In Optical fabrication and testing (p. OMD1). Optical Society of America.

Tricard, M., Dumas, P., & Menapace, J. (2008, August).Continuous phase plate polishing using magnetorheological finishing. In Optical Engineering + Applications (p. 70620V). International Society for Optics and Photonics.

10

Nanofinishing of Freeform
Surfaces Using BEMRF

Faiz Iqbal and Sunil Jha

*Department of Mechanical Engineering, Indian Institute of Technology Delhi,
New Delhi, India*

CONTENTS

10.1 Introduction

The performance of a manufactured product is primarily decided by its surface quality. A better quality surface is characterised by the level of its surface finish. The manufacturing process used to produce a surface determines its surface finish level. Some processes are capable of producing better surfaces than others. The conventional processes recognised for good surface finish are grinding, honing, lapping, buffing, super finishing and polishing.

Conventional finishing processes are labour intensive and less precise as compared to the new advanced finishing processes. Finishing forces are precisely controlled in newly developed processes. The new advanced finishing processes are capable of finishing complex geometries and difficult-to-reach surfaces up to the nanometre level. Such geometries are very difficult to finish using conventional methods.

10.2 Freeform Surfaces

Freeform surfaces or sculptured surfaces have been widely used in the automobile, aerospace, consumer products and mould industries. Freeform surfaces are usually designed to improve aesthetics and/or to meet functional requirements. The definitions of freeform or sculptured surfaces are widely discussed in Campbell and Flynn [1]. Often, they are defined as surfaces containing one or more non-planar non-quadratic surfaces generally represented by parametric models. Many different mathematically defined methods have been developed to represent freeform surfaces; some of the most commonly used methods are Bézier, B-spline and NURBS. The Bézier curve is generally defined by an array of control points P_i ($i = 0$ to n) in three-dimensional (3D) space. Similarly, a Bézier surface is defined by a grid of control points $P_{i,j}$ ($i = 0$ to n, $j = 0$ to m). A major problem with Bézier curves and surfaces is that, usually, one wants to keep the degrees low, preferably not more than 3 or 4, and hence complex shapes are to be modelled by a composition of various curves or surfaces. This requires specific control point configurations to ensure a certain order of continuity [2].

Lasemi et al. [3] have thoroughly discussed the challenges that are faced by operators who intend to machine or polish freeform surfaces that are mainly tool path generation, tool orientation identification and tool geometry selection.

Li and Liu [4] defined the B-spline surface by the following equation:

$$s(u,v) = \sum_{i=0}^{nu-1} \sum_{j=0}^{nv-1} Bi, p(u) \cdot Bj, q(v) \cdot \varphi ij,$$

where

n_u and n_v are the number of control points in the u and v directions, respectively;

φ_{ij} $(i = 0, 1, \ldots, n_u - 1, j = 0, 1, \ldots, n_v - 1) - n$ $(n = n_u \times n_v)$ are control points and $B_{i,p}(u)$ and $B_{j,q}(v)$ are normal B-splines of degree p and q in the u and v directions, respectively.

They further normalised the previous equation with suitable assumptions to determine the coordinates of a measurement point r_k on the surface, with location parameters (u_k, v_k) as follows:

$$x(k) = \sum_{i=0}^{nu-1} \sum_{j=0}^{nv-1} Bi, p(uk) \cdot Bj, q(vk) \cdot x_{ij},$$

$$y(k) = \sum_{i=0}^{nu-1} \sum_{j=0}^{nv-1} Bi, p(uk) \cdot Bj, q(vk) \cdot y_{ij}$$

and

$$z(k) = \sum_{i=0}^{nu-1} \sum_{j=0}^{nv-1} Bi, p(uk) \cdot Bj, q(vk) \cdot z_{ij},$$

where x_{ij}, y_{ij} and z_{ij} are the coordinates of the B-spline surface control points φ_{ij} [4].

NURBS, as shown in Figure 10.1, are a generalised form of B-spline curves and surfaces. The most important properties of NURBS are the following:

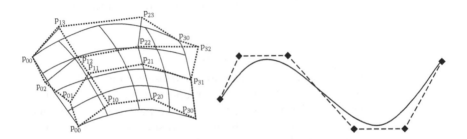

FIGURE 10.1
NURBS curves and surface. (From Zlatanova, S., Pu, S., Bronsvoort, W.F., Freeform curves and surfaces in DBMS: A step forward in spatial data integration, in Nayak, Pathan I, Garg (Eds.). Proceedings of ISPRS Commission IV Symposium on Geospatial Databases for Sustainable Development, September 27–30, 2006, Goa, India.)

i. They are piecewise rational polynomial curves and have the same continuity conditions at knots as a B-spline curve.

ii. NURBS curves are projective invariant; i.e. one can apply affine and projective transformations by applying these to the control points.

iii. NURBS curves can exactly represent conic sections, such as circles and ellipses.

iv. NURBS curves are, just like B-splines curves, locally modifiable and contained within the convex hull of their control points.

10.3 Nanofinishing Processes

Nanofinishing of freeform surfaces requires various challenges to be addressed. Before understanding these challenges, let us discuss some of the advanced finishing processes and their principles and limitations.

10.3.1 Abrasive Flow Machining

Identified in the 1960s as a method of deburring, polishing and radiusing, abrasive flow machining (AFM) uses an abrasive-laden viscoelastic polymer to flow over edges and difficult-to-reach surfaces. In AFM, two vertically opposed cylinders extrude an abrasive medium back and forth through the passage formed by the workpiece and tooling. Abrasion occurs wherever the medium passes through the restrictive passages. The key components of AFM are the machine, the tooling, types of abrasives, medium composition and process settings [5]. The three major elements of the process are the following:

a. The tooling, which confines and directs the abrasive medium flow to the areas where deburring, radiusing and surface improvements are desired.

b. The machine to control the process variables such as extrusion pressure, medium flow volume and flow rate.

c. The abrasive-laden polymeric medium, whose rheological properties determine the pattern and aggressiveness of the abrasive action. To formulate the AFM medium, the abrasive particles are blended into the special viscoelastic polymer, which shows a change in viscosity when forced to flow through restrictive passages [6].

AFM can process many selected passages on a single workpiece or multiple parts simultaneously. Inaccessible areas and complex internal passages can be finished economically and productively.

However, the biggest limitation of AFM in finishing freeform surfaces or complex geometries is the independent tooling requirement for every workpiece. An open freeform surface cannot be polished by AFM; it requires closed restrictive passages for the viscoplastic fluid to be squeezed through for the finishing action to take place. Therefore, the tooling requirement is a must for any kind of workpiece in AFM. Freeform surfaces in themselves are difficult to generate, and for AFM to finish such shapes, a mirror image of the workpiece is required to make a closed passage. This tooling requirement for AFM is difficult and expensive to create and may not be possible for all shapes of surfaces.

10.3.2 Magnetic Abrasive Finishing

Magnetic abrasion has emerged as an important finishing method for metals and ceramics. Magnetic abrasive finishing (MAF) is one such finishing method developed in the recent past to produce efficiently and economically good quality finish on the internal and external surfaces of tubes as well as flat surfaces made up of magnetic or non-magnetic materials. In this process, usually ferromagnetic particles are sintered with fine abrasive particles (Al_2O_3, SiC, CBN or diamond) to form ferromagnetic abrasive particles (or magnetic abrasive particles). However, homogeneously mixed loose ferromagnetic and abrasive particles are also used in certain applications. Figure 10.2a shows a plane MAF process in which finishing action is generated by the application of a magnetic field across the gap between the workpiece surface and the rotating electromagnet pole. The enlarged view of finishing zone (Figure 10.2b) shows the forces acting on the work surface to remove material in the form of micro/nano-chips. Force due to the magnetic field is responsible for normal indentation force causing abrasive particles penetration inside the workpiece surface while rotations of the magnetic abrasive brush (i.e. north pole) result in material removal in the form of chips [6].

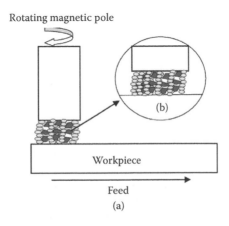

FIGURE 10.2
(a) Plane MAF and (b) finishing zone.

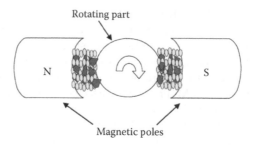

FIGURE 10.3
MAF of circular parts.

MAF can also finish circular parts and has a different mechanism of finishing as compared to flat surfaces; Figure 10.3 shows a schematic of MAF for circular parts.

The application of MAF has widely been accepted for flat surfaces and hollow tubes and an appreciably good surface finish has been achieved [7]. However, the limitations of MAF lie in finishing freeform surfaces and the complex geometries owing to its inherent limitation in generating a uniform surface finish on uneven curves and bends of a freeform surface.

10.3.3 Magnetorheological Finishing

The Center for Optics Manufacturing in Rochester, New York (USA), developed a technology to automate the lens finishing process using magnetorheological finishing (MRF). The MRF process relies on a unique 'smart fluid', known as magnetorheological (MR) fluid. MR fluids are suspensions of micron-sized magnetisable particles such as carbonyl iron, dispersed in a non-magnetic carrier medium like silicone oil, mineral oil or water.

Figure 10.4a shows a schematic representation of the dispersion of abrasive particles and CIPs in the absence of magnetic field, and Figure 10.4b shows the formation of chains by the CIPs embedding the abrasive particles between them. Because energy is required to deform and rupture the chains (Figure 10.4b), this micro-structural transition is responsible for the onset of a large 'controllable' finite yield stress [8]. When the field is removed, the particles return to their random state (Figure 10.4a) and the fluid again exhibits its original Newtonian behaviour.

In the MRF process, a convex, flat or concave workpiece is positioned above a reference surface and a thin layer of MR fluid ribbon is deposited on the rotating wheel rim, and on application of magnetic field in the gap, the stiffened region forms a transient work zone or finishing spot. Surface smoothing, removal of sub-surface damage and figure correction are accomplished by rotating the lens on a spindle at a constant speed while sweeping the lens about its radius of curvature through the stiffened finishing zone [9]. Material removal takes place through the shear stress generated as the MR

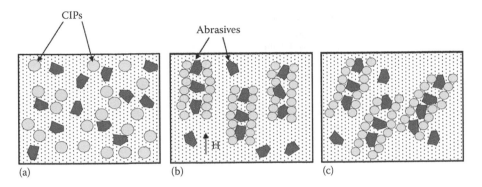

FIGURE 10.4
Schematic representation of dispersion of abrasive particles and CIPs (a) in the absence of magnetic field, (b) in the presence of magnetic field and (c) with applied shear strain γ.

polishing (MRP) ribbon is dragged into the converging gap between the part and the carrier surface. The zone of contact is restricted to a spot, which conforms perfectly to the local topography of the part. Deterministic finishing of flats, spheres and aspheric surfaces can be accomplished by mounting the part on a rotating spindle and sweeping it through the spot under computer control, such that the dwell time determines the amount of material removal.

The computer-controlled MRF process has demonstrated the ability to produce surface roughness of the order of 10–100 nm peak to valley by overcoming many fundamental limitations inherent to the traditional finishing techniques [10]. These unique characteristics make MRF the most efficient and able process for high-precision finishing of optics. MRF made finishing of freeform shapes possible for first time.

The advanced finishing processes developed in the last three to four decades are capable of nanofinishing common geometries/shapes but still have some limitations in finishing any complex 3D surface. Some processes are limited to flat and externally curved surfaces only, whereas others require complex tooling for workpiece fixture, thereby making it difficult or tedious for the users to finish freeform surfaces.

10.4 Ball End MRF

Ball end MRF (BEMRF) is an advanced finishing process that makes use of a rotating spot of stiffened MR fluid for the finishing. The MR fluid consists of carbonyl iron particles (CIPs) and abrasive particles mixed together in a base fluid medium. In the absence of a magnetic field, the fluid shows no special behaviour and particles are randomly dispersed in base fluid, but when a magnetic field is applied across the flow of the MR fluid, the CIPs

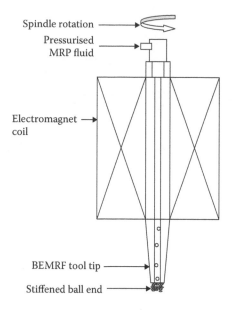

Spindle rotation

Pressurised
MRP fluid

Electromagnet
coil

BEMRF tool tip

Stiffened ball end

FIGURE 10.5
Mechanism of the BEMRF process.

form chains along the magnetic lines of force and the abrasive particles get trapped between the CIP chains. Figure 10.4a and b shows arrangement of CIP and abrasive particles when no magnetic field is applied and the structure formed in the presence of magnetic field, respectively.

In the BEMRF process, a small amount of MRP fluid is supplied at the tip of the BEMRF tool, which, under the influence of the magnetic field, turns into a stiff hemispherical ball as the CIPs form a chain along the magnetic lines of force as shown in Figure 10.5.

The abrasive particles held by carbonyl iron chains abrade the workpiece surface and finish the workpiece surface. The amount of material sheared from the peaks of the workpiece surface by abrasive particles depends on the bonding strength provided to the abrasives by CIP chains under the magnetic field [11]. The magnetic flux density in the finishing zone can be varied using the power supply, which controls the input current to the electromagnet.

10.5 Ball End MR Finishing Tool

The BEMRF tool is designed to have precise control of the process parameters during finishing. Figure 10.6 shows a schematic of a BEMRF tool.

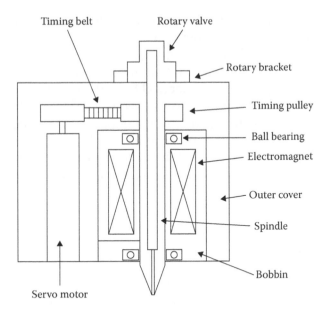

FIGURE 10.6
Schematic diagram of the BEMRF tool.

The main components of the BEMRF tool are as follows:

i. Spindle: The spindle is a cylindrical rotating magnetic core designed with a hollow axial passage at the centre to deliver MRP fluid at the tip of the tool.

ii. Tool tip: The tip of the spindle is a changeable tool that facilitates the requirement of variable-sized finishing spots. Having flexibility in the diameter of the tool tip helps in covering a wide range of work sizes for finishing.

iii. Bobbin: The bobbin is a cylindrical base for electromagnet copper winding.

iv. Electromagnet: The electromagnet is an essential component of the BEMRF tool. It is a cylindrical solenoid with copper winding around the spindle. When energised with electric current, it magnetises the spindle to produce magnetic flux density.

v. Outer cover: The outer cover forms the housing for the electromagnet. The electromagnet, when energised with an electric current up to 7–8 A, experiences a rise in coil temperature. There is a need to control the coil temperature during finishing, which is achieved by a continuous flow of low-temperature transformer oil in the outer cover completely immersing the electromagnet in it. The oil is maintained at a low temperature of around 0–5°C and is continuously circulated externally into the outer cover, thereby keeping the electromagnet temperature under control.

vi. Temperature sensor: To keep the temperature of the electromagnetunder under control, it is important to monitor it continuously. Resistance temperature detector (RTD) type temperature sensors are embedded in three layers of electromagnet winding for the same.

vii. Rotary valve: The MRP fluid from the fluid delivery system (FDS) is required to be supplied in the rotating spindle. The rotary valve is used to keep the fluid delivery tube static against the rotating spindle.

viii. Servo motor: The spindle rotary motion is achieved by coupling the spindle to a servomotor.

10.6 BEMRF Process Parameters

The BEMRF process is governed by its process and fluid parameters, which are explained in the following sections.

10.6.1 Process Parameters

The effectiveness of process parameters in the BEMRF process performance is measured by the percentage change in roughness (% ΔR_a). The following are the independent controllable parameters:

a. Current (A): The electric current in the electromagnet governs the strength of the magnetic field at the tool tip. The range of current varies from 1 A to 10 A.

   The magnetic flux density in the working gap increases with increasing current induced in the electromagnet coil as more magnetic lines of force pass through the BEMRF polishing spot. As a result, the interparticle magnetic force between the CIP in the chains is increased. Therefore, the increasing current in the electromagnet coil increases the % ΔR_a (Figure 10.7). However, this increase in % ΔR_a with increasing current is not linear and is saturated after the upper limit of current is reached.

b. Working gap (mm): In the BEMRF process a small gap is maintained between the tool tip and the work surface. This gap is filled with the MRP fluid, which forms a stiffened hemispherical ball under the influence of a magnetic field. A typical range of the working gap is 0.6 mm to 1.5 mm. The magnetic flux density is also affected by the working gap. This is because the magnetic flux density decreases with increasing distance from the electromagnet coil. Thus, the percentage

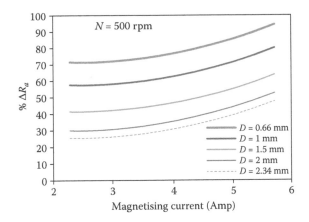

FIGURE 10.7
Effect of magnetising current on percentage reduction in R_a value at different working gaps (D).

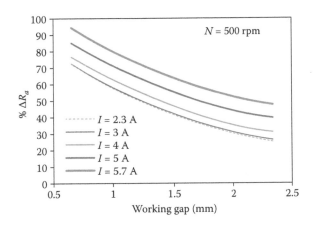

FIGURE 10.8
Effect of working gap on percentage reduction in R_a value at different magnetising currents.

reduction in R_a will reduce with increasing working gap as shown in Figure 10.8.

c. Spindle rotational speed (RPM): As the spindle of the BEMRF tool rotates on its axis along with the ball ended tip, different abrasive particles are made to move relative to the work surface. The spindle speed affects the rate of surface finish on the work surface and varies between 100 and 500 RPM. The % ΔR_a first increases with the increase in tool rotational speed as the abrasives with the CIP chains move faster across the workpiece surface and remove material. However, the % ΔR_a starts to decrease, as seen in Figure 10.9, if the tool rotational speed is increased beyond a given value. This is

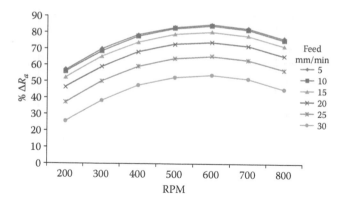

FIGURE 10.9
% ΔR_a obtained at different RPMs for different feed rates.

because at higher rotational speeds, the effect of centrifugal force acting on CIPs at a given position begins to increase beyond the magnetic force of attraction on the CIPs. The CIP chains are then not able to hold the abrasive particles strongly enough during finishing.

d. Table speed (mm/min): Similar to the spindle rotation, the table feed rate also affects the surface finish in the BEMRF process. The table feed rate is the distance travelled by the work holding table across the BEMRF tool per unit time. The table feed rate ranges from 5 mm/min up to 50 mm/min. Figure 10.9 shows that % ΔR_a reduces with an increase in table feed rate. This is because when the workpiece is moving at a faster rate under the rotating ball end, the CIP chains and abrasives have less time to finish at a particular spot. The abrasive interaction time with workpiece surface reduces with table speed.

10.6.2 Fluid Parameters

Other than process parameters, increasing the fluid composition also plays an important role in process performance. The following are fluid parameters and their effect on change in surface roughness:

a. Abrasive mesh size: The abrasive mesh size defines the size of the abrasive particles. Surfaces having initial roughness (R_a) of the order of 1 μm require abrasives of mesh size 600–800. The mesh size of abrasives for the BEMRF process varies from 600 to 2000.

b. Abrasive concentration (%vol): The MRP fluid comprises of CIPs, abrasive particles and base fluid. The percentage concentration of abrasive particles in the fluid ranges from 10% to 20% by volume.

c. CIP concentration (%vol): The CIP concentration in the MRP fluid governs the number of chains tuned under the application of the magnetic field. More chains will trap more abrasive particles between them and better finishing may be obtained. However, in all concentration-based parameters, each has an optimum concentration value range below or beyond which the fluid does not provide good results while finishing. CIP concentration in the MRP fluid ranges from 10% to 40% by volume.

d. Abrasive types: Depending on the hardness of the workpiece material, abrasives such as Al_2O_3, SiC, B_4C, diamond, etc., are used in MRP fluid.

10.7 Mechanism of Material Removal and Surface Finishing in the BEMRF Process

Surface finishing and material removal in the BEMRF process depend on duration of interaction of the abrasive particles, while abrading the work material. The electromagnet is switched on once the MRP fluid reaches the tool tip. The fluid stiffens in the shape of a ball end as shown in Figure 10.10a. The abrasive particles get trapped in between the CIP chains. Inside the non-magnetic abrasive particles, the flux density is very small for any external magnetic field intensity. Therefore, the majority of the abrasive particles are repelled from the higher gradient of the magnetised tool tip surface towards the lower gradient of magnetic flux density at the workpiece surface [12]. Those abrasive particles that are in contact with the workpiece surface are called active abrasive particles and are responsible for the material removal during the rotation of the ball ended tool on the workpiece surface. The active abrasive particles are tightly gripped by CIP chain structures toward the outer periphery of the finishing spot of the MRP fluid. When the gripped active abrasive particles have relative motion on the workpiece surface during the rotation of the finishing spot of the MRP fluid (Figure 10.10b), the peaks of the workpiece surface wear out.

Stiffened ball end ←   → BEMRF tool tip

→ Stiffened ball end rotating

(a)   (b)

FIGURE 10.10
Tool tip (a) when the electromagnet is on and (b) during surface finishing by abrasion action.

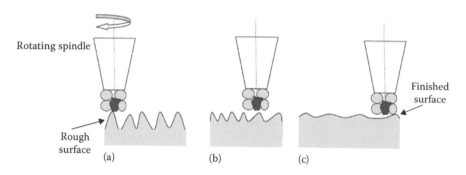

FIGURE 10.11

Mechanism of material removal in the BEMRF process: (a) gripped active abrasive particle with CIP chains approaching the initial roughness peaks of the workpiece surface, (b) updated roughness peaks after removing the first layer in the form of microchips and (c) final roughness peaks after removing the almost all its layers in the form of microchips.

Figure 10.11 shows the mechanism of material removal in the BEMRF process. The higher yield strength of the MRP fluid can be found at the highly magnetised tool tip surface. The CIP chains will be able to hold abrasive particles more firmly and strongly like a single body for a longer period of time under high shear strength. This is necessary for the effective removal of material from the workpiece surface during finishing operation. Due to higher magnetic normal forces, the abrasive particles are in more constant contact with the workpiece surface for a longer period of time, which results in deeper penetration of the abrasive particles on roughness peaks (Figure 10.11a). When an MRP fluid finishing spot with a high yield strength rotates on the workpiece surface, the high shear strengths of the gripped active abrasive particles shear the peaks of the surface in the form of microchips (Figure 10.11b) due to the abrasion by the two-body wear mechanism.

When continuous feed is given to the workpiece surface with respect to the rotation of the finishing spot of MRP fluid, the final surface finish can be achieved after wear-out of almost all layers of roughness peaks by the abrasion process, as shown in Figure 10.11c. The constrained contact and depth of penetration of active abrasive particles on the workpiece surface depend on the magnetic normal force, while the shear strength of the MRP finishing spot is responsible for the removal of material by abrasion in the form of microchips during rotation of the central tool core.

Figure 10.12 shows a schematic representation of material removal during the finishing operation by a gripped active abrasive particle with a CIP chain structure in the absence of feed (Figure 10.12a) and with feed (Figure 10.12b) to the workpiece. When the finishing is performed at a single spot on the workpiece surface without any feed relative to the MR finishing spot, there are chances of tool rotation marks appearing on the workpiece surface due to repeated interaction of the abrasive particles at the same location (Figure 10.12a). This may result in non-uniform finishing of the workpiece surface.

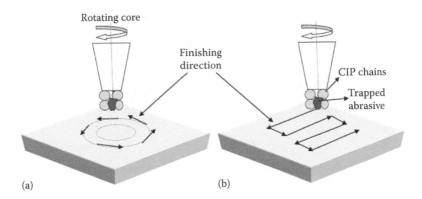

FIGURE 10.12
Schematic representation of material removal by a gripped active abrasive particle with CIPs chains in the BEMRF process (a) without any feed to the workpiece and (b) with feed to the workpiece.

When the workpiece surface is finished by giving feed in the X and Y directions relative to the rotation of MR finishing spot (Figure 10.12b), there are very little chances of getting feed marks on the surface due to the continuous change of location of the rotating active abrasive particles on the workpiece surface. This results in uniform finishing of the workpiece surface [13].

10.8 Mathematical Model for Material Removal in the BEMRF Process

The following are the assumptions made to develop the mathematical model for surface roughness and normal forces [14]:

- The magnetic field due to magnetisation of the CIPs is assumed to be insignificant/negligible.
- The indentation force of each abrasive particle is assumed to be uniform.
- Only the axial variation of the magnetic flux density is considered.
- It has been assumed that all abrasive particles and CIPs are of uniform size and there is a homogeneous distribution of abrasive particles and CIPs in the polishing fluid forming a uniform body centred chain structure.
- Magnetic flux leakages have been considered as negligible.
- While calculating the rotation speed of fluid, the fluid is assumed to be Newtonian, to solve the Navier–Stokes equation.

10.8.1 Magnetic Flux Density in the Working Gap

Magnetic flux density due to a finite solenoid at an axial point is given as [15]

$$B(x) = \frac{\mu_o i N}{2(r_2 - r_1)}\left[x_2 ln \frac{\sqrt{r_2^2 + x_2^2} + r_2}{\sqrt{r_1^2 + x_2^2} + r_1} - x_1 ln \frac{\sqrt{r_2^2 + x_1^2} + r_2}{\sqrt{r_1^2 + x_1^2} + r_1} \right], \qquad (10.1)$$

where

B is the magnetic flux density, measured in tesla, at any point on the axis of
    the solenoid.
The direction of the field is parallel to the solenoid axis.
N is the number of turns of wire per unit length of the solenoid.
r_1 is the inside radius of the solenoid.
r_2 is the outside radius of the solenoid.
x_1 and x_2 are the distances from the two ends of the solenoid, on the axis,
    as shown in Figure 10.13.
i is the current in the wire measured in amperes.
μ_o is the permeability constant.

Example 10.1

Calculate the magnetic flux density (B) due to an electromagnet at an
axial distance of 1 mm. The electromagnet has 1000 turns in a length of
200 mm. The inner and outer radii of the electromagnet are 35 mm and
135 mm, respectively. A current of 2 A is given in the electromagnet coil.

SOLUTION

Given are $n = 1000, l = 200$ mm, $r_1 = 35$ mm, $r_2 = 135$ mm, $i = 2$ A, $z = 1$ mm

$$N = n/l = \frac{1000}{200} = 5, x_1 = 1 \text{ mm and } x_2 = 201 \text{ mm}$$

$$\mu_o = 1.256 * 10^{-6}.$$

Putting all values in Equation 10.1:

$$B(x) = \frac{1.256 * 10^{-6} * 2 * 5}{2(135 - 35)}\left[201 ln \frac{\sqrt{135^2 + 201^2} + 135}{\sqrt{35^2 + 201^2} + 35} - ln \frac{\sqrt{135^2 + 1^2} + 135}{\sqrt{35^2 + 1^2} + 35} \right]$$

$$B(x) = 0.0628 * 10^{-6}[201 ln 1.577 - ln 3.856]$$

$$B(x) = 0.0628 * 10^{-6}[90.307]$$

$$B(x) = 5.671 * 10^{-6} \text{ H.A}/m^2$$

$$B(x) = 5.671 * 10^{-3} \text{ Tesla.}$$

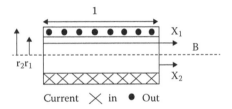

FIGURE 10.13
A finite solenoid.

The magnetic flux density produced by the solenoid will be responsible for the magnetisation of the ferromagnetic material (either itself as a workpiece or kept under a non-ferromagnetic workpiece). The magnetised ferromagnetic material will produce its own magnetic field, which will enhance the magnetic field within the gap. If the magnetic field strength (H) in the ferromagnetic material can be determined, the intensity of magnetisation (M) can be obtained.

To determine the magnetic field strength at a distance (d) from the tool tip, the following relation can be used [16]:

$$H = \frac{m}{4\pi\mu_0\mu_r d^2} , \tag{10.2}$$

where m is the magnetic pole strength developed at the tool tip, μ_r is relative permeability of the ferromagnetic material and d is the distance of ferromagnetic material from the tool tip.

The dipole moment of a coil is [15] $\mu = i * A$, where A is area of coil and i is current flowing through coil.

If there are n numbers of layers in coil, then net dipole moment will be

$$\mu_{net} = \sum_{i=1}^{n} \mu_i. \tag{10.3}$$

The pole strength can be calculated as [15]

$$m = \frac{\mu_{net}}{N}. \tag{10.4}$$

Here, N is the number of turns in unit length $\left(= \frac{n}{l} \right)$.

From the pole strength obtained in Equation 10.4 and substituting it in Equation 10.2, the magnetic field strength (H) in the ferromagnetic material can be obtained.

The relation between M and H must be known to calculate the intensity of magnetisation.

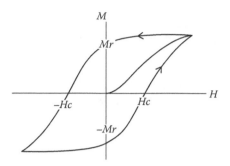

FIGURE 10.14
A typical *M–H* curve for a ferromagnetic material. (From Pasrija, K., Analysis and Mathematical Modelling of Magnetorheological Finishing Process, M. Tech thesis, Indian Institute of Technology Delhi, New Delhi, India, 2014.)

For paramagnetic and diamagnetic materials, the relation is linear, that is

$$M = X_m H. \tag{10.5}$$

But for a ferromagnetic material, the relation will be non-linear, as shown in Figure 10.14, and for any value of magnetic field strength (H), we can get the intensity of magnetisation (M).

By its definition, magnetisation, $M = \dfrac{dipole\ moment}{volume}$.

And also, dipole moment = pole strength * distance separating poles.

From the relation of the dipole moment, we can get the pole strength of the ferromagnetic material, which gets magnetised due to the field produced by the current carrying solenoid (Figure 10.15).

The magnetic flux density produced by a magnetic pole having strength m is [9]

$$B = \frac{\mu_0 m}{4\pi r^2}, \tag{10.6}$$

where m is pole strength and r is the axial distance from the pole.

Now, the magnetic flux density in the working gap can be obtained from Equations 10.1 and 10.6 as

$$\vec{B}(z) = \frac{\mu_0 in}{2(r_2 - r_1)} \left[x_2 ln \frac{\sqrt{r_2^2 + x_2^2} + r_2}{\sqrt{r_1^2 + x_2^2} + r_1} - x_1 ln \frac{\sqrt{r_2^2 + x_1^2} + r_2}{\sqrt{r_1^2 + x_1^2} + r_1} \right]$$
$$+ \frac{\mu_0 m}{4\pi(t - z)^2} - \frac{\mu_0 m}{4\pi(a + t - z)^2}, \tag{10.7}$$

where $x_1 = z$ and $x_2 = l + z$, where l is the length of the solenoid.

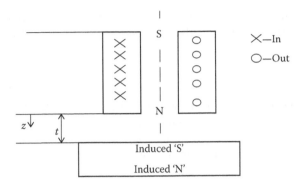

FIGURE 10.15
Setup showing induced magnetisation of ferromagnetic piece. (From Pasrija, K., Analysis and Mathematical Modelling of Magnetorheological Finishing Process, M. Tech thesis, Indian Institute of Technology Delhi, New Delhi, India, 2014.)

For a distance of 1 mm between the ferromagnetic workpiece and the tool tip and a current of 3 A in the solenoid, magnetic flux density variation from Equation 10.7 is shown in Figure 10.16.

After applying the quadratic equation on the theoretically obtained variation from Figure 10.16, the following equations are obtained:

$$B = 49.1z^2 - 177.9z + 1088$$

$$\frac{dB}{dz} = 98.2z - 177.9.$$

(10.8)

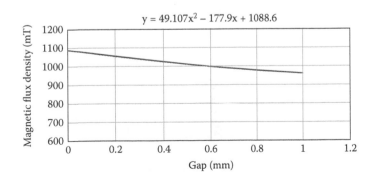

FIGURE 10.16
Theoretical variation of magnetic flux density in the gap. (From Pasrija, K., Analysis and Mathematical Modelling of Magnetorheological Finishing Process, M. Tech thesis, Indian Institute of Technology Delhi, New Delhi, India, 2014.)

This magnetic flux density in Equation 10.8 will be responsible for the magnetic force on CIPs in the MR fluid. Now, the force on the CIP is given by [16]

$$F_m = m_m B \nabla B, \tag{10.9}$$

where m is mass of CIP, $m = \rho * 4/3\pi r^3$, m is the mass susceptibility of particle (m³/kg), B is the magnetic flux density (T) in the gap and ∇B is the gradient of magnetic field in the z-direction. These CIPs will arrange themselves in a chain-like structure, holding abrasive particles between them. Hence, normal force on the abrasive can be obtained using Equation 10.9. The magnitude of normal force will decide the quality of the surface finish. It has been observed experimentally that the gradient near the tool tip is higher as compared to that near the workpiece. So, it can be said that most of the CIPs will be accumulating at the tool tip and the concentration of abrasive particles will be more near the workpiece surface. The force mentioned in Equation 10.9 will act on all the CIPs in the MR fluid.

10.8.2 MRP Fluid Micro-Structure and Active Abrasive Particles

There is an assumption that the abrasive particles participating in the material removal process are forming a half-BCC structure at the workpiece surface.

Figure 10.17 shows an exaggerated view of a single chain of CIPs, holding the abrasive particle in between. Let there be n number of CIPs in a single

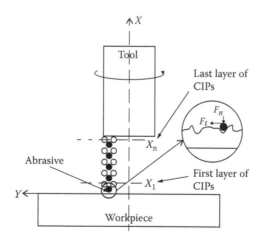

FIGURE 10.17
Exaggerated view of magnetic particles forming a chain. (From Pasrija, K., Analysis and Mathematical Modelling of Magnetorheological Finishing Process, M. Tech thesis, Indian Institute of Technology Delhi, New Delhi, India, 2014.)

chain. The two forces, normal force and tangential force, are shown in Figure 10.17, acting in mutually perpendicular directions. Normal force is responsible for the indentation and tangential force is responsible for shearing.

10.8.3 Normal Force on an Abrasive Particle

The chain structure near the workpiece surface is assumed to be half-body centred cubic; hence, for a single active abrasive particle on the workpiece surface, there are four magnetic particles surrounding it for a specific fluid concentration [13], and each magnetic particle exerts force on four abrasive particles, as can be seen in Figure 10.17. The magnetic particles experience the magnetic force when the magnetic field is applied. The force transferred to an active abrasive particle due to four surrounding magnetic particles is equal to F_m. Considering a single column chain structure, in the first layer, an active abrasive particle is touching the workpiece surface, and in the second and subsequent layers, magnetic particles are forming a BCC structure with an abrasive particle right at the centre of the BCC cell. Thus, from the first layer of magnetic particles, the $F_{m,1}$ force is transferred on the active abrasive particle, and from the second layer of the CIPs, $F_{m,2}$ and so on. Hence, from the nth layer, $F_{m,n}$ force is transferred through the chains on the active abrasive particle.

Total magnetic force, $F_{m,sum}$, on an active abrasive particle is calculated by summing the forces from all layers of magnetic particles and is given by (using Equation 10.9):

$$F_{m,sum} = F_{m,1} + F_{m,2} + \ldots + F_{m,n} = \frac{\sum_{i=1}^{n'} m(B)_i \left(\frac{\partial B}{\partial z} \right)_{imi}}{\mu_0}. \qquad (10.10)$$

Length of one BCC unit cell, $a = \dfrac{2(r_1 + r_2)}{\sqrt{3}}$,

where r_1 is the radius of CPI and r_2 is the radius of abrasive.

Distance of first layer of CIP from workpiece surface, $X_1 = r_2 + \dfrac{a}{2}$.

Distance of last layer of CIP from workpiece surface, $X_n = X_1 + (n' - 1)a$.

From this relation, for a gap of 1 mm and unit cell of 18 μm, the value of $n' = 55$, which will be used in Equation 10.9 to find the net normal force on the abrasive particle.

For a gap of 1 mm and current of 3 A, $r_1 = 6$ μm and $r_2 = 10$ μm and $a = 18$ μm. The normal force on an active abrasive can be calculated using Equation 10.9:

$$F_{m,sum} = 3.1104 * 10^{-11} \text{ N}.$$

10.8.4 Modelling of Surface Finish

The normal force (i.e. the force obtained from the net magnetic force on an abrasive; Equation 10.9) acting on an abrasive particle forces it to indent the workpiece surface. The Brinell hardness number (BHN) has a direct correlation with the depth of indentation in the material as follows:

$$BHN = \frac{F_{m,sum}}{\frac{\pi}{2} D_g \left(D_g - \sqrt{D_g^2 - D_i^2} \right)}, \qquad (10.11)$$

where $F_{m,sum}$ is the indentation force on each abrasive particle, D_i is the indentation diameter and D_g is the diameter of the abrasive particle. The BHN is measured in kgf/mm². We will obtain the value of D_i by substituting the respective values.

From the geometry of Figure 10.18a, the depth of indentation t is obtained as [17]

$$t = \frac{D_g}{2} - \frac{1}{2}\sqrt{D_g^2 - D_i^2}. \qquad (10.12)$$

For $D_g = 20$ μm and $D_i = 0.6708$ μm, the indentation depth t is calculated as follows:

$$t = \frac{(20 * 10^{-6})^2}{2} - \frac{1}{2}\sqrt{(20 * 10^{-6})^2 - (0.6708 * 10^{-6})^2}$$

$$t = 5.626 * 10^{-9} \text{ m.}$$

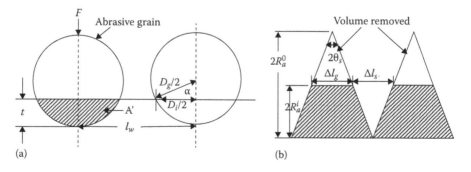

(a)   (b)

FIGURE 10.18
(a) Schematic diagram showing the penetration of abrasive grain in work material and (b) triangular shape of the irregularity.

There is an assumption that the initial surface profile is triangular, as shown in Figure 10.18b. It is assumed that the initial surface profile of the workpiece is uniformly distributed with initial surface roughness R_a^0 and the abrasives move in a perpendicular direction to the direction of lay.

The final R_a after each pass can be obtained from:

$$R_a^i = R_a^o - \frac{\sqrt{3}}{2(r_1 + r_2)} \left[\frac{D_g^2}{4} \sin^{-1} \frac{2\sqrt{t(D_g - t)}}{D_g} - \sqrt{t(D_g - t)} \left(\frac{D_g}{2} - t \right) \right]. \quad (10.13)$$

The final R_a^i obtained after each pass is used as the initial surface roughness value for estimating R_a during the consequent pass.

Example 10.2

For a given material having $BHN = 89$ kgf/mm², force due to all CIP layers $F_{m,sum} = 3.1104 * 10^{-11}$ and diameter of abrasive particle $D_g = 20$ μm, find the indentation diameter D_i.

SOLUTION

Given here are $BHN = 89$ kgf/mm², $F_{m,sum} = 3.1104 * 10^{-11}$ and $D_g = 20$ μm. Putting all values in Equation 10.11,

$$89 = \frac{3.1104 * 10^{-11}}{\frac{\pi}{2} * 20 * 10^{-6}[20 * 10^{-6} - \sqrt{((20 * 10^{-6})^2 - D_i^2)}]}$$

$$89 * \frac{\pi}{2} * 20 * 10^{-6} = \frac{3.1104 * 10^{-11}}{[20 * 10^{-6} - \sqrt{((20 * 10^{-6})^2 - D_i^2)}]}$$

$$2796.017 * 10^{-6} = \frac{3.1104 * 10^{-11}}{[20 * 10^{-6} - \sqrt{((20 * 10^{-6})^2 - (D_i)^2)}]}$$

$$55920.349 * 10^{-12} - 2796.017 * 10^{-6}[\sqrt{((20 * 10^{-6})^2 - (D_i)^2)} = 3.1104 * 10^{-11}$$

$$1.9988 * 10^{-5} = \sqrt{((20 * 10^{-6})^2 - (D_i)^2)}.$$

Squaring both sides,

$$3.9955 * 10^{-10} = 400 * 10^{-12} - D_i^2$$

$$D_i^2 = 4 * 10^{-10} - 3.9955*10^{-10}$$

$$D_i = \sqrt{(4 * 10^{-10} - 3.9955 * 10^{-10})}$$

$$D_i = \sqrt{(4.5 * 10^{-3})} = 6.708 * 10^{-7} = 0.6708 \; \mu m.$$

10.9 BEMRF Machine Tool

A dedicated machine tool for the BEMRF process is required for the following:

i. The process needs motion control so that a proper path may be followed by the BEMRF tool tip to finish the desired geometry and shapes of the workpiece.

ii. Control of the process parameters of the BEMRF process is required.

iii. A synchronised control of the motion and process parameters is needed so that effective finishing may be obtained.

iv. A dedicated user interface for the automatic control of the BEMRF process to allow the user to have a better control over the finishing process.

Considering these requirements, a five-axis computer numerical control (CNC) BEMRF machine tool is developed. The following are the four major parts of the machine tool.

10.9.1 Machining Area

The machining area consists of three linear and two rotary positioners to accomplish motion in five axes, i.e. X, Y, Z, B and C. The system uses servomotors and servo drives, which are controlled by a programmable automation controller (PAC).

The Y-axis linear positioner is mounted at the bottom, over which the X-axis linear positioner is fixed rigidly such that when the Y carriage moves, it carries the X-axis positioner along with it. The C-axis rotary positioner is mounted over the X-axis linear positioner. The workpiece holding mechanism is mounted over the C- and X-axes arrangement. This arrangement gives the workpiece linear movement in two directions, viz., X and Y and rotary motion about the Z-axis, i.e. C-axis movement. The complete five-axis arrangement is schematically shown in Figure 10.19.

10.9.2 Control Panel

The control panel has servo drives, PAC, I/O, modules and the necessary electrical components for the machine tool. Figure 10.20 shows the schematic

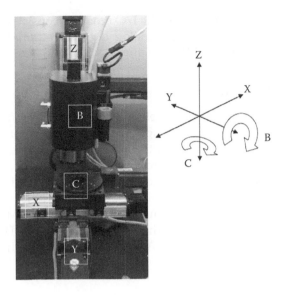

FIGURE 10.19
Five-axis arrangement of a BEMRF machine tool.

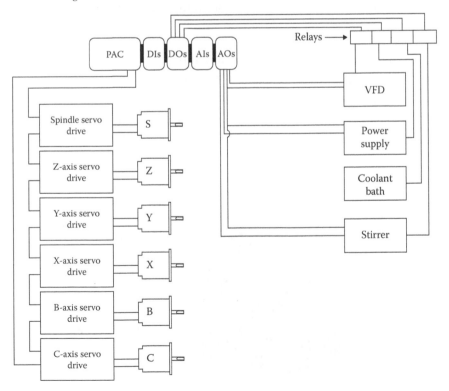

FIGURE 10.20
Control panel schematic.

representation of the control panel. The I/O modules consist of digital inputs, digital outputs, analog inputs and analog outputs required to control the process parameters of the BEMRF machine tool.

10.9.3 Fluid Delivery System

The MRP fluid plays an important role in material removal and carries heat and debris away from the polishing zone. A FDS is incorporated in the BEMRF machine tool, which is controlled to supply a desired volume of MRP fluid from the reservoir to the tool tip.

A schematic of the MRP-FDS is shown in Figure 10.21. It is prepared in a funnel-shaped reservoir with the help of a direct current (DC) speed controlled stirrer. A peristaltic pump is used for delivering the MRP fluid to the tip surface of the rotating core from the MRP fluid reservoir. The speed of the peristaltic pump is controlled by an AC variable frequency drive.

10.9.4 Graphical User Interface

A dedicated graphical user interface for the automatic control of the BEMRF machine enables the user to have the following controls:

   i. Manual control of all motion axes
  ii. Manual control of all process parameters
 iii. Automatic control of complete BEMRF process
  iv. Temperature monitoring of the BEMRF tool
   v. Automatic motion control of the five axes as per the part program developed according to the geometry of the profile to be polished

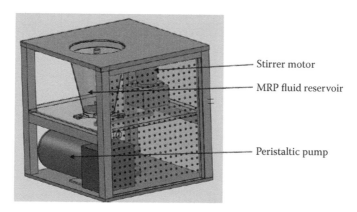

Stirrer motor

MRP fluid reservoir

Peristaltic pump

FIGURE 10.21
Schematic representation of FDS.

10.10 Applications of BEMRF

The BEMRF process is the most versatile process suitable for finishing complex 3D geometries; the only difference lies in the fluid composition for different materials. The BEMRF process finds its applications in the nanofinishing of many important materials and products; some of the application areas are polishing of different materials, which are listed here:

i. Hardened dies: Chromium-steel-based dies are widely used for moulding purposes; one example is moulds of compact fluorescent lamp (CFL) tubes, which are prepared using hardened steels. Finishing up to nanometer level is required in such cases, and finishing such hard materials is difficult from conventional finishing processes. BEMRF becomes useful in such application areas as it can finish a variety of materials from as soft as copper and aluminium to as hard as chromium steel. In case of hardened steels, diamond-powder-based fluid is prepared. Diamond, being the hardest material, facilitates the finishing of hardened steels.

ii. Metal mirrors: Metal mirrors are used in areas where high temperature conditions exist; metal mirrors are preferred because of high thermal conductivity, as they can withstand higher temperatures as compared to glass mirror coatings. Copper mirrors are thus very useful. Nanofinishing of copper is different from hardened steels as copper is ductile, and therefore, polishing fluid for nanofinishing of copper in BEMRF will have alumina (Al_2O_3) abrasives.

iii. Glass: Glass is a non-crystalline, colourless compound. It possesses a very low thermal expansion coefficient combined with excellent optical qualities and good transmittance over a wide spectral range. It is resistant to scratching and thermal shock, and one of the important applications is high-energy laser optics. This demands the glass to have a very fine surface finish. Singh et al. [18] have explained the use of BEMRF in polishing of fused silica glass and obtained good results by achieving R_a value as low as 0.146 nm from an initial value of 0.74 nm.

Other applications areas of BEMRF process are nanofinishing of compact disc (CD) moulds, gems and other precision roughness areas.

Exercises

1. A BEMRF tool has an electromagnet with 1000 turns in a length of 240 mm. The outer and inner radii of the electromagnet are 125 mm and 35 mm, respectively. The distance between the tool tip and a

ferromagnetic workpiece is 1.5 mm. For a current of 3 A in the electromagnet, calculate the magnetic flux densities in the working gap at distances of 0.5 mm and 1 mm from the tool tip.

(Magnetic permeability of free space $\mu_o = 1.256 * 10^{-6}$ H/m)

[Answer: $B = 1.77$ T for $z = 0.5$ mm, $B = 2.02$ T for $z = 1$ mm]

2. Using the value of magnetic flux density at $z = 1$ mm found in problem 1, find the normal force on an active abrasive particle due to the first layer of the CIP. The radius of a CIP is 6 μm and that of an abrasive particle is 10 μm. Values of the magnetising current and the working gap are the same as in problem 1.

[Answer: $F_{m1} = 6.754 * 10^{-12}$ N]

3. Using the normal force and particle size from problem 2, find the indentation diameter D_i and indentation depth t for a given material having $BHN = 89$ kgf/mm².

[Answer: $D_i = 0.3193$ μm, $t = 1.274 * 10^{-9}$ m]

4. Using the indentation depth found in problem 3 and initial surface roughness R_a 1 μm, calculate the ΔR_a after one pass. The size of the electromagnet, CIP and abrasive particle are same as in problems 1 and 2.

[Answer: $\Delta R_a = 5 * 10^{-5}$ μm]

References

1. Campbell, R.J., Flynn, P.J., 'A survey of free-form object representation and recognition techniques', *Computer Vision and Image Understanding* 81 (2001) 166–210.
2. Zlatanova, S., Pu, S., Bronsvoort, W.F., 'Freeform curves and surfaces in DBMS: A step forward in spatial data integration' in Nayak, Pathan I & Garg (Eds.). Proceedings of ISPRS Commission IV Symposium on 'Geospatial Databases for Sustainable Development', 27–30 September 2006, Goa, India.
3. Lasemi, A., Xue, D., Gu, P., 'Recent development in CNC machining of freeform surfaces: A state-of-the-art review', *Computer-Aided Design* 42 (2010) 641–654.
4. Li, F.Y., Liu, Z.G., 'Method for determining the probing points for efficient measurement and reconstruction of freeform surfaces', *Journal of Measurement Science and Technology* 14 (2003) 1280–1288.
5. Rhoades. L.J., 'Abrasive flow machining', *Manufacturing Engineering* 101 (1988) 75–78.
6. Taniguchi, N. 'Current status in, and future trends of ultra-precision machining and ultrafine material processing', *Annals of CIRP*, 32/2 (1983) 573–582.
7. Fox, M., Agrawal, K., Shinmura, T., Komanduri, R., 'Magnetic abrasive finishing of rollers', *Annals of CIRP*, 43/1 (1994) 181–184.

8. Jain, V.K., Jha, S., 'Nano-finishing techniques' in Mahalik, N.P. (Ed.). *Micromanufacturing and Nano-Technology*, Springer Verlag, New York (2005) 171–195.

9. COM, Magneto-rheological finishing, Article by Center for Optics Manufacturing (http://www.opticam.rochester.edu) (1998).

10. Kordonski, W.I., 'Magneto-rheological finishing', *International Journal of Modern Physics B*, 10(23&24) (1996) 2837–2849.

11. Singh, A.K., Jha, S., Pandey, P.M. 'Design and development of nanofinishing process for 3D surfaces using ball end MR finishing tool', *International Journal of Machine Tools & Manufacture* 51 (2011) 142–151.

12. Nathan, I., *Engineering Electro-magnetics*, Springer, New York, 2004 (Chapter 9).

13. Singh, A.K., Jha, S., Pandey, P.M., 'Mechanism of material removal in ball end magneto-rheological finishing process', *Wear* 302 (2013) 1180–1191.

14. Pasrija, K., 'Analysis and Mathematical Modelling of Magnetorheological Finishing Process', M. Tech thesis, Indian Institute of Technology Delhi, New Delhi, India (2014).

15. Ginder, J.M., Davis, L.C., 'Shear stresses in magnetorheological fluids: Role of magnetic saturation,' *Applied Physics Letters* 65 (26), 3410–3412.

16. Stradling, A.W., 'The physics of open-gradient dry magnetic separation'. *International Journal of Mineral Processing* 39 (1993) 19–29.

17. Jain, R.K., Jain, V.K., Dixit, P.M., 'Modeling of material removal and surface roughness in abrasive flow machining process', *International Journal of Machine Tools & Manufacture* 39 (1999) 1903–1923.

18. Singh, A.K., Jha, S., Pandey, P.M., 'Nanofinishing of Fused Silica Glass Using Ball-End Magnetorheological Finishing Tool', *Materials and Manufacturing Processes*, 27(10) (2012) 1139–1144.

11

Nanofinishing Process for Spherical Components

Jomy Joseph,[1] Ajay Sidpara,[1] Jinu Paul[1] and Vijay K. Jain[2]

[1]*Mechanical Engineering Department, Indian Institute of Technology Kharagpur, Kharagpur, India*

[2]*Department of Mechanical Engineering, Indian Institute of Technology Kanpur, Kanpur, India*

CONTENTS

11.1 Introduction

Superior form accuracy and nanofinishing requirements are two essential characteristics for components of high-precision devices. In many cases, these can be achieved only by non-traditional/advanced finishing processes. Conventional finishing processes have limitations regarding workpiece geometry, forces involved, surface integrity, surface finish requirements and efficient control of process parameters. Most of the non-traditional finishing processes, although better than the conventional processes, still find it difficult to meet the surface finish requirements of ceramic balls, ceramic rollers, optical lenses, etc., used in precision devices. This is mainly due to the difficulty to fix and grip spherical components and to get a uniform finish all over the surface.

Ceramic balls or silicon nitride (Si_3N_4) balls are preferred to traditional chrome balls or steel rolls in precision bearings and bearing tracks (Figure 11.1) because of their low thermal expansion, low thermal conductivity, low

FIGURE 11.1
Ceramic ball bearings. (Courtesy of http://www.directindustry.com/prod/cerobear/product
-61988-401645.html.)

deformation, high corrosion resistance, lightweight, high temperature hardness, high flexural strength and fracture toughness (Industrial Tectonics Inc., http://www.itiball.com/silicon-nitride-balls.php). They have applications in aircraft braking assemblies, automotive industries, windmills, etc. Ceramic balls are appropriate for components that cannot have frequent maintenance. The inert nature of ceramic balls makes them suitable for biological applications also (Thomson, http://www.thomsonprecisionball.com/ceramic-balls.html).

The performance and life of ceramic balls depend on their surface integrity (Yuan et al., 2002). Manufacturing of highly precise ceramic balls and providing them with a high-quality surface finish are challenging tasks. Similarly, ceramic rollers also require a high-quality surface finish for their functional requirements.

The conventional method for finishing ceramic balls is grinding followed by V-groove lapping (Figure 11.2), which is also used for finishing of steel balls. The balls revolve around the pad, spinning continuously, gliding and rolling against the contacting surfaces of the pad. The process employs high loads (~10 N per ball), low polishing speeds (~50 rpm) and diamond abrasive particles. Hence, it is much time consuming (Umehara et al., 2006).

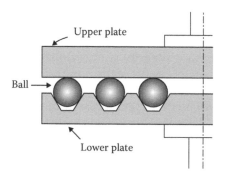

FIGURE 11.2
Schematic diagram of conventional V-groove lapping of ceramic balls.

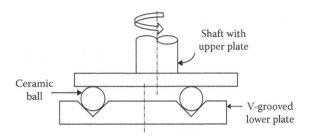

FIGURE 11.3
Eccentric V-groove mechanism.

The surface finish obtained is dependent on the rotation speed, spin speed and angle of spin axis of the ball and dimension of the V-groove. For traditional concentric V-groove lapping, the spin angle is constant and the grinding tracks on the ball surfaces are fixed circles. Therefore, a true sphere cannot be obtained unless the spin axis of the ball is changed. This situation can be changed by using a spin angle control method or setting the wheel and the V-groove eccentrically as shown in Figure 11.3. The eccentricity between the wheel and the V-groove also introduces skidding between the ball and the grinding wheel, leading to an increase in the material removal rate (MRR) and uniform surface finish (Zhang and Nakajima, 2000).

The lapping process has a high probability of surface defects under heavy loads. These surface defects act as nucleation sites for cracks, leading to brittle fractures and subsequent failure of the balls. In order to minimise cost, finishing time, surface damage, etc., an advanced well-controlled finishing process called 'magnetic fluid grinding' (also known as 'magnetic float polishing' [MFP]) was developed in Japan (Umehara and Kato, 1988).

11.2 The MFP Process

According to Rosenweig's ferrohydrodynamics theory (Rosenweig, 1966), a buoyant force acts on a non-magnetic body immersed in a magnetic fluid under the effect of a magnetic field. A magnetic fluid is a suspension of ferromagnetic particles and non-magnetic abrasive particles dispersed uniformly in a liquid medium as shown in Figure 11.4a. When the magnetic field is applied, the magnetic particles are accumulated close to the magnet, whereas the non-magnetic particles move away from the magnet and remain floated over the magnetic particles (Figure 11.4b).

Based on this principle, Tani et al. (1984) developed the MFP process. Abrasive particles are selected based on the surface finish requirements and workpiece material properties, and they are added as non-magnetic bodies into a magnetic fluid. A series of permanent magnets having opposite pole

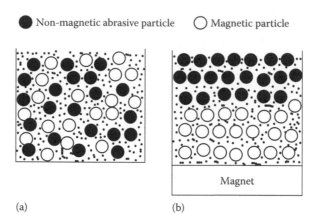

● Non-magnetic abrasive particle   ○ Magnetic particle

(a)   (b)

FIGURE 11.4
Schematic diagram of (a) uniform distribution of magnetic and non-magnetic particles and (b) separation of non-magnetic and magnetic particles under the presence of magnetic field.

direction side by side are placed at the bottom of the apparatus to provide the required magnetic field. When magnetic fluid is loaded with abrasive particles under the magnetic field, the abrasive particles float over the magnetic particles, as shown in Figures 11.4b and 11.5. The workpiece is plunged and rotated in the magnetic fluid, and the relative motion between the workpiece and the abrasive particles results in finishing of the workpiece surface. The MRR and finishing rate during the process can be controlled by varying the magnetic field strength, normal force and rotational speed of the workpiece.

The magnetic buoyant force of abrasive particles alone is very low and it is not sufficient to give an efficient finishing rate and the desired shape accuracy. As a solution to this problem, Umehara and Kato (1990) added an additional feature to the apparatus called 'float'. The finishing pressure with

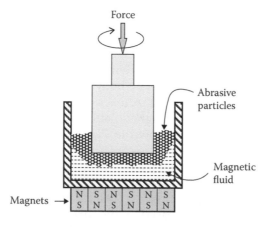

FIGURE 11.5
Finishing of cylindrical bar using magnetic fluid without float.

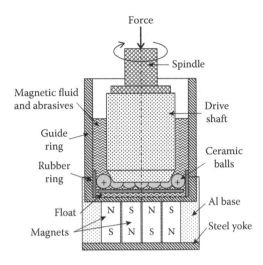

FIGURE 11.6
Finishing of ceramic balls using MFP. (From Sidpara, A., Jain, V.K., Magnetic float polishing: An advanced finishing process for ceramic balls, In *Introduction to Micromachining*, V.K. Jain, ed., Narosa Publishers, India, 2010.)

a float at a distance of 1 mm from the magnet is 20 times larger than that of without a float (Umehara, 1994). Figure 11.6 shows a schematic diagram of an MFP setup with a float. The ceramic balls are in contact with the float at the bottom, the chamber wall on the side and the drive shaft at some angle.

The salient features of the MFP process are as follows:

- Very high finish and accuracy can be obtained (Ra of ~4 nm; sphericity of 0.15–0.25 μm).
- Little or no surface damage to workpiece due to extremely low load.
- Faster process compared to conventional V-groove lapping.
- Workpiece surfaces can be processed to final finish in a single operation – no need to change the polishing equipment for roughing, semi-finishing and finishing stages.
- Low capital and running costs.

Zhang et al. (1995, 1998) suggested various modifications to the process to improve its efficiency. The use of a taper thrust float to increase the MRR proved to be effective. In this setup, the surface roughness is independent of the eccentricity. A further modification of the process (Figure 11.7) by Zhang and Nakajima (2003) for magnetic-fluid-based grinding of Si_3N_4 ceramic balls has the magnetic fluid sealed in the chamber beneath the float with a rubber membrane and provides a soft support. Grinding experiments of Si_3N_4 ceramic balls were carried out with four different types of supports such as gum sheet/magnetic fluid supports, steel sheet/magnetic fluid supports,

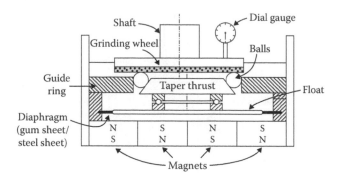

FIGURE 11.7
MFP with tapered thrust force.

TABLE 11.1

Comparison of Lapping and MFP for Finishing of Ceramic Balls

Parameters	Lapping	MFP
Abrasive particles	Diamond	B_4C, SiO_2, CeO_2, Al_2O_3 etc.
Load per ball (N)	10–100	0.5–1
Speed (RPM)	50	5000
Number of balls per batch	1000–5000	10–50

Source: Komanduri, R., *CIRP Ann. Manuf. Technol.*, 45, 509–514, 1996; Childs, T.H.C., Mahmood, S., Yoon, H.J., *Tribol. Int.*, 28, 341–348, 1995.

steel sheet/colloid fluid supports and steel sheet/air supports. Gum sheet/ magnetic fluid supports give the fastest improvement and the best sphericity during grinding amongst the four types of supports. Table 11.1 compares the conventional V-groove lapping and MFP for finishing of ceramic balls.

11.3 Process Parameters of MFP

The MFP is a tribo-chemo-mechanical polishing process where tribology, chemical action and mechanical abrasion between the magnetic fluid and the surface to be finished need to be investigated. Figure 11.8 shows the process parameters of MFP that decide the final quality of surface finish and sphericity.

11.3.1 Effect of Type, Size and Concentration of Abrasive Particles

Abrasive particles for MFP are of two types depending on their mechanical hardness and chemical activeness with the work material in the given

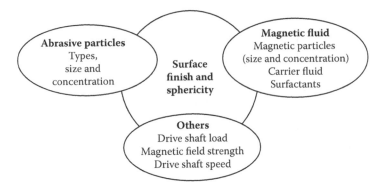

FIGURE 11.8
Process parameters of MFP.

environment. The first one predominantly has a mechanical polishing action and the other has a chemo-mechanical polishing action. Very fine diamond, boron carbide (B_4C) and silicon carbide (SiC) abrasive particles (which are harder than Si_3N_4 work material) are used for mechanical polishing with high MRRs. The material removal in this case is due to mechanical micro-fracture. Material removal by brittle fracture occurs at a microscale due to low depth of indentation, low polishing force, flexible float system and fine abrasive particles. As a result, the cracks generated at localised (top) surfaces are suppressed from propagating into the material. Consequently, subsurface damage is minimised. Final polishing of the Si_3N_4 balls using softer abrasive particles such as CeO_2 or Cr_2O_3 (that chemo-mechanically react with the Si_3N_4 work material) results in high-quality ceramic balls with superior surface finish. Figure 11.9 shows the properties of various abrasive particles (Jiang et al., 1998).

Under constant test conditions, it is observed that zirconium oxide (ZrO_2) and cerium oxide (CeO_2) are the most effective abrasive particles, followed

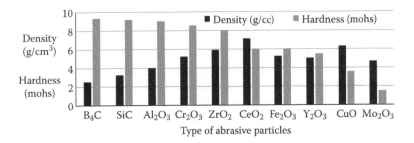

FIGURE 11.9
Properties of abrasive particles (B_4C – boron carbide, SiC – silicon carbide, Al_2O_3 – aluminum oxide, Cr_2O_3 – chromium oxide, ZrO_2 – zirconium oxide, CeO_2 – cerium oxide, Fe_2O_3 – iron oxide, Y_2O_3 – yttrium oxide, CuO – copper oxide, Mo_2O_3 – molybdenum oxide). (From Jiang, M., Wood, N.O., Komanduri, R., *Wear*, 220, 59–71, 1998.)

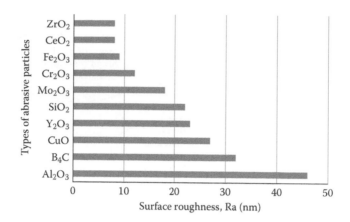

FIGURE 11.10
Effect of different abrasive particles on surface roughness.

by iron oxide (Fe_2O_3) and chromium oxide (Cr_2O_3), after 90 minutes of finishing time, as shown in Figure 11.10 (Yuan et al., 2002).

CeO_2 abrasive particles chemically react with Si_3N_4 balls, forming a SiO_2 layer over the ceramic balls. The hardness of this layer is 1/3 times that of the ceramic ball. This layer can be easily removed by subsequent mechanical polishing using softer abrasive particles without creating deep and long scratches. Chemical reactions between Cr_2O_3 and Si_3N_4 result in the formation of chromium silicate (Cr_2SiO_4) and chromium nitride (CrN) (Bhagavatula and Komanduri, 1996). As the abrasive particle size decreases, the depth and size of brittle fracture by each abrasive particle decreases, improving the surface finish and decreasing the MRR.

11.3.2 Effect of Magnetic Fluid Constituents

Magnetic fluid consists of fine ferromagnetic particles in a carrier fluid such as water or oil. Surfactants are added to reduce agglomeration of particles. The magnetic particles in the base fluid are colloidal due to Brownian motion in the presence of surfactants. It is observed that water as a carrier fluid is more effective than oil for finishing of ceramic balls with CeO_2. Oil film formed between the abrasive particle and workpiece prevents chemical reaction between them. But water-based slurry helps in the formation of a SiO_2 soft layer, which is easily removed by mechanical abrasion of abrasive particles with the workpiece.

$$Si_3N_4 + 6H_2O = 3SiO_2 + 4NH_3$$

$$Si_3N_4 + 6H_2O = 3SiO_2 + 2N_2(g) + 6H_2(g) \text{ when } T > 200°C$$

11.3.3 Effect of Field Strength, Drive Shaft Speed and Shaft Load

The magnetic field intensity should be consistent all over the chamber base in order to levitate the float uniformly and apply uniform pressure. Jiang and Komanduri (1997) studied the effect of shaft speed, shaft load and abrasive particle volume percentage on surface finish in MFP of silicon nitride (Si_3N_4) balls. It is observed that surface roughness reduces with an increase in shaft load and shaft speed but increases with an increase in abrasive particle concentration. Furthermore, it is found that the polishing force (shaft load) is most significant, followed by the polishing speed and then the abrasive particle concentration. Umehara et al. (2006) studied the effect of shaft speed, shaft load and abrasive particle volume percentage on surface finish, sphericity and MRR in MFP of large size/large batch silicon nitride (Si_3N_4) balls for hybrid bearing applications. It was observed that MRR increases with an increase in the polishing load, speed and abrasive particles volume percentage.

11.4 Analysis of MFP

It is difficult to analyse the forces acting on ceramic balls and the velocities involved during the MFP process exactly due to multiple and random contacts and motions. However, a generalised diagram of forces and velocities can be proposed, and it is shown in Figure 11.11 (Lee et al., 2009a).

Childs et al. (1994) developed a theoretical model of volume removal rate for sliding contact during MFP and it was developed based on abrasive wear law (Archard's wear law):

$$V = K W v / H,$$

where V is the volume of the material removed, K is wear coefficient, v is sliding velocity, W is contact load and H is hardness of the abraded material.

Considering the ball motion under the condition of rolling and sliding, a modified equation for the volume removal rate in the MFP was developed as:

$$V = 0.54 \left(\frac{K}{H} \right) \left(\omega_b \sin \beta + \Omega_f \right) F_m^{4/3} \left(\frac{r_b}{E} \right)^{1/3},$$

where ω_b is the angular velocity of the ball around its own axis and inclined at β to the horizontal, Ω_f is angular velocity of the float, F_m is magnetic force, r_b is ball radius, $\dfrac{1}{E} \cong \dfrac{1}{E_1} + \dfrac{1}{E_2}$, where E_1 and E_2 are the Young's moduli of the ball and float materials, respectively.

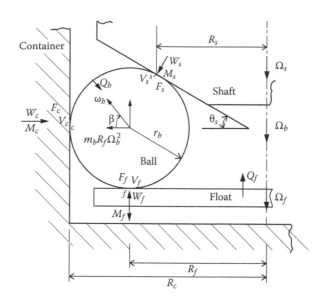

FIGURE 11.11
Force and velocity diagram of MFP (Ω_s, Ω_b and Ω_f are angular velocities of the shaft, ball, float, respectively; ω_b is angular velocity of the ball around its own axis and inclined at β to the horizontal; θ is cone slope of drive shaft; R_c, R_f and R_s are radial distances from the cell-centre line to the ball contact with the container, float and shaft respectively; F_c, F_f and F_s are friction force components due to drag at the contact points of ball with the container, float and shaft, respectively; M_c, M_f and M_s are the contact torque of the ball with the container, float and shaft, respectively; W_c, W_f and W_s are contact loads of the ball with the container, float and shaft, respectively and Q_b and Q_f are fluid drag torque on the ball and float, respectively).

According to Childs et al. (1994), skidding occurs between the balls and shaft because of higher shaft speeds. The lowest shaft speed at which skidding occurs depends on many factors – load, fluid viscosity and friction coefficient at contacts between the balls and the float, shaft and container wall. Childs et al. (1994), attributed the higher MRRs in the MFP to the higher sliding velocities when compared to V-groove lapping. However, Childs et al. (1994), model is based on a steady-state solution for the motion of the ball and the float and cannot address variable shaft speed and variable contact load conditions.

Uniform distribution of the grinding tracks over the ball surface is very important in achieving the roundness of balls during the grinding process. Hence, a dynamic analysis of the MFP system is essential. Uneven loading is a major cause of differential MRRs in MFP. These removal rates occur when skidding motions are generated between the balls and the shaft. The same skidding motions also cause grinding track marks on the surface of the balls. Lee et al. (2009a) carried out an investigation of the effects of the grinding load and shaft speed on grinding tracks with the help of a dynamic analysis of ball motion.

The sliding speeds V_s, V_c and V_f at the contacts of the ball with the shaft, container and float, respectively, can be derived from the cell dimensions and angular motions:

$$V_s = R_s (\Omega_s - \Omega_b) - r_b (\omega_{br} \cos \theta_s + \omega_{bz} \sin \theta_s), \tag{11.1}$$

$$V_c = R_c \Omega_b - r_b \omega_{bz}, \tag{11.2}$$

and

$$V_f = R_f (\Omega_b - \Omega_f) - r_b \omega_{br}. \tag{11.3}$$

The equations of motion for a ball and a float may be written by direct application of Newton's second law or by Lagrange's method as follows:

$$Q_{br} + M_c - M_s \sin \theta_s - r_b (F_s \cos \theta_s + F_f) + I_b \dot{\omega}_{br} = 0, \tag{11.4}$$

$$Q_{bz} + M_s \cos \theta_s + M_f - r_b (F_s \sin \theta_s + F_c) + I_b \dot{\omega}_{bz} = 0, \tag{11.5}$$

$$F_s - F_c - F_f - D_b - m_b \dot{\Omega}_{br} - 2m_b \Omega_b \dot{r} = 0, \tag{11.6}$$

$$W_s = \frac{W_f - m'g - m_b \ddot{z}_b}{\cos \theta_s}, \tag{11.7}$$

$$W_c = W_s \sin \theta_s + m_b r \Omega_b^2 + m_b \ddot{r}, \tag{11.8}$$

and

$$N(R_f F_f - M_f) - Q_f - I_f \dot{\Omega}_f = 0, \tag{11.9}$$

where, Q_{br} and Q_{bz} are r- and z-components of fluid drag torque on the ball respectively; I_b and I_f are moment of inertia for the ball and float, respectively; D_b is fluid drag force on a ball; m_b and m_f are mass of the ball and float, respectively; m' is the effective mass of the ball in the fluid and N is the number of balls in a cell.

If the inertia terms are set to zero, then Equations 11.4, 11.5 and 11.9 become identical with the steady-state motion obtained by Childs et al. (1994). Three cases are considered for the contact status of the ball with the container, float and shaft in Figure 11.12.

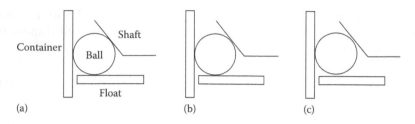

FIGURE 11.12
Contact status of the ball with the container, float and shaft. (a) Three-point contact – ball in contact with the shaft, the container and the float. (b) Two-point contact – ball in contact with the container and the float. (c) One-point contact – ball in contact with the container.

Dynamic analysis of the ball motion by Lee et al. (2009a,b) arrived at the following conclusions:

(a) The effects of variable shaft speed and variable contact load on the ball spin angle and area covered by the grinding tracks are small when the ball remains in contact with the shaft during finishing.

(b) Intermittent separation results in variation in the ball spin angle as the ball loses the driving force from the shaft.

(c) When the ball separates from the shaft, the combined effect of variable shaft speed and the variable contact load on the area covered by the grinding tracks is different from that of the variable contact load alone. As a result, it is possible to grind a truly spherical surface by MFP.

(d) When the intermittent separation occurs at the geometrical imperfections on the ball orbit, it causes a large oscillation in the ball spin angle and the ball spin speed. As a result, the effect of the imperfections in the ball orbit on the area covered by the grinding tracks is larger than that of the ball geometry.

(e) Ball–ball contacts cause a large oscillation in the ball spin angle, resulting in a uniform distribution of the grinding tracks. Hence, ball–ball contacts have a significant role in achieving a uniform distribution of the grinding tracks.

11.5 MFP for Ceramic Rollers

Bearings with ceramic rollers have the advantages of a higher degree of rigidity, higher speed capability, reduced centrifugal and inertial forces within the bearing, reduced frictional heat, less energy consumption and extended bearing service life (SKF, http://www.skf.com/group /products/bearings-units-housings/super-precision-bearings/cylindrical -roller-bearings/design-and-variants/hybrid-bearings/index.html). Ceramic rollers manufactured from silicon nitride are tougher and have a service life of up to 20 to 30 times that of welding rollers made from steel. These ceramic rollers have high thermal shock resistance, improved precision, extreme hardness, high compressive strength and high toughness (CeramTec, https://www.ceramtec.com/welding-rollers). Ceramic or ceramic-coated rollers (Figure 11.13) are widely used in plastics, leather, paper, textile, packaging and printing, steel and other industries. Compared with conventional hard chrome plating, thermal spray ceramic rollers have the advantages of high temperature resistance, abrasion resistance and erosion resistance (Murata, http://www.murataroll .com/en/products.asp?id=7).

Umehara and Komanduri (1996) successfully performed MFP of ceramic rollers. Figure 11.14 shows the details of a roller polishing arrangement. Each magnet in the magnetic ring A is magnetised in the radial direction. Magnetic ring A supports the float, as well as concentrates the abrasive particles around the centre of the chamber. Each magnet in magnetic ring B is magnetised in the longitudinal direction. Magnetic ring B pushes the float in the longitudinal direction with the help of a magnetic buoyant force. A roller holder holds the roller in a way to obtain high sliding velocity between the roller and the drive shaft leading to higher MRR and uniform surface finish.

(a)

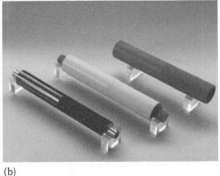
(b)

FIGURE 11.13
(a) Ceramic welding rollers (Courtesy of CeramTec, https://www.ceramtec.com/welding -rollers.) and (b) Ceramic rollers for paper and textile industries. (Courtesy of Murata, http:// www.murataroll.com/en/products.asp?id=7.)

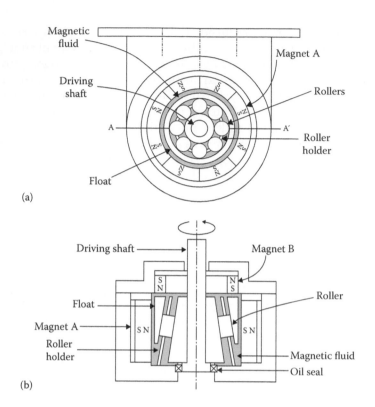

(a)

(b)

FIGURE 11.14

MFP equipment for ceramic rollers: (a) top view and (b) side section view through line A-A′ of (a). (From Umehara, N., Komanduri, R., *Wear*, 192, 85–93, 1996.)

Question Bank

1. Magnetic float polishing is generally used for
   (a) Fabrication of ceramic balls
   (b) Finishing of ceramic balls
   (c) Finishing of metallic balls
   (d) Fabrication of metallic balls

2. Ceramic balls are generally made of
   (a) Silicon nitride
   (b) Aluminium alloys
   (c) Stainless steel
   (d) Titanium alloys

3. Magnetic float polishing works on the principle of
   (a) Magnetic levitation force
   (b) Gravitational force
   (c) Newton's law
   (d) Centrifugal force

4. Ceramic balls are in contact with
   (a) Float at the bottom
   (b) Chamber wall on the side
   (c) Drive shaft at some angle
   (d) All of these

5. State the important properties of silicon nitride balls.

6. Write a few applications of ceramic balls.

7. What are the problems of conventional V-groove lapping during ceramic ball finishing?

8. How is the magnetic float polishing better than the conventional V-groove lapping?

9. What are the abrasive particles used for finishing and which are the most preferred abrasive particles?

10. What is the role of a surfactant in ferrofluid?

11. Explain the principle of magnetic levitation force.

12. What are the advantages of use of float in the MFP process?

13. Draw a schematic diagram of the MFP process and label its important components.

14. What are the important process parameters of the MFP process?

15. Explain schematically the force and velocity components of the MFP process.

16. What are the major issues of the MFP process?

17. What are the advantages of eccentric V-groove lapping over conventional V-groove lapping process?

18. Explain the material removal mechanism in MFP.

19. Explain the effect of shaft speed and shaft load on the surface finish achieved during MFP.

20. Draw a schematic diagram explaining MFP of ceramic cylinders.

21. Explain how a taper thrust float is better than conventional float in MFP.

References

Bhagavatula, S. R., Komanduri, R. (1996) *Philosophical Magazine*, A-74 (4), 1003–1017.

CeramTec. Increased process reliability in longitudinal tube and pipe welding. https://www.ceramtec.com/welding-rollers.

Childs, T. H. C., Mahmood, S., Yoon, H. J. (1994) The material removal mechanism in magnetic fluid grinding of ceramic ball bearings. *Proceedings of the Institution of Mechanical Engineers, Part B: Journal of Engineering Manufacture*, 208(1), 47–59.

Childs, T. H. C., Mahmood, S., Yoon, H. J. (1995) Magnetic fluid grinding of ceramic balls. *Tribology International*, 28(6), 341–348.

http://www.directindustry.com/prod/cerobear/product-61988-401645.html.

Industrial Tectonics Inc. http://www.itiball.com/silicon-nitride-balls.php.

Jiang, M., Komanduri, R. (1997) Application of Taguchi method for optimization of finishing conditions in magnetic float polishing (MFP). *Wear*, 213, 59–71.

Jiang, M., Wood, N. O., Komanduri, R. (1998) On chemo-mechanical polishing (CMP) of silicon nitride (Si_3N_4) work material with various abrasives. *Wear*, 220, 59–71.

Komanduri, R. (1996) On material removal mechanisms in finishing of advanced ceramics and glasses. *CIRP Annals–Manufacturing Technology*, 45(1), 509–514.

Lee, R., Hwang, Y., Chiou, Y. (2009a) Dynamic analysis and grinding tracks in the magnetic fluid grinding system. Part, I. Effects of load and speed. *Precision Engineering*, 33, 81–90.

Lee, R., Hwang, Y., Chiou, Y. (2009b) Dynamic analysis and grinding tracks in the magnetic fluid grinding system Part II. The imperfection and ball interaction effects. *Precision Engineering*, 33, 91–98.

Murata. Ceramic mirror finish roller, satin finish roller. http://www.murataroll.com/en/products.asp?id=7.

Rosenweig, R. E. (1966) Fluid magnetic buoyancy. *AIAA Journal*, 4(10), 1751–1758.

Sidpara, A., Jain, V. K. (2010) Magnetic float polishing: An advanced finishing process for ceramic balls, In *Introduction to Micromachining* (Ed. V. K. Jain), Narosa Publishers, India.

SKF. Hybrid bearings. http://www.skf.com/group/products/bearings-units-housings/super-precision-bearings/cylindrical-roller-bearings/design-and-variants/hybrid-bearings/index.html.

Tani, Y., Kawata, K., Nakayama, K. (1984) Development of high-efficient fine finishing process using magnetic fluid. *Annals of the CIRP*, 33(1), 217–220.

Thomson. Thomson precision balls. http://www.thomsonprecisionball.com/ceramic-balls.html.

Umehara, N. (1994) Magnetic fluid grinding – A new technique for finishing advanced ceramics. *Annals of the CIRP*, 43(1), 185–188.

Umehara, N., Kato, K. (1988) A study on magnetic fluid grinding – 1st report: The effect of the floating pad on removal rate of Si_3N_4 balls. *Transaction of The Japan Society of Mechanical Engineers (JSME)*, 54, 1599–1604 (in Japanese).

Umehara, N., Kato, K. (1990) Principles of magnetic fluid grinding of ceramic balls. *International Journal of Applied Electromagnetics in Materials*, 1, 37–43.

Umehara, N., Komanduri, R. (1996) Magnetic fluid grinding of HIP-Si$_3$N$_4$ rollers. *Wear*, 192, 85–93.

Umehara, N., Kirtane, T., Gerlick, R., Jain, V. K., Komanduri, R. (2006) A new apparatus for finishing large size/large batch silicon nitride (Si$_3$N$_4$) balls for hybrid bearing applications by magnetic float polishing (MFP). *International Journal of Machine Tools & Manufacture*, 46, 151–169.

Yuan, J. L., Lu, B. H., Lin, X., Zhang, L. B., Ji, S. M. (2002) Research on abrasives in the chemical-mechanical polishing process for silicon nitride balls. *Journal of Materials Processing Technology*, 129, 171–175.

Zhang, B., Nakajima, A. (2000) Spherical surface generation mechanism in the grinding of balls for ultraprecision bearings. *Proceedings of the Institution of Mechanical Engineers, Part J-Journal of Engineering Tribology*, 214, 351–357.

Zhang, B., Nakajima, A. (2003) Dynamics of magnetic fluid support grinding of Si$_3$N$_4$ ceramic balls for ultraprecision bearings and its importance in spherical surface generation. *Precision Engineering*, 27, 1–8.

Zhang, B., Umehara, N., Kato, K. (1995) Effect of the eccentricity between the driving shaft and the guide ring on the behavior of magnetic fluid grinding of ceramic balls. *Journal of the Japan Society for Precision Engineering*, 61(4), 586–590.

Zhang, B., Uematsu, T., Nakajima, A. (1998) High efficiency and precision grinding of Si$_3$N$_4$ ceramic balls aided by magnetic fluid support using diamond wheels. *JSME International Journal, Series C*, 41(3), 499–505.

Section V

Hybrid Nanofinishing Processes

12

Chemomechanical Magnetorheological Finishing (CMMRF)

Prabhat Ranjan,[1] **R. Balasubramaniam,**[1] **Vinod K. Suri**[2] **and Vijay K. Jain**[3]

[1]*Bhabha Atomic Research Center Bombay, Mumbai, India*

[2]*Bhabha Atomic Research Center Bombay and MGM Institute of Health Sciences, Mumbai, India*

[3]*Department of Mechanical Engineering, Indian Institute of Technology Kanpur, Kanpur, India*

CONTENTS

12.1 Introduction

Surface roughness is one of the important parameters to enhance the performance and efficiency of engineering components. Nanofinishing is a manufacturing process that reduces surface roughness upto a few nanometres and sub-nanometres in some cases. The process has impact on various fields of engineering, science and biomedical applications like precision bearings, biomedical chips and implants, general optics, laser optics, X-ray optics, microelectronic fabrication, micro-fluidics, etc., to improve their functionality.

Recent applications in the engineering and biomedical fields need a wide variety of advanced materials to meet improved functionalities. Nanofinishing of these materials poses challenges to researchers and industries. To develop nanofinishing technology for such advanced materials, various techniques are developed and implemented towards the realisation of products at the industrial level. The following properties of workpiece materials pose challenges for nanofinishing processes:

1. *Hardness*: Workpiece hardness resists the indentation of abrasive particles in the workpiece material thereafter removal of the material during the finishing process, and it reduces the finishing rate. To overcome this challenge, hard and stiff tools with the appropriate mechanism to soften the workpiece surface prior to material removal are essential.

2. *Ductility and malleability*: This property inhibits the use of abrasive-based finishing processes and leads to defects like scratches, pits, etc., on the finished surface. To overcome this problem, generally, path controlled processes that are deterministic in nature are used. A diamond turn machining is one of such processes employed to generate nanofinished surface on ductile materials.

3. *Brittleness*: Fracture toughness [$K_{IC} = \sigma_c(\pi c)^{0.5} = (2\gamma E)^{0.5}$ as per Griffith's theory, where σ_c is critical stress, c is the crack length, γ is the surface energy per unit area and E is the Young's modulus of elasticity] dictates the generation of surface and sub-surface damages. To avoid such surface damages, ductile mode of machining is preferred while finishing brittle materials.

Nanofinishing processes are carried out by various methods, basically mechanical abrasion, magnetic field assisted mechanical abrasion, chemical reaction assisted mechanical abrasion and ion beam energy assisted machining. Of these, chemomechanical polishing (CMP) is one of the nanofinishing processes that work on the principle of chemical assisted mechanical abrasion. In this process, a chemical reaction is used for creating a passivated superficial layer on the surface followed by random abrasive polishing to remove this passivated layer (softer than parent material) using a polishing pad. CMP is extensively used in integrated circuit (IC) industries to generate surface finish of atomic level. However, CMP is not capable of finishing geometries other than circular and flat wafers. When a magnetic field along with magnetorheological fluid (MR fluid) is used to impart stiffening force (holding and cutting forces) to the abrasive particles, as in magnetorheological finishing (MRF), the process gets a higher level of flexibility and it can be extended to 3D components as well as to any geometry. When both CMP and MRF are combined, a new hybrid finishing process, named chemomechanical magnetorheological finishing (CMMRF), is obtained (Jain et al., 2010; Ranjan et al., 2013). This process has the capability to finish any material, either ductile or brittle, in the scale of few nanometres to sub-nanometres.

In general, the nanofinished surface will have the following improved surface properties:

1. *Wettability*: Wettability of the nanofinished surfaces is improved due to the increased effective contact area. Improved wettability has applications in micro-contact printing, bio-fouling, DNA immobilisation, cell growth and tissue engineering.

2. *Bearing ratio*: Surface roughness with a negatively skewed profile consists of a large number of valleys as compared to peaks. These surfaces have oil- or fluid-retaining capability and find applications in the field of automobiles, tribology, biomedical implants and cutting tools.

3. *Surface profile with atomic level of surface finish*: Surfaces that are atomically aligned (local radius of curvature tends to infinity at microscopic scale, i.e. $1/r \rightarrow 0$) exhibit optical properties with least scattering phenomena. The ratio of 'surface Gibb's free energy' to 'volume Gibb's free energy' becomes inversely proportional to surface curvatures (Mullin, 2001), which becomes extremely low (tends to zero). Hence, atoms on such surfaces lessen their surface Gibb's free energy to make it stable and inert, and hence, the corrosion resistivity of the surface increases. Moreover, atomically finished surfaces mitigate micro-cracks, micro-valleys, nano-valleys and nano-peaks, which enhance other mechanical properties like hardness, fatigue strength, wear resistance and low friction.

12.2 Mechanisms

CMMRF has been developed for surface finishing of engineering materials. The process combines the essential aspects of the CMP and MRF processes. Chemical reactions associated with CMP are used to improve the surface finish, whereas MR fluid and magnets of the MRF setup are used to control the magnitude of the abrading forces acting on the workpiece for nano-abrasion as well as to control flexibility for finishing non-planer surfaces.

12.2.1 Nano-Abrasion

In abrasive-based surface finishing by ductile mode abrasion, under precisely controlled abrasion in which material removal rate becomes a few nanometres per 'stroke of abrasion', surface finish (*Ra* value) is achieved of the order of a few nanometres. Such precise abrasion is called nano-abrasion. A schematic of a typical ductile mode nano-abrasion is presented in Figure 12.1. Material removal in this mode occurs due to shear action, which basically depends on the mechanical properties of the workpiece.

Nano-abrasion can be further classified as

1. Mechanical abrasion
2. Atomic scale abrasion

12.2.2 Mechanical Abrasion

In this process, material removal takes place by shear deformation in the form of small chips, as illustrated in Figure 12.1, and the chip does not get detached completely from the parent material.

When two solid objects/particles (small in size) like an abrasive particle and the chip come closer, they are attracted and form secondary bonding. The secondary bond formation depends on the attractive force due to the atomic interaction force between the abrasive particle and the chip with

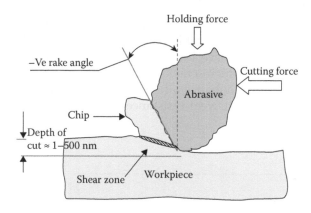

FIGURE 12.1

Ductile mode nano-abrasion: −ve rake angle applies compressive force on cutting zone. This mode is able to generate surface finish without crack layers and with un-cut chip thickness of few nanometres.

respect to their gravitational force; hence, the size of the chip becomes vital while forming a strong bonding. The inter-atomic force depends on shape as well as size (Wang and Chen, 2012). In general, the magnitude of inter-atomic forces varies on the scale of micro-Newton forces that can be investigated by the surface forces apparatus. Nano-Newton forces can be easily measured by atomic force microscopy, and femto-Newton forces are measured by a method using internal reflection microscopy and looking at particles under Brownian motion (Fewkes et al., 2015). These inter-atomic forces are caused by a secondary bonding force or van der Waals bonding force. Generally, they are weak in comparison to the primary force (ionic, covalent or metallic). Secondary bonding forces arise from atomic or molecular dipoles. The secondary bonding force between two dissimilar surfaces can be described by the Lennard-Jones potential, which deals with the interaction between two atoms (Callister, 2000). This force depends on inter-atomic distance (inter-atomic separation), as shown in Figure 12.2, wherein the repulsive force arises between atoms at a short range and the attractive force occurs at a longer range. The repulsive force is caused by Paulie exclusion principles, which inhibit overlapping of the two electron clouds (to avoid same quantum numbers of different atoms). The attractive force is called the van der Waals force. It arises from the distortion of the electron cloud of one molecule or atom by the presence of the other.

The secondary bonding force plays a vital role in removing the chip, as shown in Figure 12.3. The secondary bonding force between the abrasive

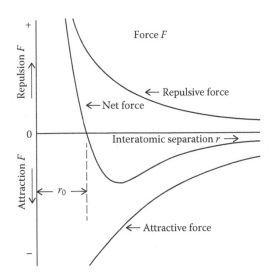

FIGURE 12.2
The repulsive force dominates at $r < r_O$ and the attractive force becomes vital at $r > r_O$. At equilibrium, inter-atomic separation becomes equal to r_O (0.3–0.5 nm), in which net force becomes zero. (From Raghavan, V., *Material Science and Engineering*, 5th ed., PHI Learning Private Limited, Delhi, India, 2013.)

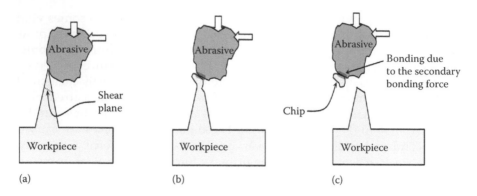

FIGURE 12.3
Abrasion stages for material removal are presented from left to right as follows: (a) interaction of abrasive particle on peak of workpiece irregularity in which shear plane is indicated with dotted line; (b) partially sheared off chip over the peak. In this stage, the abrasive particle and chip are joined due to the secondary bonding force and (c) the shear plane has become thinner and the material resisting force is less than the secondary bonding force between abrasive and chip, which helps in the removal of the chip.

and the chip depends on the shape and size of the abrasive particle as well as the chip, and the chip is removed if the secondary bonding force is higher than material resisting force on the shear plane.

Nano-abrasion on rough surface occurs when shear stress in the shear zone becomes more than the yield shear stress. There are factors that help in efficient abrasion:

- Cutting force and holding force: These come from the stiffness or hardness of the polishing medium, which transfers force from the machine tool to abrasive particles during relative movement between the polishing medium and the workpiece. For nano-abrasion, cutting and holding forces need to be precisely controlled. Hence, soft and flexible finishing medium or polishing pad is recommended.
- Shape of the surface profile: The shape and size of local features (at microscopic level) on the surface profile exhibit shear stress on the shear zone during nano-abrasion. A sharp and triangular-shaped peak on the surface shows more shear stress under the same working parameters like depth of cut and cutting and holding force; this mechanism is presented in Figure 12.4.

Material resisting force (Fsr) along the shear plane and resultant force (R) acting on the workpiece can be written as

$$Fsr = \tau_y wb \text{ and } R = (Fc^2 + Fv^2)^{0.5},$$

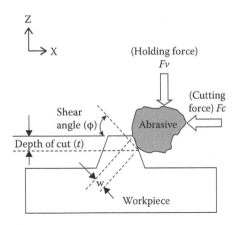

FIGURE 12.4
Schematic for mechanical abrasion.

where $w = t/\sin \phi$; b = width of abraded material along the y axis; Fc = cutting force; Fv = holding force;

$$\text{Force acting along the shear plane} = Fsa = R\cos\{\tan^{-1}(Fv/Fc) + \phi\};$$

$$\text{Shear stress developed along the shear plane } (\tau_\phi) = Fsa/wb \quad (12.1)$$

$$\text{Material removal condition or criterion for} \\ \text{shear deformation can be } \tau_\phi \geq \tau_y. \quad (12.2)$$

$$\text{Therefore, } \tau_y wb \leq (Fc^2 + Fv^2)^{0.5} \cos\{\tan^{-1}(Fv/Fc) + \phi\}. \quad (12.3)$$

Otherwise, t will be reduced due to sliding and rolling of the abrasive particles with respect to the work surface and it will reduce Fsr until Equation 12.2 is satisfied. Hence, material removal takes place under precise abrasion until sharp features on the surface become smooth.

Example 12.1

Calculate the depth of material removal per stroke by a single alumina abrasive particle during mechanical-based abrasion on brass work material. Use the following parameters for the calculation: cutting force = 10 μN; vertical holding force = 10 μN; shear angle = 15°; shear strength of brass = 80 MPa; abrasive diameter = 10 μm; assume width of abraded material = 0.8 times of abrasive diameter.

SOLUTION

After applying the given parameters to Equation 12.3, the equation can be modified as follows:

$$80 \times 10^6\, wb = 10^{-6}(10\sqrt{2})\cos(\tan^{-1}1 + 15) = 10^{-6}(10\sqrt{2}) \cos 60° = 10^{-5}/\sqrt{2}.$$

After placing $b = 0.8 \times 10$ µm, W can be computed:

$$W = 10^{-6}/(64\sqrt{2})m = 1/90 \text{ µm, whereas } t = W \sin \varphi = 2.88 \text{ nm.}$$

Hence, material removal per stroke due to single abrasive particle is 2.88 nm.

12.2.3 Atomic Scale Abrasion

There is another mechanism of nano-abrasion in which the abrasive particle reacts with the workpiece surface and forms chemical bonding. This chemical bond becomes stronger than the chemical bond between the atoms of the parent material. This mechanism is responsible for creating cold welding of an abrasive with the work surface. The chip in the form of 'atomic cluster' is dislodged from the workpiece due to relative movement between workpiece and abrasive particles, as shown in Figure 12.5.

Note: There are few 'material combinations' of 'abrasive and work material' in which they interact to form new chemical bonds towards 'atomic scale abrasion'. Example: Silica as abrasive reacts with various engineering materials like silicon, iron, sapphire, etc. Si-O-Si is a newly formed chemical bond between silica abrasive and silicon as a workpiece. This bond has higher strength than an Si-Si bond because of the higher electronegativity of oxygen (3.44) as compared to silicon (1.90).

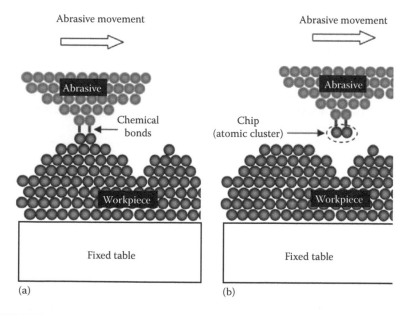

FIGURE 12.5
(a) Chemical reaction and bonding between abrasive particle and workpiece and (b) dislodgment of 'cluster of atoms' from peak of the workpiece.

The extent of mechanical and chemical interaction depends on the 'potential energy per atom' of the surface atoms. Low potential energy per atom brings physical, chemical and mechanical stability or inertness of the specific atom. This potential can be formulated and conceived using suitable atomic models like the Morse Potential, the Born Mayer Potential, the Tersoff Potential or the Embedded Atom Potential. The Morse Potential, suitable for cubic-crystalline metals, can be presented as follows:

$$V_{ij} = D_0 \left[e_{ij}^{-2\alpha(r-r_0)} - 2e_{ij}^{-\alpha(r-r_0)} \right] = -D_0\, e_{ij}^{-2\alpha(r-r_0)} \left[2e_{ij}^{\alpha(r-r_0)} - 1 \right] \quad r < r_c , \quad (12.4)$$

where V_{ij} = Morse potential on the ith atom due to jth atom, r = inter-atomic distance, r_c = cut-off distance (≈ 2 nm) and D_0, α, and r_0 are constants determined on the basis of physical properties of the material.

On the basis of the previous potential function, the total potential on the ith atom can be determined by the addition of potentials due to their neighbour atoms as follows:

$$V_i = -\sum_j D_0\, e_{ij}^{-2\alpha(r-r_0)} \left[2e_{ij}^{\alpha(r-r_0)} - 1 \right]. \quad (12.5)$$

V_i can be expressed as P.E./atom computed using molecular dynamics simulation (MDS). Figure 12.6 was generated using an MDS in which the effects of sharing atoms on P.E./atom and surface adsorptions are presented.

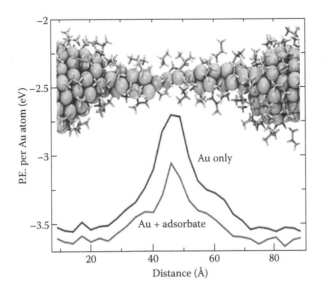

FIGURE 12.6
Effect on potential per atom and adsorption due to variation of sharing atoms. (From French, W.R., Iacovella, C.R., Cummings, P.T., *J. Phys. Chem.*, 115, 18422–18433, 2011.)

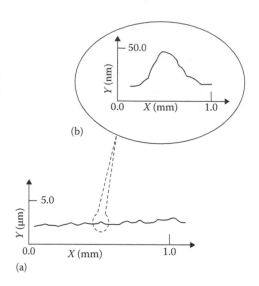

FIGURE 12.7
(a) Surface profile and (b) magnified view of local portion on the surface profile.

On the basis of previous studies, nano-abrasion can be concluded according to the 'shape and size at microscopic level on the workpiece surface profile' to maintain material removal or polishing of the engineering materials.

According to Equation 12.1, a sharp peak subjected to a small depth of cut will develop higher shear stress (τ_φ), and accordingly, a sharp peak, as shown in Figure 12.7b, is amenable for nano-abrasion. Apex atoms of a sharp peak as shown in Figures 12.5 and 12.6 become highly unsaturated as compared with core atoms. These unsaturated atoms on sharp edges of the surface profile are highly susceptible for atomic level abrasion as well.

Hence, both aspects of nano-abrasion like mechanical abrasion and atomic scale abrasion depend on shape and size of surface profile at microscopic scale. Thus, nano-abrasion has the ability to remove sharp features on surface profiles; the abrasion rate (material removal) tends to zero or stops, as shown in Figure 12.8. However, further abrasion needs more abrasion force (indentation as well as cutting force) which may lead to damage the surface integrity in form of scratches, degenerated layers, nano-crack layer, etc.

12.2.4 Advantages of Nano-Abrasion

The following advantages are associated with nano-abrasion assisted finishing processes:

1. Surface finish with asymmetric surface profile that shows negative skewness or a convex surface profile. These surfaces pose a high bearing ratio, which is suitable for the automobile and bearing industries.

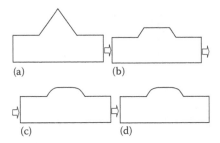

FIGURE 12.8
(a) initial surface topography at microscopic scale, (b) plateau formation after few cycles of nano-abrasion, (c) removal of all sharp features after many cycles of nano-abrasion and (d) no significant change on the surface but existence of stable feature (convex geometry) yields saturated or critical surface roughness.

2. This process also improves the optical properties of the surface.
3. It yields an atomic level of surface finish if the initial surface profile is started with nanometric variations on the surface.
4. This process needs less variety of consumables and tool.
5. It does not create side effects like contamination and damages on the surface.

12.2.5 Demerits of Nano-Abrasion

This process removes material at the nanometric scale with a mechanical abrasion technique. Hence, there are a few demerits that are associated with the process:

1. Extremely low polishing rate amongst all finishing processes except focus ion beam machining.
2. Poor 'critical surface roughness' if the initial surface has a very poor surface finish.
3. Excess abrasion forces induce surface and sub-surface damages like scratches, digs, cracks, material integrity (metallurgical changes along depth), etc.

To progress further abrasion in a tactile way, it is recommended to dissolve or make the stable features (smoothened peak) mechanically softer than the workpiece material. This technique reduces the stable region, and the roughness height is reduced accordingly with minimal defects on the final finished surface.

12.2.6 Chemical Reaction on Rough Surfaces

The P.E./atom of surface atoms themselves is different from that of atoms from within the core. This difference in a distinct energy condition (elevated

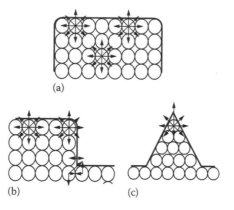

FIGURE 12.9
Schematic representation of field of forces on surfaces of different shapes: (a) plane surface, (b) edge and (c) corner. (From Hebda, M., Wachal, A., *Tribology*, WNT, Warsaw, Poland, 1980.)

energy) of the surface atom enhances adsorption activity (Figure 12.6). There are various types of surfaces as per the variation on atomic arrangement, as shown in Figure 12.9. Atoms at the surfaces of solids have a very limited freedom of movement. Saturation of the forces of adhesion between them depends on other atoms in their vicinity. A smaller number of sharing atoms reduces adhesion forces, and hence, the surface energy becomes higher. The microscopic feature on the surface, like Figure in 12.9c, exhibits very high P.E./atom, and this is why the atom becomes highly unstable and suitable for vigorous chemical reaction with surrounding species.

The properties of surface atoms are different from those of interior atoms or molecules, due to fewer bonds linking to their nearest neighbour atoms or molecules as compared with their interior counterpart. However, chemical potential is also dependent on the radius of the curvature (generated by nano-abrasion) of a surface, which enhances the rate of chemical reaction or depth of chemically passivated layer. The relationship between the chemical potential and the surface curvature is illustrated as follows.

According to Fick's law of diffusion, it is considered that the chemical from a fluid transfers to a spherical solid particle, as shown in Figure 12.10, in which dn atoms from the fluid are transferred to a particle or surface with a radius of R. The volume change in spherical particle, dV, is equal to the atomic volume, Ω, times dn as follows (Cao, 2004):

$$dV = 4\pi R^2 dR = \Omega dn. \tag{12.6}$$

The change in chemical potential per atom on the surface ($d\zeta$) is

$$d\zeta = \gamma dA/dn = 2\gamma\Omega/R = 2(K.E. + V_i)\rho_a\Omega/R, \tag{12.7}$$

where ρ_a = atomic surface density and γ = surface energy per unit area.

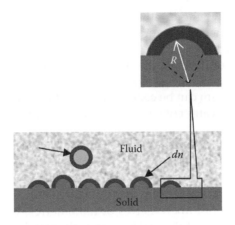

FIGURE 12.10
Schematic shows surface passivation on a solid surface as well as particles by transferring *dn* atoms from the fluid's chemical to the surface of solid materials. Dark regions are passivation layers on particle and the solid surface. (From Cao, G., *Nanostructures and Nanomaterials: Synthesis, Properties & Applications*, Imperial College Press, 2004.)

As per the Arheneous theory, the change in chemical potential will define the intensity of the chemical reaction rate. Hence, rate of chemical reaction on the work surface would be proportional to $\left(e^{\frac{d\zeta - Ea}{kT}} \right)$. Thus,

$$\text{depth of passivation layer} = \int k'e^{\frac{d\zeta - Ea}{kT}}\, dt = \int k1 e^{\frac{d\zeta}{kT}}\, dt = \int k1 e^{\alpha \frac{(K.E._i + V_i)}{R}}\, dt,$$

(12.8)

where *Ea* and *K*1 are the activation energy and constant for chemical reaction kinetics, respectively, which will depend on chemical composition and temperature. $K.E._i$ and V_i are the kinetic energy and the potential energy of the *i*th atom on the work surface, respectively. *k* is Boltzmann constant, α is the dimensional constant and *R* is the local radius of curvature on the work surface.

Initially, sharp peaks on the surface will get chemically reacted and removed as 'sharp peak' intensifies the chemical reaction. After removal of sharp peaks, the P.E./atom on the surface would be approximately uniform throughout the surface atoms. Hence, the depth of the passivation layer will be modified as per the following equation:

$$\text{depth of passivation layer} = \int k1 e^{\frac{\alpha 1}{R}}\, dt,$$

(12.9)

where $\alpha 1 = \alpha(K.E. + V_i)$, which is another constant.

Equation 12.9 shows the depth of passivated layers, a function of work material property, chemical composition, temperature and radius of curvature on the work surface.

As illustrated in Figure 12.10, the chemical (the fluid) attacks vigorously on highly irregular surface such as a convex surface with a small radius of curvature. This mechanism can be conceived using Equation 12.9. An extremely irregular surface indicates convexity with $R \to 0$, which yields an extremely high reaction rate.

The chemical reaction converts micro and nanometric features into a soft passivation layer, which is amenable for mechanical abrasion. The rate of chemical reaction on an uneven surface is dependent on the size of the local variation (radius of curvature) of the surface. The convex feature on the surface is of the order of micrometres or nanometres, which is subjected to chemical reaction. This is followed by a suitable abrasion process that shears off the convex features and flattens it up to atomic scale. As $R \to \infty$ a further chemical reaction is stopped (as per Equation 12.9), and hence, mechanical abrasion is also mitigated, as shown in Figure 12.11.

The passivated layers need to be removed by mechanical abrasion such that the depth of abrasive penetration is equal to the thickness of the passivated layer. There are a few techniques that are being used for mechanical abrasion as follows:

1. Mechanical abrasion using a soft lapping pad: This is also known as abrasive-free polishing. In this process, a lapping pad swipes or removes the soft passivated layer without inducing any damages (stress, scratches or cracks) on the surface. However, pad

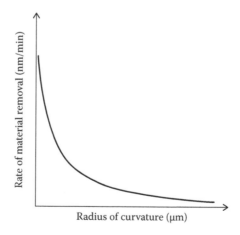

FIGURE 12.11
Material removal rate due to chemical reaction with varying radius of curvature on the surface profile. This plot was prepared on the basis of Equation 12.9.

degradation with time and poor polishing rate are deficiencies of the process.

2. Mechanical abrasion using an abrasive assisted lapping pad: This is a kind of three-body abrasion. This process removes the passivated layer with better abrasion rate.

3. Hydrodynamic-based abrasion: In this process, hydrodynamic pressure transfers in the form of abrasion force on the work surface through the abrasive.

4. Magnetorheological-based abrasion: Magnetic force is transferred from the magnetic fluid to the abrasive to abrade and removes the passivated layer. This process exhibits a better material removal rate as well as control on abrasion forces.

There are a few aspects that are needed for efficient abrasion to generate surface finish of atomic scale.

1. *Micro-patterns on polishing pad*: This is needed to hold and transfer the holding as well as cutting force on the abrasive particle. This feature is available on a micro-textured lapping pad and magnetically stiffened MR fluid (magnetic fluid). In case of micro-textures of the lapping pad, they are degraded with time due to abrasion effect and chemical reaction on the pad, which need to be conditioned after some extent of degradation.

2. *Flexible pad*: Flexibility on a polishing pad is required to finish three-dimensional (3D) surfaces like freeform surface. A soft lapping pad and magnetic fluid are eligible media to this requirement.

3. *High surface area on polishing pad*: This aspect is needed to enhance the secondary bonding force between the debris/chip (which is generated during the finishing process) and the polishing pad. During nano-abrasion, chips are formed and removed, as demonstrated in Figure 12.3b and c. This phenomenon is possible with a magnetically stiffened magnetic fluid in which soft spherical micro-particles form the polishing pad with high surface energy. The magnetically stiffened polishing pad exhibits better flexibility (ability to be deformed as per workpiece surface topography at macroscopic as well as microscopic scale). This flexibility increases the possibility to access debris from the surface of the workpiece at the macroscopic as well as the microscopic scale. However, the high surface energy of the pad adheres and carries away the debris from the surface of the workpiece. In general, micro-particles (diameter ≈1–5 μm) of the magnetic fluid exhibit high surface energy density as compared to macro-particles or flat surfaces. A high-surface-energy tool acts as a magnet to attract any micro- and nano-particles using van der Walls force of interaction.

On the basis of these three aspects, it is necessary to combine MR fluid with the CMP process.

Example 12.2

Calculate the percentage change in chemical reaction rate on a local point of a surface profile when the radius of curvature (*RoC*) is changed from 10 μm to 100 μm due to finishing operation.

SOLUTION

Rate of chemical reaction rate when *RoC* is 10 μm (C1) $\propto e^{\frac{d\zeta 1 - Ea}{kT}}$;

Rate of chemical reaction rate when *RoC* is 100 μm (C2) $\propto e^{\frac{d\zeta 2 - Ea}{kT}}$;

$$C2/C1 = e^{\frac{d\zeta 2 - d\zeta 1}{kT}} = e^{(2\gamma\Omega 10^6)\left(\frac{0.01 - 0.1}{kT}\right)} ;$$

% change in chemical reaction rate

$$= (C1 - C2)100/C1 = 100\left\{1 - e^{(2\gamma\Omega 10^6)\left(\frac{0.01 - 0.1}{kT}\right)}\right\}\% .$$

12.2.7 Magnetorheological Fluid

Magnetic + rheo (flow) + logical fluid = MR fluid

MR fluid is manageable fluid whose rheological properties like viscosity, strain rate and yield stress are controlled under varying magnetic fields. This fluid comprises carbonyl iron particles (CIPs) as a carrier medium (Phule, 2001). The fluid behaves like a Newtonian fluid. The behaviour of the fluid is changed under the influence of a magnetic field. The fluid behaviour under the magnetic field becomes similar to Bingham plastic fluid. Transformation of the MR fluid behaviour from Newtonian to Bingham plastic occurs due to the magnetic field (Jolly et al., 1996) under a specific value of time, which is known as the response time. In general, the response time is a few milliseconds to create a chain of CIPs. Moreover, the response time becomes a few seconds to form thick layers of MR fluid (Tao, 2001). In general, applications of MR fluid have been classified into four major categories. These classes are expressed on the basis of operation, like shear mode, valve mode and squeeze mode or a combination (Carlson and Jolly, 2000). Amongst all these modes of operation, the squeeze mode can produce the highest stress and pressure.

To combine MR fluid with the CMP process, a flexible polishing pad is formed with the help of MR fluid and a permanent magnet. In CMMRF, the MR fluid forms a polishing pad and works on squeeze as well as shear mode.

Towards pad formation, MR fluid does not align along a stable path of magnetic field initially, and it shows little higher magnetic energy, which is stored in the MR fluid due to external magnetic field. After a few seconds, MR fluid flows to minimise its magnetic energy and forms a flexible pad, as shown in Figure 12.12.

To realise the forces (cutting force and holding) on the abrasive, it is necessary to know the stiffness of magnetically stiffened MR fluid along the cutting direction as well as the vertical direction. The stiffness of the fluid can be computed using Kelvin's formulation (Bakuzis et al., 2005) on MR fluid, which is listed in Equations 12.10 through 12.12:

Force density on MR fluid under magnetic field $(B) = \nabla(B.B)/(2mu)$,   (12.10)

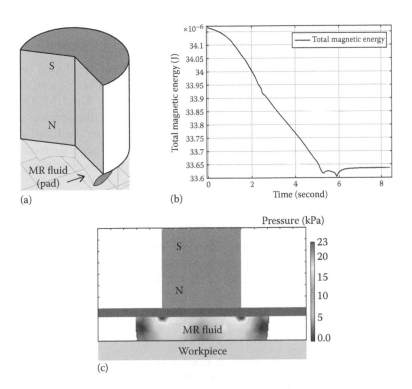

(a)   (b)

(c)

FIGURE 12.12
Results obtained by finite element analysis using Equation 12.10 and the Navier-Stokes equation: (a) formation of MR-fluid-based polishing pad under the presence of permanent magnet. (b) The curve shows that the magnetic energy is stabilised within 5 seconds. It indicates the time required to form the polishing pad. (c) Pressure builds in the polishing pad while finishing at a fixed working gap between the magnet and the workpiece.

where *mu* = magnetic permeability of the MR fluid, *B* = magnetic flux density and ∇*B* = vector gradient of magnetic flux density.

In case of cylindrical magnet, the stiffening force along the radial (cutting direction) and vertical directions are presented as follows (Bakuzis et al., 2005):

$$\text{Radial force } (Fr) = (\partial Vm/\partial z * \partial^2 Vm/\partial r \partial z + \partial Vm/\partial r * \partial^2 Vm/\partial r^2) * mu \quad (12.11)$$

and

$$\text{Vertical force } (Fz) = (\partial Vm/\partial z * \partial^2 Vm/\partial z^2 + \partial Vm/\partial r * \partial^2 Vm/\partial r \partial z) * mu, \quad (12.12)$$

where *Vm* is magnetic potential at a specified location along radial as well as vertical direction. Thus, *Vm* can be *Vm(r,z)*.

12.3 Chemomechanical Magnetorheological Finishing

A hybrid finishing process, namely, CMMRF, has been developed by combining the constructive aspects of CMP and MRF exclusive of their hostile effects while surface finishing at the nanometric scale. The CMMRF process is suitable for a wide range of engineering materials like semiconductor, ceramic, metal and alloys. It has various advantages over the CMP as well as the MRF process, like excellent polishing rate, ultra-high surface finish, improved process repeatability, polishing pad life, process flexibility, etc. A schematic of the process is presented in Figure 12.13.

12.3.1 Mechanism of CMMRF

During material removal in the CMMRF process, two actions simultaneously take place, namely, a chemical reaction and magnetic assisted mechanical abrasion. The schematic of the CMMRF mechanism is illustrated in Figure 12.14.

There are essential attributes that are associated with the mechanism of CMMRF:

- Effect of chemical reaction: A chemical reaction does take place on the superficial surface of the workpiece. The chemical reaction makes a passivated layer on the surface to protect it against further chemical reaction. However, this passivated layer shows altered mechanical properties that are favourable for nano-abrasion.
- Effect of magnetic field: Nano-abrasion depends on abrasion force, which comes from the stiffening force of the MR fluid under the

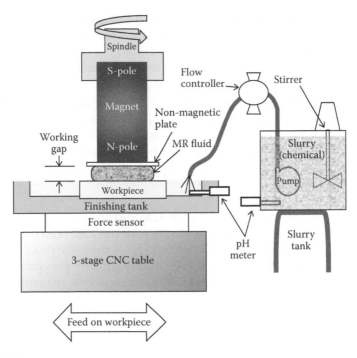

FIGURE 12.13
Schematic of the CMMRF process. (From Ranjan, P., Balasubramaniam, R., Suri, V.K., *Int. J. Precis. Technol.*, 4, 230–246, 2014.)

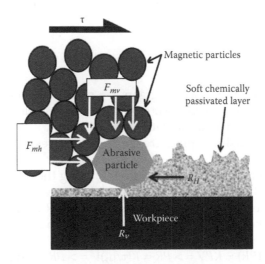

FIGURE 12.14
Material removal mechanism of the CMMRF process. Here, R_v = reaction force in the vertical direction; R_H = reaction force in the horizontal direction and F_{mv} and F_{mh} are magnetic force on CIPs along the vertical and horizontal directions, respectively. (From Ranjan, P., Balasubramaniam, R., Suri, V.K., *Int. J. Precis. Technol.*, 4, 230–246, 2014.)

influence of magnetic field as presented in Equations 12.10 through
12.12. This nano-abrasion force depends on the intensity of the magnetic flux density as well as its gradient. In the case of a permanent
magnet, abrasion force depends on working gap, grade of the magnet and magnetic permeability of the MR fluid.

- Texturing of the polishing pad: It is an essential attribute of the CMP
  process to hold and transfer the abrasion force on the abrasive. In
  CMMRF, CIPs are aligned along the magnetic lines of field in which
  abrasive particles are entangled in between the CIPs. Hence, pad
  conditioning has been automatically achieved by implementation of
  MR-fluid-based polishing pad.

To understand the mechanism of the CMMRF process, it is required to
know the mathematical models associated with the CMP and MRF process.
Preston (1927) has developed a polishing model of the CMP process in which
the polishing pressure and relative velocity are directly proportional to the
polishing rate (it can also be expressed like abrasion rate, nanofinishing rate
or material removal rate):

$$\text{Polishing Rate } (PR) = KPV, \tag{12.13}$$

where K = constant for chemical reaction; P = polishing pressure on abrasive;
V = polishing velocity.

In the MRF process, MR fluid acts as a Bingham plastic fluid under the
influence of a magnetic field (Figure 12.14) which can be written as shear
stress (Jain et al., 2011):

$$\tau = \tau_o + \mu(\dot{Y}) = \tau_o + \mu(V/g), \tag{12.14}$$

where $\dot{Y}$ = strain rate, g = working gap; τ = shear stress on the MR fluid at
the finishing plane; τ_o = yield shear strength of the MR fluid at a given magnetic field and μ = dynamic viscosity of the MR fluid.

In MR fluid, polishing pressure and shear stress can be correlated
(Figure 12.15).

According to force equilibrium conditions, along the direction of the fluid
flow, $\tau = \mu_k N$, and along normal to the work surface, $N = P$, where μ_k is the
friction factor between the work surface and the MR fluid. From Equations
12.13 and 12.14, the polishing pressure can be written as follows:

$$\text{Polishing Rate } (PR) = KPV => PR = K(\tau/\mu_k)V => PR = K[\tau_o + \mu(V/g)](V/\mu_k). \tag{12.15}$$

In the CMMRF process, chemical passivation plays a vital role to finish
engineering materials to the optical level. Hard and brittle materials like

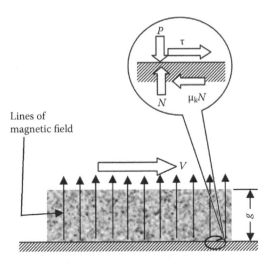

FIGURE 12.15
Schematic of MR fluid while in working condition.

silicon, quartz, sapphire, etc., need a mechanism for efficient finishing, as shown in Figure 12.16a, which can be expressed as the thinner passivation technique. In this mechanism, the depth of abrasive indentation becomes equal to the depth of the passivated layer by increasing the indentation force. The stress developed by indentation force becomes higher than the yield stress of the passivated layer, but it has to be lower than the yield stress of the parent material. Thus, the entire passivated layers are swept away, which result in improved polishing rate without damaging the parent material.

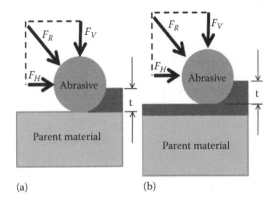

FIGURE 12.16
(a) Thinner layer of passivation for a hard workpiece $[\sigma_y(\Pi/4)(2Dt) < F_V]$. (b) Thicker layer of passivation for a soft workpiece $[\sigma_y(\Pi/4)(2Dt) = F_V]$. Here, D = mean diameter of the abrasive; σ_y = yield strength of the passivating layer in compression and F_V, F_H and F_R are the cutting force, holding force and resultant force generated by the magnetism, respectively.

To finish ductile and malleable materials like metals and alloys, another technique, named as 'thicker layer passivation', works, which is shown in Figure 12.16b. In this process, abrasive indentions always become less than or equal to the depth of the chemically passivated layer. This mechanism is used to protect the parent materials against abrading action directly on the surface of the parent material. This technique enables a very high level of surface finish with very minimal level of surface defects.

Example 12.3

Find the percentage change in the polishing rate of the CMMRF process if the friction factor of MR fluid is increased by 15%.

SOLUTION

Let us consider a friction factor of first MR fluid = μ_{k1} and friction of second fluid = μ_{k2}, where μ_{k2} can be written as $\mu_{k2} = \mu_{k1}(1.15)$.

Polishing rate due to first MR fluid $(PR1) = K[\tau_o + \mu(V/g)](V/\mu_{k1})$;

Polishing rate due to second MR fluid $(PR2) = K[\tau_o + \mu(V/g)](V/\mu_{k2})$;

$PR2/PR1 = \mu_{k1}/\mu_{k2} = 1/1.15$; Increasing in polishing rate = $1 - PR2/PR1 = 0.15/1.15 = 0.13$.

Thus, polishing rate will increase by 13%.

12.3.2 Preparation of the CMMRF Process

The CMMRF process has three major units, which can be expressed on the basis of Figure 12.13 as slurry system, finishing zone and controller.

Slurry system: This unit comprises the slurry of the MR fluid and essential chemicals with the fluid circulation mechanism as shown in Figure 12.13. The MR fluid consists of CIPs, abrasive particles, dispersion medium (water is preferred for CMMRF) and additives to avoid particle agglomeration and sedimentation. Chemicals are used for passivating the work surface. In some cases like finishing of metals, corrosion inhibitors are used to avoid particle dissolution as well as corrosion of finishing surface. MR fluid is placed between the magnet and the work material to form the polishing pad, whereas chemicals are continuously circulated and it is ensured to submerge the work material.

Finishing zone: This unit has a 'provision to retain the slurry' and 'spindle with polishing head'. There is an add-on like force transducer or dynamometer to monitor polishing pressure online.

Controller: This arrangement maintains the required movement of the polishing head with respect to the work material, working gap, polishing speed (in form of rotational speed of the spindle) and feed using computer numerical control (CNC)-based controller. However, there are other parameters that are related to slurry, like pH value and zeta potential, and they need to be monitored and controlled.

12.3.3 CMMRF Parameters

The CMMRF process has a few parameters to control the process, which can be expressed as MR fluid, magnetic field in working zone, chemicals, rotational speed, feed to the workpiece and finishing time.

> *MR fluid*: MR fluid is directly involved in nano-abrasion in the CMMRF process. The magnetic permeability of the MR fluid leads to higher viscosity and higher yield shear stress under a specified magnetic field, which results in better abrasion and better polishing rate. The magnetic permeability depends on the type and size of magnetic particles, dispersed in the base medium of MR fluid. There are two types of magnetic iron particles (shown in Figure 12.17): electrolytic iron particles and CIPs. CIPs confirm similar permeability and much better stability against particle sedimentation with respect to the electrolytic iron particle. However, there are some more attributes of MR fluid like abrasive size and its concentration. An abrasive with a bigger diameter (mean diameter) yields a better polishing rate, but the final surface roughness saturates at poor surface finish (higher roughness value), whereas a smaller abrasive particle polishes the surface at a lower polishing rate but yields better surface finish (low surface roughness).

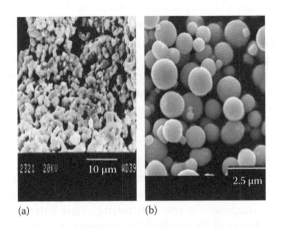

(a)   (b)

FIGURE 12.17
(a) Scanning electron microscope image of electrolytic iron powder, manufactured by electrolytic reduction of iron oxide; particle size ranges from 10 μm to 150 μm (according to their mesh number). (b) CIPs manufactured by thermal decomposition of iron pentacarbonyl. It shows spherical shape and their size that ranges from 1 to 2 μm. (Courtesy of BASF, http://www.monomers.basf.com.)

Magnetic field (due to magnet or working gap): An Intensified magnetic field increases mechanical abrasion, which is favourable for hard materials to be finished. But it damages the workpiece in the form of scratches, pits and degenerated layers on ductile and malleable material like aluminium and copper. Hence, a low magnetic field is needed while finishing such soft materials. The magnetic field can be controlled by varying the magnetic strength of magnet (permanent or electromagnet) as well as working gap.

Chemicals: Chemicals have two roles during the CMMRF operation, viz., formation of a chemically passivated layer on the work surface and separating particles with the help of suitable additives and zeta potential of the chemical. The pH value plays a significant role in controlling the passivation rate. For finishing (polishing) of semiconductor and ceramics, a basic medium with a pH value in the range of 10 to 11 is used as the optimal value, whereas in the case of metals and alloys, it is recommended to use an acidic pH value (pH in range of 3–5). There are some metals like stainless steel (SS) and aluminum that are stained while finishing with acidic medium; hence, they need basic chemicals.

Rotational speed: This allows the number of cutting strokes for nano-abrasion per unit time, which increases polishing rate and yields better surface finish. On the other hand, rotational speed beyond some value starts hitting the work surface in an unwanted way of nano-abrasion, which creates defects in the form of pits and particle embedment. Hence, there is some value of rotational speed in which a higher polishing rate and better surface finish are achievable. This value is also called the optimum value, which varies in the range of 200 to 300 rpm, and it depends on the size of the polishing pad.

Feed: Feed is being used to have a uniform polishing rate on the entire surface of workpiece as well as degree of randomness on the finishing process itself. A high degree of randomness yields a better surface finish by removing the amplitude of peaks, which are generated due to the combined effect of feed and rotational speed. A lower feed always gives a better surface finish. A range of feed rate of 1–5 mm/min is the recommended value to obtain nanometric surface finish.

Finishing time: Keeping all other parameters constant, increasing the finishing time reduces the surface roughness value. After some time, the surface roughness starts saturating. This time to saturate the surface roughness is also called the saturation time. Hence, surface roughness reduces by increasing the finishing time until saturation time. The surface roughness obtained after saturation time is also called 'critical surface roughness'. The recommended finishing time is always higher than or equal to the saturation time.

12.3.4 Workpiece Material and Chemical Reaction

There are chemicals that are incorporated with the MR fluid to conduct the CMMRF operation on various work materials.

Materials that are suitable for the CMMRF process are silicon (mono-crystalline as well as poly-crystalline), copper and its alloys, SS (all grades of non-magnetic SS such as SS304, SS316, SS310, etc.), sapphire, tungsten, silicon nitride, aluminum, etc.

There is a wide list of abrasives in which some are chemically reactive to the work materials and some of them are inert, which do only mechanical abrasion.

Silicon carbide (SiC): Silicon carbide is used for mechanical action (abrasion) for many work materials like tungsten, sapphire, copper, steel, silicon nitride, etc. It causes surface and sub-surface damages.

Alumina (Al_2O_3): Alumina is chemically unstable in acidic as well as alkali media, which is suitable for silicon, quartz, copper, silicon nitride, tungsten, etc. The surface damage in this case is less as compared to SiC abrasive.

Ceria (CeO_2): Ceria reacts chemically with silicon as well as silicon dioxide directly and forms a passivation layer as a complex of Si-O-Ce, which is softer than ceria itself. This abrasive is suitable for silicon, quartz and silicon nitride to generate atomistic surface without surface damages.

Silica (SiO_2): A silica particle is chemically stable in an alkali environment and has the capability of 'atomic scale abrasion' for various engineering materials like silicon, quartz, silicon nitride, steel, sapphire, etc.

Boron carbide (B_4C) and diamond (C): These are extremely hard abrasives with sharp edges, which are suitable for all types of engineering materials to enhance material removal rate at the coarse level of finishing operation. These abrasives cannot be used for fine finishing because they lead to scratches and degenerated layers during finishing operation.

There are some more abrasives for silicon, quartz and silicon nitride as chemical-reactive-abrasives, like chromium oxide (Cr_2O_3), zirconium oxide (ZrO_2), iron oxide (Fe_2O_3), yttrium oxide (Y_2O_3), copper oxide (CuO) and molybdenum oxide (Mn_2O_3).

The chemical of the CMMRF slurry reacts with the work material and acts as a dispersion medium, which has the ability to carry abrasive and iron particles without agglomeration as well as sedimentation. The agglomeration and sedimentation of the particles in the slurry are inhibited by increasing the zeta potential of the slurry in the range of ±30 to ±60 mV. There are constituents to improve zeta potential, which are also called additives. Glycine (NH_2CH_2COOH), hydrogen peroxide (H_2O_2), glycerol ($C_3H_5(OH)_3$), polyacrylic acid and colloidal silica with pH = 10–12 are being used as additives, whereas benzotriazole (BTA) was in demand for incorporation in the finishing of metals and alloys as a corrosion inhibitor. The corrosion inhibitor also mitigates the dissolution of iron particles in the chemicals.

Chemical reaction for the formation of passivated layer of engineering materials:

For silicon and quartz: The following chemical reactions are suitable to formulate soft passivation layer on surface of silicon and quartz.

Chemical reactions on silicon:

$$Silicon(Si) + 3H_2O = H_2SiO_3 \text{ (soluble silicate)} + 4\,H^+ + 4e^- \tag{12.16}$$

$$2CeO_2 + Si \rightarrow Ce_2O_3 + SiO \tag{12.17}$$

$$Ce_2O_3 + 1/2O_2 \rightarrow 2CeO_2 \tag{12.18}$$

$$CeO_2 + Si + O_2 \rightarrow CeO_2.SiO_2$$
$$\text{(soft material which can be easily removed).} \tag{12.19}$$

Chemical reactions on quartz:

$$Quartz(SiO_2) + H_2O = H_2SiO_3 \text{ (soluble silicate)} \tag{12.20}$$

$$SiO_2 + CeO_2 \rightarrow CeO_2.SiO_2$$
$$\text{(soft material which can be easily removed).} \tag{12.21}$$

For copper: Aqueous ammonia in MR fluid introduces basicity (pH > 7) of the fluid and forms complexes during chemical reaction. A superficial layer of chemically passivated copper converts from cuprous to cupric because the cupric state is more stable than the cuprous form; this effect can be conceived with the following electronic equations:

$$Cu = Cu^{2+} + 2\,e;\ E^0 = -0.52\,V;\ Cu = Cu^{2+} + e^-,\ E^0 = -0.15\,V \tag{12.22}$$

$$2Cu^+ = Cu^{2+} + 2e;\ E^0 = 0.367\,V;\quad K = [Cu^{2+}]/[Cu^+] \approx 10^6. \tag{12.23}$$

Copper forms chemical passivation by using the following chemical reactions with acid and aqueous ammonia with glycine.

Reaction with diluted acid:

$$3Cu + 8HNO_3 \rightarrow 3Cu(NO_3)_2 + 4H_2O + 2NO \qquad (12.24)$$

$$2Cu + 2H_2SO_4 + O_2 \rightarrow 2CuSO_4 + 2H_2O. \qquad (12.25)$$

Reaction with aqueous ammonia and glycine:

$$Cu + NH_4^+ + 2OH^- \rightarrow Cu(OH)_2 + NH_4^+ \qquad (12.26)$$

$$Cu^{2+} + glycine \rightarrow Cu\text{-}glycine\ complex + OH^\bullet \qquad (12.27)$$

$$3Cu + 4OH^\bullet \rightarrow Cu_2O + CuO + 2H_2O. \qquad (12.28)$$

Thus, chemical reactions make soft-brittle (hydroxide and oxide) layers on the Cu workpiece surface, which can be mechanically removed by flexible magnetic polishing pad.

For SS: To achieve a nanometric surface finish, the following issues are involved during CMMRF.

- Chemical composition of SS.
- Major alloying elements are Fe, Cr and Ni.
- Formation of chemical passivation layer due to Cr.
- Need to break inert layer of Cr_2O_3 by mechanical action (mechanical abrasion).
- Chemical reaction upon Fe, Cr and Ni to make their hydroxides.
- With alkali media, Fe, Cr and Ni are converted in to their hydroxides.
- Magnetic assisted mechanical abrasion of the hydroxides in nanometric domain.

For sapphire (α-Al_2O_3): Sapphire is the alpha phase of monocrystalline alumina, which performs the following chemical reactions to make soft passivation layers for CMMRF.

Slurry should be alkaline as per the following reactions:

$$Al_2O_3 + H_2O = 2AlO(OH) \qquad (12.29)$$

$$Al_2O_3 + 3H_2O = 2Al(OH)_3 \qquad (12.30)$$

$$Al_2O_3 + 2OH^- = 2AlO_2^- + H_2O \qquad (12.31)$$

$$Al(OH)_3 + OH^- = AlO_2^- + 2H_2O. \qquad (12.32)$$

In case of acidic media, alum is ion formed, which is difficult for cleaning as follows:

$$Al_2O_3 + 6H^+ = 2Al^{3+} + 3H_2O \qquad\qquad (12.33)$$

$$Al(OH)_3 + 3H^+ = Al^{3+} + 3H_2O \qquad\qquad (12.34)$$

$$AlO(OH) + 3H^+ = Al^{3+} + 2H_2O. \qquad\qquad (12.35)$$

Hence, it is recommended to use a chemical with a pH value ≈10–12.

For tungsten (W): Slurry should have an oxidant for passivation, such as H_2O_2 and $Fe(NO_3)_3$. Here, H_2O_2 is decomposed on the surface and produces reactive hydroxyl radical. Ferricyanide etchant is also used to make passivation on W as follows:

$$W + 6Fe(CN)_6^{-3} + 4H_2O \rightarrow WO_4^{-2} + 6Fe(CN)_6^{-4} + 8H^+ \qquad (12.36)$$

$$W + 6Fe(CN)_6^{-3} + 3H_2O \rightarrow WO_3 + 6Fe(CN)_6^{-4} + 8H^+. \qquad (12.37)$$

For silicon nitride (Si_3N_4): Chemical reactions on silicon nitride are presented in the following chemical reactions.

$$Si_3N_4 + 6H_2O \rightarrow 3SiO_2 + 4NH_3 \qquad\qquad (12.38)$$

Reactions with abrasives, like Fe_2O_3, Cr_2O_3, ZrO_2 and CeO_2 are as follows:

$$Si_3N_4 + Fe_2O_3 \rightarrow SiO_2 + FeO + FeSiO_3/FeO.SiO_2 + Fe_4N + N_2 \quad (12.39)$$

$$Si_3N_4 + 2Cr_2O_3 \rightarrow 3SiO_2 + 4CrN \qquad\qquad (12.40)$$

$$Si_3N_4 + ZrO_2 \rightarrow SiO_2 + ZrSiO_4/ZrO_2.SiO_2 + ZrN + N_2 \qquad (12.41)$$

$$Si_3N_4 + CeO_2 \rightarrow SiO_2 + CeO_{1,2,3} + CeO_{1,8,3} + Ce_3O_2 + N_2. \qquad (12.42)$$

TABLE 12.1

List of Chemicals for CMMRF of Engineering Materials

Sr. No.	Work Material	MR Fluid and Chemical Media
1	Mono silicon	CIPs, silica, H_2O, glycerol, CeO_2
2	Poly silicon	CIPs, silica, H_2O, glycerol, CeO_2
3	Copper	CIPs, alumina, H_2O, glycerol, NH_4OH, BTA
4	Cu-be alloy	CIPs, alumina, H_2O, glycerol, NH_4OH, BTA
5	Steel (316&304)	CIPs, alumina, H_2O, glycerol, NH_4OH, BTA
6	Sapphire	CIPs, silica, H_2O, glycerol, NH_4OH
7	Tungsten	CIPs, SiC, H_2O, H_2O_2, ferricyanide
8	Silicon nitride	CIPs, silica, H_2O, glycerol, CeO_2
9	Aluminum	CIPs, silica, H_2O, glycerol, H_2O_2, NH_4OH, BTA

All these chemicals are presented in Table 12.1, in which work materials are placed with their suitable chemicals for CMMRF operation.

12.3.5 Case Study

For surface finishing on mono-crystalline silicon (m-Si), the following activities have been carried out to execute CMMRF operation on m-Si.

Preparation of MR fluid: The fluid encompasses solid particles and a dispersion medium. For solid particles, CIP:CeO_2:Al_2O_3 particles have been mixed in a volumetric proportion of 6:3:1, respectively. Liquid chemicals are prepared by mixing deionised water and glycerine with a proportion of 26:1 (v/v), respectively. Solid mixture and liquid chemicals are now mixed with a ratio of 4:6 (v/v) and stirred for 1 hour to maintain homogeneous dispersion of solid particles in the liquid medium.

Preparation of polishing pad: A polishing pad arrangement using non-magnetic cover and plates as shown in Figure 12.12 is carried out in which MR fluid is put underneath the non-magnetic plate. After a few seconds (4–10 seconds), the MR fluid stiffens and forms a polishing pad. However, the non-magnetic plate is used to retain the fluid in the magnetic domain, which is created by using permanent magnet.

Now, the finishing process is carried out (Figure 12.13). The polishing pad is squeezed against the work material (m-Si) to maintain the working gap ($\approx$1 mm). During the finishing operation, continuous flow of the slurry is ensured using suitable flow control. After finishing for 8 hours with different levels of parameters, a parametric graph is generated, as shown in Figure 12.18, which shows optimal surface finish at optimum working gap.

As per previous studies, it is obvious that CMMRF employs a flexible polishing pad, which can finish a 2D as well as a 3D surface. There are some samples of copper alloys that are optically finished, as shown in Figure 12.19. It articulates the capabilities of the process for 2D as well as 3D surface finishing.

The process has the capability to generate atomic-level surface finish on a wide range of engineering materials. A few surfaces are shown in Figure 12.20.

A comparison amongst CMP, MRF and CMMRF processes is presented in Table 12.2.

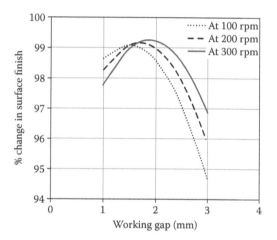

FIGURE 12.18
Effect of working gap and rotation speed on CMMRF of m-Si.

FIGURE 12.19
CMMRF processed on (a) Oxygen-free high thermal conductivity (OFHC) copper substrate, flat surface, (2D) and (b) bronze coin (3D patterns).

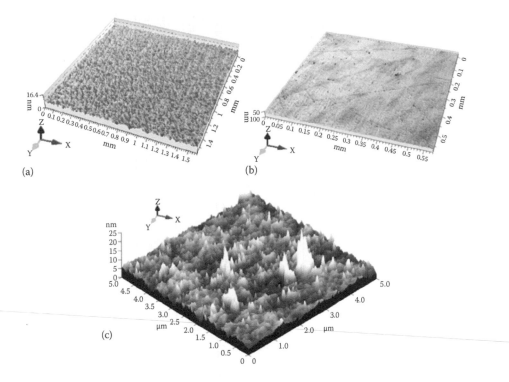

FIGURE 12.20
3D Surface topography of CMMRF processed materials: (a) SS (SS316) with surface finish Ra = 0.6 nm, (b) Cu-substrate with surface finish Ra = 2.5 nm and (c) polycrystalline silicon with surface finish Ra = 1.0 nm.

12.4 Applications of CMMRF

The CMMRF process has the capability to generate surface roughness at atomic and nanometric scales. The surface at the atomic scale exhibits a high level of stability against chemical, physical and mechanical properties, and this is why the surface shows enhanced properties in the following areas:

- Wear resistance (surface wear occurs due to peaks removal during friction on the surface)
- Fatigue strength and endurance limit (due to stress concentration on valleys)
- Corrosion resistance (chemical stability)
- Enhanced optical properties (reduced light scattering phenomena)
- Reduction on friction factor (relative movement between solid and fluid)

TABLE 12.2

Comparison amongst the CMP, MRF and CMMRF Processes

Sr. No.	Basis for Comparison	CMP	MRF	CMMRF	Suitability
1	Polishing rate	$\approx$10 nm/min	$\approx$1 nm/min	$\approx$10 nm/min	CMP and CMMRF
2	Final surface finish (*Ra*) on metal and alloys	<1.0 nm	$\approx$10 nm	<1.0 nm	CMP and CMMRF
3	Final surface finish (*Ra*) on semiconductor and ceramics	$\approx$0.2–0.4 nm	$\approx$1.0 nm	$\approx$0.4–0.5 nm	CMP and CMMRF
4	Surface complexity (2D and 3D)	Flat-circular wafers (2D)	2D as well as 3D surfaces	2D as well as 3D surfaces	MRF and CMMRF
5	Type of materials to process	All materials	Not suitable for malleable materials	All materials	CMP and CMMRF
6	Surface defects (scratch-dig)	–	–	–	CMP and CMMRF
7	Residual stress	Negligible	Significant stress	Negligible	CMP and CMMRF

Applications on specific need in various engineering domains are briefly discussed as follows.

High-speed bearing components: Hydrodynamic bearings like journal-bearing that works on the principle of thin layer lubrication dynamics. This bearing needs a high surface finish to reduce friction factor and surface wear with excellent bearing efficiency. However, there are other types of bearings like ball bearing and roller bearing that need high surface finish. The surface roughness at the atomic scale and form accuracy at the nanometric scale bring improved performance such as increased rolling speed range and life of the bearings and less noise, which lead towards finally making noiseless bearings.

Biomedical implants: Knee prosthesis and hip prosthesis need atomic scale of surface roughness to maintain smooth movement and reduced wear on moving implants like knee-femoral component and femoral head of hip implant. In this view, patient comfort and implant life increase with such surfaces.

Corrosion resistance: Surface finish at the atomic scale makes it highly stable and chemically inert, and this is why corrosion resistance improves drastically.

Optical lens and mirrors: High surface finish reduces light scattering phenomenon, and this is why the nanometric surface brings better reflection quality on mirrors and better transmission quality on lenses.

LASER optics: A metal mirror with a nanometric surface finish improves reflection efficiency. However, the heat conduction quality of metal brings a cooling effect on the mirror to avoid thermal damages to the mirror.

X-ray optics: Surface finish of the order of a few angstroms (1–4 Å) with form error of the order of a few nanometres satisfy the need of X-ray optics towards lesser scattering and better focusing efficiency of the X-ray.

Linear accelerator (LINAC) cavity: LINAC needs higher quality factor (Q-factor) for better efficiency of the accelerator. The Q-factor increases by reducing surface stresses and surface roughness with minimal amount of magnetic susceptibility. Hence, CMMRF-finished copper alloy is suitable for LINAC cavity.

Micro-fluidics: In micro-fluidic devices, a small amount of fluid needs to flow (with required value of flow rate) through a micro-channel under the specified pressure difference across the channel. The pressure difference for the required flow rate depends on the channel's surface roughness. Low surface roughness reduces the pressure difference. The surfaces of micro-fluidic devices are not plain, rather they are complex in three dimensions (combination of land and groove area). Hence, these land and grooves need to be polished up to the nanometric level, which is possible by the CMMRF process.

Micro-fabrication: To improve the extent of miniaturisation as well as compactness between micro-structures, nanofinishing operation is needed before and during micro-fabrication.

12.5 Remarks

The CMMRF process has the capability to obtain nanofinishing and sub-nanofinishing for various engineering materials. It can generate a sub-nanometric surface finish in the case of monocrystalline engineering materials like silicon, quartz and sapphire. On the other hand, CMMRF has the capability to produce nanometric surface finish on polycrystalline materials like SS, OFHC copper, structural aluminum, etc. Thus, the concluding remarks are as follows:

- It has the ability to finish brittle, hard and soft materials due to the combined effect of chemical passivation and mechanical abrasion on the finishing surface.
- The mean surface roughness value on silicon substrate is less than 0.5 nm, which declares process capability.
- Implementation of CMMRF for other materials can be carried out with their appropriate chemicals to execute chemical reaction and formation of passivating layer.
- CMMRF slurry for other materials should have a balanced amount of additives to maintain slurry stability throughout the finishing process by monitoring of pH and zeta potential.
- The process has flexibility to finish 2D as well as 3D surfaces; hence, it has wide application areas in the field of automobiles, biomedical, optics, micro-fluidics and micro-electronics.

12.5.1 Future Research Scope

Since the process has been designed independently to finish brittle as well as ductile materials, all engineering materials can be processed with the opti-mised chemical composition of the MR fluid and polishing chemicals. Few parameters have been optimised, like working gap, rotational speed and feed and pH value. Yet, there are other parameters that need optimisation, like concentration of corrosion inhibitor and additives. Further, the following issues also need attention of researchers:

- MDS to study chemical reaction kinetics and nano-abrasion.
- Study and enhancement of polishing pad flexibility for efficient finishing of 3D surfaces.

12.5.2 Unsolved Problems

1. For a diamond turned surface that has a uniform surface profile, peak to valley distance on surface roughness is equal to 10 nm and wavelength of the roughness is 1.0 µm. Establish the relationship between the abrading force (for nano-abrasion) and material removal depth.

2. In a brass workpiece, an alumina abrasive is used for mechanical abrasion. Calculate the required cutting force and holding force for pit-free surface and depth of cut = 2 nm/abrasive-stroke. Assume the following parameters, shear strength = 70 MPa, diameter of abrasive = 2.0 µm, shear angle = 0°.

3. Derive a relationship between material removal and local 'radius of curvature' on workpiece surface during CMMRF. Assume chemical reaction rate is equal to material removal rate.

4. Calculate maximum cutting force and holding of a CMMRF polishing pad. Magnetic field is given as follows: flux density $(B) = 0.1i + 0.2j + 0.6k$ Tesla and gradient $(\mathrm{grad}B) = 0.8i + 0.6j + 0.2k$ kilo Tesla/m. Abrasive diameter is equal to 50.0 µm and the magnetic permeability of MR fluid is equal to 2000 N/T^2-m^3.

5. Calculate the polishing rate in CMMRF using the following parameters: constant for chemical reaction $(K) = 1$, shear yield strength of MR fluid at 0.6 T = 10 kPa, apparent viscosity = 0.5 Pa-s, relative velocity = 10 mm/s, working gap = 1.0 mm and friction factor between MR fluid and work surface = 0.2.

6. Draw a characteristic curve between surface roughness and relative velocity for CMMRF.

References

Bakuzis, A.F., Chen, K., Luo, W., Zhuang, H., 'Magnetic body force', *International Journal of Modern Physics B*, 19 (2005) 1205–1208.
BASF, http://www.monomers.basf.com.

Callister, W.D., *Fundamentals of Materials Science and Engineering*, 5th edition, John Wiley & Sons, Inc., Wiley, New York (2000).

Cao, G., *Nanostructures and Nanomaterials: Synthesis, Properties & Applications*, Imperial College Press, London (2004).

Carlson, D.J., Jolly, M.R., 'MR fluid, foam and elastomer devices', *Mechatronics*, 10 (2000) 555–569.

Fewkes, C.J., Tabor, R.F., Dagastine, R.R., 'Sphere to rod transitions in self assembled systems probed using direct force measurement', *The Royal Society of Chemistry 2015, Soft Matter*, 11 (2015) 1303–1314.

French, W.R., Iacovella, C.R., Cummings, P.T., 'The influence of molecular adsorption on elongating gold nanowires', *The Journal of Physical Chemistry*, 115 (2011) 18422–18433.

Hebda, M., Wachal, A., *Tribology*, WNT, Warsaw, Poland (1980) (in Polish).

Jain, V.K., Ranjan, P., Suri, V.K., Komanduri, R., 'Chemo-mechanical magneto-rheological finishing (CMMRF) of silicon for microelectronics applications', *CIRP Annals-Manufacturing Technology*, 59 (2010) 323–328.

Jain, V.K., Sidpara, A., Sankar, M.R., Das, M., 'Nano-finishing techniques: A review', *Proceedings of the Institution of Mechanical Engineers, Part C: Journal of Mechanical Engineering Science*, 226 (2011) 327–346.

Jolly, M.R., Carlson, J.D., Munoz, B.C., 'A model of the behaviour of magnetorheological materials', *Smart Material Structure*, 5 (1996) 607–614.

Mullin, J.W., *Crystallization. Chemical, Petrochemical & Process*, Elsevier Butterworth-Heinemann, London (2001) 1–600.

Phule, P.P., 'Magnetorheological (MR) fluids: Principles and applications', *Smart Materials Bulletin*, 2001 (Issue 2) (2001) 7–10.

Preston, F., 'The theory and design of plate glass polishing machines', *Journal of the Society of Glass Technology*, 11 (1927) 214–256.

Raghavan, V., *Material Science and Engineering*, 5th edition, PHI Learning Private Limited, Delhi, India (2013).

Ranjan, P., Balasubramaniam, R., Suri, V.K., 'Development of chemo-mechanical magnetorheological finishing process for super finishing of copper alloy', *International Journal of Manufacturing Technology and Management*, 27 (2013) 130–141.

Ranjan, P., Balasubramaniam, R., Suri, V.K., 'Modelling and simulation of chemo-mechanical magnetorheological finishing (CMMRF) process', *International Journal of Precision Technology*, 4 (2014) 230–246.

Tao, R., 'Super-strong magnetorheological fluids', *Journal of Physics: Condensed Matter*, 13 (2001) R979–R999.

Wang, C., Chen, S.-H., 'Factors influencing particle agglomeration during solid-state sintering', *Acta Mechanica Sinica*, 28(Issue 3) (2012) 711–719.

13

Electrochemical Grinding

Divyansh S. Patel, Vijay K. Jain and J. Ramkumar

Department of Mechanical Engineering, Indian Institute of Technology Kanpur, Kanpur, India

CONTENTS

13.1 Introduction

Electrochemical grinding (ECG) is a hybridised process that is a combination of electrochemical machining (ECM) and mechanical grinding processes. The ECG process is applicable for shaping or grinding an electrically conductive material. In mechanical grinding, the ground surface consists of micro-burrs and lays. The addition of electrochemical (EC) anodic dissolution, in which metal is removed from the anode in clusters of atoms/molecules, produces burr- and lay-free surfaces during ECG. Heat generation during this process is much lower as compared to mechanical grinding because a major part of material removal takes place due to electrolytic anodic dissolution. Thus, no thermal residual stresses and heat-affected zones are obtained during the ECG process. The process requires an abrasive-laden grinding wheel, which is bonded with an electrically conductive material. The grinding wheel is connected to the negative terminal (cathode) and the workpiece to the positive terminal (anode) of a direct current (DC) power source. The tool and workpiece are separated by a gap of a few tens micrometres. In this gap, electrolyte is supplied through a jet, which is recycled after removing debris and reaction products (chips and precipitates formed during EC dissolution) through a filter. Redox (reduction-oxidation) reactions take place between the tool and workpiece, and simultaneously, active abrasive particles in the machining zone start removing metal from the workpiece through erosion. In general, only 5%–10% of material removal takes place by mechanical action, while EC dissolution is responsible for 95%–90% of the material removal [1].

The recent trend shows that advanced material processing has become a challenge to industries because of outstanding mechanical properties of materials such as high strength-to-weight ratio, high melting point, high ductility, etc. Industries are also facing challenges in fulfilling the desired precision, accuracy and surface integrity of the products in the present day's customer-driven market.

13.2 EC Reactions

An electrolytic cell is formed by providing polarity to two electrodes (grinding wheel and workpiece) and employing a conductive liquid (electrolyte) in between, which results in oxidation and reduction at the anode and cathode, respectively. The cell and anodic reactions are shown in Figure 13.1.

An electrolyte consists of an aqueous solution of chemical salt(s) and additives. Salts such as $NaNO_3$, $NaCl$, $NaOH$, etc., when added in water, are

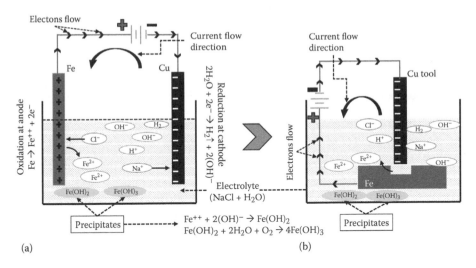

FIGURE 13.1

Schematic diagram of an electrolytic cell showing redox reactions and ion motions: (a) direction of electrons flow, cations, anions and precipitates generated in redox reaction in an electrolytic cell and (b) conversion of an electrolytic cell into an EC machining process.

disassociated into cations and anions. The motion of these ions causes the flow of electricity through the electrolyte and makes the solution electrically conductive. The conductivity of the electrolytes can be controlled by increasing or decreasing the electrolyte concentration and temperature. When a highly conductive electrolyte is desired, strong acids such as sulphuric acid (H_2SO_4), hydrochloric acid (HCl) or perchloric acid ($HClO_4$) are added in the aqueous solution. When these acids are added into an aqueous solution, H^+ ions are added to the solution. The mobility of H^+ ions is highest as compared to any other ion because of a higher transport number. The transport number is the fraction of current carried by a specific ion in the electrolyte. Thus, addition of H^+ ions to an electrolyte increases the conductivity and, hence, the rate of anodic dissolution.

Figure 13.1a shows a schematic diagram of an electrolytic cell, as electric the circuit is completed with the motion of ions (Na^+ and Cl^-). The potential between the two electrodes causes the flow of current from the cathode to the anode. Cations (Na^+) move towards the cathode (negative electrode) and anions move towards the anode (positive electrode) due to opposite polarity attraction (as shown in Figure 13.1a). Simultaneously, oxidation (at the anode) and reduction (at the cathode) take place due to potential difference, and atoms (Fe) from iron electrode (anode/workpiece) are converted into ions (Fe^{++}) by releasing two electrons.

Figure 13.1b shows the converted electrolytic cell to an ECM process, where EC anodic dissolution is achieved through concentrating the current density at the desired location. The cathode is shaped in such a way that the negative shape can be replicated on the anode. The tool is fed towards the workpiece to achieve the desired shape and size.

13.3 Electrochemical Grinding

Figure 13.2a shows a schematic diagram of the mechanism of the ECG process, in which the tool (shown in Figure 13.1b) is replaced by an electrically conductive grinding wheel. The grinding wheel is rotated as well as fed in the desired direction for the facing or grinding operation. The feed of the wheel is constrained by the rate of material removal due to the EC reaction and due to mechanical abrasion to maintain a minimum inter-electrode gap (IEG) so that short-circuiting is avoided. When EC anodic dissolution starts, an oxide layer is formed (while using a certain type of electrolyte) on the workpiece surface, which works as an insulator. This oxide layer is removed by forcing the electrolyte between the IEG with a high velocity. Figure 13.2b shows a magnified schematic view of the ECG process. The machining area is categorised into three zones (zone I, zone II and zone III). In zone I, material removal takes place due to pure EC anodic dissolution. The IEG keeps on varying along the electrolyte flow direction. Rotation of the wheel enhances the circulation and drawing of the electrolyte through the narrow machining gap. As EC reactions in zone I occur, reaction products such as gases, precipitates, oxide layer, etc., mix in the electrolyte and change its conductivity. In fact, the presence of certain sludge, which is normally electrically conductive and temperature gradient, to some extent, increases the conductivity of the electrolyte. But the presence of gases generated and contaminated precipitates decrease the conductivity, which generally lead to the decreased value of net conductivity of the electrolyte [2].

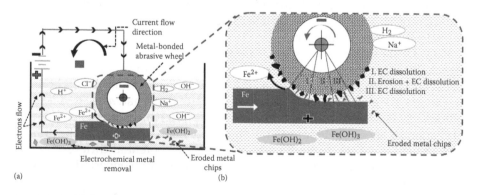

(a)   (b)

FIGURE 13.2
(a) Schematic diagram of mechanism of ECG process and (b) magnified schematic view showing different zones of ECG process.

The rotational speed of the ECG wheel and electrolyte jet velocity force this electrolyte into zone II. The electrolyte conductivity reduces in zone II and the IEG becomes smaller than in zone I. Further, if the IEG is smaller than the protrusion height of the abrasive particles, then removal of a small amount of the workpiece material by erosion takes place. As a result of the reduction in the IEG, the volume of gases in the gap is reduced; hence, the material removal rate (MRR) due to anodic dissolution is increased. The abrasive particles of the grinding wheel remove the non-reactive (oxide) passivation layer if generated due to EC reactions. Most of the metal oxides formed during EC reactions are insoluble in water and electrically non-conductive. The formation of this layer acts as an insulator and becomes an obstacle in anodic dissolution. Hence, removal of the oxide layer is an essential step in the ECG process. Thus, in zone II, material removal takes place due to mechanical action in the form of removal of the passivation layer and workpiece material in the form of micro/nano-chips or in the form of metal ions that react in the electrolyte and form reaction products. However, at the same time, material removal takes place due to EC anodic dissolution also.

In zone III, metal removal takes place solely through EC dissolution because the abrasive particles are disengaged from the work surface. Electrolyte pressure is released in this zone; hence, bubbles formation reduces the electrolyte conductivity. As a result, the EC anodic dissolution rate decreases. However, the current density at the discontinuity is much higher; hence, the scratches and burrs produced in zone II dissolve.

In ECG, metal removal depends on the gap between the wheel and workpiece. The gap between the two electrodes influences the amount of electrolyte and current carrying capacity. When the grinding force is low, metal removal is due to EC dissolution only. In general, an aqueous solution of salts or bases is used as an electrolyte in ECG. EC dissolution in the presence of such electrolytes results in a passive layer on the anode surface. This layer increases resistance to the flow of current between the wheel and workpiece; hence, material removal due to EC reactions decreases. As the wheel is fed towards the workpiece, the gap becomes smaller, which leads to an increase in grinding force. As the grinding force is higher, abrasives remove the passive layer by an erosion process from the workpiece; hence, it allows the EC processes to take place on the fresh (virgin) workpiece surface. The existence of these two material removal processes increases net material removal by a significant amount. Figure 13.3 shows the results of an experimental study carried out by Kuppuswamy [3], indicating separately the contribution of material removal through mechanical action and EC action. When machining with an Al_2O_3 grinding wheel, the amount of mechanically removed material ($\approx 5 \times 10^{-4}$ g/min) does not change much while increasing the grinding force. However, material removed by EC dissolution varies on increasing the grinding force and attains an optimum value. Hence, the combined effect of both mechanisms results in the highest MRR.

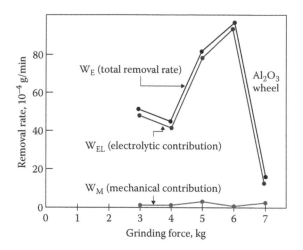

FIGURE 13.3
Contribution of mechanical and EC material removal in ECG while using AL_2O_3 wheel. (From Kuppuswamy, G., *Tribol. Int.*, 9, 29–32, 1979.)

ECG can be performed in internal as well as external surfaces. It can be used for a formed surface, flat surface or the face of a cylindrical surface. Hence, it can be classified as follows [4]: EC cylindrical grinding, EC form grinding, EC surface grinding, EC face grinding and EC internal grinding.

13.3.1 Power Supply for ECG

An ECG machine can operate on a continuous DC or pulsed DC power supply. Both have their own merits and demerits.

1. Continuous DC power (DC-ECG)

    The EC machining operation is fast enough when continuous DC power supply is employed since the amount of material removal depends upon the magnitude of current flowing between the cathode and anode.

2. Pulsed DC power (P-ECG)

    The P-ECG supply provides a pulsating energy flux because the current is supplied in small segments at the desired frequency. This implies proper execution of the EC dissolution by providing a pulse-off time and comparatively low MRR. When the pulse is off, the sludge produced due to EC reactions is flushed out from the machining zone. The pulsed EC process enhances surface integrity. In pulsed EC finishing, the surface roughness (R_z) value can be reduced from 3 μm to 1.22 μm in 4 min [7]. It has been observed that the P-ECG supply in ECG is an appropriate means for enhancing control over the process variables and repeatability. In addition, P-ECG reduces overcut significantly [8].

In the pulsed power operation mode, the factor that regulates the current is duty cycle, which is defined as follows:

Duty cycle = [Pulse on time/(Pulse off time + Pulse on time)] × 100%

Further, duty cycle is regulated by the voltage pulse width, which consequently increases the mean current [8]. Figure 13.4 shows a graph of the linear trend of mean current on varying the duty cycle in ECG with tool and die steel and stainless steel 304 as the workpiece material. An increase in pulse width results in increased value of the duty cycle, which consequently increases the mean current. This type of plot can be used to determine the sum of overvoltage and electrode potential in ECM and ECG. The trend line is extended to the horizontal axis (intersection of the trend line with the abscissa); the point corresponds to the condition where the current is initiated. Electrode potential corresponds to a duty cycle of 0.145 for tool and die steels, and for the stainless steel, electrode potential corresponds to a duty cycle of 0.084, as shown in Figure 13.4.

13.3.2 Comparison of ECG with Other Finishing/Grinding Processes

Table 13.1 shows a comparison of the parameters, advantages, MRR and machinable materials of ECG and other finishing operations.

13.3.3 Variants of ECG

The ECG operation can be performed in several ways to achieve the required material removal from the workpiece. The wheel of an ECG machine can be replaced by a belt or some other type of tool on the basis of the desired applications. Some variants of ECG in practice are as follows:

1. Belt-type ECG

   The process finishes flat and cylindrical workpieces using an electrically conductive flexible belt laden with abrasive particles. The belt is tightened on a couple of pulleys placed at a certain distance. The assembly of the system is made in such a way that the workpiece always remains in contact with the composite belt at a point where the pulley provides backing to the belt [9,10]. A schematic diagram of the belt type of ECG is shown in Figure 13.5a. The electrolyte is supplied in between the abrasive-particles-laden belt and the workpiece throughout the machining zone with the help of centrifugal force.

TABLE 13.1

Comparison of Other Finishing/Grinding Processes with ECG

Processes	Multi-Operations	Specific Energy	Merits and Demerits	MRR	Workpiece Type
			Performance Parameters		
ECG	• Grinding • Machining	Moderate	• High surface finish, No surface defects, low accuracy • 0.005 mm tolerance, • $Ra: \approx 1.01~\mu m$	• High	• Electrically conductive
Chemomechanical polishing (CMP)	• Only finishing	Moderate	• Finish: ≈0.5 nm • Mirror-like surface finish	• Lower than ECG	• Silicon wafer • Metals
Lapping/honing	• Only finishing	Moderate	• Moderate surface finish, mirror-like surface finish • High precision and accuracy • $Ra: 13–1500$ nm	• Lower than ECG	• Generally for metals
Grinding	• Machining • Finishing	High	• Metallurgical defects, heat affected zone (HAZ), low roughness, lay, burrs • $Ra: \approx 25–6000$ nm	• Lower than ECG	• Any material softer than the abrasive particles
Abrasive flow finishing (AFF)	• Finishing • Mirror-like surface finish	Very high	• High finishing rates • Best $Ra: \approx 50$ nm • More suitable for cylindrical surfaces	• Lower than ECG	• Any material softer than the abrasive particles
Magnet abrasive finishing (MAF)	• Finishing	High	• Low finishing rates • Best $Ra: \approx 7.6$ nm • Mirror-like surface finish	• Lower than ECG but higher than AFF	• Metals • Internal/external finishing of cylindrical surface • More suitable for non-magnetic materials
Magnetorheological finishing (MRF)	• Finishing	Very high	• Low finishing rates • $Ra: \approx 0.8$ nm • Mirror-like surface finish	• Lower than other finishing processes	• Freeform surfaces • Convex glasses • More suitable for non-magnetic materials

Source: Jha, S., Jain, V.K., Nanofinishing techniques, in: *Micromanufacturing Nanotechnol.*, 2006, 171–195; Jain, V.K., Sidpara, A., Sankar, M.R., Das, M., *Proc. Inst. Mech. Eng. Part C J. Mech. Eng. Sci.* 226, 327–346, 2011.

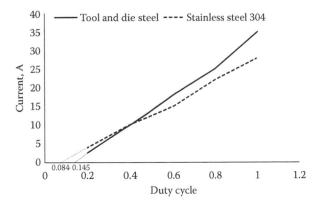

FIGURE 13.4
Variation of mean current with duty cycle.

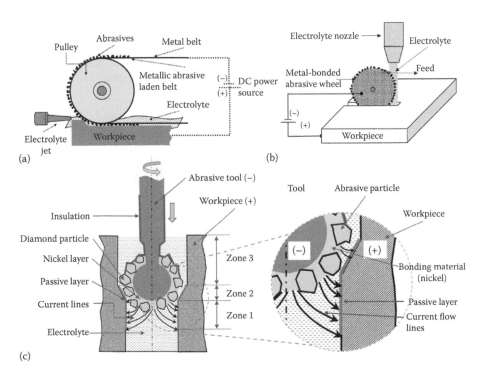

(a)

(b)

(c)

FIGURE 13.5
Variants of ECG: (a) belt-type ECG, (b) EC cut-off grinding and (c) ECG boring.

The process of the belt-type ECG is well accepted in commercial applications because MRR is higher as compared to the conventional ECG (schematic diagram shown in Figure 13.2). In addition, the process provides uniform ground surface with good surface finish.

2. EC cut-off grinding

This machine works on the principle of ECG, which can cut off composite materials consisting of metals and non-metals. The process requires a lower cutting force than the conventional machining processes and results in a low surface roughness value due to the effect of EC dissolution [11]. Figure 13.5b shows a schematic diagram of the EC cut-off grinding operation.

3. ECG boring

This type of machining setup (Figure 13.5c) can enlarge micro-holes on difficult-to-machine materials of about 500 μm diameter depending upon the tool accuracy [12]. Figure 13.5c shows a schematic diagram of the ECG boring (ECGB) operation, where the pre-existing micro-hole size is enlarged. In ECGB, a metal rod with a spherical end coated with diamond abrasive particles is used as an ECG tool to enlarge a hole of D_0 to $(D_0 + \Delta D_0)$ diameter.

In this process, the material removal occurs in two zones, and in some cases, in three zones:

Zone 1–Material is removed totally by EC dissolution. Due to EC reactions, a passivation layer is formed on the inner surface of the hole if a passive electrolyte ($NaNO_3$) is used.

Zone 2–Metal is removed through mechanical erosion and EC dissolution both. When the tool is fed in side the hole, the gap between abrasive particles and workpiece is reduced. Hence, removal of non-reactive oxide layer (if formed) and workpiece material takes place due to erosion by the sharp edges of abrasive particles. Further, a small amount of the material is removed by EC anodic dissolution in the micro/nano-electrolytic cell formed between the ECGB tool bonding material and the workpiece.

Zone 3–No material removal takes place since half of the surface area of the spherical end of the tool is insulated. This implies that there is no electric current flow between the workpiece and the tool in this zone. If the tool is not insulated, EC dissolution takes place and material removal starts in zone 3 also.

13.3.4 Fishbone Diagram

Figure 13.6 shows the cause-and-effect diagram of the ECG process.

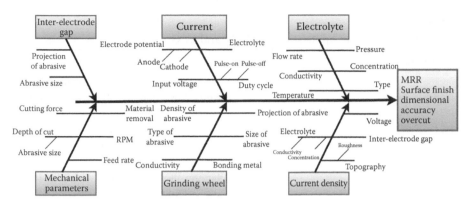

FIGURE 13.6
Cause-and-effect diagram of ECG.

13.4 Grinding Wheel

Ceramic is the major constituent of vitrified grinding wheels. The melting temperature of ceramics is significantly higher than that of metallic substances, which is meant to make the wheels conductive. Impregnation is a technique that induces the property of electrical conductivity to a vitrified wheel but needs a lot of post-processing steps (drying, firing, etc.). In impregnation, the wall of the pores in the wheel is coated by metallic substances and there is a tendency to have a non-uniform internal conductivity of the wheel. This leads to variations in performance of the wheel [13].

Amero et al. [13] invented a technique of manufacturing vitrified bonded wheels with excellent electrical conductivity. In addition, they can retain their shape better than conventional vitrified wheels and are comparable to resin bonded wheels. Four steps in manufacturing such wheels are as follows: (a) A suitable conductive substance (metal) is converted into a fine powder. (b) The fine powder of the metal is mixed into ceramic bonding material. (c) Abrasive particles are combined to this blend (metal powder and ceramic bonding material). (d) The whole combination is then moulded into a desired shape and fired to a temperature below the melting temperature of metallic powder.

Manufacturing of the wheel used in ECG is an essential task because the main purpose of the abrasive particles is to maintain a uniform gap between the electrically conductive wheel and the workpiece throughout the rotation. In addition, secondary material removal by the abrasive particles takes place by removing the oxide layer formed during electrolysis [14] and material removal from the workpiece in general. Metal (bronze) bonded diamond composite wheels are used to grind electrically conductive ceramics [15].

TABLE 13.2

Electrolytes and Their Applications for ECM and ECG

Electrolyte	Application
NaCl	Steel, cast iron, nickel and cobalt alloys
$NaNO_3$	Ferrous alloys, titanium alloys, white cast iron
HCl	Nickel alloys
NaCl + H_2SO_4	Nickel alloys
10% HF + 10% HCl +10% HNO_3	Titanium alloys
NaOH	Tungsten carbide
KCl, $NaNO_3$, NaCl	Titanium alloys, aluminium alloys, Cu alloys
$NaClO_3$	Hardened steel, grey cast iron
NaCl/KCl, NaOH	Molybdenum

Source: Chouhan, A.S., Micro tool fabrication using electro chemical micro machining, M. Tech thesis, Indian Institute of Technology Kanp, 2009.

13.5 Electrolytes

Sodium chloride is an efficient electrolyte for grinding ferrous, nickel and cobalt alloys. But in some specific cases, NaCl is not recommended since it is highly aggressive towards anodic dissolution as well as corrosion. The problems of rust and tolerance control have more concern when NaCl is used as an electrolyte. $NaNO_3$ (Sodium nitrate) is therefore employed to many alloys and tungsten carbide. Additives are added to $NaNO_3$ such as rust inhibitor and chelating agent (which forms bond with single metal ion) [16]. Some of the electrolytes and their applications are shown in Table 13.2.

13.6 Material Removal in ECG

This section discusses material removal in ECG in three parts: Section 13.6.1 discusses the mathematical formulation of a linear MRR for a special case of ECG of tungsten carbide (WC)-Co. Section 13.6.3 discusses the effect of hydrogen gas produced during the process on MRR. Section 13.5.3 describes the effect of applied voltage, feed rate and depth of cut on MRR.

13.6.1 Material Removal Rate Analysis

It is assumed that in ECG, mechanical and EC removal processes take place independently, hence total volumetric material removal will be [1]

$$m = m_m + m_{ec}$$ (13.1)

where m_m = material removed by mechanical action (grinding) process and m_{ec} = material removed by EC action.

m_m can be obtained by experimentally performing a grinding operation without supplying electrolyte and electrical power to the system to have a comparison between theoretical and experimental MRRs due to grinding only. Overall MRR can be achieved by weight difference methodology. MRR due to ECM can be obtained by subtracting experimental MRR due to grinding from the overall experimental MRR. Determination of the theoretical MRR (m_{ec}) due to ECM can be obtained from the Faraday's laws of electrolysis as follows:

$$m_{ec} = \alpha I, \tag{13.2}$$

where I is current and α is a proportionality constant, which can be defined by Faraday's second law:

$$\alpha = \frac{M}{FZY}, \tag{13.3}$$

where F is Faraday's constant ($F = 96500$ A-s), M is atomic mass, Z is valency of the anode material and Y is density.

The efficiency (η) of the EC material removal can be expressed as

$$\eta \equiv \frac{m - m_m}{\alpha I}. \tag{13.4}$$

13.6.2 EC Analysis

To understand the mechanism of ECG process, let us consider the analysis of machining of WC-Co. EC dissolutions of carbide and cobalt are determined [1] as follows.

Anodic dissolution of cobalt takes place as follows:

$$Co \rightarrow Co^{+2} + 2e^- \ (z_{Co} = 2). \tag{13.5}$$

Anodic dissolution of tungsten carbide takes place as follows:

$$WC + 4H_2O \rightarrow WO_3 + Co + 4H_2 + 8e^- \ (z_{WC} = 8). \tag{13.6}$$

Volumetric MRR according to Faraday's law is shown in Equation 13.2. The linear (one-dimensional) MRR for a part of cross-sectional area A can be determined by dividing m_{ec} by the cross-sectional area A in Equation 13.2.

Hence, specific etching velocity $MRR_{l,e}$ is given as

$$MRR_{l,e} = \alpha \frac{1}{A} = \alpha J, \tag{13.7}$$

where J is current density.

Assumption: Cobalt and tungsten carbide are both considered at the same level at the surface as EC dissolution starts. Cobalt and tungsten carbide are considered to behave as parallel electrical conductors from the surface to a small active depth h into the material.

The total current is the sum of the currents through cobalt (I_{Co}) and tungsten carbide (I_{WC}):

$$I = I_{Co} + I_{WC}. \tag{13.8}$$

The voltage V_h drop across this active zone is

$$V_h = I_{Co}R_{Co} = I_{WC}R_{WC}, \tag{13.9}$$

where R_{Co} and R_{WC} are the resistances of both phases:

$$R_{Co} = \frac{\rho_{Co}h}{A_{Co}} \tag{13.10}$$

and

$$R_{WC} = \frac{\rho_{WC}h}{A_{WC}}, \tag{13.11}$$

where ρ_{WC} and ρ_{Co} are the resistivities of the WC and cobalt phases, respectively.

The total cross-sectional area is then

$$A = A_{Co} + A_{WC}. \tag{13.12}$$

Substituting Equations 13.10 and 13.11 in Equation 13.9, it can be shown [1]:

$$V_h = I_{Co}\frac{\rho_{Co}h}{A_{Co}} = I_{WC}\frac{\rho_{WC}h}{A_{WC}}$$

and

$$V_h = I_{Co}\frac{\rho_{Co}}{\rho_{WC}} = I_{WC}\frac{A_{Co}}{A_{WC}}.$$

Defining $\dfrac{\rho_{Co}}{\rho_{WC}} = \beta$ (ratio of resistivity of cobalt to WC) and $\dfrac{A_{WC}}{A_{Co}} = \varepsilon$ (ratio of area of WC to cobalt),

$$V_h = I_{Co}\beta = I_{WC}\frac{1}{\varepsilon}.$$

Substituting I_{Co} and I_{WC} into Equation 13.8, we get

$$I_{Co} = \frac{I}{1+\beta\varepsilon}, \tag{13.13}$$

$$I_{WC} = I\frac{\varepsilon\beta}{1+\varepsilon\beta}, \tag{13.14}$$

$$\beta \equiv \frac{\rho_{Co}}{\rho_{WC}} \simeq 0.30, \tag{13.15}$$

and

$$\varepsilon \equiv \frac{A_{WC}}{A_{Co}}. \tag{13.16}$$

From Faraday's first law (Equation 13.2), the initial volumetric dissolution rate of each phase can now be written as [1]

$$m_{ec,Co} = \alpha_{Co}I_{Co} = \alpha'_{Co}I, \tag{13.17}$$

$$m_{ec,WC} = \alpha_{WC}I_{WC} = \alpha'_{WC}I, \tag{13.18}$$

$$\alpha'_{Co} \equiv \frac{\alpha_{Co}}{1+\varepsilon\beta} \tag{13.19}$$

and

$$\alpha'_{WC} \equiv \frac{\alpha_{WC}\varepsilon\beta}{1+\varepsilon\beta}.$$

(13.20)

In terms of specific etching velocities from Equation 13.7,

$$MRR_{l,Co} = \alpha'_{Co}(1+\varepsilon)\frac{I}{A} = \alpha'_{Co}(1+\varepsilon)J$$

(13.21)

and

$$MRR_{l,WC} = \alpha'_{WC}\left(\frac{1+\varepsilon}{\varepsilon}\right)\frac{I}{A} = \alpha'_{WC}\left(\frac{1+\varepsilon}{\varepsilon}\right)J.$$

(13.22)

The ratio of etching velocity of cobalt to tungsten carbide is

$$\frac{MRR_{l,Co}}{MRR_{l,WC}} = \frac{\varepsilon\alpha'_{Co}}{\alpha'_{WC}}.$$

(13.23)

Cemented carbide consisting of 94% WC and 6% Co by weight results in ratio of area $\varepsilon = 8.83$. For this material, the ratio of the initial specific etching velocity of cobalt to that of tungsten carbide in Equation 13.23 is equal to 7.2.

The previous analysis shows that the initial specific etching of cobalt is much faster than that of WC. But this does not continue throughout the process because the higher removal rate of cobalt results in a decreased value of I_{Co}. To satisfy Equation 13.8, the current I_{WC} must increase consequently. The decreasing current in the cobalt and increasing current in WC results in a decreased value of the specific etching velocity of cobalt and increased value of the specific etching rate of WC. At some point, a condition arrives at which both specific etching rates become equal [1].

Beyond that, the machining rate of both phases reaches the same specific etching velocity. The specific etching velocity of both phases at the steady-state condition can be calculated. $MRR^*_{l,CO}$ and $MRR^*_{l,WC}$ denote the steady-state specific etching velocities of cobalt and tungsten carbide, respectively.

$$MRR^*_{l,Co} = \frac{\alpha_{Co}I^*_{Co}}{A_{Co}}$$

(13.24)

$$MRR^*_{l,WC} = \frac{\alpha_{WC}I^*_{WC}}{A_{WC}}$$

(13.25)

For parallel electrical current flow,

$$I^* = I^*_{WC} + I^*_{Co}. \tag{13.26}$$

Solving Equations 13.24 through 13.26 and combining with Equations 13.12 and 13.16, we get the following:

$$I^*_{CO} = \left(\frac{\alpha_{WC}}{\alpha_{WC} + \varepsilon \alpha_{Co}} \right) I \tag{13.27}$$

and

$$I^*_{WC} = \left(\frac{\varepsilon \alpha_{Co}}{\alpha_{WC} + \varepsilon \alpha_{Co}} \right) I. \tag{13.28}$$

From Equations 13.27 through 13.28 and 13.24 through 13.25, the specific etching velocities at the steady state (or equilibrium condition) are

$$MRR^*_i = MRR^*_{i,Co} = MRR^*_{i,WC} = \left[\frac{\alpha_{Co}\alpha_{WC}(1+\varepsilon)}{\alpha_{WC} + \varepsilon \alpha_{Co}} \right] i = \alpha^* J, \tag{13.29}$$

where $\alpha^* = \left[\dfrac{\alpha_{Co}\alpha_{WC}(1+\varepsilon)}{\alpha_{WC} + \varepsilon \alpha_{Co}} \right] J.$

In this analysis, the passivation of WC due to oxide film (Equation 13.6) is accounted. Such oxide layer should minimise the current density in the WC phase and henceforth further encourages selective etching of the cobalt phase. Removal of this oxide layer by mechanical grinding action during ECG should minimise the passivation.

13.6.3 Effect of Hydrogen Bubbles on MRR

Measurement of the mechanical grinding forces acting on the workpiece can be done using a strain gauge dynamometer. There are two forces involved in the mechanical grinding process: (1) vertical force and (2) horizontal force. In a pure mechanical grinding process, these forces are affected by feed, depth of cut and rotation per minute (RPM) of the wheel. But in ECG, it was experimentally found that the vertical force decreases as the current density increases. Figure 13.7 shows the variation of mechanical MRR with grinding force and current density [18]. Figure 13.7a shows that as the mechanical grinding force increases, the mechanical MRR increases, which is obvious due to the increased rate of erosion. But when the current density increases, it has been recorded that mechanical MRR decreases, as shown

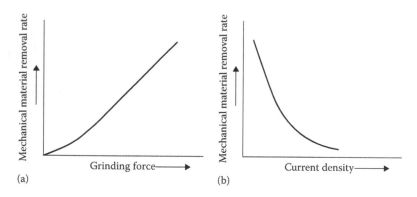

FIGURE 13.7
Variation of mechanical MRR with (a) grinding force and (b) current density.

in Figure 13.7b. As the current density increases, the linear MRR due to EC dissolution increases, which results in an increased IEG. Hence, there is a reduction in mechanical force on the workpiece exerted by the ECG wheel. Further, the force in the upward direction exerted by bubbles increases as the bubble volume fraction increases due to increase in current density [18]. This would also contribute to a decrease in MRR due to erosion.

13.6.4 Effect of Input Parameters on MRR

The major amount of material removal takes place due to EC dissolution; hence, MRR mainly depends on the amount of current flowing between the electrodes. The trends of variation in MRRs are shown in Figure 13.8 [14]. MRR increases as applied voltage increases according to Faraday's law of electrolysis (Figure 13.8a) in all the grades of diamond-impregnated and aluminium oxide metal bonded wheels. Figure 13.8b shows an increase in MRR with increasing feed rate. Figure 13.8c shows that an increase in the set depth of the cut slightly increases the MRR. Approximately a linear trend of variation is observed in all the grades of wheels [14].

The MRR also marginally increases with increasing grit size, with variation in applied voltage and feed rate. But when the depth of cut is increased, the MRR is higher in the case of smaller grit size. This is because the depth of cut is set according to the reference of the outer periphery of the wheel having protruded grits. Hence, when the depth of cut is set with a finer grit wheel, it leads to a reduction in the effective depth of cut and IEG. Therefore, more current flows between the wheel and workpiece, resulting in a higher MRR. Generally, low-concentration wheels lead to higher current produced due to the lower number of abrasive particles per unit area. It results in a higher apparent area for EC dissolution. Hence, a marginally higher MRR is obtained as compared to higher concentrated (with abrasive particles) wheels.

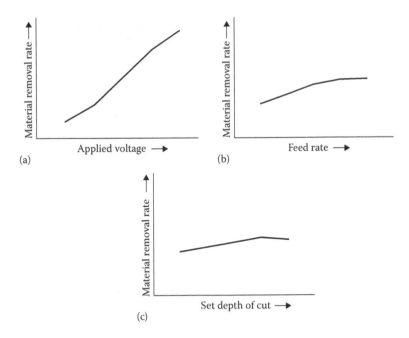

FIGURE 13.8
Variation of MRR on varying (a) applied voltage, (b) feed rate and (c) depth of cut.

13.7 Overcut in ECG

In conventional ECG, the overcut can be reduced to a minimal value (almost zero) by employing a pulsating input voltage. In ECG, the overcut is influenced by all the input parameters, such as input voltage, current density, feed rate, depth of cut, wheel characteristics and electrolyte properties. Almost all tool design processes for ECM tools are aimed to minimise overcut in ECM. These tool design processes with certain modifications are applicable for ECG. Hence, it is very important to comprehend the relationship between process input parameters and overcut to enhance the process efficiency [8]. The effects of some important input parameters on overcuts are discussed in the following sections.

13.7.1 Input Voltage

Input voltage is a significant parameter that controls the magnitude of overcut observed during ECG of slots, cavities, holes, etc. As the input voltage is increased, the rate of EC dissolution of the anode increases. Figure 13.9a shows the relationship between the overcut and the average applied voltage to compare direct P-ECG and DC-ECG during grinding of tool and die steel [8]. It can be seen that at lower voltages, P-ECG supply results in a smaller

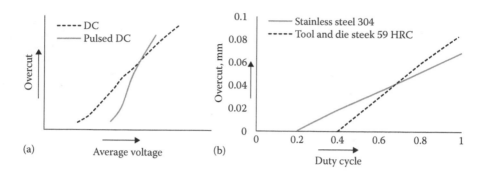

FIGURE 13.9
Variation in overcut. (a) Relationship between overcut and input voltage in ECG with pulsed and continuous DC power supply. (b) Relationship between duty cycle and overcut for stainless steel and tool and die steel workpieces.

overcut (0.03 mm) compared to smooth DC-ECG. At lower voltages, a greater area of the wheel remains in contact with the workpiece than at higher voltages. When the voltage is increased, EC dissolution dominates the mechanical removal and leads to a higher value of overcut. In P-ECG, when pulse-on time (or duty cycle) is smaller, there are higher chances of a larger area of abrasive grits eroding the workpiece.

13.7.2 Duty Cycle in Pulsed ECG

Figure 13.9b shows the relationship between an overcut and a duty cycle in a pulsed ECG using a diamond wheel for grinding of tool and die steel and stainless steel [8]. A duty cycle of 1 represents continuous ECG and less than 1 duty cycle represents pulsed ECG. The results discussed earlier correspond to the zone where only EC machining takes place. The figures show that duty cycles of 0.2 and 0.4 are responsible for initiating machining/overcut in the tool-die steel and stainless steel workpieces, respectively. Hence, the trend of overcutting on varying duty cycles of the power source suggests that a low duty cycle corresponds to a lesser overcut and a higher accuracy during ECG operation, but at the cost of machining time. Geva et al. [19] presented experimental observations of overcuts on varying voltages at different depths of cut with different carbide grades (K20, K30 and K40). During grinding with a K20 wheel at 0.00 mm depth of cut, when the voltage is varied from 5 V to 10 V, it results in a linear increment in overcut from 5 to 6 μm. If the depth of cut is increased to 0.75 mm, it leads to an increase in overcut significantly (30–40 μm). A low depth of cut lowers the interaction of the cathode and anode, hence the low material removal, which results in a low overcut. Using K30 and K40 grades, it is observed that a lower depth of cut results in a higher overcut.

13.8 Surface Roughness in ECG

The trends observed in the variation of surface roughness by varying various operating parameters are shown in Figure 13.10 [14]. When material removal takes place only due to EC dissolution, there is no particular influence of wheel parameters. When the input voltage is increased beyond 10 V, wheel variables (feed, speed and depth of cut) come into action. Figure 13.10a shows that surface roughness (*Ra* value) increases upon increasing the applied voltage. In general, the surface roughness (1.27–1.78 µm) at 10 V is better than that at 15 V. Machined surfaces at 5 V and 10 V show evidence of mechanical material removal in the form of scratches. Mechanical material removal is greater at 5 V than that at 10 V. In a low-voltage ECG operation, wheel parameters such as grit size and grit type affect the machined surface roughness more than EC dissolution. Larger abrasive particles penetrate deeper, hence resulting in greater roughness. Increasing the feed rate shows a considerable improvement in surface finish (by decreasing the roughness value) because it reduces the time of interaction of stray cutting (Figure 13.10b). Figure 13.10c shows the variation of surface roughness with increasing depth of cut. The effect is not very predictive based on these results because of the erratic nature of variation [14].

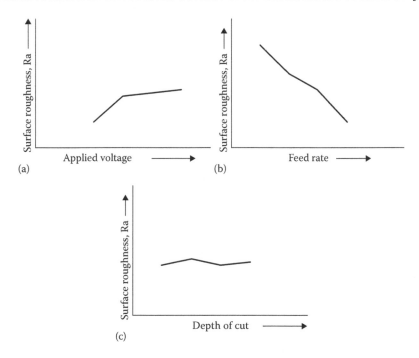

FIGURE 13.10
Variation of surface roughness with (a) increasing applied voltage, (b) feed rate and (c) depth of cut.

Different types of wheels lead to distinct machined surface characteristics. The harder the material (diamond abrasive wheel), the deeper is the crater formed. Hence, a surface machined with a Norelek wheel has a lower surface roughness value (better surface finish) than that with a diamond abrasive wheel. The hardness and sharpness of the abrasive particles also play an important role in deciding the surface integrity of the machined part. Diamond abrasive particles are harder as well as sharper than aluminium oxide abrasive particles. The quality of impregnation of the wheel of ECG may also affect the quality of the machined surface up to a considerable extent.

According to the experimental investigations of Geva et al. [19], during ECG of WC-Co, the surface roughness value increases with increasing input voltage and the surface roughness decreases with increasing depth of cut. In a certain range, when ECG is performed at 5 V applied voltage with 0.75 mm depth of cut, the roughness value achieved is 2 µm. If the voltage is increased to 8 V with the same depth of cut, the roughness value attained is higher (6 µm) due to increased electric potential and current density. But a further increment in applied voltage leads to a decreasing trend of roughness value. In general, the net effect of increasing input voltage increases the roughness of the machined surface. A passivation layer formed during the EC dissolution results in differential material removal. Thus, the stability of the passivation layer also plays an important role in deciding the selective dissolution [19]. In the active region, increasing voltage results in a high potential between the electrodes, and increased current density leads to rapid oxidation and reduction. Hence, higher values of overcut and roughness are attained. In passive regions, the values of roughness and overcut are low.

13.9 Application of Magnetic Field and Specific Energy

Btyumenfel'd and Gol'dfeld [20] demonstrated the relation of the magnetic field with the conductivity of aqueous solutions of an electrolyte. Simultaneous action carried by the magnetic field and electric potential leads to a change in motion (transport number) of ions; hence, conductivity is changed.

$$\text{Specific energy} = \text{Spindle power/total metal removal}$$

In the mechanical grinding process, the specific energy (W/mg-min) required is comparatively higher than in other conventional machining processes. In the case of ECG, specific energy substantially reduces because of EC dissolution of the anode, and hence, resistance to metal removal also decreases due to passive layer formation and its removal. When a magnetic field is

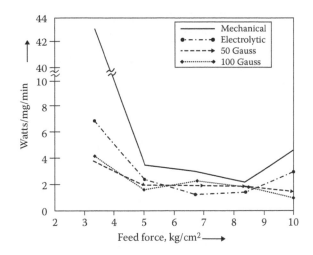

FIGURE 13.11
Variation of specific energy with feeding force. (From Kuppuswamy, G., *Wear*, 54, 257–272, 1979.)

applied to the grinding zone, the conductivity of the electrolyte is enhanced, which results in decreased specific energy. At lower and higher feed force conditions, when ECG is assisted with a suitable magnetic field, it results in increased MRR. Figure 13.11 shows the effect of magnetic field on specific energy with varying feed forces [21].

Faraday efficiency is defined as the electrolytic portion of the MRR divided by the theoretical electrolytic removal rate. In ECG, a diamond wheel on P20 carbide specimen leads to a Faraday efficiency of 85%, which is evidence of effective electrolysis and implies a higher MRR. When ECG is performed with the assistance of a magnetic field, increased Faraday efficiency (more than 100%) is recorded.

13.10 Wear of Cathode in ECG

Wear of a grinding wheel (cathode) occurs in the ECG process in a considerable amount, which should be accounted for increasing the quality of the ground surfaces. The wear of the metal bonded grinding wheel (abrasive tool) is determined by measuring loss in mass or degradation of outer layer and change in mass of particles or change in volume of abrasive particles. Different mechanisms of wear are categorised in Figure 13.12 [22].

The mechanism of wear of abrasive particles encompasses the concurrent effect of the following processes: mechanical abrasion, chemical and EC dissolution, adhesion, diffusion, cracking, spalling of particles and

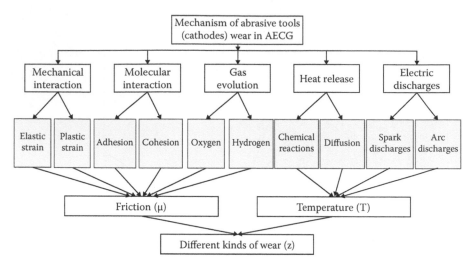

FIGURE 13.12
Mechanisms of wear of abrasive wheels in ECG. (From Zaborski, S., Łupak, M., Poro, D., *J. Mater. Process. Technol.*, 149, 414–418, 2004.)

metallic binder. These factors act in different ways in different modes of material removal in ECG. In a mechanical material removal zone, the mechanism of material removal is dependent on the coefficient of friction, μ. In the EC dissolution zone, the wear mechanism depends on the temperature gradient generated due to chemical reactions, EC reaction, diffusion, sparking or arc discharge.

Forms of abrasive wear:

1. The wear of the abrasive wheel, which occurs as a consequence of the heat generated in the machining zone during the process, is mainly affected by adhesion and diffusion. As a result of that, the hardness of the abrasive particles decreases and the resistance of the cutting edge reduces. In the IEG, the temperature of the zone crucially affects the wear mechanism of abrasive particles.

2. Cracking of the abrasive particles and spalling of top layers of particles occur under high stress due to friction.

3. Chemical wear is caused by a corrosive environment and high current density. Hydrogen gas generated due to EC reactions destruct the abrasive particles and metallic binder, revealing the spalling of particle fragments and the metallic binder [22].

Figure 13.13 shows more specific type of wear and the dominant form of the wear.

Figure 13.14 shows the difference in three different bonded wheels. This results in rubbing of the workpiece during grinding. In resin bonded wheels,

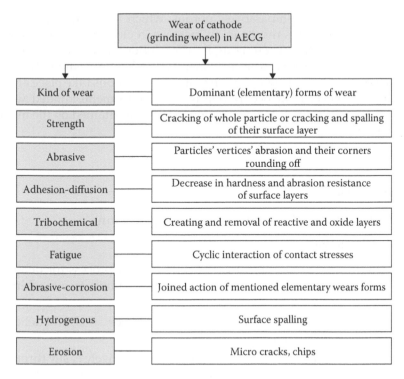

FIGURE 13.13
Types of abrasive tool (cathode) wear in ECG. (From Zaborski, S., Łupak, M., Poro, D., *J. Mater. Process. Technol.*, 149, 414–418, 2004.)

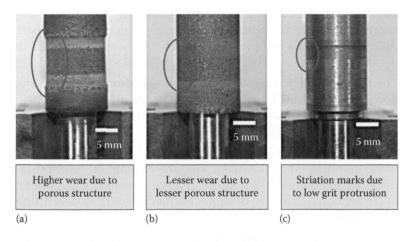

FIGURE 13.14
Wear scars on (a) resin bond, (b) electroplated type A and (c) metal bond tools. (From Curtis, D.T., Soo, S.L., Aspinwall, D.K., Sage, C., *CIRP Ann. Manuf. Technol.*, 58, 173–176, 2009.)

the bond structure is comparatively porous, which results in a higher wear rate (Figure 13.14a). This results in an increase in electrolyte reachability in the IEG, which implies encouraging EC dissolution rather than mechanical material removal. Porosity in electroplated-type wheels is comparatively less than that of resin bonded wheels; hence, less EC reactions take place and less resin-bonded wheel wear occurs (Figure 13.14b). In metal bonded wheels, grit protrusion is limited; hence, striation marks are obtained on the metallic bonded surface (Figure 13.14c) [23].

13.11 Advantages, Limitations and Applications

13.11.1 Advantages

1. Conventional grinding process requires a post-process for deburring the machined surface, but ECG provides a burr-free machined surface.

2. ECG results in a stress-free ground surface; hence, small and delicate products can be ground effectively.

3. Hard-to-grind materials can be ground comparatively easily through the ECG process, since the MRR is independent of the hardness and temper of metals. ECG is moreover a 'cold' process as the temperature at the wheel–workpiece interface does not rise beyond 100°C [24].

4. In the grinding process, the required intense specific cutting energy leads to a very high temperature rise. EC ground parts are without metallurgical damage (such as work hardening, structural change, micro-cracks, etc.) and without any change in mechanical properties (hardness) because comparatively much less heat is generated during the ECG process.

5. The life of EC grinding wheel is higher as compared to a conventional grinding wheel; hence, the cost of production is lower. Further, frequent wheel dressing is not required since wheel loading and glazing are rare in ECG [25].

6. Surface finish and tolerance precision are higher than in conventional grinding since the major amount of material removal takes place due to EC anodic dissolution. The valleys and crests generated in conventional grinding disappear when using EC-assisted finishing or grinding [7].

7. The ECG process has a higher MRR since it has a hybrid mechanism of material removal. A larger depth of cut can be employed to increase the production rate. Metal removal from the anode per unit of abrasive particle is always greater than in conventional grinding [15].

13.11.2 Limitations

1. A material that is electrochemically reactive can only be ground by ECG.
2. ECG performs effectively only when the workpiece is electrically conductive.
3. Generated overcuts due to un-controlled ECM by the side face of the wheel limit the ECG process; hence, the accuracy is affected [24].
4. Maintaining least IEG for ECG to happen requires very precise control during the feed.

13.11.3 Applications

The ECG process is suitable for grinding exotic alloys, carbides and other hard-to-machine materials. ECG is generally employed in the space and nuclear industries for special purposes, such as grinding of heat- and stress-sensitive materials, form grinding, face or peripheral grinding, etc. [24]. Metals such as steels, Haste alloy, aluminium (Al), copper (Cu) and Inconel alloys can be effectively machined by ECG. Machining of other materials such as nickel/titanium, cobalt alloys, Rene 41, rhenium, rhodium, stelllite, vitalium, zirconium and tungsten is also possible [25].

13.12 Electrolytic In-Process Dressing Grinding

Electrolytic in-process dressing (ELID) [26] is a hybrid process in which electrolytic anodic dissolution is employed for dressing of the metal bonded grinding wheel and mechanical grinding is subjected to material removal from the workpiece (electrically conducting, non-conducting or both). The ELID grinding process also forms an electrolytic cell where anodic dissolution is confined to the surface of a metal bonded grinding wheel. In the ELID grinding process, EC dissolution takes place on a metal bonded grinding wheel and material is removed through erosion from the workpiece, whereas in ECG, EC dissolution takes place on the (only electrically conducting) workpiece. In general, the ELID process is preferred to a normal grinding process in which wheel loading and wheel wear rate are higher due to the mechanical properties of the workpiece. The ELID system consists of a metal bonded grinding wheel, an electrode that covers 1/6th to 1/4th of the periphery of the wheel, a DC (pulse or continuous) power supply and electrolyte as a conductive medium between the electrodes, as illustrated in Figure 13.15.

The grinding wheel is connected to the positive terminal, and the electrode is connected to the negative terminal of the DC power source. Hence,

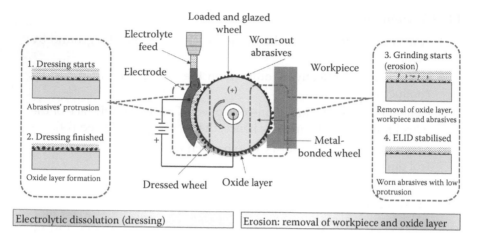

FIGURE 13.15
Illustration of an ELID grinding system and mechanism of material removal.

the metal bonded grinding wheel becomes an anode (where oxidation takes place) and the electrode becomes a cathode (where reduction takes place). The wheel and electrode are retained at a small gap of 0.1 to 0.3 mm for maintaining an electrolyte band in between. It leads to the occurrence of electrolysis (redox reactions) due to electric potential across the electrolyte [26].

The process starts with grinding of the workpiece material, and metal removal occurs due to erosion through the metal bonded wheel. This erosion phenomenon results in the wear of abrasive particles; hence, the protrusion of abrasive particles is reduced. Another part of the system that consists of an electrolytic cell starts electrolytic dissolution of the metal bonded wheel and results in a fresh surface having abrasives of higher protrusion (trued wheel). Due to the redox reactions, the solid metal converts into metal ions and forms metal oxide, and hence, the wheel surface is oxidised. Thus, an oxide layer is deposited during the electrolysis on the surface of the wheel, which reduces further anodic dissolution. In the next step, the oxide layer is removed from the wheel surface due to erosion, and trued wheel removes material from the workpiece in the form of chips.

ELID is classified into four major groups based on the workpiece to be ground and the nature of the job, even though the working principle of in-process dressing is same for all the methods. The four groups are listed as follows [26]:

1. ELID-I–Electrolytic in-process dressing
2. ELID-II–Electrolytic interval dressing
3. ELID-III–Electrolytic electrode-less dressing and
4. ELID-IIIA–Electrolytic electrode-less dressing using alternate current

From the experimental studies of the process, it can be stated that ELID grinding is an efficient process for producing a surface with nanometre level surface finish on various materials such as metals, glasses and ceramics. However, this technique needs fundamental analysis to produce the desired results. The robustness of the process can be increased by integrating the advanced motion controlling devices and intelligent processors.

Review Questions

1. What is the working principle of ECG?

2. What are the three distinct zones in ECG? Explain with the help of a schematic diagram.

3. Why is a pulsed power supply better for increasing accuracy?

4. What is duty cycle? What range of duty cycle will result in higher accuracy of machining?

5. What are the variants of the ECG process?

6. Draw a schematic of the ECGB operation and describe the different machining phases.

7. Derive an equation for a linear MRR by considering ECG of a tungsten carbide cobalt (WC-Co) workpiece.

8. Explain the effect of hydrogen bubbles produced during the ECG operation. How it is related with MRR?

9. What is overcut in ECG? How can it be minimised?

10. How does the surface roughness during an ECG operation depend on applied voltage, feed rate and depth of cut? Explain with the help of graphs.

11. What happens if a magnetic field is applied to the ECG operation? Describe the changes in specific energy and MRR.

12. What are the mechanisms of grinding wheel wear? Describe the type of wheel wear occurring in the ECG process.

13. Write down the merits and demerits of the ECG process.

14. Draw a schematic of ELID and describe how it is different from ECG.

15. Why is ELID called as a hybrid process? What are the different mechanisms involved in material removal during ELID?

Nomenclature

A	Cross-sectional area
A_{Co}	Cross-sectional area of cobalt phase

A_{WC}	Cross-sectional area of carbide phase
F	Faraday's constant ($F = 96500$ A-s),
h	Active layer depth (interelectrode gap)
I	Current
I_{Co}	Current in cobalt phase
I_{WC}	Current in tungsten carbide phase
J	Current density
m	Total volumetric material removal
m_m	Material removed by grinding process
m_{ec}	Material removed by electrochemical process
M	Molecular weight
$MRR_{l,e}$	Specific etching velocity
MRR_l^*	Specific etching velocity at steady state
$MRR_{l,Co}, MRR_{l,Co}^*$	Initial and steady state specific etching velocities into cobalt phase
$MRR_{l,WC}, MRR_{l,WC}^*$	Initial and steady state specific etching velocities into tungsten carbide phase
V_h	Voltage drop in active layer
Z	Valency of anode material
z_{WC}, z_{Co}	Valences of tungsten carbide and cobalt phases
α^*	Proportionality constant for initial steady state etching
$\alpha, \alpha_{Co}, \alpha_{WC}$	Proportionality constants for Faraday's First Law
$\alpha_{Co}', \alpha_{WC}'$	Proportionality constant for initial steady state etching
β	Ratio of resistivity of cobalt to resistivity of tungsten carbide
γ	Density of anode material
ε	Ratio of area of tungsten carbide to cobalt
η	Efficiency of the EC material removal
ρ_{Co}, ρ_{WC}	Resistivity of cobalt and tungsten carbide

References

1. R. Levinger, S. Malkin, Electrochemical grinding of WC-Co cemented carbides, *ASME J. Eng. Ind.* 101 (2015) 285–294.
2. A.K. Chouksey, Modelling of electrolyte conductivity during electrochemical machining, M. Tech. Thesis, IIT Kanpur, Kanpur (UP), India (2015).
3. G. Kuppuswamy, Wheel variables in electrolytic grinding, *Tribol. Int.* 9 (1976) 29–32. doi:10.1016/0301-679X(76)90067-0.
4. V.K. Jain, *Advanced machining processes*, New Delhi: Allied Publisher, 2002.
5. S. Jha, V.K. Jain, Nanofinishing techniques, in: *Micromanufacturing and Nanotechnology*, (Ed. N. P. Mahalik), Berlin Heidelberg, New York: Springer, 2006: pp. 171–195.

6. V.K. Jain, A. Sidpara, M.R. Sankar, M. Das, Nano-finishing techniques: A review, *Proc. Inst. Mech. Eng. Part C J. Mech. Eng. Sci.* 226 (2011) 327–346. doi:10.1177/0954406211426948.

7. N. Ma, W. Xu, X. Wang, B. Tao, Pulse electrochemical finishing: Modeling and experiment, *J. Mater. Process. Technol.* 210 (2010) 852–857. doi:10.1016/j.jmatprotec.2010.01.016.

8. A.F. Tehrani, J. Atkinson, Overcut in pulsed electrochemical grinding, *Proc. Inst. Mech. Eng. Part B J. Eng. Manuf.* 214 (2000) 259–269.

9. R.M. Bell, Belt type electro-chemical (or electrolytic) grinding machine, United States Pat. Office. (1969) 3,448,023.

10. R.M. Bell, Belt type electrolytic grinding machine, United States Pat. Off. (1964) 17–20.

11. M. Yoshino, T. Shirakashi, T. Obikawa, E. Usui, Electrolytic cut-off grinding machine for composite materials, *J. Mater. Process. Technol.* 74 (1998) 131–136.

12. D. Zhu, Y.B. Zeng, Z.Y. Xu, X.Y. Zhang, Precision machining of small holes by the hybrid process of electrochemical removal and grinding, *CIRP Ann. Manuf. Technol.* 60 (2011) 247–250. doi:10.1016/j.cirp.2011.03.130.

13. J.J. Amero, Vitrified bonded wheel for electrochemical grinding containing coductine metal and a thermoset polymer filler, United States Pat. Off. 3,535,832 (1965) 2–6.

14. A. Geddam, C.F. Noble, An assessment of the influence of some wheel variables in peripheral electrochemical grinding, *Int. J. Mach. Tool Des. Res.* 11 (1971) 1–12. doi:10.1016/0020-7357(71)90043-6.

15. T.M.A. Maksoud, A.J. Brooks, Electrochemical grinding of ceramic form tooling, *J. Mater. Process. Technol.* 55 (1995) 70–75.

16. Electrochemical grinding, http://osp.mans.edu.eg/s-hazem/NTM/ECG.html (2015).

17. A.S. Chauhan, Micro tool fabrication using electro chemical micro machining, M. Tech. thesis, Indian Institute of Technology Kanpur (2009).

18. R.R. Cole, An experimental investigation of the electrolytic grinding process, *J. Eng. Ind.* 83 (1961) 194–201.

19. M. Geva, E. Lenz, S. Nadiv, Peripheral electrochemical grinding of sintered carbides–Effect on surface finish, *Wear* 38 (1976) 325–339.

20. L.A. Btyumenfel'd, M.G. Gol'dfeld, The effect of magnetic field on the electrical conductivity of water and aqueous solutions of electrolytes, *Inst. Chem. Physics, Acad. Sci. USSR* 9 (1968) 379–384.

21. G. Kuppuswamy, An investigation of a magnetic field on electrolytic diamond grinding, *Wear* 54 (1979) 257–272.

22. S. Zaborski, M. Łupak, D. Poro, Wear of cathode in abrasive electrochemical grinding of hardly machined materials, *J. Mater. Process. Technol.* 149 (2004) 414–418. doi:10.1016/j.matprotec.2004.02.015.

23. D.T. Curtis, S.L. Soo, D.K. Aspinwall, C. Sage, Electrochemical superabrasive machining of a nickel-based aeroengine alloy using mounted grinding points, *CIRP Ann. Manuf. Technol.* 58 (2009) 173–176. doi:10.1016/j.cirp.2009.03.074.

24. A. Tiwari, G.K. Lal, Dimensional analysis and process optimization of electrochemical grinding, M. Tech. Thesis, IIT Kanpur (1983).

25. Everite, Electrochemical grinding process review, www.everite.com/capabilities /burr-free-cutting-machining/electrochemical-grinding-process-overview /(n.d.).

26. F.K. Patham, The electrolytic in process (ELID) grinding, in *Introduction to Micromachining*, 2nd edition (Ed: V.K. Jain), New Delhi: Narosa Publishing House (2014).

14

Electrochemical Magnetic Abrasive Finishing

K.B. Judal[1] and Vinod Yadava[2]

[1]*Government Engineering College Patan, Gujarat, India*

[2]*Motilal Nehru National Institute of Technology Allahabad, Uttar Pradesh, India*

CONTENTS

14.1 General

The quality of surface has great influence on the functional properties of major engineering parts, viz., wear resistance, power loss in friction, fatigue life, etc. Therefore, modern manufacturing industries demand high-quality surfaces and also high efficiency of the machining process to meet the present demand of the market. Traditional machining consists of a single process that cannot satisfy the current demand for both high quality and high efficiency simultaneously. With the development in recent engineering materials, the conventional methods of abrasive finishing, viz., grinding, honing and lapping, may not satisfy the demand of high-quality surface with high efficiency because of their limits of performance. Thus, an advanced abrasion-based hybrid process has to be developed that integrates advanced abrasion-based machining process with a non-abrasive process to meet the demand of finishing industries.

Researchers have combined ultrasonic or electrochemical elements with magnetic abrasive finishing (MAF). Ultrasonic-assisted MAF is one of the hybrid processes that integrate ultrasonic vibrations and MAF for nano-level finishing of plane surfaces in a short time [1]. When electrochemical action is combined with MAF, the process so developed is called electrochemical MAF (EMAF). EMAF can be applied to finish difficult-to-machine electrically conductive advanced engineering materials with high efficiency. During processing, first, electrochemical reaction produces a passive film (oxide layer) on the surface of the workpiece [2]. The flexible magnetic abrasive brush (FMAB) formed in MAF removes the passive layer formed, particularly from the peaks of the surface irregularities, which exposes fresh metal for further electrochemical dissolution (ECDi). Thus, EMAF produces desired surface finish efficiently due to abrasion–passivation synergism. This chapter describes in depth the understanding of EMAF process mechanism, EMAF apparatus and the effect of process variables on the performance of the EMAF process and modelling of the EMAF process using finite element method (FEM).

14.2 EMAF Process and Mechanism of Machining

EMAF is an advanced abrasion-based hybrid finishing process in which, first, ECDi produces a micron-thickness passive film on the workpiece surface, whose hardness is less than that of the parent material, which is easily removed by FMAB formed in the MAF process [2]. Also, due to the magnetic field in the electrode gap, traditional electrochemical polishing enhances the dissolving velocity of peak points and side faces [3]. This hybrid process led to the improvement in surface finish and machining efficiency due to the synergetic effect of both constituent processes and the presence of magnetic field in the inter-electrode gap (IEG) during electrochemical reaction. In the EMAF process, the material is removed from the surface of the workpiece due to the combined effect of MAF and ECDi. The combination of these two processes leads to increased passivation and/or abrasion. The material removed in the EMAF process is considered as the sum of material removal due to MAF, ECDi and synergistic material removal due to combination of MAF and ECDi. The removal mechanism of each component is discussed in the following sub-sections.

14.2.1 Mechanism of MAF

During MAF, the workpiece is kept between the N-Pole and S-Pole of a magnet, as shown in Figure 14.1. The working gap between the workpiece and the magnet is filled with magnetic abrasive particles (MAPs). MAPs are prepared by mixing ferromagnetic particles (FPs) (e.g. iron particles, steel grit, etc.) with abrasive particles (SiC, Al_2O_3, Cr_2O_3, diamond powder, etc.). MAPs can be used as bonded, unbonded or loosely bonded [4]. The MAPs join each other along the lines of magnetic force and form an FMAB between the workpiece and the magnetic pole. The gradient of the magnetic field in the

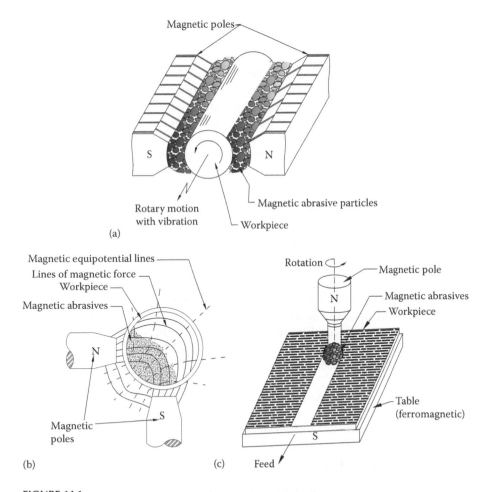

FIGURE 14.1
Applications of MAF to different configurations of workpiece. (a) External cylindrical MAF. (From Shinmura, T., Takazawa, K., Hatano, E., Aizawa, T. *Bull. Jpn. Soc. Prec. Eng.*, 19, 54–55, 1985.) (b) Internal cylindrical MAF. (From Shinmura, T., Aizawa, T. *Bull. Jpn. Soc. Prec. Eng.*, 23, 37–41, 1989.) (c) Plane MAF. (From Shinmura, T., Aizawa, T., *Bull. Jpn. Soc. Precis. Eng.*, 23, 239–239, 1989.)

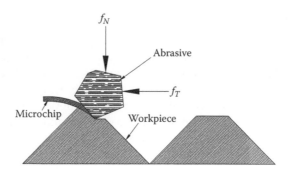

FIGURE 14.2
Material removal from the peaks.

working gap presses the MAPs on the surface and cutting edge of abrasive is capable of infinitesimal cutting. The relative motion between the FMAB and the workpiece is responsible for material removal due to abrasion. The material will be removed particularly from the peaks of the workpiece surface in the form of microchips, as illustrated in Figure 14.2. Valleys of the surface irregularities are not easily accessible. Hence, peaks are truncated and the surface roughness is reduced.

14.2.2 Mechanism of ECDi

During the ECDi process, the gap between the copper electrode and the workpiece is filled with electrolyte. The electrode (tool) is connected to the cathode while workpiece is connected to the anode of the direct current (DC) power source. The material is removed atom by atom from the workpiece kept at anodic potential. The surface of the workpiece is dissolved according to Faraday's law of electrolysis. Depending on the operating conditions and the metal–electrolyte combinations, different anodic reactions take place. During processing of ferrous alloy workpiece with $NaNO_3$ as an electrolyte, the reactions that occurred can be represented as follows [8].

$$\text{Anode (workpiece): } Fe \rightarrow Fe^{+2} + 2e^-$$

$$\text{Cathode (electrode): } 2H_2O + 2e^- \rightarrow 2(OH)^- + H_2 \uparrow$$

Electrochemical reactions:

$$NaNO_3 \rightarrow Na^+ + (NO_3)^-$$

$$Fe^{++} + 2(OH)^- \rightarrow Fe(OH)_2$$

$$4Fe(OH)_2 + 2H_2O + O_2 \rightarrow 4Fe(OH)_3 \downarrow \text{ (sludge)}$$

The rate of these reactions depends on the ability of the system to remove the reaction products from the IEG as they are formed and the supply of electrolyte to the IEG (keeping all other parameters constant). The amount of any substance dissolved is proportional to the electrical current that is passed through the electrolyte (I_e). The volume of material dissolved in specified time (T_m) is given by [9]

$$V = \frac{I_e T_m M}{n \rho F},$$ (14.1)

where F represents Faraday's constant and M, n and ρ represent atomic weight, valency of oxidation and density of the anode material, respectively.

14.2.3 Abrasion–Passivation Synergism

Many factors are responsible for the rate of abrasion and electrochemical passivation. In ECDi, a passive layer formed on the workpiece surface (Figure 14.3a) is not broken by low current densities applied in this compound process. The local abrasion of the passive film in MAF leads to abrasion-assisted dissolution and rapid removal of the locally depassivated metal surface, followed by further passivation (Figure 14.3b). The extra material removed in the EMAF process (in addition to the sum of material removal due to MAF and ECDi when they are applied independently) contributes to an increase

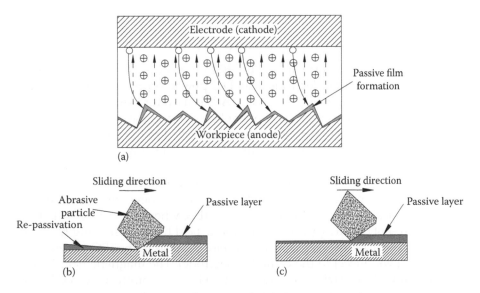

(a)

(b)　　　　(c)

FIGURE 14.3
(a) Passive film formation. (b) Complete passive film removal. (c) Partial removal of passive film.

in ECDi due to quick removal of passive film and change in abrasion rate due to change in mechanical properties of the workpiece surface. The extra material removal (MR) can be represented as

$$MR_{(EXTRA)} = MR_{(abrasion\text{-}assisted\ dissolution)} + MR_{(passivation\text{-}assisted\ abrasion)}. \tag{14.2}$$

If the passive film is completely removed by abrasion (Figure 14.3b), then the new surface will be exposed to the electrolyte and rapid ECDi takes place, which enhances the abrasion-assisted dissolution. If the passive film is partially removed (Figure 14.3c), then it reduces further dissolution and the contribution of abrasion-assisted dissolution decreases [10]. The passivation-assisted abrasion takes place due to a change in the mechanical surface properties (viz. hardness, wear resistance, etc.) of the workpiece. The thickness of the new surface formed due to electrochemical passivation and its new surface hardness can be calculated, which is incorporated in the modelling of MAF to consider the effect of passivation-assisted abrasion. Thus, in the case of EMAF, the oxide layer formed on the peaks and front side of the surface profile of workpiece is quickly removed by MAPs in the MAF. The increase in ECDi depends on the active wear area of the workpiece surface exposed to the electrochemical reaction and the time interval during which the unit length of the workpiece passes through the electrochemical zone of reaction under the electrode. The active wear area depends on the indentation made by abrasive particles, which in turn depends on magnetic flux density (MFD). Extra material removed due to abrasion-assisted dissolution can also be expressed in terms of increased current efficiency in ECDi due to abrasion.

14.2.4 Mechanism of Material Removal in EMAF

Figure 14.4a shows the principle of the EMAF process. Here, the electrolyte supply system, copper electrode and DC power supply are prepared for ECDi. The pair of magnetic poles and MAPs is used for MAF. During experimentation, the IEG between the copper electrode and the workpiece is filled with electrolyte. The copper electrode is connected to the negative terminal (cathode), while the workpiece is connected to the positive terminal (anode) of the DC power source. A micron-thick passive layer is created on the workpiece surface due to ECDi. The passive layer is removed gently by FMAB during the MAF process, which further exposes the new metal surface. The electrochemical reaction takes place in the IEG, where the magnetic field is already present (Figure 14.4b). The magnetic field lines are moving from the N-pole to S-pole of the magnet, whereas electric field in the IEG passes from the anode (workpiece) to the cathode (electrode). The magnetic field is perpendicular to the electric field. The negative ions are jointly activated by the Lorentz force and the electrical field force. The Lorentz force (N) acting on

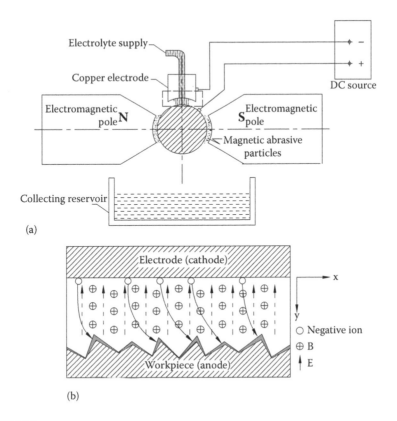

FIGURE 14.4
(a) Process principle of EMAF. (b) Electrochemical reactions in IEG. (From Yan, B.H., Chang, G.W., Cheng, T.J., Hsu, R.T. *Int. J. Mach. Tools Manuf.* 43, 13 1355–1366, 2003.)

the ions due to the combined effect of electric and magnetic field is given as [3]

$$\vec{F} = q(\vec{E} + \vec{v} \times \vec{B}),$$

(14.3)

where q and $\vec{v}$ represent charge (C) and velocity (m/s) of the ions, respectively. $\vec{E}$ and $\vec{B}$ represent vector of electric field intensity (V/m) and magnetic field density (T), respectively. By using Newton's second law of motion,

$$\vec{a} = \frac{\vec{F}}{m} = \frac{q}{m}(\vec{E} + \vec{v} \times \vec{B}) = \frac{d\vec{v}}{dt},$$

(14.4)

where $\vec{a}$ is the acceleration of ion, m is the mass of ion and t is the time.

By considering a two-dimensional (2D) domain in the x–y plane as shown in Figure 14.4b, the equations of ionic movement are given as [3] follows:

$$v_x = \left(v_{x(0)} - \frac{U}{B.y_e} \right) \cos \frac{qB}{m} t + v_{y(0)} \sin \frac{qB}{m} t + \frac{U}{m.y_e} \tag{14.5}$$

and

$$v_y = -\left(v_{x(0)} - \frac{U}{B.y_e} \right) \sin \frac{qB}{m} t + v_{y(0)} \sin \frac{qB}{m} t, \tag{14.6}$$

where $v_{x(0)}$ and $v_{y(0)}$ represent the components of initial flow velocities of electrolyte in the x and y directions, respectively; U is the electrode potential and y_e is the IEG.

The position of ions in the electrode gap while moving from the cathode to the anode is given by

$$x = \frac{m}{qB} \left(v_{x(0)} - \frac{U}{B.y_e} \right) \sin \frac{qB}{m} t - \frac{m.v_{y(0)}}{qm} \cos \frac{qB}{m} t + \frac{U}{B.y_e} t + \frac{m.v_{y(0)}}{qB} + x_0 \tag{14.7}$$

and

$$y = \frac{m}{qB} \left(v_{x(0)} - \frac{U}{B.y_e} \right) \cos \frac{qB}{m} t + \frac{v_{y(0)}}{qB} \sin \frac{qB}{m} t + \frac{m.U}{qB^2.y_e} t - \frac{m.v_{x(0)}}{qB} + y_0. \tag{14.8}$$

x_0 and y_0 represent the initial positions of the ions in the electrode gap. From Equations 14.7 and 14.8, it is observed that the ions move in a complex cycloidal path. The angle of approach of ions moving towards the anode depends on the electrode gap, potential and MFD in IEG. Under appropriate conditions, the possibility of the electrolytic ions reaction with the peak points and the front side of the irregularities is more than that of their reaction with valley points. A thick passive film is produced on peak points and front sides of the irregularities. The passive film at the valley points and the rear side of the irregularities is thin due to lower chances of reaction. During the process, the passive film produced on the peak points is removed by FMAB, but it remains in the valleys, which is inaccessible to the abrasives. As soon as a new surface appears after the soft layer is removed, a new passive layer is immediately produced and it is then removed again. Repeating the process very quickly removes the surface irregularities, and the surface roughness is quickly reduced.

14.3 Configurations of EMAF

EMAF can be applied to external cylindrical, internal cylindrical and plane surfaces like MAF. When EMAF is applied to external cylindrical surfaces of rotational jobs, it is called the cylindrical electrochemical MAF (C-EMAF) process (Figure 14.4a). When EMAF is applied to internal surfaces (Figure 14.5a) and plane surfaces (Figure 14.5b), it is known as internal-EMAF and plane-EMAF, respectively.

14.4 EMAF Apparatus

The EMAF apparatus consists of various systems/arrangements as discussed in the following.

1. Magnetic source: During EMAF, the abrasion pressure (cutting force) on the workpiece surface can be controlled by a magnetic field in the working gap. A magnetic field can be generated by either a permanent magnet or an electromagnet. Magnetic flux can be changed by changing current to the electromagnet or by varying the air gap in the magnetic circuit in the case of a permanent magnet. The magnetic flux produced by the magnetic source can be transferred and concentrated in the working gap by means of cores, poles and yoke.

2. Magnetic core, poles and yoke: The magnetic cores, poles and yoke are generally made of material having high magnetic permeability. The arrangement of the poles and yoke is different for different configurations of EMAF. The pole ends are designed in such a way that they maintain a uniform working gap with the workpiece. In order to machine different shapes and sizes of workpiece on the same setup and to maintain uniform image contact over the work surface, a unique pair of exchangeable pole inserts for different shapes/sizes can be fixed over the pole base (Figure 14.6). For finishing cylindrical workpieces of different diameters with a uniform image contact set of unique pole inserts for each diameter

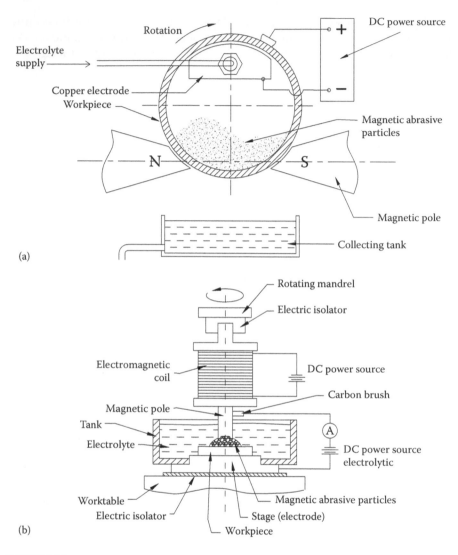

FIGURE 14.5
Configurations of EMAF: (a) internal-EMAF and (b) plane-EMAF.

(Figure 14.6b) can be used with same magnetic pole base as shown in Figure 14.6a. The actual photograph of pole inserts used during the finishing of cylindrical workpieces in electrochemical magnetic abrasive machining (EMAM) is shown in Figure 14.6c. These pole inserts facilitate replacement with new ones at low cost if they are worn out due to continuous abrasion [11].

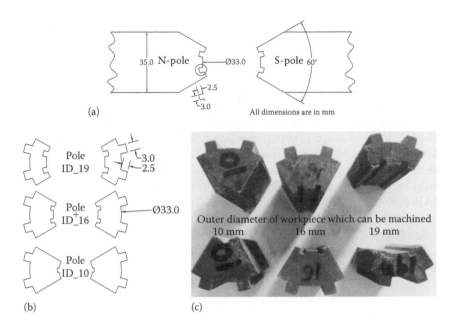

FIGURE 14.6
Magnetic poles: (a) pole base, (b) pole inserts and (c) photograph of pole inserts.

3. MAPs: MAPs used may be bonded or unbonded. Bonded magnetic abrasives are prepared by sintering FPs and abrasive particles at very high pressure and temperature in inert gas atmosphere and then crushing and sieving to a specific size. Unbonded magnetic abrasives (UMAs) are prepared by simple mechanical mixture of FPs and abrasive powder [4]. Steel or iron grit may be used as FPs. Aluminium oxide (Al_2O_3), silicon carbide (SiC), chromium oxide (Cr_2O_3) and diamond powder can be used as abrasive particles.

4. Electrolyte supply system: This consists of chemical metering pump, electrolyte reservoir, pipe and pipe fittings and disposal arrangement. The chemical metering pump supplies electrolyte into the IEG at specified flow rate, which falls into a collecting reservoir after electrolytic reaction.

5. Electrodes and power supply system: During the EMAF process, cathode and anode electrodes are important. Tool electrode and workpiece are connected to cathode and anode of the DC power source, respectively, by some suitable arrangement. Hollow copper electrode can be used as a tool electrode during cylindrical-EMAF and internal-EMAF, which serves the dual purpose of supplying electrolyte in the IEG and electrolytic reaction. The shape of the cathode electrode ensures smooth flow of electrolyte in the electrode gap without separation to perform electrolysis at high relative velocity due to workpiece

rotation. In case of plane-EMAF, magnetic pole serves the purpose of cathode electrode [12]. A DC power source with suitable voltage range and current rating is used for electrolytic dissolution.

During finishing, the working gap and IEG are set properly. A suitable mass of MAPs is filled into the working gap. The MAPs get excited due to the magnetic field in the working gap and form the FMAB. The surface of the workpiece to be finished is kept in contact with the FMAB. Then, passive electrolyte solution is allowed to flow through the IEG by an electrolyte supply system and an electrolytic current is switched on. Some suitable arrangement should be provided to give relative motion between the workpiece and the FMAB. For effective finishing, the IEG should be clear with a continuous path of electrolyte between the workpiece and the cathode electrode. The electrolyte collected in the reservoir after the electrolytic reaction is discarded, because it contains products of the electrochemical reaction and abrasives. The finishing operation can be continued for some time. The amount of material removed during EMAF can be calculated by measuring the mass of the workpiece before and after finishing. The final surface roughness can be measured using surface roughness tester.

14.5 Process Parameters in EMAF

There are large number of variables in EMAF regarding electrolysis, MAF and MAPs in addition to the geometry and material properties of the workpiece and magnetic poles. The parameters that influence the performance of the EMAF process can be categorised as follows:

1. Electrolysis-related parameters
   a. Type of electrolyte
   b. Electrolyte concentration
   c. Flow rate in the gap
   d. Current
2. MAF-related parameters
   a. Magnetic flux in the working gap or current to electromagnet
   b. Working gap
   c. Relative velocity between workpiece and magnetic poles
3. MAP-related parameters
   a. Type of MAPs, viz., bonded, loosely bonded, unbounded
   b. Size of FP/MAP

    c. Size of abrasives

    d. Type of abrasives, viz., SiC, Al_2O_3

4. Workpiece geometry and properties

    a. Shape, viz., cylindrical, plane, internal features

    b. Hardness

    c. Chemical composition of workpiece

5. Magnetic pole geometry and properties

    a. Shape of poles

    b. Material of poles

    c. Internal features in poles viz. Slot, groove, etc.

14.6 Performance Characteristics and Their Evaluation

In order to assess the suitability of any production, it is essential to know the performance characteristics of the process. For EMAF, these characteristics are as follows:

1. Material removal
2. Surface finish

14.7 Effect of Process Parameters on the Performance of the EMAF Process

Various researchers tried to study the influence of various process parameters on the performance of the combined electrochemical-magnetic abrasive finishing process. Kim and Choi [13] proposed a magneto-electrochemical abrasive polishing system that includes a magnetic field and an electrolytic abrasive polishing system and observed that magnetic field increases the finishing efficiency by accelerating and stirring the electrolytic ions movement. They pointed that the electrolytic current density, abrasive pressure and feed rate should be lower in the final finishing step than in the initial step to simultaneously obtain high production efficiency and a good surface finish. Kim et al. [14] developed non-woven abrasive pads to use with magneto-electrolytic-abrasive polishing process, which remove oxide membrane formed during electrolytic polishing effectively and no scratches left on the workpiece surface. They have also tested the polishing ability of SiC and Al_2O_3 non-woven abrasive pads on difficult-to-cut materials of Cr-coated rollers. Yan et al. [2] developed an EMAF process, which was a compound finishing process of the electrolytic process and MAF. They studied the effect of process parameters such as electrolytic current, electrode gap and rate of workpiece revolution on finishing characteristics such as surface roughness (R_a) and material removal (MR) during finishing of SKD11 tool steel. They compared the finishing characteristics of MAF and EMAF. Figure 14.7 illustrates the finishing results of EMAF and MAF at 0.5 A electrolytic current. It shows that EMAF always removes more material and yields a better surface roughness than MAF. The effect of IEG on the finishing results of EMAF is shown in Figure 14.8. It indicates that 3 mm IEG has a better surface finish and the more material removal than the 5-mm gap does. This suggests that the IEG should not be too wide for the EMAF to ensure a superior finish and to save power.

El-Taweel [15] developed a response surface model for material removal rate and surface roughness by considering MFD, applied voltage, tool feed rate and workpiece rotation as input parameters during hybrid electrochemical turning (ECT) with MAF of 6061 Al/Al_2O_3 composite. This author observed that increasing both the applied voltage and the tool feed rate leads to an increase in machining efficiency and improves the surface roughness significantly. The increase in machining efficiency and surface quality during combined ECT-MAF was 147.6% and 33% respectively with respect to traditional ECT. Liu et al. [16] focused on the design of the composite tool for EMAF, which serves the purpose of an electrode during electrolysis and a magnetic pole during MAF. As one possible solution, they have suggested a hollow structure of magnetic pole.

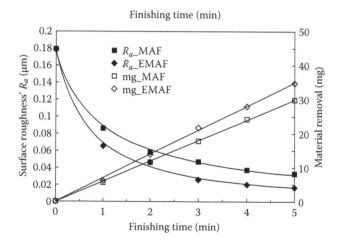

FIGURE 14.7
Variation in the surface roughness and material removal with finishing time for the EMAF and MAF processes (speed: 500 RPM, electrode gap: 3 mm, electrolytic current: 0.5 A). (From Yan, B.H., Chang, G.W., Cheng, T.J., Hsu, R.T., *Int. J. Mac. Tools Manuf.* 43, 1355–1366, 2003.)

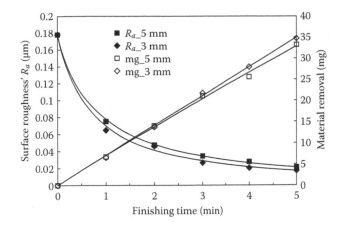

FIGURE 14.8
Effects of IEG on surface roughness and material removal with finishing time for EMAF (speed: 500 RPM, electrolytic current: 0.5 A). (From Yan, B.H., Chang, G.W., Cheng, T.J., Hsu, R.T., *Int. J. Mac. Tools Manuf.* 43, 1355–1366, 2003.)

They have also analysed the structure parameters of the hollow tool on the abrasive brush formation during EMAF. Judal and Yadava [10] developed the *MR* and surface roughness (R_a) model for machining of non-magnetic stainless steel during the cylindrical-EMAM process as a function of current to electromagnet, electrolytic current, workpiece rotational speed and frequency of vibration. But the maximum MFD established in the working gap in case of non-magnetic stainless steel (AISI-304) was 0.32 T. They have observed that the workpiece rotational speed and electrolytic current have a considerable effect on *MR* as well as R_a. During the machining of magnetic materials in a cylindrical-EMAM process, the role of magnetic abrasion dominates because the MFD generated in the working gap greatly depends on the magnetic nature of the workpiece material [17]. They also conducted experiments to investigate the performance of the cylindrical-EMAM process during processing of AISI-420 grade steel by adopting one parameter at a time approach. The effects of electrolytic current and current to electromagnet on synergistic *MR* and R_a were discussed [18].

14.8 Modelling of Material Removal during EMAF

The total material removed during the EMAF process is the sum of material removal during MAF and ECDi, when they are applied independently, and an extra amount of material removed due to abrasion–passivation synergism. In this section, an attempt is made to model the material removal during cylindrical EMAF process using UMAs by combining these three different phenomena by theoretical-cum-experimental approach. A magnetic potential (ϕ) distribution in the 2D domain can be calculated using FEM to find the forces acting on the FPs at the workpiece–brush interface to find the contribution of MAF only. An empirical relation can be developed to consider the effect of ECDi and abrasion–passivation synergism on material removal considering area of fresh metal exposed to an electrochemical reaction, electrolytic current density and frequency of depassivation. It is then corrected based on experimental results for material removal due to ECDi and synergistic material removal due to combination of ECDi and MAF. The step-by-step procedure to model material removal during EMAF is explained in the following sections.

14.8.1 Modelling of Magnetic Potential during MAF

The physical phenomenon of the MAF process can be described by the boundary value problem. The MAF process is complex and unpredictable in nature when UMAs are used. It also involves an intricate geometry of solution domain, as represented in Figure 14.9a, and composite material properties. Under such conditions, it would be very difficult to obtain the solution of the governing equation using analytical methods. It can be solved numerically either by FEM or the finite difference method. The finite difference method is difficult to apply on complex and irregular geometries, whereas the FEM is more flexible with respect to geometry of the problem and material properties [19]. Therefore, the FEM can be used to obtain the approximate numerical solution of a complicated MAF problem.

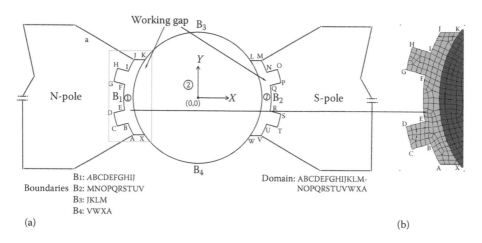

FIGURE 14.9
(a) Solution domain of MAF. (b) Portion of discretised domain. (From Judal, K.B., Yadava, V., *J. Mater. Process. Technol.*, 213, 2089–2100, 2013.)

The following assumptions are made to make the problem mathematically feasible:

1. The solution domain is considered two dimensional, i.e. $\dfrac{\partial \phi}{\partial Z} = 0$, where ϕ represents the magnetic scalar potential.

2. All magnetic field lines are passing from the N-pole to the S-pole via a working gap and magnetic workpiece as shown in domain.

3. Leakage of magnetic flux from the domain is neglected and the magnetic poles, cores and yoke are considered saturated uniformly throughout the cross-section.

4. The MAPs are closely packed in the working gap between the workpiece and the poles.

5. The relative permeability of UMAs is calculated by considering relative volume fractions of FPs and abrasive particles.

To determine the distribution of magnetic forces, the governing equation of the process is expressed in terms of the magnetic potential, which is primary variable. The steady-state Laplacian form of the governing equation within a 2D domain (Figure 14.9a) can be written as [21]

$$\frac{\partial}{\partial X}\left(\mu_r \frac{\partial \phi}{\partial X} \right) + \frac{\partial}{\partial Y}\left(\mu_r \frac{\partial \phi}{\partial Y} \right) = 0 \text{ in the domain,} \tag{14.9}$$

where μ_r represents the relative permeability of material in the domain and X and Y represent the global coordinates.

Boundary conditions: There are total four numbers of boundaries: B_1, B_2, B_3 and B_4.

1. Essential boundary conditions

The scalar magnetic potential ϕ is specified on boundaries B_1 and B_2. Here, there is no current source in the domain, so the problem is solved based on reduced scalar potential strategy. If the magnetic field is generated by means of electromagnet having N_m number of turns through which I_m current is flowing, then

$$\phi = k_m N_m I_m \text{ on } B_1, \tag{14.10}$$

where k_m represents the efficiency of electromagnet to consider the loss of magnetic flux at different current to electromagnet. The value of k_m can be determined by numerical computation of MFD generated at particular point in the working gap and experimental measurement of MFD at same location by gauss meter at different current to electromagnet.

$$\phi = 0 \text{ on } B_2 \qquad (14.11)$$

2. Natural boundary conditions

In the case of a non-magnetic workpiece, all magnetic field lines may not pass through the workpiece, but some field lines may flare and pass from the N-pole to the S-pole through air, while for magnetic workpiece, the field lines follow the less reluctant path. On boundaries B_3 and B_4, the normal derivative of the scalar potential is zero.

$$\frac{\partial \phi}{\partial n} = 0 \text{ on } B_3 \text{ and } B_4 \qquad (14.12)$$

The FEM [22] is used to evaluate the magnetic potential distribution within the solution domain. A part of the discretised solution domain is shown in Figure 14.9b. Galerkin's finite element approach is applied for solving the governing equation (Equation 14.9) and boundary conditions (Equations 14.10 through 14.12). For each typical element in the solution domain, an elemental equation has been developed. The elemental equation over a typical element is given by

$$[K]^e \{\phi\}^e = 0, \qquad (14.13)$$

where $[K]^e$ is the elemental coefficient matrix and $\{\phi\}^e$ is the magnetic scalar potential vector of the respective element. The elemental coefficient matrix is given by

$$[K]^e = \int_{A_e} \mu_r [B]^{e^T} [B]^e \, dX \, dY. \qquad (14.14)$$

The $[B]^e$ represents the matrix of derivatives of shape functions for a corresponding element. Then, the global coefficient matrix $[GK]$ is obtained after assembling the elemental coefficient matrices over all the elements. The solution leads to a global equation as given in

$$[GK]\{\phi\} = \{0\}. \qquad (14.15)$$

Equation 14.15 is a set of linear algebraic equations. These equations are solved by a simultaneous equation solver after imposing essential boundary conditions. The solution leads to nodal values of the magnetic scalar potential (ϕ) at all nodes in the domain.

14.8.2 Calculation of Cutting Forces during MAF

From the nodal values of ϕ, the values of the magnetic field intensity (H) and gradients of the magnetic field intensity $\left(\dfrac{dH}{dX} \text{ and } \dfrac{dH}{dY}\right)$ are calculated at Gauss points near the workpiece contact surface in region (1) (Figure 14.9a) to compute the forces acting on the FP along the X and Y directions. The forces acting on an FP in the magnetic field are given by [23]

$$F_X = \chi_m \mu_0 VH\left(\frac{dH}{dX}\right) \tag{14.16}$$

and

$$F_Y = \chi_m \mu_0 VH\left(\frac{dH}{dY}\right), \tag{14.17}$$

where magnetic field intensity, $H = -\nabla\phi$, μ_0 = permeability of free space, V = volume of the FP,

$$\chi_m = \text{susceptibility of MAPs.}$$

Here, the MAPs are a simple mechanical mixture of FP and abrasives. The relative permeability of MAPs (μ_m) can be calculated as [19]

$$\mu_m = \alpha\mu_{FP} + (1 - \alpha)\mu_{ABR} \text{ and } \chi_m = \mu_m - 1, \tag{14.18}$$

where α = volume fraction of FPs in UMAs, μ_{ABR} = relative permeability of abrasives and μ_{FP} = relative permeability of FPs.

The contact surface between the workpiece and FMAB is represented by an arc of circles, which is the common portion of regions (1) and (2) (Figure 14.9a). The values of H and its derivatives are more accurate at Gauss points than at nodes. Therefore, the approximate length of the mess element is selected as a function of size of FP, so that the centre of each FP at contact surface approximately lies on the Gauss point.

The cutting forces F_N and F_T acting on steel grit can be calculated by resolving the forces, F_X and F_Y, along normal N–N and tangent T–T directions as shown in Figure 14.10a [24].

$$F_N = F_X \cos\theta + F_Y \sin\theta \tag{14.19}$$

$$F_T = F_X \sin\theta + F_Y \cos\theta \tag{14.20}$$

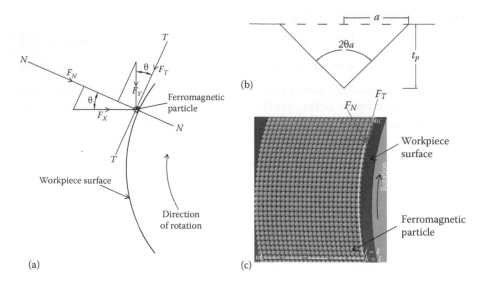

FIGURE 14.10
(a) Forces acting on the FP during MAF. (b) Indentation by wedge-shaped abrasive particle.
(c) Grid of FPs at contact surface. (From Judal, K.B., Yadava, V., *J. Mater. Process. Technol.*, 213, 2089–2100, 2013.)

Both F_N and F_T are spatial in nature as their magnitude and direction vary with the location of the FP in a 2D system.

14.8.3 Calculation of Material Removal Due to MAF Only

Abrasives are randomly distributed in the UMAs. It is assumed that the primary cutting action is performed by abrasive particles, and FPs provide abrasion pressure due to normal force. The number of abrasive particles that take part in abrasion action beneath each FP may vary from one instance to another. Based on the probability statistics function, the actual number of contacting grains may vary from 3.8% to 18% of the total number of abrasive particles per unit area [25]. Considering these aspects, the number of abrasive particles under each FP is selected by a random permutation of numbers.

Let n_a be the number of abrasive particles under the FP at an instance as captured from random permutation.

A normal force acting on single abrasive particle responsible for indentation is given by

$$f_N = \frac{F_N}{n_a}.$$ (14.21)

The abrasive particles have angular cutting edges, which can be considered as wedge shape with included angle $2\theta_a$ [26]. The indentation depth of such wedge-shaped abrasive at a particular instance can be determined from Figure 14.10b:

$$f_N = H_{mt}\pi\alpha^2, \tag{14.22}$$

where $a = t_p \tan\theta_a$, H_{mt} = hardness of workpiece surface, α = radius of projected area of indentation and t_p = depth of penetration of abrasive particle in workpiece surface.

The cross-sectional area of penetration by abrasive particle is given by

$$A_a = t_p^2 \tan\theta_a. \tag{14.23}$$

Figure 14.9c shows the grid of FPs at the contact surface of the cylindrical workpiece within the FMAB during MAF. Each FP provides normal force on the abrasive particle beneath it. The normal force, $f_N(j)$, acting on the abrasive particle depends on the location of the FP in the 2D system (X–Y plane) and the number of abrasive cutting edges (n_a) simultaneously penetrating the workpiece surface. Let nl and nc be the number of FPs simultaneously in contact with the surface along the workpiece length and circumferential direction, respectively, as shown in Figure 14.10c. By considering the workpiece surface as a uniform triangular profile without statistical distribution, the total amount of material removed by MAF (V_{MAF}) from the surface within a given machining time can be calculated as follows:

$$V_{MAF} = K_{maf}\, v_{rel}\, T_m \left(1 - \frac{R_a}{R_a^0}\right) \sum_{k=1}^{nl} \sum_{j=1}^{nc} \left(\frac{n_a\, f_N(j)}{H_{mt}\pi \tan\theta_a}\right), \tag{14.24}$$

where K_{maf} is the constant for MAF, which depends on the number of factors as MAF can be considered as a case of three-body abrasion system; v_{rel} represents the relative velocity of abrasive particles at the workpiece surface and R_a^0 and R_a represent initial and final arithmetic mean surface roughness, respectively.

$$K_{maf} = \frac{k'k_0 k_1 k_2 k_3 k_4}{k_5} \tag{14.25}$$

Wang and Wang [26] have recommended various criteria to consider different factors and their values in the case of three-body abrasion. In the case of UMAs, the values of different factors in Equation 14.25 are given in Table 14.1 along with consideration.

TABLE 14.1

Value of Different Factors while Calculating the K_{maf}.

Factor	Consideration	Value
k'	Proportionality constant, which is more or less constant	0.03
k_0	The hardness of the workpiece and the steel grit is approximately same.	0.50
k_1	Abrasives are closely packed between the steel grit and the workpiece surface.	1.0
k_2	Material of the workpiece is steel.	1.1
k_3	Electrolyte-containing water is always present in the form of lubrication between abrasive and workpiece surface.	0.67
k_4	Ductility of the workpiece material.	0.5
k_5	Relative hardness of workpiece material and abrasives; here, the hardness of the SiC abrasive is two times greater than the hardness of the workpiece.	1.0

Source: Wang, Y.L., Wang, Z.S., *Wear* 122, 123–133, 1988.

14.8.4 Calculation of Material Removal Due to ECDi Only

During ECDi, the material is removed atom by atom from the workpiece kept at the anodic potential. The surface of the workpiece is dissolved according to Faraday's laws of electrolysis. Depending on the operating conditions and metal-electrolyte combinations, different anodic reactions take place. In order to consider different dissolution rates of various species in the workpiece material, Faraday's law is modified. The material removed during ECDi (V_{ECDi}), when it is applied independently, is given by [9]

$$V_{ECDi} = \frac{\eta_c I_e T_m}{\rho F \sum_{i=1}^{m} \frac{x_i n_i}{M_i}}. \tag{14.26}$$

The dissolution efficiency (η_c) is the ratio of actual material removed to the theoretically calculated and is obtained by performing set of experiments on the developed setup for electrochemical dissolution process independently. Here, x_i, n_i and M_i represent weight fraction, valence of oxidation and atomic weight, respectively, of the ith species in an alloy having a total m number of species.

14.8.5 Calculation of Extra Material Removed Due to Abrasion-Assisted Dissolution

Many factors are responsible for the dependence of abrasion and electrochemical passivation as explained in the previous section. Adler and Walters [27] investigated that the passive (oxide) film did not affect the mechanical surface properties (scratch hardness and wear resistance) of stainless steel.

Therefore, in the case of a stainless steel workpiece, the change in mechanical surface properties can be neglected for simplicity.

During evaluation of the increased process performance under the combined effects of MAF and ECDi, it is assumed that abrasive particles slide on a workpiece surface kept at the anodic potential during electrochemical reaction. The abrasion action of the abrasive particles leads to the destruction of passive film formed and then repassivation starts [28]. The average anodic current during abrasion–passivation system can be written as given in the following [29]:

$$I_a = K_r\, f_{Nave}^{0.5} H_m^{-0.5} \int_0^{1/f} i_p\, dt.f,$$ (14.27)

where K_r represents the proportionality factor, which depends on the fraction of the wear track area being effectively depassivated, the probability that abrasion takes place on the already depassivated area and the number of abrasive particles in contact; f_{Nave} is mean normal force acting on the abrasive particle; i_p represents repassivating current density and f is the frequency of depassivation.

The previous relation can be modified for the EMAF system under consideration based on experimental results for material removal due to ECDi and synergistic material removal due to a combination of ECDi and MAF.

The relationship for current efficiency, which considers the material removal due to ECDi and abrasion–passivation synergism, is written as [20]

$$\eta_{ecap} = \frac{I_a}{I_e}.$$ (14.28)

The volume of material removed due to ECDi and abrasion-assisted dissolution ($V_{ECD} + V_{EXTRA}$) can be predicted by replacing η_c in Equation 14.26 by η_{ecap}.

The total volume of material removed from the workpiece during the EMAF process in a given machining time is given by

$$V_w = V_{MAF} + V_{ECD} + V_{EXTRA}.$$ (14.29)

14.9 Modelling of Surface Roughness during EMAF

Surface roughness after finishing for a certain period of time during EMAF can be predicted as a function of total volume of material removed during the process by considering regular initial surface roughness profile and

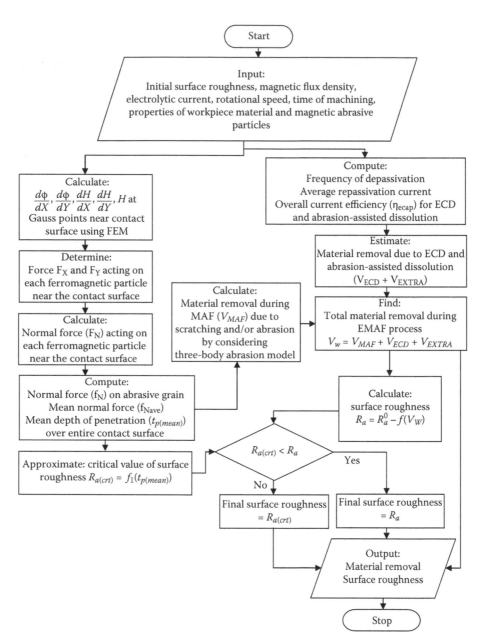

FIGURE 14.11
Flowchart of EMAF process model. (From Judal, K. B., Yadava, V. *J. Mater. Process. Technol.*, 213, 2089–2100, 2013.)

assuming a major amount of material removal from the peaks of the surface only. In addition, when surface roughness reduces to critical surface roughness, further material removal does not contribute to improvement in surface finish. The flowchart of the EMAF process model is shown in Figure 14.11.

14.10 Conclusions

EMAF is an advanced abrasion-based finishing process that constitutes ECDi and MAF. It is still in its infancy. The performance of the process is the simultaneous effect of MAF, ECDi and abrasion–passivation synergism. This process has reduced the time required to achieve the desired surface finish on difficult-to-machine materials. With the increase in MFD, material removal increases, but each value of the MFD is associated with a critical value of surface roughness beyond which further reduction under given conditions is not possible. The developments in this process will be useful to modern grinding industries particularly involved in finishing of advanced engineering materials.

Review Questions

1. Explain the EMAF process with a schematic diagram.
2. Describe the mechanism of material removal during the EMAF process.
3. Explain the basic systems of the EMAF apparatus.
4. List various process parameters that affect the performance of the EMAF process.
5. Derive expressions for normal and tangential cutting forces acting on the FP at an instance during MAF of cylindrical surfaces.

    Ans.: (Hint: Refer to Section 14.8.2)

Objective Questions

1. EMAF is the combination of
    a. electrochemical machining and abrasive finishing.
    b. ECDi and MAF.
    c. electromagnetic machining and abrasive finishing.
    d. None of the above.
2. In EMAF, ECDi is responsible for
    a. passive film formation.

   b. surface roughness reduction.

   c. material removal.

   d. All of the above.

   e. None of the above.

3. Passivation-assisted abrasion indicates an

   a. increase in abrasion due to change in mechanical properties due to passivation.

   b. increase in dissolution due to higher surface area exposed to ECDi.

   c. increase in material removal due to higher current efficiency.

   d. None of the above.

4. Abrasion-assisted passivation is due to

   a. increased ECDi due to higher fresh surface area exposed by abrasion.

   b. an increase in abrasion due to change in mechanical properties due to passivation.

   c. higher thickness of passive film formation.

   d. an increase in MFD.

5. The increased rate of finishing in EMAF is due to

   a. the cycloidal path of ions in IEG.

   b. peaks of surface irregularities are more closer to cathode electrode.

   c. greater interaction of FMAB with peaks of the irregularities.

   d. All of the above.

6. The average anodic current during EMAF depend/s on the

   a. number of abrasive particles in contact with workpiece surface.

   b. normal force acting on the abrasive particle.

   c. frequency of de-passivation.

   d. All of the above.

7. The material removal during EMAF process is

   a. equal to the sum of material removal during MAF and ECDi.

   b. greater than the sum of material removal during MAF and ECDi.

   c. less than the sum of material removal during MAF and ECDi.

   d. difficult to predict.

     Answers: 1. (b), 2. (d), 3. (a), 4. (a), 5. (d), 6. (d), 7. (b)

References

1. Mulik, R.S. and Pandey, P.M. Ultrasonic assisted magnetic abrasive finishing of hardened AISI52100 steel using unbonded SiC abrasives. *International Journal of Refractory Metals and Hard Materials* 29 (2011), 68–77.

2. Yan, B.H., Chang, G.W., Cheng, T.J., and Hsu, R.T. Electrolytic magnetic abrasive finishing. *International Journal of Machine Tools and Manufacture* 43(13) (2003), 1355–1366.

3. Fang, J.C., Jin, Z.J., Xu, W.J. and Shi, Y.Y. Magnetic electrochemical finishing machining. *Journal of Materials Processing Technology* 129(1–3) (2002), 283–287.

4. Jain, V.K., Kumar, P., Behera, P.K., and Jayswal, S.C. Effect of working gap and circumferential speed on the performance of magnetic abrasive finishing process. *Wear* 250 (2001), 384–390.

5. Shinmura, T., Takazawa, K., Hatano, E., and Aizawa, T. Study on magnetic abrasive process – Process principle and finishing possibility. *Bulletin Japan Society of Precision Engineering* 19(1) (1985), 54–55.

6. Shinmura, T. and Aizawa, T. Study on internal finishing of non-ferromagnetic tubing by magnetic abrasive machining process. *Bulletin Japan Society of Precision Engineering* 23(1) (1989), 37–41.

7. Shinmura, T. and Aizawa, T. Study on magnetic abrasive finishing process – Development of plane finishing apparatus using a stationary type electromagnet. *Bulletin Japan Society of Precision Engineering* 23(3) (1989), 239–239.

8. Jain, V.K. *Introduction to Micromachining*, Narosa Publishers, India, 2009.

9. McGeough, J. *Micromachining of Engineering Materials*, Marcel Dekker, New York, 2002.

10. Judal, K.B. and Yadava, V. Cylindrical electrochemical magnetic abrasive machining of AISI-304 stainless steel. *Materials and Manufacturing Processes*, 28(4) (2013), 449–456.

11. Judal, K.B. and Yadava, V. Electrochemical magnetic abrasive machining of AISI304 stainless steel tubes. *International Journal of Precision Engineering and Manufacturing* 14(1) (2013), 37–43.

12. Judal, K.B., Yadava, V., and Mishra, L. Plane electrolytic magnetic abrasive finishing: Development and experimentation, International Conference on Advancements and Futuristic Trends in Mechanical and Materials Engineering, Kapurthala, India (2013), 319–323.

13. Kim, J.D. and Choi, M.S. Development of the magneto-electrolytic-abrasive polishing system (MEAPS) and finishing characteristics of a Cr-coated rollers. *International Journal of Machine Tools and Manufacture* 37(7) (1997), 997–1006.

14. Kim, J.D., Xu, Y.M., and Kang, Y.H. Study on the characteristics of magneto-electrolytic abrasive polishing by using the newly developed nonwoven-abrasive pads. *International Journal of Machine Tools and Manufacture* 38 (1998), 1038–1043.

15. EI-Taweel, T.A. Modelling and analysis of hybrid electrochemical turning – Magnetic abrasive finishing of 6061 Al/Al$_2$O$_3$ composite. *International Journal of Advanced Manufacturing Technology* 37 (2008), 705–714.

16. Liu, G.Y., Guo, Z.N., Li, Y.B., and Liu, J.W. Composite tools design for electrolytic magnetic abrasive finishing process with FEM. *Advanced Materials Research* 325 (2011), 536–541.

17. Judal, K.B. and Yadava, V. Experimental Investigations into electrochemical magnetic abrasive machining of cylindrical shaped non-magnetic stainless steel workpiece. *Materials and Manufacturing Processes* 28(4) (2013), 1095–1101.

18. Judal, K.B. and Yadava, V. Experimental investigations into cylindrical electrochemical magnetic abrasive machining of AISI-420 magnetic stainless steel. *International Journal of Abrasive Technology* 5 (2012), 315–331.

19. Jayswal, S.C., Jain, V.K., and Dixit, P.M. Modeling and simulation of magnetic abrasive finishing process. *International Journal of Advanced Manufacturing Technology* 26 (2005), 477–490.

20. Judal, K.B. and Yadava, V. Modeling and simulation of cylindrical electrochemical magnetic abrasive machining of AISI-420 magnetic steel. *Journal of Materials Processing Technology* 213(12) (2013), 2089–2100.

21. Chandrupatla, T.R. and Belegundu, A.D. *Introduction to Finite Elements in Engineering*. PHI Learning Private Limited, New Delhi, 2008.

22. Reddy, J.N. *An Introduction to the Finite Element Method*. Tata McGraw-Hill Publishing Company Limited, New Delhi, 2005.

23. Smolkin, M.R. and Smolkin, R.D. Calculation and analysis of the magnetic force acting on a particle in the magnetic field of separator. Analysis of the equation used in the magnetic methods of separation. *IEEE Transactions on Magnetics* 42 (2006), 3682–3693.

24. Judal, K.B. and Yadava, V. Modeling and simulation of cylindrical electrochemical magnetic abrasive machining process. *Machining Science and Technology* 18(2) (2014), 221–250.

25. Hou, Z.B. and Komanduri, R. Magnetic field assisted finishing of ceramics – Part III: On the thermal aspect of magnetic abrasive finishing (MAF) of ceramic rollers. *ASME Journal of Tribology* 120(4) (1998), 660–667.

26. Wang, Y.L. and Wang, Z.S. An analysis of the influence of plastic indentation on three-body abrasive wear of metals. *Wear* 122 (1988), 123–133.

27. Adler, T.A. and Walters, R.P. Corrosion and wear of 304 stainless steel using a scratch test. *Corrosion Science* 33 (1992), 1855–1876.

28. Watson, S.W., Friedersdorf, F.J., Madsen, B.W., and Cramer, S.D. Methods of measuring wear-corrosion synergism. *Wear* 181–183 (1995), 476–484.

29. Mischler, S., Debaud, S., and Landolt, D. Wear-accelerated corrosion of passive metals in tribocorrosion systems. *Journal of the Electrochemical Society* 145 (1998), 750–758.

15

Electro-Discharge Diamond Grinding

Vinod Yadava

Mechanical Engineering Department, Motilal Nehru National Institute of Technology, Allahabad, Uttar Pradesh, India

CONTENTS

15.1 Hybrid Machining Technology

The emergence of advanced engineering materials, having remarkable technological characteristics in terms of high strength temperature resistance, high hardness, high wear resistance, high toughness, etc., plays an important role in modern manufacturing industries, especially in aircraft, automobile, tool and die, medical and electronics. Shaping these materials with stringent design requirements such as high precision, complex shapes and high surface quality is inevitable to put them in use. To exploit these difficult-to-machine advanced engineering materials with new challenges, many advanced machining processes have been developed, but they have their own strengths and weaknesses [1]. Technological improvement of advanced machining processes can further be achieved by combining two or more different mechanisms of material removal simultaneously on the material being processed [2]. The machining processes so developed are called hybrid machining processes (HMPs). These processes are developed to exploit the potential advantages and to restrict the disadvantages associated with an individual constituent process. Usually, the performance of HMP is better than the sum of the performance of constituent processes with the same parameter settings. In these processes, besides the performance from individual component processes, an additional contribution may also come from the interaction of the component processes [2].

Practically feasible HMPs have been developed either combining erosion-based electro-discharge action with erosion-based electro-chemical action or combining abrasive abrasion action with erosion-based electro-discharge action or electro-chemical action [3,4]. The former category of machining processes can be called erosive HMPs (EHMPs) and the later category of processes can be called as abrasive HMPs (AHMPs) (Figure 15.1).

EHMPs are conceived to overcome the major limitation of electro-chemical machining (ECM) and electro-discharge machining (EDM) in which the tool and workpiece are required to be electrically conducting. Electro-chemical spark machining (ECSM) is such a type of HMP in which the phenomenon of electro-chemical discharge is employed for material removal from electrically non-conductive materials. In ECSM, the discharge takes place (in the form of desirable sparks) between the tool and the surrounding electrolyte in the vicinity of the electrically non-conducting material workpiece. The ECSM process has been successfully applied for the machining of soda lime glass, borosilicate glass, quartz, glass fibre reinforced plastics and ceramics [5–7].

AHMPs are developed [3,7] by combining abrasion-action-based abrasive machining with erosion-action-based ECM or EDM and are correspondingly called electro-chemical abrasive machining (ECAM) and electro-discharge abrasive machining, respectively. When a metal bonded abrasive tool electrode is used in the form of a disc (wheel), then the corresponding processes

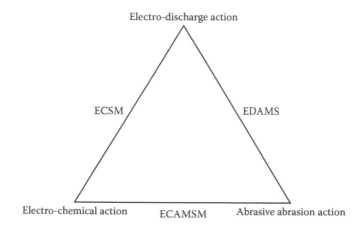

FIGURE 15.1
Hybrid machining processes.

are called as electro-chemical abrasive grinding (ECAG) and electro-discharge abrasive grinding (EDAG), respectively.

The ECAG process is developed by combining the electro-chemical grinding and abrasive grinding (AG) processes in such a way that both the processes occur simultaneously. In this process, material is removed by the combined effect of electro-chemical dissolution and abrasive abrasion [3]. For this purpose, a metal bonded abrasive wheel is used, which is connected to the negative polarity while the workpiece is connected to the positive polarity. The bond material of the wheel is responsible for anodic dissolution when electric current is applied. In this process, most of the material is removed due to electro-chemical dissolution (90% to 95%), while abrasive abrasion is only responsible for removal of the oxide film [3]. The performance of the ECAG process is better than that of the conventional grinding and ECM processes. In this process, the abrasive particles are always in contact with the workpiece and established the gap through which the electrolyte can easily flow out, resulting in improvement in the machining performance.

The EDAG process is developed by combining the features of electro-discharge grinding (EDG; use of a disc-shaped tool electrode without abrasives) and AG (metal bonded abrasive tool) processes for machining of the electrically conductive difficult-to-machine materials [8–10]. In this process, a metal bonded abrasive wheel is used for spark erosion and abrasive abrasion purposes. The spark is generated between the metallic bonding of the wheel and workpiece when a direct current (DC) pulse power supply is applied between them. The major advantages of this process are higher material removal rate (MRR), better surface finish, low grinding forces, low specific energy, continuous self-dressing of the grinding wheel and removal of micro-cracks and recast layer from the workpiece surface [10]. The process is more acceptable in modern industries due to the non-hazardous, chemically non-reactive and eco-friendly nature of dielectric used in EDAG process as compared to electrolytic-based AHMP such as ECAG. Hence, researchers have highly concentrated their research activities toward the EDAG process for machining of hard and brittle electrically conductive materials. However, the EDAG process is experimentally feasible only with metal bonded diamond abrasives; hence, it is known by the name of the electro-discharge diamond grinding (EDDG) [10].

15.2 The EDDG Process and Mechanism of Machining

The metallic disc used in the EDG process is replaced with a metal bonded diamond wheel. Generally, brass or bronze is used as the bonding material.

The wheel–workpiece interaction and mechanism of material removal of EDDG process are shown in Figure 15.2. In this process, the spark is generated between the bonding material and the workpiece when the DC pulse power supply is applied between them. Due to spark energy, the workpiece surface becomes softer, which is removed by diamond abrasives at low grinding forces. In this process, the workpiece is simultaneously influenced by electrical discharges and abrasive abrasions [3,10]. Therefore, the material is removed by a combined effect of spark erosion of EDG and abrasive abrasion of diamond grinding (DG).

In the EDDG process, the protrusion heights of abrasive particles are kept more than the gap between the bonding material of the wheel and workpiece surface. Here, the diamond abrasive particles are always making contacts with the workpiece surface, but short-circuit phenomenon during pulse on-time of the process is avoided because of the non-electrical conductivity of diamond abrasives. On the other hand, the high thermal conductivity of the diamond particles assists to regain the ambient temperature after each spark discharge.

The EDDG process has the potential to remove the problems related to the DG as well as the EDG process. In this HMP, the spark generated during the EDG process is utilised for softening the workpiece material because of which reduction in the grinding forces occurs while the abrasive abrasion assists to remove the recast layer with micro-cracks on the workpiece surface. The advantages of the EDDG process are reduction in grinding forces, reduction in thermal residual stresses, elimination of wheel loading and glazing problems, elimination of re-solidified layer and micro-cracks by effective abrasive action and achievement of higher MRR and better surface finish. Further, EDDG suffers from several inherent problems such as continuous loss of diamond abrasives, entrapment of chips between the workpiece

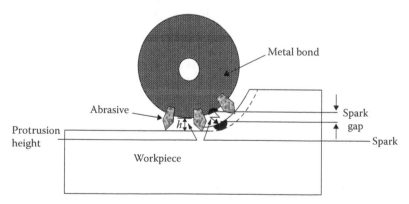

FIGURE 15.2
Material removal in the EDDG process.

and bond material, accumulation of debris particles into the inter-electrode gap (IEG) and abnormal arcing or short-circuiting due to the accumulation of debris and chips into IEG.

In the EDDG process, abrasive abrasion and electro-discharge erosion can be controlled by adjusting wheel feed rates and electro-discharge pulse parameters. The improved machining performance in EDDG is due to continuous in-process dressing and de-clogging of the grinding wheel. Consequently, the grinding wheel can maintain its grinding ability without becoming dull. Since the contributions of abrasive abrasion and electro-discharge erosion are adjustable, the EDDG can be used either in DG-dominant mode, with a relatively less effect of electrical discharge to acquire a reduced heat-affected surface layer, or in an EDG-dominant mode, with a relatively less effect of diamond abrasion to reduce the machining force, or in a well-balanced state between the DG and EDG. Applications of this process include machining of components made of advanced engineering materials such as Al-SiC, Inconel, titanium alloy, tungsten carbide, high-speed steel (HSS), cemented carbide, polycrystalline diamond, etc.

15.3 Variants of EDDG Process

The EDDG process was developed in the USSR during the late 1970s. But in the late 1980s, its applications gained momentum in various industries where there is a need to grind parts made of difficult-to-machine materials. The EDDG process is in its early stage of development, and literature available is scarce. Grodzinskii and Zubotaya [11,12] have done extensive experimental work on EDDG. But they all are more exploratory in nature. The role of electrical discharges on grinding forces, grinding wheel wear and geometrical accuracy while grooving and cutting off of cemented carbide and few other advanced ceramics have been experimentally studied by Aoyama and Inasaki [9]. Rajurkar et al. [13] have reported the characteristics of EDM-grinding hybrid process, which they call abrasive EDG. Koshy et al. [10], Choudhary et al. [14], Singh et al. [15,16], Yadav and Yadava [17] and Agarwal and Yadava [18,19] have done many experimental studies on EDDG.

The variants of EDDG can be categorised into four major categories depending upon its use for cylindrical workpiece and prismatic workpiece: electrical discharge diamond peripheral surface grinding (EDDPSG), electrical discharge diamond face surface grinding (EDDFSG), electrical discharge diamond peripheral cylindrical grinding (EDDPCG) and electrical discharge diamond face cylindrical grinding (EDDFCG) (Figure 15.3).

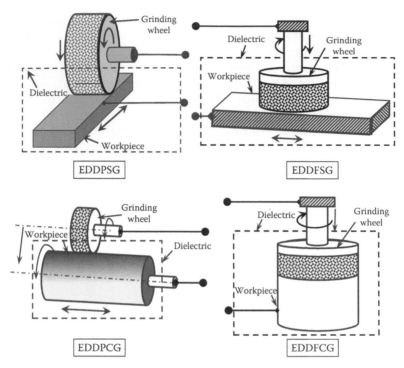

FIGURE 15.3
Configurations of the EDDG process.

15.4 Development and Performance Study of EDDG

Here, the peripheral surface grinding variant of EDDG, which is called EDDPSG, is discussed for machining flat rectangular workpieces. In EDDPSG, electrical sparks occur between the peripheral surface of the metal bonded DG wheel and the flat surface of rectangular workpiece, and abrasion takes place by diamond abrasives whose protrusion height is greater than the IEG. The rotation of the grinding wheel improves the machining performance due to synergetic interaction and also due to the effective flushing between the working gaps. With all the procurement of components and accessories, the EDDPSG attachment was fabricated by Agarwal [18]. The fabricated setup was fitted to the ram of a ZNC-EDM machine. Figure 15.4 shows the schematic diagram of the EDDPSG fabricated attachment. This attachment is fitted after replacing the original tool holder of a ZNC-EDM machine. The circuit used in the setup consists of many sections like relay control section, output section, drive section, etc. The drive section is connected to the variable port, which is used to vary the speed of the workpiece. The relay section is connected to the input of

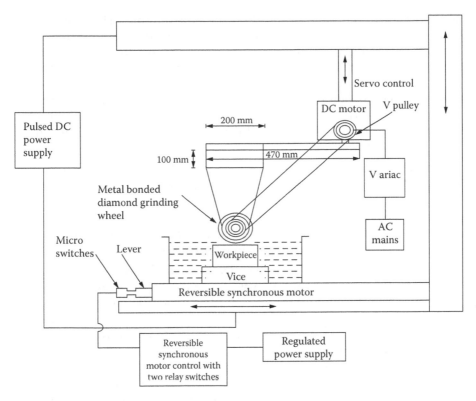

FIGURE 15.4
Schematic diagram of the EDDPSG setup.

reversible synchronous motor. Limit switches have been attached to the relay control section.

In the EDDPSG setup, the grinding wheel is driven with the help of variable-speed DC motor through a belt pulley arrangement. The speed of the motor is varied by changing the supply voltage with the help of a variac. For imparting surface grinding motion, the lead screw of the machine table is driven by using a reversible synchronous motor. Since automatic to and fro motion of the table motor should automatically rotate both in the clockwise and anti-clockwise directions as and when required, a reversible synchronous motor control circuit has been designed using a relay switch, two limit switches and regulated power supply.

The DC motor is a very important part of the EDDPSG attachment, which is used to drive the shaft with the help of a belt. It is located on a horizontal flat plate. According to the requirements, we can control the wheel speed with the help of an auto transformer (variac). To fulfil the requirement of smooth power transmission, a DC motor of 1.25 kW power and 7200 RPM (no load) is used. This DC motor provides speed to the shaft in the range of 500–2500 RPM with the help of variac. The lead screw of the *X*-axis of the ZNC-EDM machine table is driven by the use of a reversible synchronous motor. The motor can be rotated clockwise and anticlockwise direction as per the requirement.

After conducting experiments on the developed setup for the performance study of the EDDPSG process, the following conclusions are drawn [18].

- The combination of electro-discharge peripheral surface grinding (EDPSG) and diamond peripheral surface grinding (DPSG) improves the performance parameters MRR and R_a during machining of Al-SiC and Al-Al$_2$O$_3$ MMC.

- EDDPSG experiments on Al-SiC and Al-Al$_2$O$_3$ MMC indicate that the MRR increases with an increase in current, wheel speed, work-piece speed, pulse on-time and depth of cut while it decreases with an increase in duty factor. The R_a increases with an increase in current, duty factor, depth of cut, pulse on-time and workpiece speed and decreases with an increase in wheel speed.

- It was observed that during the machining of Al-SiC, the MRR was increased by 10 times and R_a was reduced to half when wheel speed was varied between 1000 and 1400 RPM under the range of current applied. An increase in MRR by a factor of 2.2 and in R_a by a factor of 1.2 was also noted with an increase in depth of cut from 20 μm to 40 μm and workpiece speed of 3 mm/s to 5 mm/s. If current is 24 A, then the MRR was found to be almost the same if the duty factor is changed from 0.49 to 0.82. It was found that the MRR increases more than eight times and R_a by two times when the pulse on-time is changed from 50 μs to 150 μs.

- Micrograph examinations of machined surfaces indicate that they are affected more at high current as compared to low current at the same values of other parameters.

- It was observed that during machining of Al-Al$_2$O$_3$, the MRR increased by more than eight times and R_a reduced by more than three times when the wheel speed was varied from 800 RPM to 1400 RPM. An increase in MRR by a factor of 1.5 and in R_a by a factor of 1.2 was also noted with an increase in depth of cut from 10 μm to 30 μm and workpiece speed from 3 mm/s to 5 mm/s. If the current is 24 A, then the MRR increases by 1.6 times and R_a by 1.5 times if the duty factor is changed from 0.49 to 0.82. It was found that the MRR increases more than six times and R_a by more than three times when the pulse on-time is changed from 50 μs to 150 μs [18].

Another variant of EDDG, called EDDFCG, was designed and developed by Singh [15]. The setup consists of an electrically conductive metal bonded DG wheel, motor, shaft, V-belt and bearing. Figure 15.5 shows a line sketch of the EDDFCG setup assembled on a ZNC-EDM machine. The shaft is a rotating element of the attachment, which is held between the two bearings, and

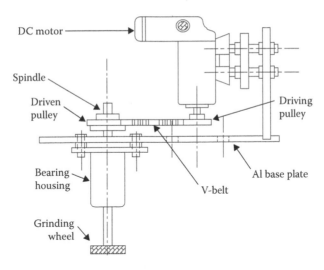

FIGURE 15.5
Schematic diagram of the EDDFCG setup.

is used to rotate the metal bonded DG wheel. One end of the shaft is used to mount the wheel and the other end was for mounting V-pulley. Here, a steel shaft of 15-mm diameter was used to rotate the wheel up to 3000 RPM. The V-belt of a trapezoidal cross section is used to transmit power from the driver to the driven pulley. A bearing is used to support the movement of the shaft. It permits a relative motion between the contact surfaces of the members while carrying the load. Selection of the bearing needs the weight of the shaft, grinding wheel and pulley. A motor is used to drive the shaft of the attachment with the help of a belt and is placed on a 170 mm × 60 mm vertical flat plate. In the EDDFCG attachment, smooth power transmission is required because fluctuation in the speed affects the spark. A single-phase motor of 0.5 kW capacity was selected to provide the speed to the shaft in the range of 500–3000 RPM. A metal bonded DG wheel is mounted on the shaft with the help of collet nut.

After conducting experiments on the developed setup for the performance study of EDDPSG process, the following conclusions were drawn [15,16].

- EDDFCG experiments on HSS workpiece indicate that the MRR increases with an increase in wheel speed, current and pulse on-time but decreases with an increase in duty factor. The wheel wear rate (WWR) and R_a increase with an increase in wheel speed, current, pulse on-time and duty factor.

- EDDFCG experiments on WC-Co composites indicated that the MRR increases with an increase in current and wheel speed, while it decreases with an increase in pulse on-time for higher pulse on-time (above 100 μs). The WWR and R_a increase with an increase in wheel speed and current.

15.5 Experimental Modelling and Performance Prediction of EDDG

In experimental modelling, the model parameters are determined on the basis of measured values followed by the mathematical method for developing the input–output relation. Artificial neural network (ANN) modelling is a new experimental-based modelling technique used for predicting the performance of a machining process as a function of input parameters. ANN derives its computing power first through its massively parallel-distributed structure and, second, its ability to learn and therefore generalise. Generalisation refers to the neural network (NN) producing reasonable outputs from inputs not encountered during training (learning). Regression models use linear combinations of variables and therefore are not adept at modelling non-linear complex interactions.

ANN, being non-parametric, makes no assumption about the model and is capable of letting the data speak for itself. The training of the network is repeated for many such data until the network reaches a steady state, where there are no further significant changes in synaptic weights. Non-linearity, which is a highly important property, particularly if the mechanism responsible for the generation of an input signal is inherently non-linear. In the present chapter, the description of applicability of a multi-layer feed-forward ANN for prediction of process performance of EDDPSG is presented. A multi-layer feed-forward ANN consists of three parts: input layer, hidden layer and output layer (Figure 15.6). The neurons between the layers are connected by the links having synaptic weights. The error back-propagation training algorithm is based on weight updates so as to minimise the sum of squared error for K-number of output neurons.

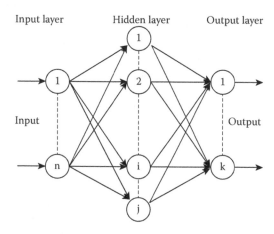

FIGURE 15.6
Multi-layer feed-forward ANN.

With reference to our context, the input layer contains six neurons (wheel speed, current, depth of cut, pulse on-time, workpiece speed and duty factor), while the output layer contains two neurons, MRR and R_a. The training process ends when the specified goal of mean square error or maximum number of epochs is achieved. Before training and validation, the total input and output data are normalised for increased accuracy and speed of the network. The total number of exemplar in the data set is 55. The training of ANN for 55 input–output patterns has been carried out using programming in an NN toolbox accessible in MATLAB software.

The present chapter deals with modelling the EDDPSG process and parametric analysis through a developed ANN model for two workpieces: Al-SiC and Al-Al$_2$O$_3$. It is apparent that an ANN model can be reliably used for predicting output responses such as MRR and R_a in close conformity to the actual experimental data. The parametric analysis has been carried out by developing surface plots for MRR and R_a with current, pulse on-time, duty factor, wheel speed, workpiece speed and depth of cut as input process parameters. The conclusions drawn after the study are as follows [18,19]:

- The ANN back-propagation algorithm with six inputs, two outputs and one hidden layer with 26 neurons (for Al-SiC workpiece) and 29 neurons (for Al-Al$_2$O$_3$ workpiece) has been found suitable to establish the process model. After proper training, the model is capable of predicting the response parameters. The number of hidden layer neurons and the learning factors employed are found to be appropriate. There subsist highly non-linear relationships between the MRR, R_a and the machining conditions. This justifies the use of ANN to develop the model.

- The MRR can be high if the current, wheel speed, speed of workpiece and depth of cut are chosen from their high ranges and duty factor and pulse on-time from the lower range. To have a lower R_a value on the workpiece, a lower value of pulse current, duty factor, depth of cut, speed of workpiece and pulse on-time and higher wheel speed should be chosen.

- The maximum absolute percentage error during training was found to be around 1.09 and 0.99 (for Al-SiC workpiece) and 1.01 and 0.98 (for Al-Al$_2$O$_3$ workpiece) for MRR and R_a, respectively. The minimum absolute percentage error was found to be 0.01 and 0.01 (for Al-SiC workpiece) and 0.03 and 0.01 for (for Al-Al$_2$O$_3$ workpiece) for MRR and R_a, respectively. Maximum absolute percentage error during testing was around 4.04 and 3.33 (for Al-SiC workpiece) and 3.94 and 3.93 (for Al-Al$_2$O$_3$ workpiece) for MRR and R_a, respectively. The minimum absolute percentage error was around 1.03 and 0.19 (for Al-SiC workpiece) and around 0.05 and 0.29 (for Al-Al$_2$O$_3$ workpiece) for MRR and R_a, respectively.

- During machining of Al-SiC, the MRR becomes higher but R_a also increases by a factor of 1.5 if the depth of cut is changed from 20 μm to 40 μm. At a higher range of current, the MRR becomes almost the same if the duty factor changes from 0.492 to 0.817.
- During machining of Al-Al$_2$O$_3$, the MRR increases by more than seven times when the pulse on-time is changed from 50 μs to 150 μs under the range of current investigated, while R_a increases by more than 3.2 times.

15.6 Thermal Modelling and Temperature Simulation of EDDG

EDDPSG integrates DPSG and EDPSG. Heat energy generated during EDDPSG can cause undesirable effects such as micro-cracks, generation of critical residual stresses and metallurgical damage that may affect the surface integrity of finished components. An accurate prediction of thermal history (or transient temperature distribution) in any machined component is a prerequisite for a reliable prediction of surface integrity. A configuration of a typical EDDPSG is shown in Figure 15.7. Here, L is workpiece length, V_f is feed velocity, V_s is wheel speed, D is the diameter of metal bonded DG wheel, B is the width of workpiece and H is height of the workpiece. Two electrodes of the EDDPSG, namely the workpiece and the metal bonded DG wheel, are dipped in a liquid dielectric as shown in Figure 15.7.

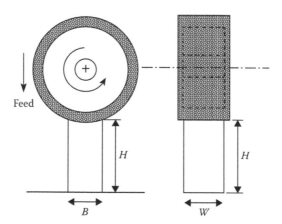

FIGURE 15.7
Schematic of the EDDPSG process.

The finite element method (FEM) can be used for determining the temperature distribution in the workpiece due to EDDPSG. The temperature distribution in the whole workpiece domain due to EDDPSG is obtained by superposing the temperature distributions due to DPSG and EDPSG. The workpiece is simultaneously subjected to heating due to electrical sparks generated between the bond and workpiece and heat generated due to abrasion by abrasive grains having a protrusion height more than the IEG. Hence, the transient temperature distribution due to the EDDG process is considered as a superposition of the transient temperature due to both grinding heat source as well as EDM heat source. Temperature distribution due to grinding is determined by considering the whole domain as a two-dimensional (2D) boundary-initial value problem, whereas the temperature distribution due to sparking action is treated as an axisymmetric boundary-initial value problem. Due to the random and complex nature of the EDDG process, the following assumptions are made to make the problem mathematically tractable [2,20].

1. The workpiece material is homogeneous and isotropic.
2. The properties of the workpiece material are temperature independent.
3. Only a fraction of grinding as well as discharge energy is dissipated as heat into the workpiece.
4. The protrusion height of all the grains is equal and remains constant throughout the operation.
5. The length scales of the grinding action domain and the discharge action domains are different.

Assumptions 1–3 are valid for both DPSG as well as EDPSG, whereas assumption 4 is valid for DPSG only. Assumption 5 is used for superposition.

15.6.1 Thermal Modelling of DPSG

In a DPSG operation, which is a constituent process of EDDPSG, it is assumed that the total grinding energy is converted into heat within a small grinding zone, which leads to a temperature rise in the grinding zone. To find the temperature distribution in the workpiece, the heat flux generated during DPSG is considered as a rectangular heat flux over the workpiece top surface (Figure 15.8). Here, B and H are the width and height of the workpiece enclosed by the boundary surfaces S1, S2, S3 and S4. Its value is calculated by using $q_{wg} = F_{wg} U_o V_f$, where F_{wg} is energy partition due to grinding, V_f is feed given to the workpiece and U_o is specific grinding energy. Here, the value of F_{wg} is chosen as 0.08.

The physical phenomenon of DPSG can be described by the energy equation (which is a differential equation) and appropriate boundary and initial

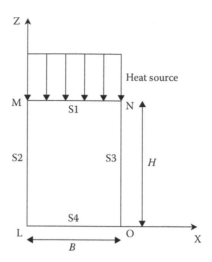

FIGURE 15.8
Thermal model of DPSG.

conditions. The governing equation used to describe the temperature field due to DPSG is governed by a thermal diffusion-convection differential equation as

$$\rho C_s \frac{\partial T}{\partial t} = \frac{\partial}{\partial X}\left(k\frac{\partial T}{\partial X}\right) + \frac{\partial}{\partial Z}\left(k\frac{\partial T}{\partial Z}\right) \text{ in the enclosed}$$

domain LMNO (Figure 15.8),   (15.1)

where T is temperature, t is time, ρ is density, k is thermal conductivity and C_s is the specific heat capacity of the workpiece material in solid state.

Energy transferred to the workpiece as heat input serves as a thermal boundary condition on the top surface, S1 (Figure 15.8). The bottom surface of the workpiece is assumed to be sufficiently away from the top surface so as to remain at its initial temperature throughout the grinding pass. The heat loss to the coolant on the surfaces S1, S2 and S3 (Figure 15.8) is modelled using convective boundary condition. Thus, the boundary conditions are as follows:

$$\left.\begin{array}{lll} q = -q_w + h_c T & on & S1, S2, S3 \\ \text{where} & & \\ q_w = q_{wg} + h_c T_0 & on & S1 \\ q_w = h_c T_0 & on & S2, S3 \\ T = T_0 & on & S4 \end{array}\right\} \text{ when } t > 0.$$ (15.2)

Here, h_c is convective heat transfer coefficient, T_o is ambient temperature (i.e. dielectric temperature) and q_{wg} is heat flux supplied to the workpiece due to grinding. The initial temperature, T_i, can be taken as normal room temperature, T_o, of the dielectric in which the workpiece is completely dipped. Thus, the initial condition becomes

$$T_i = T_o \quad \text{at} \quad t = 0. \tag{15.3}$$

The FEM is used to find the temperature distribution in the workpiece due to DPSG.

15.6.2 Thermal Modelling of EDPSG

The governing equation applicable for heating a workpiece due to a single spark (Gaussian distributed) is assumed to be axisymmetric, i.e. $\dfrac{\partial T}{\partial \theta} = 0$. Therefore, the temperature field is governed by a thermal diffusion differential equation:

$$\rho C_s \frac{\partial T}{\partial t} = \frac{1}{R} \frac{\partial}{\partial R} \left(k \frac{\partial T}{\partial R} \right) + \frac{\partial}{\partial Z} \left(k \frac{\partial T}{\partial Z} \right) \text{ in domain EFGH (Figure 15.9),} \tag{15.4}$$

where R and Z are coordinate axes.

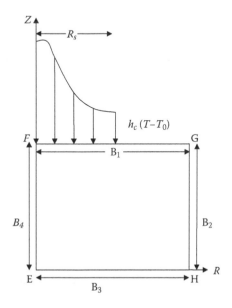

FIGURE 15.9
Axisymmetric thermal model of EDPSG.

The boundary and initial conditions can be written considering energy transferred to the workpiece as heat input serves as a thermal boundary condition on the top surface B_1 (Figure 15.9). Heat loss to the coolant on the surface B_1 is modelled using the convective boundary condition. Further, the boundaries B_2 and B_3 are at such a large distance that there is no heat transfer across B_2 and B_3 (Figure 15.9). Similarly, there is no heat transfer across the axis of axisymmetric, i.e. across B_4 (Figure 15.9). The heat transfer across any boundary is given by

$$q = -q_w + hT,$$

where

$$
\left.
\begin{array}{rll}
q_w = -q_{ws} & \text{for} & R \leq R_s \\
= h_c T_o & \text{for} & R \geq R_s \\
= 0 & \text{for} & \text{off-time} \\
h = 0 & \text{for} & R \leq R_s \\
= -h_c & \text{for} & R \geq R_s \\
= 0 & \text{for} & \text{off-time}
\end{array}
\right\} \quad \text{for } t > 0 \text{ on } B_1
\tag{15.5}
$$

$$q = 0, \quad \text{on} \quad B_2, B_3, B_4 \ (R = 0). \tag{15.6}$$

Here, q_{ws} is the quantity of heat flux entering into the workpiece due to spark and R_s is the radius of spark. The initial temperature, T_i, can be taken as the temperature of the dielectric in which the workpiece is completely dipped:

$$T_i = T_o \quad \text{at} \quad t = 0. \tag{15.7}$$

Here also, the Galerkin weighted residual method [22] is used to convert the governing differential equation into algebraic equations by using approximate weighting functions. The heat flux q_{ws} (R) at a radius (R) can be calculated using the following equation:

$$q_{ws}(R) = \frac{4.45 F_{ws} U_b I}{\pi R_s^2} \exp\left\{-4.5\left(R/R_s\right)^2\right\}, \tag{15.8}$$

where U_b is breakdown (discharge) voltage, I is current and F_{ws} is the energy partition (fraction of input heat going into the workpiece) due to EDM. In the present work, the value of F_{ws} is taken as 0.08. Also, the latent heat required for phase change is accounted for by modifying the expression for specific heat of the workpiece (C_s). Considering the enthalpy before and after the phase change, the latent heat of melting (L_m) and the latent heat of evaporation (L_v) are incorporated into modified specific heats, C_l and C_v, in the vicinity of the melting temperature, T_m, and the boiling temperature, T_v, respectively [20]:

$$C_l = C_s + \frac{L_m}{2\Delta T} \quad \text{for } T_m - \Delta T \leq T \leq T_m + \Delta T \qquad (15.9)$$

and

$$C_v = C_l + \frac{L_v}{2\Delta T} \quad \text{for } T_v - \Delta T \leq T \leq T_v + \Delta T, \qquad (15.10)$$

where C_l and C_v are specific heats of workpiece material in liquid and vapour states, respectively. ΔT is the temperature rise.

15.6.3 Temperature Simulation Using the Theory of Superposition

According to this method, the resulting temperature (T) due to both DPSG and EDPSG is obtained by superposition of the two temperatures as follows:

$$T = T_0 + T_g + T_s, \qquad (15.11)$$

where T_g is temperature rise due to DPSG and T_s is temperature rise due to EDPSG.

The EDDPSG is used for an HSS workpiece of width 6 mm, thickness 6 mm and height 20 mm (Figure 15.10a) with a grinding wheel of thickness 6.5 mm at the top surface of the workpiece. Top surface discretisation and locations of sparks (points●) on the workpiece surface are shown in Figure 15.10b. The superposition technique is used to obtain the temperature distribution in the workpiece due to EDDPSG. First, the temperature distribution in the cuboid shape workpiece due to DPSG is obtained by using plane thermal modelling of DPSG. Let us consider a plane $M_1M_2M_3M_4$ parallel to the Y–Z plane (Figure 15.10a). It is assumed that the temperature along the thickness of the workpiece (Y-axis) at a specified depth in the given plane, say $M_1M_2M_3M_4$, is the same during DPSG. For example, the temperature at any point along the line M_1–M_2 on the top surface of the workpiece will be the same during DPSG. In the same way, the temperature at any point along the line PP$_1$ (Figure 15.10a) will also be the same but of different magnitude as compared to that along M_1–M_2 in the plane $M_1M_2M_3M_4$. Using this procedure, nodal temperatures in the 3D domain of a cuboid shape workpiece are obtained at any specified grinding time, which is an integral multiple of the time required for the grinding wheel to move downward by one element length (0.5 mm).

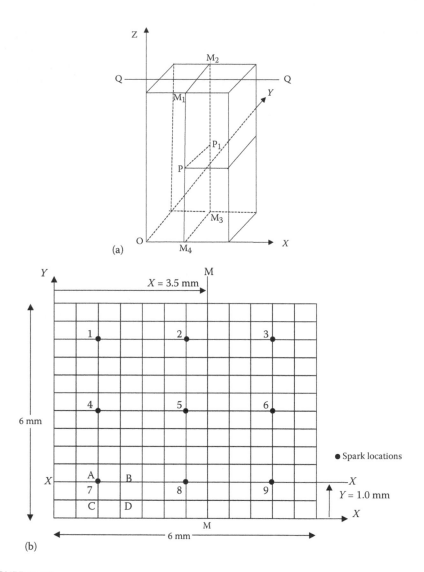

FIGURE 15.10
(a) Cuboid shape workpiece whose top surface is machined using EDDPSG. (b) Discretised workpiece top surface used for superposition of temperature.

It has been assumed that the spark strikes at the workpiece top surface at selected points (Figure 15.10b). It has also been considered that only one spark takes place at a time (say, at point 1) and the next spark at the same location is repeated in sequence after sparking at all other selected locations (2 to 9 in Figure 15.10b) are over. The grinding time for one element movement in the downward direction is completely matched with the total time of sparking at all other nine locations. For this matching of time, 24 times sparking is needed at each of the nine spark locations with on-time of 75 μs and off-time of 150 μs. In this way, at each spark location, heat will be supplied for 75 μs, and again, it will be repeated after 1950 μs (sum of the cycle time of eight sparks of eight locations plus off-time of the spark at present location).

The rectangular parallelepiped domain of 16R × 16R × 8R around a spark could be reasonably considered as a domain independent of the effect of heat from the neighbouring sparks (assuming that the temperature distribution is evaluated after 0.05 s of grinding). Nine such solid blocks are assumed to be fitted within a single solid block of dimension 48R length along the X-axis, 48R width along the Y-axis and 8R depth from the top surface opposite to Z-axis.

Figure 15.11 shows the top surface temperature distribution after superposition of DPSG and EDPSG temperature distributions. The figure shows that the peak temperatures are observed at the nine spark locations.

The temperature distribution in the workpiece during EDDPSG can be simulated by means of the FEM. This reduces the need for costly and time-consuming measurements to determine workpiece temperature. The proposed FEM approach takes care of the different length scales of the constituent process of EDDPSG (DPSG and EDPSG) by using the superposition theorem. Thus, from the plane temperature analysis of DPSG and the axisymmetric temperature analysis of EDPSG, it is possible to predict the temperature distribution in the cuboid shape workpiece due to EDDPSG. A sharp increase in temperature is observed around spark locations at the top surface of the workpiece. Further, high-temperature gradients, which introduce the thermal stresses in the workpiece, are found within a thin surface layer near the spark location. This shows

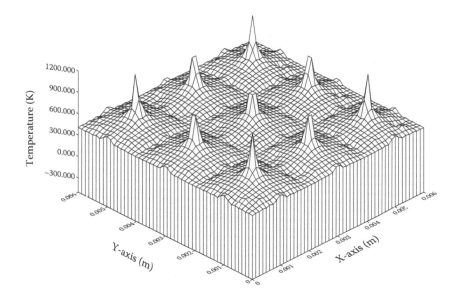

FIGURE 15.11
Temperature peaks on the top surface of a workpiece during EDDPSG process for t_{on} = 75 μs, t_{off} = 150 μs after t = 0.05 s.

that temperature-softening effects are limited to a thin layer around the spark flux. The model can predict the temperature distribution in the workpiece, including the effects of duty cycle, current, feed and time of machining.

15.7 Conclusions

The present chapter dealt with the concept of hybrid machining followed by a detailed description of the development and modelling of the EDDG process. Experimental modelling and performance prediction studies have been conducted on Al-SiC and Al-Al$_2$O$_3$ workpieces. The feasibility study of the EDDG process has been experimentally tested to study the effect of input parameters considering current, pulse on-time, duty factor, wheel speed, workpiece speed and depth of cut on MRR and R$_a$. A three-layer feed-forward ANN has been found suitable to generalise the system characteristics by predicting values close to the actual values. A three-dimensional temperature distribution in the workpiece has been successfully achieved using 2D analysis of the constituent processes of EDDG to the study thermal effects on the workpiece.

Example 15.1

Prove that the heat flux due to a Gaussian distributed spark heat source effective up to 3 sigma limit can be given by the expression

$$q_{ws}(R) = \frac{4.45 F_{ws} U_b I}{\pi R_s^2} \exp\left\{-4.5\left(\frac{R}{R_s}\right)^2\right\}.$$

SOLUTION

The Gaussian distributed spark heat flux profile in the ξ and η coordinate system is given in the following:

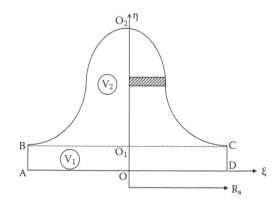

Consider the effect of spark up to 6-σ limit, i.e. $6\sigma = 2R_s$ or $\sigma = R_s/3$. The heat flux q_{ws} (R) at a radius (R) can be calculated if maximum intensity q_o at the axis of a spark and its radius (R_s) is known.

$$q_{ws}(R) = q_o \exp\left\{-4.5\left(\frac{R}{R_s}\right)^2\right\} \tag{15.12}$$

At $\xi = 0$, $\eta = q_0$ and at $\xi = R_s$, $\eta = q_0\,e^{-4.5}$. Now from Equation 15.12,

$$\ln\frac{\eta}{q_0} = -4.5\frac{\xi^2}{R_s^2} \quad \text{or} \quad \xi^2 = -\frac{R_s^2}{4.5}\ln\frac{\eta}{q_0}. \tag{15.13}$$

If it is assumed that the total power of each pulse is to be used only by one spark, then the total volume of the spark is given by

$$V = F_{ws}U_b I = V_1 + V_2, \tag{15.14}$$

where

Volume after rotating OO_1CD about OO_2 $V_1 = \pi R_s^2 q_o e^{-4.5} = 0.011\pi R_s^2 q_o$
$$\tag{15.15}$$

Volume after rotating O_1O_2C about O_1O_2 $V_2 = \int_{q_0 e^{-4.5}}^{q_0} \xi^2\, d\eta = 0.2086\pi R_s^2 q_o.$
$$\tag{15.16}$$

Putting Equations 15.15 and 15.16 in Equation 15.14,

$$F_{ws}U_b I = 0.011\pi R_s^2 q_0 + 0.2086\pi R_s^2 q_0 \tag{15.17}$$

or

$$q_o = \frac{4.45 F_{ws}U_b I}{\pi R_s^2} \tag{15.18}$$

Putting the value of q_0 from Equation 15.18 into Equation 15.12, we get

$$q_{ws}(R) = \frac{4.45 F_{ws}U_b I}{\pi R_s^2} \exp\left\{-4.5\left(\frac{R}{R_s}\right)^2\right\}, \tag{15.19}$$

where U_b is breakdown (discharge) voltage and I is current.

Review Questions

1. With the help of a neat sketch, explain the mechanism of material removal in the EDDG process.

2. Explain the working of the different feasible configurations of EDDG.

3. Explain the working of EDDG in grinding-dominant mode and discharge-dominant mode. How can EDDG be used in the balanced mode?

4. Explain why only diamond abrasives are practically suitable for the EDAG process. Also, clearly explain the specification of a grinding wheel used for the EDDG process.

5. Compare DG, EDG and EDDG in terms of input parameters and output performance parameters.

6. What modifications are required to be made in a conventional EDM machine to use it for different configurations of EDDG?

7. Why is determination of temperature distribution in the workpiece required during EDDG? Develop an expression, one for each, for the determination of heat flux on the workpiece surface due to grinding using an equilateral triangular heat flux distribution and due to EDG using Gaussian heat flux distribution effective up to two-sigma limit only.

8. Explain the procedure of finding residual stresses in the workpiece after finding the temperature distribution in the workpiece.

9. How can different experimental modelling techniques be applied for modelling the EDDG process?

10. Write the applications and limitations of the EDDG process.

WRITE TRUE OR FALSE

1. EDDG can be used at any low and high RPM of grinding wheel.

2. In EDDG, the spark is generated between the diamond abrasive and the bonding material of the wheel.

3. To apply EDDG, the protrusion height of abrasives should be more than the IEG.

4. In EDDG, a resin bonded diamond abrasive wheel will give better performance than a metal bonded wheel.

5. The major problems of EDDG are wheel loading and wheel clogging.

6. The wheel used in EDDPSG is same as that used in EDDFSG in terms of specification and performance.

7. Differential equations are used for experimental modelling of the EDDG process.

8. Superposition theorem can be applied for determining the thermal stresses in the workpiece due to EDDG.

9. Wire-EDDG is better option for making circular cross-section hole in the MMC workpiece.

10. EDDG can be used only in EDG dominant mode.

FILL IN THE BLANKS

1. In EDDG process the electro-discharge takes place between _____ and abrasive abrasion takes place between _____.

2. In HMP, the performance of the process is the sum of their _____ processes and _____.

3. ECSM is a _____ type of HMP, whereas EDDG is a _____ type of HMP.

4. In EDG, the process of sparking takes place between _____ and _____.

5. EDDPSG is used for machining _____ workpiece, and EDDFCG, for _____.

6. In EDDPSG, the grinding wheel is driven with the help of _____ through the _____ arrangement.

7. The input parameters of EDDG are _____ and the output parameters are _____.

8. A three-dimensional temperature distribution in the workpiece during EDDG can be obtained using _____ analysis of _____ and _____ analysis of _____ along with _____ theorem.

9. The equation used for heat flux to workpiece due to discharge action in thermal modelling of EDDG is _____.

10. The heat balance equation at boundary S1 can be written as _____, whereas at S2, it can be written as _____ in thermal modelling of diamond surface grinding.

Multiple Choice Questions

1. EDDG can be used for the machining of
   a. conducting materials only.
   b. non-conducting materials only.
   c. both conducting and non-conducting materials.
   d. None of these.

2. In EDDG, the spark is generated between
    a.   the abrasive and the workpiece.
    b.   the bonding material and the dielectric.
    c.   the bonding material and the electrolyte.
    d.   the bonding material and the workpiece.

3. In EDDG, the protrusion height of the abrasive grain is
    a.   equal to IEG.
    b.   more than IEG.
    c.   half of the IEG.
    d.   None of these.

4. Critical voltage between electrodes for generation of spark is EDDG is dependent on
    a.   abrasive size.
    b.   protrusion height.
    c.   IEG.
    d.   All of the above.

5. For imparting surface grinding motion in EDDG, the lead screw of machine table is driven by
    a.   DC motor.
    b.   reversible synchronous motor.
    c.   servo motor.
    d.   None of these.

6. For precision machining of the end surface of a cylindrical work-piece, which one of the following configurations of EDDG is used?
    a.   EDDPSG
    b.   EDDPCG
    c.   EDDFCG
    d.   EDDFSG

7. Grinding forces are reduced during EDDG due to
    a.   softening of wheel by spark.
    b.   softening of workpiece by spark.
    c.   wheel de-clogging.
    d.   None of these.

8. Which one of the following is not related to EDDG characteristics?
    a.   Elimination re-solidified layer
    b.   Reduction in grinding forces

    c. Higher MRR

    d. Generation of wheel glazing and loading problems

9. A combination of electro-chemical action and electro-discharge action leads to

    a. ECAM.

    b. ECSM.

    c. EDDG.

    d. None of these.

10. ECAFCG combines the features of

    a. ECPSG and APSG.

    b. ECFCG and ACG.

    c. ECPCG and APCG.

    d. None of these.

References

1. Jain V.K., *Advanced Machining Processes*, Allied Publisher, Bombay, India, 2001.
2. Yadava V., Finite Element Analysis of Electro-Discharge Diamond Grinding, PhD thesis at Indian Institute of Technology Kanpur, Kanpur, India, 2002.
3. Yadav R.N., Some Investigations on Slotted-Electrical Discharge Diamond Grinding, PhD thesis at Motilal Nehru National Institute of Technology Allahabad, Allahabad, India, 2014.
4. Yadav R.N. and Yadava B., Experimental study of erosion and abrasion based hybrid machining of hybrid metal matrix composite, *International Journal of Precision Engineering and Manufacturing*, Vol. 14, No. 8, pp 1293–1299 (2013).
5. Panda M.C., Thermal Finite Element based Intelligent Modeling and Optimization of Electro-Chemical Spark Machining Process, PhD thesis at Motilal Nehru National Institute of Technology Allahabad, Allahabad, India, 2010.
6. Panda M.C. and Yadava V., Finite element prediction of material removal rate due to traveling wire electrochemical spark machining, *International Journal of Advanced Manufacturing Technology*, Vol. 45, pp. 506–520 (2009).
7. Panda M.C. and Yadava V., Intelligent modeling and multi-objective optimization of die sinking electro-chemical spark machining process, *Materials and Manufacturing Processes*, Vol. 27, No. 1, pp. 10–25 (2012).
8. Kozak J. and Kazimierz E.O., Selected problems of abrasive hybrid machining, *Journal of Materials Processing Technology*, Vol. 109, pp. 360–366 (2001).
9. Aoyama T. and Inasaki I., Hybrid machining – Combination of electrical discharge machining and grinding, Proceedings of the 14th North American Manufacturing Research Conference, SME (1986), 654–661.

10. Koshy P., Jain V.K., and Lal G.K., Mechanism of material removal in electrical discharge diamond grinding, *International Journal of Machine Tools and Manufacture*, Vol. 36, No. 10, pp. 1173–1185 (1996).
11. Grodzinskii E.Y., Grinding with electrical activation of the wheel surface, *Machines and Tooling*, Vol. 50, No. 12, pp. 10–13 (1979).
12. Grodzinskii E.Y. and Zubotaya L.S., Electrochemical and electrical discharge abrasive machining, *Soviet Engineering Research*, Vol. 2, No. 3, pp. 90–92 (1982).
13. Rajurkar K.P., Wei B., Kozak J., and Nooka S.R., Abrasive electro-discharge grinding of advanced materials, *Proceedings of the 11th International Symposium of Electro-Machining (ISEM-11)* (1995) 863–869.
14. Choudhary S.K., Jain V.K., and Gupta M., Electrical discharge diamond grinding of high-speed steel, *Machining Science and Technology*, Vol. 3, No. 1, pp. 91–105 (1999).
15. Singh G.K., Electro-Discharge Diamond Face Grinding: Development, Modeling and Optimization, PhD thesis at Motilal Nehru National Institute of Technology Allahabad, Allahabad, India, 2011.
16. Singh G.K., Yadava V., and Kumar R., Diamond face grinding of WC-Co composite with spark assistance: Experimental study and parameter optimization, *International Journal of Precision Engineering and Manufacturing*, Vol. 11, No. 4, pp. 509–518 (2010).
17. Yadav S.K.S. and Yadava V., Experimental investigations to study EDDCG machinability of cemented carbide, *Materials and Manufacturing Processes*, Vol. 28, No. 10, pp. 1077–1081 (2013).
18. Agarwal S.S., Some Investigations on Surface-Electrical Discharge Diamond Grinding of Metal Matrix Composites, PhD thesis at Motilal Nehru National Institute of Technology Allahabad, Allahabad, India, 2013.
19. Agarwal S.S. and Yadava V., Development, experimental investigation and modeling of surface-electrical discharge diamond grinding of Al-SiC metal matrix composite, *International Journal of Abrasive Technology*, Vol. 5, No. 3, pp. 223–244 (2012).
20. Yadava V., Jain V.K., and Dixit P.M., Temperature distribution during electro-discharge abrasive grinding, *Machining Science and Technology – An International Journal*, Vol. 6, No. 1, pp. 97–127 (2002).
21. Balaji P.S. and Yadava V., Three dimensional thermal finite element simulation of electro-discharge diamond surface grinding, *Simulation Modeling Practice and Theory*, Vol. 35, pp. 97–117 (2013).
22. Reddy J.N., *An Introduction to the Finite Element Method*, 3rd Edition, McGraw-Hill, Inc., New Delhi, 2005.

16

Fine Finishing of Gears by Electrochemical Honing Process

Neelesh Kumar Jain and Sunil Pathak

Discipline of Mechanical Engineering, Indian Institute of Technology Indore, Madhya Pradesh, India

CONTENTS

16.1 Introduction

A gear is a modified form of a wheel. It is one of the basic machine elements used for transmitting power and/or motion between two parallel shafts (cylindrical gears, i.e. spur and helical), intersecting shafts (conical gears, i.e. straight and spiral bevel) and non-parallel and non-intersecting shafts (i.e. hypoid gears and worm and worm-wheel). Gear drives are preferred for various power and/or motion transmission purposes due to their compactness and higher reliability. Gears are extensively used in the fields of automobiles, avionics, ship-building, construction machinery, agricultural machinery, machine tools, biomedical applications, domestic appliances, miniature and micro systems and products, etc. Continuous requirements of gears and advancements in these fields have compelled the need to manufacture accurate, high-quality and high-reliability gears.

The surface characteristics of gears have a significant influence on their service life, operating performance and characteristics related to wear, noise and motion transfer as presented in Table 16.1. Surface characteristics of gears have two major components, namely, (i) surface quality, which includes surface finish, micro-geometry (i.e. form and location errors) and wear characteristics; and (ii) surface integrity, which encompasses microstructure, micro-hardness and residual stresses.

TABLE 16.1

Surface Characteristics That Affect the Various Aspects of Gears

To Improve Service Life and Operating Performance	To Reduce Noise and Transmission Error
• Reduce parameters of surface roughness (i.e. R_a, R_{max} and R_z) on the tooth flank	• Eliminate micro-geometry errors of tooth flank profile
• Improve wear characteristics by reducing sliding coefficient of friction and friction forces	• Remove sharp corners from gear tooth
• Improve the microstructure	• Increase the wear resistance
• Increase the fatigue life	

Errors in surface characteristics lead to the premature failure of gears. To prevent premature failure of gears, careful consideration is required for the interrelationship between the following factors: (i) shape or geometry of gear tooth, (ii) forces (static and dynamic) on gear tooth, (iii) motion of gear tooth, (iv) gear material, (v) physical and chemical characteristics of lubricant, (vi) operating environment and (vii) surface quality and surface integrity of gear tooth. The first six items are related to design and applications of the gears, whereas the surface quality and surface integrity of the gears depend on manufacturing and finishing of the gears.

The surface quality of gear teeth manufactured by the generative process is not good and most often fails to meet the requirement of the end users. Poor surface quality of a gear tooth is the major source of noise generation, errors in motion transfer, excessive wear and backlash between the meshing gears. Consequently, gear teeth should be smooth and geometrically accurate for efficient motion transmission, noiseless operation, longer service life and better operating performance and to improve the load carrying capacity. As depicted in Figure 16.1, an appropriate gear finishing process should fulfil two major goals [1], namely, (i) improvement in surface quality and reduction in form errors to maximise load capacity and (ii) flank modifications and improving surface integrity to minimise the running noise. This can be ensured by a suitable combination of gear finishing and gear surface properties enhancing processes. Figure 16.2 shows the surface characteristics of the gears that can be improved by an appropriate gear finishing process.

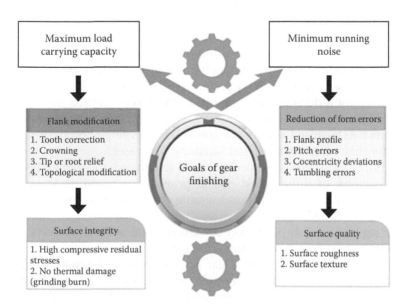

FIGURE 16.1
Goals of gear finishing.

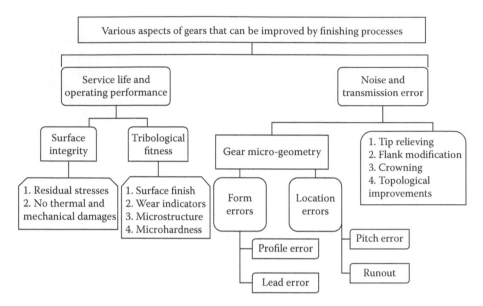

FIGURE 16.2
Surface characteristics of gears that can be improved by finishing of gears.

The most commonly used conventional processes for finishing gears are gear shaving, gear honing, gear burnishing, gear grinding and gear lapping. These processes have limited capabilities and suffer from their inherent limitations. Table 16.2 presents the capabilities, applications, advantages and limitations of these processes. Gear shaving can correct the indexing error and is able to modify the gear flank profile but it can be used only for unhardened gears or gears with hardness up to 40 of Rockwell C grade. Gear honing can be used for hardened gears for modifying their micro-geometry. It also generates a cross-hatch lay pattern on the tooth flank, which provides better lubrication during operation but it has a very limited tool life. Gear burnishing is a localised cold-working operation, in which some undesirable effects such as localised surface stresses can be developed on the gear tooth. It is faster than other conventional finishing processes.

Gear burnishing process is specially used to enhance the surface finish and micro-geometry of unhardened helical gears. Gear grinding is expensive, complicated and may result in undesirable effects such as grind burns and transverse grinding lines.

Grinding burns is a type of thermal damage that adversely affects the surface integrity of gears whereas transverse grinding lines lead to an increase in noise and vibrations of the gears. Gear lapping finishes the mating gears in a conjugate pair only and can rectify only minute deviations from the desired gear tooth profile. It is a slow process, and if performed for a longer duration, it affects the accuracy of the gear teeth profile in a detrimental manner. Gear

TABLE 16.2

Summary of Conventional Finishing Processes of Gears

Gear Finishing Processes	Capability	Applicability	Limitations
Gear shaving	Corrects error in index, helical angle, tooth profile and eccentricity. Increases load-carrying capacity of gears. Improves tooth surface smoothness. Reduces gear noise. Eliminates the danger of tooth end load concentration in service. Effectively correct spacing errors.	Widely used for teeth of straight or helical toothed external spur gears and worm wheels of moderate size.	Only for gears either having hardness up to 40 Rockwell C scale or unhardened.
Gear grinding	Very accurate method. Most frequently used gear finishing method.	Frequently used to finish tooth profiles of different types of gears of hard material that has been heat-treated to a high hardness level after gear cutting.	Relatively very expensive and complicated. Form grinding is very time-consuming.
Gear lapping	Corrects minute heat-treatment distortion errors of involute profile, helix angle, spacing and eccentricity in hardened gears. Gear tooth contact substantially improves by lapping.	Normally used for spur, helical, bevel, spiral bevel and hypoid gears. Usually employed on those gears that have been shaved and hardened.	Only corrects minute deviations from the desired gear tooth profiles. Longer lapping cycles may affect accuracy of the involute profile in a detrimental manner.
Gear honing	Improves noise characteristics of the gears. Produces cross-hatch lay pattern on the finished surface, which is useful for lubricating oil retention.	Can be used for hardened gears.	Limited life of honing gear tool.
Gear burnishing	Improves surface integrity and fatigue life of the gears. Faster than other conventional gear finishing methods.	Used for helical gears.	Can be used for unhardened gears only. It is a localised cold-working operation, with some undesirable effects such as localised surface stresses and non-uniform surface characteristics.

grinding and gear lapping are the most commonly used commercial processes for finishing gears.

It can be concluded that non-overlapping and limited capabilities and inherent limitations of the conventional processes of gear finishing do not allow a single process to improve all the surface characteristics of any gear material simultaneously and without inducing any adverse effect. Most of the time, a combination of conventional finishing processes is required to achieve the required surface quality, which can become very time consuming and laborious and affects the requirement of high productivity. These limitations can be overcome by developing a non-contact, material-hardness-independent, more productive, more economical and sustainable gear finishing process.

Electrochemical honing (ECH) is a hybrid super-finishing process that combines the capabilities and advantages of electrochemical machining (ECM) with mechanical honing and simultaneously overcoming their individual limitations. The main capabilities of the ECM process include the capability to machine/finish material of any hardness, production of stress-free and crack-free surface, higher material removal rate (MRR) and no tool wear. The main capabilities of honing are the ability to correct the geometric errors and controlled generation of functional surfaces. The main limitation of the ECM process is passivation of the anodic workpiece surface by the metal oxides formed due to the evolution of oxygen gas at the anode during its electrolytic dissolution. This anode passivation prohibits further electrolytic dissolution of the workpiece. The major limitations of honing process include limited life of the honing tool, low productivity, incapability of finishing a hardened workpiece and possibility of mechanical damage (i.e. micro-cracks, hardness alternation and plastic deformation) to the workpiece material. This makes ECH an ideal choice to explore as an alternative, superior and economical process for gear finishing.

16.2 Gear Finishing by ECH

The mechanism of material removal in ECH is based on the interaction between electrolytic actions with mechanical abrasion. It is reported in the literature that more than 90% of material removal occurs through electrolytic action, while the remaining 10% is removed by abrasive honing action [2]. The process and working principle of ECH can be better explored by understanding the process principle of ECM and mechanical honing.

16.2.1 Process Principle of ECM

ECM is controlled dissolution of the anodic workpiece through an electrolysis process governed by Faraday's law of electrolysis. It can be considered as reverse of the electroplating process with certain modifications. In this process, an approximately complementary image of a cathodic tool is reproduced on the anodic workpiece, dissolving it electrolytically but without any deposition in the cathode. In this process, no electrolyte is consumed except for flow losses and vaporisation. It can be used for machining any electrically conductive material irrespective of its hardness. Details of the ECM process, its equipment and other aspects are available in standard texts [3–6]; therefore, they are not reproduced here.

Many process variables affect the intensity and uniformity of electrolytic dissolution. These include change in the valency of electrochemical dissolution and preferential valency mode of electrochemical dissolution during the process, evolution of hydrogen gas at the cathode and oxygen gas at the anode, bubbles and voids formation, electrolyte conductivity and its dependence on electrolyte temperature, variation of temperature along the electrolyte flow path, over-potential, effect of passivity owing to metal oxide film and so on.

16.2.2 Process Principle of Mechanical Honing

Mechanical honing is a controlled, low-velocity, abrasive finishing process primarily used to generate functional characteristics for a variety of geometries and to improve the surface quality of the workpiece material. In this process, the material is removed by the shearing action of the bonded abrasive grains of the honing stones or sticks, and only a fraction of millimetres of material is removed from the workpiece [7]. Honing requires simultaneous rotary and reciprocating motions of the honing tool to generate a cross-hatch lay pattern, which enhances the lubrication and tribological properties of the functional surfaces used in various engineering applications [7,8]. The tangent of half of the cross-hatch angle is given by the ratio of reciprocating speed to rotary speed expressed in surface metre per minute (smpm). In a majority of the applications, a cross-hatch angle of 30° is common, but in practice, it varies from 20° to 45°. Commonly used abrasive materials are aluminium oxide, silicon carbide, cubic boron nitride (CBN) and diamond. Aluminium oxide is commonly used to rough-hone steel. Silicon carbide works well on cast iron, bronze, beryllium, some nonmetallic materials and to fine-finish steel. CBN is used on most steels and their alloys as well as for cast iron. Diamond is used for cast iron and tungsten carbide. Abrasive mesh size ranges from 70 to 1200. Generally, honing pressure in the range of 0.5 to 3 MPa is used. However, honing is more often controlled by feed-out or stick-out pressure compared to gauge pressure. Insufficient pressure leads to a lower MRR, while excessive pressure results in a rougher finish as the abrasive sticks break down frequently. As honing is a low-speed operation in which material is removed without an increase in workpiece temperature, unlike in grinding, there is no thermal damage to the honed surface [9]. A surface finish in terms of average surface roughness R_a in the range of 0.25 to 0.4 µm can be obtained easily and a finish of less than 0.2 µm can be achieved and reproduced with extra care.

16.2.3 Process Principle of ECH for Finishing of Gears

Material removal in ECH is based on a controlled anodic dissolution of workpiece (anode) with the tool as the cathode in an electrolyte cell. To initiate the material removal of workpiece gear in ECH, generally, a low direct current (DC) voltage in the range of 8–30 V is applied between the anode and cathode, separated by a very small value of inter-electrode gap (IEG) (usually in the range of 0.1–1.0 mm), which is maintained by a specially designed cathode tool. Current density of the order of 10–100 A/cm^2 and volumetric MRR in the range of 0.1–1,000 mm^3/min are achieved. Electrolyte should be such that the material removed from the workpiece must not deposit on the cathode tool. Products of the electrochemical reaction are flushed away from the IEG using an aqueous solution of the salt-based electrolyte such as NaCl, NaClO$_3$, NaNO$_3$ or their combination supplied through the IEG at a flow rate in the range of 10–40 lpm. This helps in maintaining the clean environment in IEG, thus helping electrochemical dissolution to continue. The electrolyte flow also takes away the heat generated due to the passage of current and possibly due to electrochemical reactions. A passivating layer of metal oxide is generated on the workpiece due to the evolution of oxygen at the anode. It prohibits its further dissolution. The thickness of this protective layer is more at valleys as compared to that on the peaks. The honing tool selectively removes this passivating layer by remaining in continuous contact with the workpiece, enabling electrolytic dissolution and mechanical honing to take place simultaneously [7–10].

The actual working principle of ECH of gears is explained by Chen et al. [10] with the help of a schematic of finishing chamber arrangements as depicted in Figure 16.3, in which the workpiece gear 1 is clamped between centres of the work table, which is reciprocating axially as indicated by the arrowhead 3. The cathode in the ECM process should be electrically conductive

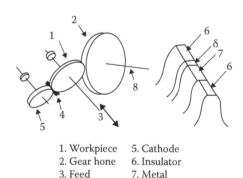

1. Workpiece	5. Cathode
2. Gear hone	6. Insulator
3. Feed	7. Metal
4. Electrolyte	8. Crossed axes

FIGURE 16.3
Design of cathode tool and working principle of ECH of gears. (From Chen, C., Liu, J., Wei, G., *Ann CIRP* 30, 103–106, 1981.)

to produce electrolysis action, but in ECH of gears, the cathode gear is in constant mesh with the workpiece gear, which will cause a short-circuit during the process. Therefore, to avoid the short circuiting, the cathode gear consists of a gear (7) made of a conducting material sandwiched between two insulating gears (6). There is difference of δ (i.e. IEG) between the gear profiles of the conducting gear and the insulating gears. The cathode has the same involute profile as the workpiece. The axis of the shaft on which the cathode is mounted is parallel to the axis of workpiece gear. A full stream of electrolyte is supplied to the gap δ, and a DC current is passed through the gap. During the process of material removal from the tooth flank, the electrolyte forms a metal oxide protective film on the workpiece gear tooth surface, which protects the surface from being further removed. This oxide layer on the tooth surface of the workpiece gear is scraped by the honing gear when it comes in contact with a cross-axis arranged honing gear (2).

There are four important advantages of ECH that make it superior to the conventional finishing processes:

1. Theoretically, there is no tool wear because there is no physical contact between the anode (i.e. workpiece gear) and the cathode (cathode gears) tool.
2. The finishing of the workpiece gear is independent of its material mechanical properties, i.e. hardness, brittleness, strength and ductility.
3. Uniformity of material removal can be achieved as the material removal is due to its anodic dissolution.
4. Ability of producing stress-free surfaces and crack-free smooth surfaces.

16.3 Controlling Anodic Dissolution by Mass Transport Phenomenon

The mass transport phenomenon plays a very important role in controlling the anodic dissolution of workpiece material and in achieving the desired level of surface quality because it affects the current density distribution at macroscopic and microscopic levels. Controlled dissolution by mass transport smooths the surface because the IEG at the peaks of a surface is smaller than the IEG at the valleys, giving more current density and more material removal at peaks than the valleys (this is referred to as a levelling effect). Anodic levelling can be achieved by three types of non-uniform current distributions, i.e. primary current distribution (Ohmic), secondary current distribution (kinetic) and tertiary current distribution due to non-uniform local mass transport [11]. In ECH, the effect of secondary current distribution is negligible if sufficient IEG and electrolyte flow velocity are maintained. Therefore, primary and tertiary current distributions mainly contribute towards smoothing of the gear surface. Moreover, the influence of charge transfer kinetics is negligible and differences due to grain orientation, grain boundaries, dislocations or small inclusions will not play any significant role because of the controlled limiting current by mass transport. As a result, material removal at the molecular level results in a very smooth surface topography (brightening effect) [6,11]. Variations in the temperature and hydrogen bubble formation have been reported to significantly affect the electrolyte conductivity and, consequently, the evenness of the electrolytic dissolution. The IEG has a tremendous influence on the intensity and uniformity of electrolytic material removal. In the electrolytic material removal process, hydrogen is engendered on the cathode and acts as a layer of bubbles mixed with electrolyte, and this layer thickens as the electrolyte flows downstream. The layer moves towards the anode surface at a speed of 0.05–0.1 m/s. Electrolyte pressure, flow velocity, density and duration of current application have been reported as the decisive parameters in defining the size and distribution of hydrogen bubbles. To avoid this problem and achieve improvement and evenness in material removal, higher values of electrolyte pressure and flow rate and lower values of current density and current duration are desirable. To minimise uneven electrolytic dissolution, a lower value of temperature variation and higher electrolyte velocity in the machining zone are preferable [5,6,11]. Thus, by controlling the previously mentioned parameters, a better control over the process can be achieved.

16.4 Equipment for Gear Finishing by ECH

Equipment for ECH of gears consists of mainly four subsystems, namely, (a) DC power supply unit, (b) electrolyte supply and cleaning system, (c) ECH tool and drive system and (d) work-holding system.

16.4.1 DC Power Supply Unit

A DC power supply unit with the capability of supplying an output voltage in the range of 0–100 V, current in the range of 10–110 A and with programmable options for setting pulse-on time and pulse-off time is generally used to supply the current in the IEG. It comprises of three parts: programmable high-power DC supply, pulse generator and pulse controller with the power switch unit. A power switch apparatus is primarily assessed by switch time, control mode. The pulse controller has the facility to modify the voltage, current, pulse-on time (T_{on}), pulse-off time (T_{off}) and, consequently, the duty cycle (ξ),

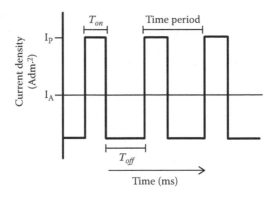

FIGURE 16.4
Pulse current waveform.

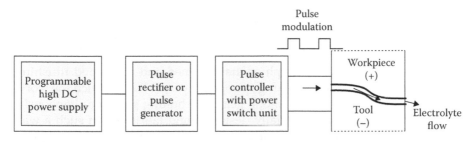

FIGURE 16.5
Components of pulsed power supply used for finishing the bevel gears by PECH.

i.e. the ratio of pulse-on time to sum of pulse-on and pulse-off times. The pulse phenomenon is used while using the pulsed-ECH (PECH) process only. A simple pulse waveform is presented in Figure 16.4, while Figure 16.5 depicts the components of the pulsed power supply used for finishing bevel gears by ECH. The positive terminal of the power supply was connected to the shaft supporting the workpiece gear, while the negative terminal was connected to the cathode gears through carbon brush and slip ring assembly.

$$Duty\ cycle\ (\xi) = \frac{T_{on}}{T_{on} + T_{off}} 100\% \tag{16.1}$$

16.4.2 Electrolyte Supply and Cleaning System

The electrolyte, besides facilitating electrochemical functions, carries away the heat and the products of machining from the finishing zone. The desirable properties of the electrolyte include high electrical conductivity, low viscosity, high specific heat, chemical stability, resistance to formation of stiff passivating film on work surface, non-corrosiveness, non-toxic, low cost and easy availability. The most commonly used electrolytes are sodium chloride and sodium nitrate. Most ECH machines require an electrolyte tank with a storage capacity of 190 to 380 litres. An electrolyte supply and cleaning system consists of a pump, filters, piping and control valves, heating and cooling coils, pressure and flow rate measuring devices and storage and settling tanks. Electrolyte is pumped through a stainless steel pump to the IEG to imitate the electrolytic dissolution of the workpiece material. Filtration is usually accomplished with coarse cartridge filters made of anti-corrosive material such as stainless steel and monel. Magnetic separators or centrifugal filters can also be used. Piping and tanks are made of stainless steel, glass-fibre-reinforced plastics, plastic-lined mild steel or similar other anti-corrosive materials.

16.4.3 ECH Tool and Drive System

16.4.3.1 Finishing Chamber for Finishing Spur and Helical Gears by ECH

Figure 16.6a shows a photograph of the designed and developed finishing chamber arrangement for finishing spur gear by ECH [12] in which the workpiece gear meshes with the cathode gear and honing gear simultaneously. The cathode gear is to be specially designed and fabricated in such a way that its electrically conducting portion never touches the anodic workpiece gear, thus avoiding short-circuiting and, at the same time, it maintains the required IEG between the anode and cathode gears. For the cylindrical gears (spur and helical gears), this is ensured by sandwiching a conducting layer between two non-conducting layers and undercutting the conducting layer by an amount equal to the required IEG as compared to the non-conducting layers. The axes of shafts of the workpiece and cathode gears are parallel to each other for finishing the spur and helical gears (Figure 16.6a and b). The workpiece gear is provided controllable rotary motion by means of a DC motor while the cathode and honing gear rotate by virtue of their tight engagement with the workpiece gear. Since the entire face width of the cathode gear is not electrically conducting, a reciprocating motion is also given to the workpiece gear by means of a programmable stepper or servo-motor to ensure finishing of the entire face width of the cylindrical gear [12,13]. The majority of finishing or material removal is done by the ECM action, which occurs between the anodic workpiece gear and the cathode gear, while the role of the honing action, which takes place between the workpiece and the honing gear, is just to selectively remove the passivation layer of metal oxide formed on the workpiece gear. All three gears should have the same involute profile and the same module.

For spur gear finishing by ECH (Figure 16.6a), a helical gear is made as honing gear; it can either be an abrasive impregnated gear or a gear having

(a)

(b)

FIGURE 16.6
(a) Photograph of the tooling system for finishing of spur gear by ECH. (From Naik, L.R., Investigations on precision finishing of gears by electrochemical honing, MTech thesis, IIT Roorkee, 2008.) (b) Photograph of the tooling system for finishing of helical gear by ECH. (From Misra, J.P., Precision finishing of helical gears by electrochemical honing (ECH) process. MTech thesis, IIT Roorkee, 2009.)

hardness more than the workpiece gear. It is mounted on a floating stock to ensure dual flank contact between the honing and workpiece gear. The honing gear having a cross-axis arrangement with the workpiece gear is mounted in such a way so as to reduce the tooth surface contact and the pressure required for finishing [12]. Helical gear finishing by ECH (Figure 16.6b) does not require cross-axis arrangement because honing and cathode gears have an opposite helix angle to that of the workpiece gear; i.e. if the workpiece gear is right-handed helical gear, then the cathode and honing gears will be left-handed, and vice versa [13].

All three gears (i.e. workpiece gear, cathode gear and honing gears) are mounted on a special type of axle made of stainless steel. Brackets are used for holding the axles of the cathode and honing gears. Bakelite has been used as bracket material for its electrical insulation and corrosion resistance properties. The entire tooling system with axles and brackets is enclosed in a machining chamber made of perspex for better visibility and corrosion resistance. The finishing chamber also has provisions for supplying fresh electrolytes, for removal of used electrolyte and for escape of gases generated during the ECH process. The finishing chamber is connected to the cast iron frame using four brass screws. Thus, it is attached to the machine column of a bench drilling machine using a swivel arrangement for ease in loading/unloading. The swivel system can slide on the machine column to achieve axial positioning of the workpiece gear with respect to the honing and cathode gears. After the proper alignment and positioning of the workpiece, the worktable can be locked in position using a lever [12,13].

16.4.3.2 Finishing Chamber for Finishing of Bevel Gears by ECH

Reciprocation of a conical gear (i.e. bevel and hypoid gears) along its axis of rotation during its finishing by ECH is not possible due to the continuously varying module along its face width. This problem was solved by envisaging a novel concept of using twin complementary cathode gears and their meshing arrangement with the workpiece and honing gears (Figure 16.7a) [14]. It ensured finishing the entire face width of the bevel gear without requiring reciprocating motion and simultaneously maintained the IEG required for the electrolytic dissolution. In this arrangement, one of the cathode gears (3) has an insulating layer of metalon sandwiched between two conducting layers of copper, whereas the other complimentary cathode gear (4) has a conducting layer of copper sandwiched between two insulating layers of metalon. The conducting layer is undercut by 1 mm as compared to the insulating layers to maintain the IEG between the cathode and anode gears, which is required for the anodic dissolution by the electrolytic action. The workpiece gear (1), honing gear (2) and cathode gears (3 and 4) are mounted in such a way that their axes of rotation are perpendicular to each other. The photograph of the designed and developed cathode gears are shown in Figure 16.7b. The workpiece gear is given rotary motion by the spindle of a drilling machine, while other gears rotate due to tight meshing with it. Electrolyte (5) is supplied to IEG and a DC voltage is applied across it. The electrolytic dissolution takes place between the

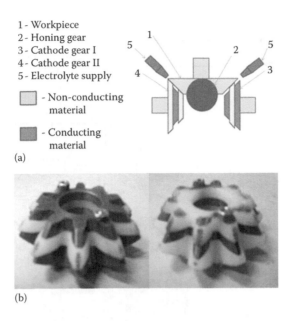

1 - Workpiece
2 - Honing gear
3 - Cathode gear I
4 - Cathode gear II
5 - Electrolyte supply

☐ - Non-conducting material

■ - Conducting material

(a)

(b)

FIGURE 16.7
(a) Concept of twin-complementary cathode gear for bevel gear finishing by ECH; (b) photograph of the complementary cathode gears. (From Shaikh, J.H., Experimental investigations and performance optimization of electrochemical honing process for finishing the bevel gears, PhD thesis, IIT Indore, 2014.)

workpiece and cathode gears, while honing action occurs simultaneously between the workpiece and honing gears. During the electrolytic dissolution process, the evolution of oxygen takes place at the anode, forming a passivating metal oxide layer on the workpiece gear surface, prohibiting its further electrolytic dissolution. The honing gear removes this passivating layer and exposes a fresh surface for further finishing by the electrolytic dissolution. This synchronous finishing by electrolytic dissolution and honing results in enhanced surface quality of the workpiece gear with high productivity.

16.4.4 Work-Holding System

The work-holding system consists of a fixture and an electrolyte chamber and is attached to the foundation with proper insulating arrangements. The worktable and fixturing design considerations include corrosion resistance and strength to take over the machining torque without deforming the workpiece. The fixture and the electrolyte chamber are made of stainless steel or perspex. Automatic gauging devices, such as air gauge, are often built into the ECH equipment.

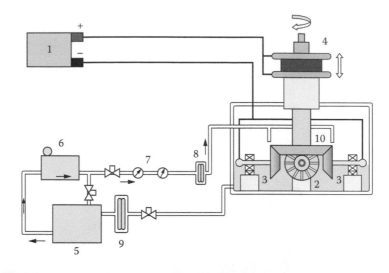

(a)

(b)

FIGURE 16.8

Experimental setup for finishing of straight bevel gears by ECH: (a) schematic diagram: (1) DC power source, (2) honing gear, (3) cathode gears, (4) carbon brush and slip ring assembly, (5) electrolyte storage tank, (6) stainless steel electrolyte supply pump, (7) flow meter and pressure gauge (8) first-stage filter with magnetic separator, (9) second-stage filter with magnetic separator and (10) workpiece gear; (b) photograph. (From Shaikh, J.H., Experimental investigations and performance optimization of electrochemical honing process for finishing the bevel gears, PhD thesis, IIT Indore, 2014.)

An innovative experimental setup for finishing the bevel gears by ECH was designed and developed by Shaikh et al. [14], whose schematic diagram is depicted in Figure 16.8a and its photograph in Figure 16.8b. This setup has four subsystems, namely, (i) DC power supply system; (ii) electrolyte supply, cleaning and re-circulating system; (iii) machining chamber housing workpiece, cathode and honing gears and (iv) a machine frame to support the machining chamber and to provide motion to the workpiece gear.

A DC power supply system has the capacity of an output voltage in the range of 0–100 V and current in the range of 10–110 A, and it can be operated either as a constant current source or as a constant voltage source. The positive terminal of the power supply was connected to the stainless steel shaft supporting the workpiece bevel gear the while negative terminal was connected to the two complementary cathode gears through a carbon brush and slip ring assembly. The electrolyte supply, cleaning and re-circulating system was designed to supply the filtered electrolyte to the machining chamber and re-circulate it back to the storage tank. A rotary pump made of stainless steel and capable of developing a wide range of pressures and flow rates was used to supply an aqueous mixture of NaCl and $NaNO_3$ as the electrolyte. Filtration was achieved by using two double-stage magnetic and stainless steel mesh filters provided in the electrolyte flow path. Electrolyte pressure and flow rate measuring devices and flow control valve were employed at the pump outlet. The electrolyte temperature was maintained by a heating element fitted with a precise temperature controller. Rotary motion to the workpiece gear was provided by a DC motor fixed on the frame of a drilling machine of 38 mm drilling capacity. This motor has a controller to vary the rotary speed continuously in the range of 30–1500 rpm. The machining chamber was fabricated using perspex sheets to provide better visualisation of the ECH process and better strength-to-weight ratio. Pedestal-type ball bearings were used to mount and support the stainless shafts on which honing and two cathode gears were mounted. Metalon blocks were used to support and mount the bearings due to their corrosion resistance, electrical insulation and higher strength-to-weight ratio. The work table of the drilling machine (size 400 mm × 400 mm) was used to mount the machining chamber [14].

16.5 Process Parameters of ECH

As ECH is a hybrid process of ECM and the mechanical honing process, its process parameters include the parameters related to both ECM and conventional honing in addition to some parameters related to the workpiece and tooling. A typical range of different ECH process parameters are presented in Table 16.3.

TABLE 16.3

Different Process Parameters of ECH and Their Typical Values

Parameters Related to ECM	Parameters Related to Mechanical Honing	Parameters Related to Workpiece and Tooling
Power-supply related parameters	• Speed of rotation (30–80 rpm)	• Electrochemical characteristics of the work material
• Voltage (V) (6 to 30 V)	• Speed of translation (0–18 m/min)	• Speed of rotation (20–100 rpm)
• Current (I) (100 to 3000 A)	• Honing pressure (0.5–3 MPa)	• Electrical conductivity of workpiece material
• Current density (J) (12 to 47 A/cm²). However, current density in the range of 0.5–2 A/cm² is also extensively used for better surface integrity results.	• Abrasive-related parameters	• Tool design
• Pulse-on time T_{on} (1 ms to 4 ms)	✓ Type (Al₂O₃, SiC, CBN, diamond)	
• Pulse-off time T_{off} (2 ms to 9 ms)	✓ Size (70–1200 mesh size)	
• Duty cycle, λ (10%–66%)	✓ Feed mechanism	
Electrolyte-related parameters		
• Type and concentration		
• Pressure (0.5 to 1.0 MPa)		
• Temperature up to 38°C		
• Flow rate up to 10–50 L/min		
• Electrolyte tank capacity (190–380 L)		
• IEG (0.25 to 1 mm)		

Source: Jain, V.K., *Advanced Machining Processes*. Allied Publishers, New Delhi; El-Hofi, H., *Fundamentals of Machining Processes*. McGraw-Hill Companies, New York, 2005; Dudley, D.W., *Gear Handbook*, McGraw-Hill Publishing Company, New York, 1962; Jain, N.K., Naik, L.R., Dubey, A.K., Shan, H.S., *Proc. Inst. Mech. Eng. Part B J. Eng. Manuf.*, 223, 665–681, 2009; Naik, L.R., Investigations on precision finishing of gears by electrochemical honing, MTech thesis, IIT Roorkee, 2008; Chen, C., Liu, J., Wei, G., *Ann. CIRP*, 30, 103–106, 1981; Pathak, S., Jain, N.K., Palani, I.A., *Int. J. Electrochem. Sci.*, 10, 1–18, 2015; Naik, L.R., Investigations on precision finishing of gears by electrochemical honing, MTech thesis, IIT Roorkee, 2008; Misra, J.P., Precision finishing of helical gears by electrochemical honing (ECH) process. MTech thesis, IIT Roorkee, 2009; Shaikh, J.H., Experimental investigations and performance optimization of electrochemical honing process for finishing the bevel gears, PhD thesis, IIT Indore, 2014.

16.6 Effects of Process Parameters on the Performance of ECH for Finishing of Gears

16.6.1 Effects of Current Density and Voltage

Either a constant DC power supply (in ECH) or a pulsed DC power supply (in PECH) can be used. A constant DC power supply provides continuous current or voltage, whereas a pulsed power DC supply is used to supply current or voltage in the form of short pulse with specific pulse-on time and pulse-off time. The MRR is proportional to the applied voltage. The linear MRR, Q_l (i.e. rate of change of IEG), and current density, J, in the IEG can be evaluated by using the following equations derived from the Faraday's laws of electrolysis:

$$Q_l = \frac{dY}{dt} = \eta \left(\frac{M}{Z} \right) \frac{J}{\rho F} - f \tag{16.2}$$

$$J = \frac{K(V_o - \Delta V)}{Y} \tag{16.3}$$

$$\Rightarrow Q_l = \frac{dY}{dt} = \eta \left[\left(\frac{M}{Z} \right) \frac{K(V_o - \Delta V)}{\rho F} \right] \frac{1}{Y} - f, \tag{16.4}$$

where η is current efficiency, M/Z is electrochemical equivalent of the workpiece material (M: atomic weight [g] and Z is valency of dissolution), K is electrolyte conductivity (Ω^{-1} mm^{-1}), V_o is pulse voltage (V), ΔV is over-voltage (V), Y is IEG (mm), ρ is density of the workpiece material (g/mm^3), F is Faraday's constant (96,500 C) and f is tool feed rate (mm/s).

From Equation 16.4, it can be seen that on increasing the value of voltage the current density will increase, which leads to a higher MRR. At the start of the ECH process, the workpiece surface has more irregularities (in terms of high peaks of roughness), and hence, the rate of electrochemical dissolution of workpiece material is high. But after a few cycles, due to continuous electrochemical dissolution, the irregularities from the workpiece surface are reduced, which increases the IEG, thereby decreasing the MRR. Therefore, the surface finish of the workpiece material increases with the increase in current up to a certain level and then starts decreasing. Also, higher voltage and electrolyte concentration increase the amount of current and the number of free ions available for the conduction of current through IEG and help in increasing the MRR.

16.6.2 Effects of Pulse-On Time and Pulse-Off Time

The introduction of T_{on} and T_{off} in an ECH process may slow down the overall process in comparison to traditional constant DC power supply. But the process becomes more effective. Lower pulse-on time restricts current only to anodic dissolution of workpiece gear material and sufficient time is not available for generating a large amount of reaction products such as sludge and gas bubbles, and whatever reaction products are generated, they are flushed away from the narrow machining zone during the pulse-off time. This ensures a clean environment in the IEG without any clogging. Also, in the PECH process, the introduction of pulse-on time decreases the concentration polarisation effect (i.e. formation of the diffusion layers of the ions at the workpiece gears surface), which acts as a barrier and restricts the anodic dissolution. Thus, from the obtained trends, it was depicted that a higher value of pulse-on time will lead to more anodic dissolution of workpiece gear material and generating more quantity of electrolytic reaction products and deteriorates the micro-geometry of the gears due to increased MRR and poor flushing of the reaction products. It is noticeable that the better results in the micro-geometry of the bevel gears can be achieved by lesser and uniform MRR only. A higher pulse-off time adversely affects the micro-geometry of bevel gears due to reduced finishing time available for the anodic dissolution of workpiece gear, while lower values of pulse-off time leads to poor flushing of the reaction products.

16.6.3 Effects of Electrolyte Parameters

Electrolyte is the most important element of the ECH system. Generally, an aqueous solution of an appropriate electrolyte is supplied through the IEG at the required flow rate. The main functions of the electrolyte flow are (a) to provide ideal conditions for the electrochemical dissolution of the workpiece material without depositing it on the cathodic tool; (b) to flush away the products of ECM formed due to electrochemical reactions (i.e. metal hydroxides and reaction gases); (c) to take away the heat generated due to the passage of current and may be due to electrochemical reactions, thus maintaining a constant temperature in the IEG; (d) to not allow concentration of ions on the electrodes and (e) not to allow oxygen evolved at the anode to passivate it by forming metal oxides.

In ECH, generally, salt-based electrolytes (i.e. $NaCl$, $NaNO_3$, $NaClO_3$, KCl, KNO_3, etc.) or their combination are supplied through the IEG. Additives such as sodium tartrate, glycerin and glycol are used to get better results. The MRR or improvement in the surface finish increases by using a high concentration of electrolytes due to the fact that the conductivity of the electrolyte depends on electrolyte concentration. Conductivity increases with concentration as more ions are available in the solution for electrolytic dissolution. Therefore, the concentration of electrolyte should be high but not greater than 10%–15% to get a better passivation effect. The current density and passivation effect of electrolyte greatly depend on electrolyte temperature. The electrolyte conductivity is very much sensitive towards electrolyte temperature and increases with it, resulting in a higher current density. Generally, an electrolyte flow rate in the range of 10–50 lpm is used to supply in the IEG for finishing of gears by ECH. Lower values of electrolyte flow rate are usually preferred so as to provide the electrolyte sufficient time to build up the required ion concentration and for allowable electrolyte temperature rise, which are required for the electrolytic dissolution of the anodic workpiece material, whereas higher electrolyte flow rate leads to more amount of fresh electrolyte, which maintains the electrolyte temperature almost constant. This decreases the ion concentration, which reduces the finishing action and results in higher values of roughness parameters of the finished surface.

A gravimetric aqueous solution of $NaNO_3$ and NaCl is the most commonly used electrolyte for finishing the gears by the ECH process due to the following reasons:

- NaCl has high corrosive nature, but its conductivity is very stable over a wide range of pH.
- Sodium nitrate ($NaNO_3$) is less corrosive but creates a strong passivation layer on the workpiece, due to which less MRR is found in comparison with NaCl. Due to less dissolution of anodic workpiece material, better smoothness can be achieved. Also, the electrolyte reactions require higher voltage when using strong passivating electrolyte. Thus, to maintain a less corrosive environment and stable conductivity in the finishing chamber, a gravimetric aqueous solution of $NaNO_3$ + NaCl is used.

16.6.4 Effects of Honing-Related Parameters

In ECH of gears, most of the material is removed by their anodic dissolution under electrolytic action, and abrasive bonded honing gear is normally used to brush away the thin and insulating metal oxide micro-film generated on the workpiece due to electrolytic passivation. Therefore, fine-grain abrasives such as sponge rubber and polyvinyl alcohol (PVA) sponge rubber are used to meet the finishing requirement. A honing gear with hardness value higher than the workpiece gear is used for removing the metal oxide layers, and Pathak et al. [15] have explained the process sequence and mechanism of gear finishing by the PECH process as depicted in Figure 16.9. It can be seen from Figure 16.9 that during the electrolytic dissolution

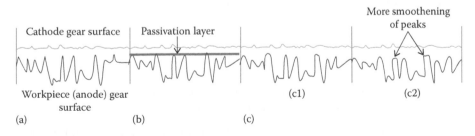

FIGURE 16.9
Sequence of finishing the bevel gear by the PECH process: (a) before finishing by PECH, (b) after electrolytic dissolution and formation of metallic oxide passivation layer and (c) honing action removing passivation layer: (c1) after honing action using an unhardened honing gear, and (c2) after honing action using a hardened honing gear. (From Pathak, S., Jain, N.K., Palani, I.A., *Int. J. Adv. Manuf. Technol.*, 2015.)

of the workpiece surface, a metallic oxide passivation layer is formed on its surface, prohibiting its further electrolytic dissolution. Use of an unhardened honing gear (hardness similar to workpiece gear) in the PECH process removes a very small amount of material from some of the highest peaks on the workpiece surface while removing the passivation layer, whereas use of a hardened honing gear (30% harder than the workpiece gear) removes material from almost all the peaks on the workpiece surface while scrubbing the passivation layer. This leads to comparatively more reduction in maximum surface roughness, which also helps in subsequent electrolytic dissolution to remove some material even from the valley, thus giving a better and uniform surface finish. They have concluded that a honing tool can also participate in material removal process although it must be optimised according to the need and system design.

16.6.5 Effect of Workpiece-Related Parameters

The ECH process is based on the electrolysis principle, and thus, the workpiece material should be electrically conducting in nature. Since ECH is noncontact process, it does not have a material hardness limitation. The rotating speed of the workpiece was found to have a significant effect on surface roughness produced in the ECH process. A very high rotary speed provides a lesser amount of time per cycle available for finishing of workpiece gear surface by electrolytic action and increases the amount mechanical honing. This results in less material being removed from the surface peaks, which yields lower values of percentage variations in surface roughness parameters. On the other hand, a very low rotary speed provides excessive time for electrolytic dissolution of the workpiece gear surface, which also yields a poor surface finish due to the sporadic breakdown of anodic film, selective dissolution of anodic workpiece gear material, flow separation and formation of eddies and hydrogen gas evolution. Therefore, an optimum value of rotary speed exists that corresponds to an to optimum combination of electrolytic dissolution and mechanical honing.

16.7 Review of Recent Work on ECH of Gears Using Constant DC Power Supply

Finishing of gears by their anodic dissolution to enhance the operating performance and productivity of mechanical honing process has been reported [16–21]. The tooling system for ECH was developed by Ann [16,17], who also reported that the ECH is a much faster process than the conventional honing process. The comparative study of the MRR for various components of unlike materials using the combined action of mechanical honing and electrolytic dissolution is presented for confirming the productiveness of the ECH process and reported that finishing the components by ECH can provide high productivity [20,21]. Use of ECH for finishing the *hardened helical* gear was explored for first time by Capello and Bertoglio [22]. Their technological innovation consisted of removing material from tooth face of the anodic helical gear having involute profile, 17 teeth and module of 2.5, mating with a specially designed cathodic helical gear tool having with 64 teeth. The test bench was built to obtain reciprocating and rotary motion of the electrodes with a controlled inter-electrode gap. Their experimental results confirmed process feasibility and need for designing the electrode tools as a function of the electrochemical parameters. Though, their results also showed that the helix and involute profiles obtained were not acceptable but it just confirmed feasibility of using ECH for gear finishing. Wei et al. [23] used a current-control method by varying the intensity of the electric field to control the intensity of electrolytic dissolution step less along the full profile of the gear using a newly developed gear-shaped cathode in the field-controlled ECH (FC-ECH) of gears. As adjusting the field intensity can be done during the process of finishing, it is not only very easy to produce any sort of tip or root relief, but the accuracy of the tooth profile can also be greatly improved provided that the errors in the tooth profile of all the teeth are nearly the same. Results of the involute profile tester, before and after FC-ECH, were presented by the authors [24,25]. This is considered as the first effort in controlling the accuracy of gear profile in the ECH process. He et al. [24] used the electrolysis time-control method to correct the gear tooth profile errors very efficiently in the process that they called slow-scanning field controlled ECH (SSFC-ECH) of gears. They also mentioned the superiority of the time control method over the current control method. They used a gear-shaped cathode that meshes with the workpiece gear during machining and which is exposed as a cathode pole only on a strip on the tooth flank. By slowly varying the centre distance between the cathode and the workpiece gear by a stepper motor, the electrolysis zone sweeps over the tooth flank from root to tip, as shown in Figure 16.10. Figure 16.11 depicts the variation in the material removal distribution under different discharge times. They established a relationship between material removal and gear tooth profile as $H(s) = E(s) + C$, where $H(s)$ is metal removal function, $E(s)$ is gear tooth profile error function, s is arc length along tooth flank and C is base metal removal.

An online profile error measuring device was delivered and a mathematical model was developed for calculating the required discharging time. Several ground gears with typical profile errors were subjected to trial machining on this working principle and the results showed that it corrects the profile errors very efficiently. Yi et al. [25] described the electrochemical gear tooth profile-modification theory. They mentioned a new process of axial modification for carbonised gears and investigated the current density distribution n in the gear teeth. Their test result indicated that both current and processing periods are principal parameters to affect the volume of crown and the amount of modification. They created a

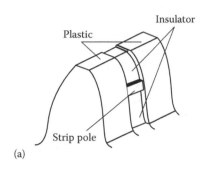

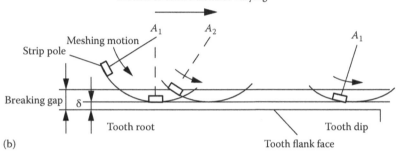

FIGURE 16.10
(a) Gear tooth of cathode. (b) Scanning motion of strip pole in SSFC-ECH. (From He, F., Zhang, W., Nezu, K., *Jpn. Soc. Mech. Eng. Int. J Ser C*, 43, 486–491, 2000.)

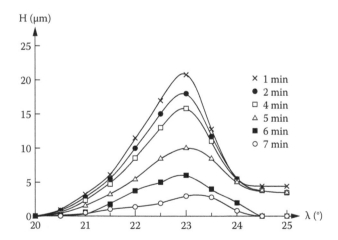

FIGURE 16.11
Metal removal distribution under different discharge time. (From He, F., Zhang, W., Nezu, K., *Jpn. Soc. Mech. Eng. Int. J Ser C*, 43, 486–491, 2000.)

new method for electrochemical tooth-profile modification based on real-time control and established a mathematical model of the electrochemical profile modification process using an artificial neural network [26]. They mentioned that one method for improving the load-carrying capacity of a gear is to shape the teeth in the direction of the gear tooth profile and gear tooth direction. The employment of gear tooth-profile modification and gear tooth crowning does not change the geometric dimensions and the smoothness of the gear.

Naik et al. [27] investigated the ECH of spur gears made of EN8 using different combinations of $NaNO_3$ and $NaCl$ as the electrolyte and EN24 as the honing gear material and reported an improvement of up to 80% and 67% in average surface roughness (R_a) and maximum surface roughness (R_{max}) values, respectively. Experimental investigations to the surface finish of helical gears made of EN8 were conducted by Misra et al. [28] to study the effects of voltage, electrolyte composition, electrolyte concentration and rotating speed of workpiece gear on the surface finish using EN24 as honing gear material and electrolyte as an aqueous solution of $NaCl$ and $NaNO_3$ in a ratio of 3:1 by weight. They reported that the effects of voltage, electrolyte concentration and rotating speed of the workpiece gear were found to have a significant influence on the surface finish of the workpiece gears. They also reported that the microstructure and surface texture of the ECH finished gear surfaces were analysed using scanning electron microscope (SEM) micrographs (Figure 16.12). It is evident from Figure 16.12 that the surface roughness present in the work surface before ECH is significantly reduced by the ECH process. The surface texture becomes more uniform with a glazed appearance. From Figure 16.12a and b, it is clear that before ECH, the work surface contains a lot of scratches and micro-cracks. The SEM micrographs of Figure 16.12c and d reveal that the deep scratches, micro-burrs and pits are reduced significantly in the ECH finished work surface. However, the ECH process was not found adequate to eliminate macro-surface defects, as clearly seen in Figure 16.12c and d.

Shaikh et al. [29] developed a novel concept of twin complementary cathode gears and introduced for simultaneous fine finishing of all the teeth of a bevel gear eliminating the need to provide reciprocating motion to the workpiece gear. The concept of machining chamber arrangements along with the cathode gears arrangement is shown in Figure 16.6. The authors identified applied voltage, electrolyte composition, electrolyte temperature and rotary speed of the workpiece as important ECH parameters influencing the microgeometry and surface finish of the straight bevel gears. They also developed theoretical models of the MRR and surface roughness of the bevel gear finished by ECH and experimentally validated them [30]. The developed a theoretical model for MRR and surface roughness of straight bevel gears finished by ECH, presented by the following equations:

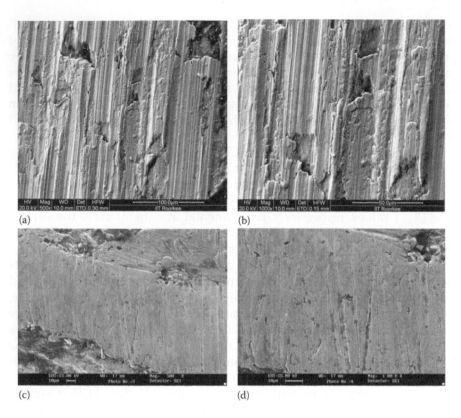

FIGURE 16.12
SEM photographs of the electrochemically honed (ECHed) helical gear tooth surface: (a) before ECH at 500×, (b) before ECH at 1000×, (c) after ECH at 500× and (d) after ECH at 1000×. (From Misra, J.P., Jain, N.K., Jain, P.K., *Proc. Inst Mech. Eng. Part B J. Eng. Manuf.*, 224, 1817–1830, 2010.)

$$
MRR_{ECH}\left(\frac{mm^3}{s}\right)
$$

$$
= (\eta ETF_w|F\rho)(K_e(V-\Delta V)|Y)\left(4r_b\left[arcsin\frac{\sqrt{4D_w^2+(w-w_t)^2}}{4r_b tana}\right]tana+2L_r+W_t+W_b\right)
$$

$$
+\frac{(KF_nN_sT)(L_i+L_r)(W_t+W_b)}{H} \tag{16.5}
$$

$$
R_{zECH}(\mu m)=R_{zi}+\frac{10^{-3}ft(2K-1)MRR_{ECH}}{TF_w\left(4r_b\left[arcsin\frac{\sqrt{4D_w^2+(w-w_t)^2}}{4r_b tana}\right]tana+2L_r+W_t+W_b\right)},
$$

$$
\tag{16.6}
$$

where *MRR* is material removal rate, R_z is maximum depth of surface roughness, η is current efficiency, *E* is electrochemical equivalent of the workpiece material (g), *T* is the number of teeth of the workpiece gear, *F* is Faraday's constant (= 96,500 C), F_n is total normal load acting along the line of action (N), F_w is face width of the bevel gear (mm), K_e is electrical conductivity of the electrolyte (Ω^{-1} mm^{-1}), ρ is density of the anodic work material (g/mm^3), *V* is applied voltage (V), ΔV is total voltage loss in the IEG (V), r_b is the radius of the base circle (mm), D_w is the working depth of gear tooth (mm), *W* is width at the base of the tooth (mm), W_b is the width of bottom land (mm), W_t is the width of top land (mm), L_i is the length of the involute profile (mm) and L_r is the length of arc of fillet at the root (mm).

16.8 Review of Recent Works on ECH of Gears under Pulsed-Power Supply

Datta and Landolt [31–33] used pulsed-ECM to achieve higher dissolution of nickel using aqueous solution of NaCl in a flow channel cell. They reported that applying repetitive short current pulses helps to dissolve the materials electrochemically at very high instantaneous current density (up to 330 A per cm^2) with good surface finish and high dissolution efficiency without any need of using high electrolyte flow velocity. The use of a pulsed current minimises the temperature and concentration gradients parallel to the electrodes and facilitates the precision with which the shape of the tool can be reproduced on the workpiece by ECM. Zhai [34] studied the effects of numerical control technology in pulse electrochemical finishing (PECF) of special purpose gears to achieve the mirror-like surface of gear teeth. The use of numerical control technology leads to better control over gear dividing devices and feed movements and also provides a better understanding of numerical simulations of the electric potential distribution. Under optimised parameters, a mirror-like surface of gear teeth of module as 3 mm, number of teeth of 33 and width of teeth of 15 mm, it was reported that

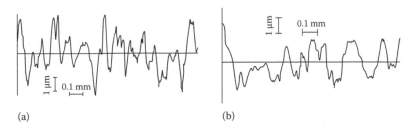

(a)   (b)

FIGURE 16.13
Surface roughness of the spiral bevel gears before and after PECF. (a) R_z = 7.13 μm, (b) R_z = 4.32 μm. (From Ning, M., Xu, W., Xuyue, W., Zefei, W., *Int. J. Adv. Manuf. Technol.*, 54, 979–986.)

the surface roughness value R_a could be reduced from more than 1.6 µm to less than 0.10 µm and surface reflectivity value could be increased from less than 30% to about 80%–90% by finishing the gear for 5 minutes by PECF. Ning et al. [35] developed an experimental setup of PECF for finishing the spiral bevel gears to achieve the improvements in surface roughness and micro-geometric accuracy. They reported that only one gear tooth was finished at a time, and for this, they used a cathode cutter, which rotates and passes through the tooth space of the workpiece gear; after reaching the full depth, the cutter withdraws and the gear is indexed for finishing the next tooth. Their experimental results showed considerable improvements in the maximum depth of surface roughness (R_z); i.e. its value decreases from 7.13 µm to 4.32 µm and showed some improvements in the micro-geometry of the gears. Figure 16.13 depicts the changes in the surface roughness profile before and after PECF [35].

PECF was used by Pang et al. [36] to achieve better surface topography of the cylindrical gears by using the scanning cathode; they reported that the tooth surface roughness R_a decreased after PECF from 3.9 µm to 0.35 µm. The tooth profile modification by PECF is achieved with an uneven IEG and profile lead modification can be achieved by using a variable moving cathode tool. Misra et al. [37] used PECH for finishing spur gears and studied the effects of the composition and temperature of electrolyte on the surface finish. They also found a ratio of 3:1 of $NaCl$ and $NaNO_3$ as optimum electrolyte composition and 30°C as optimum electrolyte temperature.

In their most recent work, Pathak et al. [38] used PECH to improve the micro-geometry of straight bevel gears by studying the effects of pulse-on time, pulse-off time and finishing time. They automated the workpiece engagement and disengagement process using a stepper motor and controller system. They yielded the percentage improvement in the average values of single pitch error (PIf_p), adjacent pitch error (PIf_u), cumulative pitch error (PIF_p) and total runout (PIF_r) as 34.22%, 39.58%, 13.34% and 18.88%, respectively, by using the optimised parameters, i.e. pulse-on time of 2 ms, pulse-off time of 4.5 ms and finishing time of 6 minutes. This improvement in the micro-geometry of bevel gears improves the gear quality in Deutsche Normen (DIN) standards up to DIN 7, which is the same as that given by the bevel gear grinding process. Figure 16.14 shows the variation in percentage improvement in the average values of single pitch error (PIf_p), adjacent pitch error (PIf_u), cumulative pitch error (PIF_p) and total runout (PIF_r) with respect to pulse-on time, pulse-off time and finishing time.

Figure 16.15a and b present the reports generated by the computer numerical control (CNC) gear metrology machine for single pitch error and adjacent pitch error for left-hand flank of the straight bevel gear before and after its finishing by PECH [38]. Figure 16.16a and b presents the same for the cumulative pitch error and total runout. It can be inferred from Figure 16.15a and b that percentage improvement in single pitch error (PIf_p) and percentage improvement in adjacent pitch error (PIf_u) for left-hand flank are 60.75% and 65.5%, respectively. From Figure 16.16, the percentage improvements

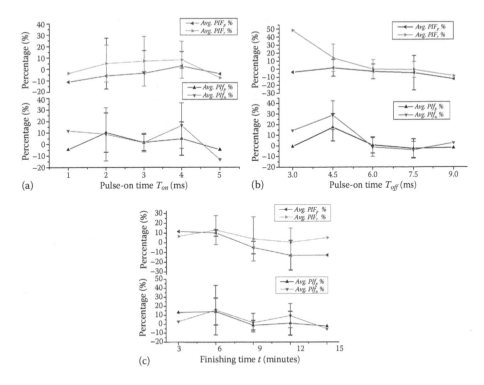

(a) Pulse-on time T_{on} (ms)

(b) Pulse-on time T_{off} (ms)

(c) Finishing time t (minutes)

FIGURE 16.14

Variation in percentage improvements in single pitch error (PIf_p), adjacent pitch error (PIf_u), cumulative pitch error (PIF_p) and total runout (PIF_r) with (a) pulse-on time, (b) pulse-off time and (c) finishing time. (From Pathak, S., Jain, N.K., Palani, I.A., *Mater. Manuf. Process.*, 29, 1461–1469, 2014.)

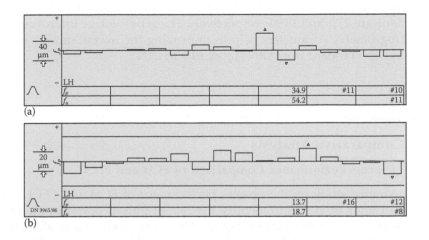

FIGURE 16.15

Single pitch error, f_p, and adjacent pitch error, f_u, of bevel gear tooth for the identified optimum PECH parameters: (a) before PECH and (b) after PECH. (From Pathak, S., Jain, N.K., Palani, I.A., *Mater. Manuf. Process.*, 29, 1461–1469, 2014.)

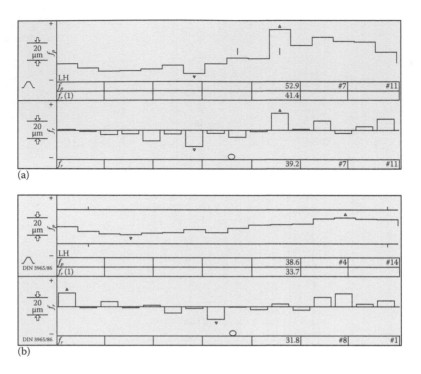

FIGURE 16.16
Cumulative pitch error, F_p, and total runout, F_r, of straight bevel gear tooth for the identified optimum PECH parameters: (a) before PECH and (b) after PECH. (From Pathak, S., Jain, N.K., Palani, I.A., *Mater. Manuf. Process.*, 29, 1461–1469, 2014.)

obtained in cumulative pitch error (PIF_p) and in total runout (PIF_r) for left-hand flank are 27% and 18.9%, respectively. These percentage improvements in microgeometry parameters help in improving the overall quality of the gears in DIN standard from DIN standard 9 to DIN7.

16.9 Comparative Analysis

16.9.1 Process Performance Comparison of ECH and PECH

A comparative study on the process performance of ECH and PECH was done by Pathak et al. [39] to identify the role of pulsed power in the ECH process for finishing straight bevel gears. The conclusion of the work reports that the application of pulse power supply in ECH helps in simultaneous improvements in the surface finish and microgeometry of the bevel gears by a significant amount, thus enhancing their service life and working performance. However, use of pulsed power supply may take a longer time to achieve the desired quality of the bevel gears, thus affecting process productivity of PECH as compared to ECH.

Figure 16.17a and b shows the SEM images of micro-structure of the tooth flank surface reported by Shaikh et al. [29] before finishing and after finishing by the ECH. Figure 16.17c and d depicts the same SEM images obtained by Misra et al. [40]. It is clearly visible from these images that the surface roughness and scratches that are generated on the tooth flank during the cutting operation are visible even after finishing them by the ECH process. Moreover, in Figure 16.17b, some milky spots can also be seen, which occur due to strong electrolytic action while using continuous voltage, which generates more oxygen bubbles and due to insufficient flushing time being available in case of ECH these bubbles are not flushed away effectively, thus leaving the milky white spots on the surface of the tooth flank. From Figure 16.17d, it is visible that the surface layer becomes quite dark after finishing by ECH; this is due to the re-deposition of the reactant, metallic hydroxide, over the surface because of insufficient flushing of the sludge products. Figure 16.18a and b depicts the

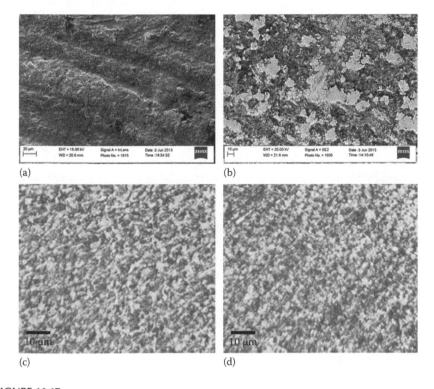

(a) (b)

(c) (d)

FIGURE 16.17
SEM images of gear tooth flank surface at 500× magnification (a) before finishing and (b) after finish by ECH (from Shaikh, J.H., Jain, N.K., Venkatesh, V.C., *Mater. Manuf. Process.*, 28, 1117–1123, 2013) and (c) before finishing and (d) after finish by ECH (from Misra, J.P., Jain, P.K., Dwivedi, D.K., Mehta, N.K., *Proc Eng.*, 64, 1259–1266, 2013), obtained using the identified optimum input parameters by these authors.

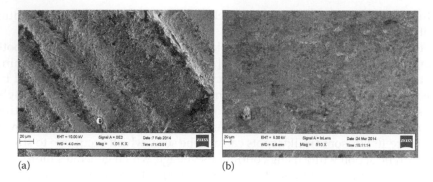

(a)   (b)

FIGURE 16.18
SEM images of gear tooth flank surface at 500× magnification of the (a) before finishing and (b) gear finished by PECH using the identified optimum input parameters. (From Pathak, S., Jain, N.K., Palani, I.A. *Mater. Manuf. Process.*, 30, 836–841, 2015.)

SEM micrographs of bevel gears before and after finishing by PECH. It can be seen from Figure 16.18b that the surface obtained after finishing by PECH is very smooth and there are no scratches, pitting or milky spots visible on the finished tooth flank surface. This smoothing is due to the proper flushing of sludge products during the pulse-off time, and due to restricted re-deposition of the reactants. Such surface smoothening by PECH will results in quieter operation as well as enhanced service life of the bevel gears.

16.9.2 Comparative Study of ECM, Honing and ECH

A comparative study of ECM, honing and ECH for finishing bevel gear was performed by Shaikh et al. [41]. It is clear from Table 16.4 that the improvement in surface finish in terms of PIR_a and PIR_{max} is much less in honing due to a much lower MRR, whereas in the case of ECM, it is more, but ECH gives a much higher improvement in surface finish in terms of PIR_a and PIR_{max}. This can be explained by the fact that the honing or scrubbing action owing to the absence of abrasives does not remove the material from the workpiece surface required to finish the irregularities on the surface. ECM has a higher MRR, which results in a slight improvement in the surface finish, whereas ECH, being a combination of honing and ECM, gives better surface finish and removes more material by more amount as compared to its constituent processes.

TABLE 16.4

Comparison of Bevel Gear Finishing by ECM, Honing and ECH

Items	ECM	Mechanical Honing	ECH
PIR_a	12.7%	−8.5%	34.6%
PIR_{max}	12.3%	0.6%	35.1%
MRR	0.06 mm³/s	0.05 mm³/s	0.31 mm³/s

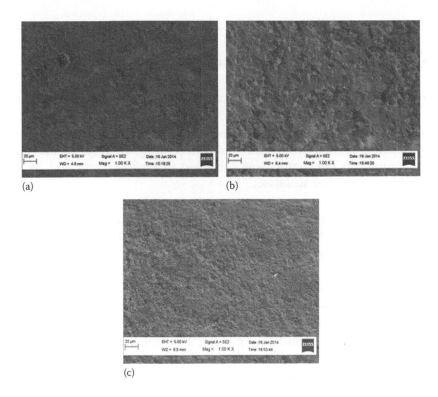

FIGURE 16.19
SEM images of bevel gear tooth flank surface at 1000× after finishing by (a) honing, (b) ECM and (c) ECH. (From Shaikh, J.H., Jain, N.K., Pathak, S., *Proc. Inst. Mech. Eng. Part B J. Eng. Manuf.*, 2015.)

Figure 16.19a–c depicts the SEM images of the similar surfaces of the gear tooth flank finished by honing, ECM and ECH, respectively. It can be seen from Figure 16.19a that the surface roughness produced during the teeth cutting operation on the gear tooth flank surface and the surface defects are not smoothened appreciably by mechanical honing, whereas the gear tooth flank surface finished by ECM (Figure 16.19b) shows more surface roughness due to etching and selective dissolution and presence of the metal oxide layer, while the similar surface finished as compared to the surface finished by ECH (Figure 16.19c). Therefore, it can be concluded that the hybridisation of ECM and honing gives better surface quality as compared to the individual processes.

16.10 Advantages

- Gears having any hardness (but electrically conducting) can be finished by ECH.

- ECH can achieve surface roughness up to 50 nm and tolerances of ±0.002 mm.
- ECH has the capability to simultaneously improve the surface finish and micro-geometry of the gears.
- ECH is a highly productive process as it can finish a gear as fast as within 2 minutes.
- ECH produces surfaces with the distinct cross-hatch lay pattern required for oil retention, surfaces with compressive residual stresses required for the components subjected to cyclic loading and completely stress-free surfaces.
- ECH not only produces high-quality surface finish and surface integrity but also has the ability to correct errors/deviations of gears in terms of form errors (i.e. deviations in lead and profile) and location errors (i.e. pitch deviations and runout) for cylindrical and conical gears.
- It takes less finishing time as compared to ECM and mechanical honing. ECH can finish materials up to 5–10 times faster than mechanical honing and four times faster than internal grinding.
- There is much less heat, thus making it suitable for the processing of parts those are susceptible to heat distortions.
- ECH automatically deburrs the work surface.
- ECH has also been reported to improve the geometry/shape such as out-of-roundness or circularity, taper, bell-mouth hole, barrel-shaped hole, axial distortion, boring tool marks, etc., for cylindrical surfaces.
- Abrasive sticks/tools have increased life due to very limited role of mechanical honing.
- As a by-product, sharp or burred edges of a cross holes or other intersections, which break into the bore, are automatically deburred as the main hole is being finished by ECH.

16.11 Limitations

- ECH can be used for finishing electrically conductive materials only.
- ECH is more costly than mechanical honing due to its electrical, fluid handling elements, need for corrosion protection, costly tooling and longer setup time. This makes ECH more economical for longer production runs than for tool room and job-shop conditions.

16.12 Applications

- ECH is an ideal choice for superfinishing, improving the surface integrity and increasing the service life of the critical components such as gears, internal cylinders, transmission gears, carbide bushings and sleeves, rollers, petrochemical reactors, moulds and dies, gun barrels, pressure vessels, etc., which are made of very hard and/or tough, wear-resistant materials, most of which are susceptible to heat distortions.

- ECH can be used to finish various hard-to-finish materials such as cast tool steels, high-alloy steels, carbide, titanium alloys, Incoloy, stainless steel, Inconel and gun steels.

- ECH is widely used in automobile, avionics, petrochemical, power generation and fluid power industries.

16.13 Summary

ECH is found to be superior to conventional gear finishing processes. This process is found to be successful in the hybridisation of ECM and mechanical honing, overcoming their individual limitations such as generation of hydrogen bubbles and metal oxide layers over the workpiece surface. ECH emerged as one of the most promising technologies to provide simultaneous improvements in surface quality and surface integrity of the gears. The comparative study reveals that majority of the material removed from the workpiece gear was due to their anodic dissolution under electrolytic action while the role of honing is limited to scrub the metal oxide layer only. This process has been developed for cylindrical and conical gears, while opportunities for processing herringbone, hypoid, worm and worm gear are yet to be explored.

Questions

1. What are design considerations for cathode gear in gear finishing by ECH?
2. Why does passivation of anodic workpiece take place in ECM-based processes?
3. What is the role of honing gear in the ECH process?
4. How is hydrogen evolution at the cathode avoided in ECM-based processes?
5. How is the IEG maintained in finishing cylindrical gears by ECH?
6. Why is the honing gear shaft inclined while finishing spur gears but not while finishing helical gear by ECH?

7. Explain the effect of constant DC power and pulsed DC power on ECH process performance.

8. What are the challenges in finishing of conical gear by ECH?

9. What is concept of complimentary cathode gears with reference to finishing conical gears?

10. Why is reciprocating motion required to be imparted to the workpiece gear while finishing cylindrical gears by the ECH process?

Nomenclature

F	Faraday's constant (96,500 C)
f	Tool feed rate (mm/s)
f_p	Single pitch error (µm)
F_p	Cumulative pitch error (µm)
F_r	Runout error (µm)
f_u	Adjacent pitch error (µm)
M	Atomic mass (g)
K	Electrolyte conductivity (Ω^{-1} mm^{-1})
R_a	Average surface roughness (µm)
R_{max}	Maximum surface roughness (µm)
R_z	Depth of surface roughness (µm)
t	Finishing time (minutes)
T_{off}	Pulse-off time (ms)
T_{on}	Pulse-on time (ms)
V	Voltage (V)
V_o	Pulse voltage (V)
ΔV	Over voltage (V)
V_{peak}	Peak voltage
Y	Valency of dissolution
Z	Coefficient of friction
μ	Current efficiency
η	Density of the workpiece material (g/mm^3)

Abbreviations

ECM	Electrochemical machining
ECF	Electrochemical finishing
ECH	Electrochemical honing

PECH	Pulsed-electrochemical honing
PIR_a	Percentage improvement in average surface roughness
PIR_{max}	Percentage improvement in maximum surface roughness
PIR_z	Percentage improvement in depth of surface roughness
PIf_p	Percentage improvement in single pitch error
PIf_u	Percentage improvement in adjacent pitch error
PIF_p	Percentage improvement in cumulative pitch error
PIF_r	Percentage improvement in total runout
MRR	Material removal rate

References

1. B. Karpuschewski, H. J. Knoche, M. Hipke, (2008) Gear finishing by abrasive processes, *Annals of CIRP*, 57(2):621–640.
2. N. K. Jain, L. R. Naik, A. K. Dubey, H. S. Shan, (2009) State of art review of electrochemical honing of internal cylinders and gears. *Proceedings of the Institution of Mechanical Engineers Part B: Journal of Engineering Manufacture*, 223(6):665–681.
3. J. F. Wilson, (1971) *Practice and Theory of Electrochemical Machining.* John Wiley, New York.
4. G. F. Benedict, (1987) *Nontraditional Manufacturing Processes.* Marcel Dekker Inc., New York.
5. V. K. Jain, (2002) *Advanced Machining Processes.* Allied Publishers, New Delhi.
6. H. El-Hofi, (2005) *Fundamentals of Machining Processes.* McGraw-Hill Companies, New York.
7. D. W. Dudley, (1962) *Gear Handbook.* McGraw-Hill Publishing Company, New York.
8. Sunil Pathak, (2016) Investigations on the Performance Characteristics of Straight Bevel Gears by Pulsed Electrochemical Honing (PECH), PhD thesis, *Discipline of Mechanical Engineering*, IIT Indore, India.
9. L. R. Naik, (2008) Investigations on precision finishing of gears by electrochemical honing. MTech thesis, IIT Roorkee.
10. C. Chen, J. Liu, G. Wei, (1981) Electrochemical honing of gears: A new method of gear finishing. *Annals of CIRP*, 30:103–106.
11. S. Pathak, N. K. Jain, I. A. Palani, (2015) On surface quality and wear resistance of straight bevel gears by pulsed electrochemical honing process. *International Journal of Electrochemical Sciences*, 10(11):1–18.
12. L. R. Naik, (2008) Investigations on precision finishing of gears by electrochemical honing. MTech thesis, IIT Roorkee.
13. J. P. Misra, (2009) Precision finishing of helical gears by electrochemical honing (ECH) process. MTech thesis, IIT Roorkee.
14. J. H. Shaikh, (2014) Experimental investigations and performance optimization of electrochemical honing process for finishing the bevel gears. PhD thesis, IIT Indore.

15. S. Pathak, N. K. Jain, I. A. Palani, (2016) Effect of honing gear hardness on surface quality and micro-geometry improvement of straight bevel gears in PECH process. *International Journal of Advanced Manufacturing Technology*, 85(9):2197–2205. (doi: 10.1007/s00170-015-7596-y).
16. R. Eshelman, (1963) Electrochemical honing reports ready for production jobs. *Iron Age*, 124.
17. Ann, (1965) Honing gets electrolytic help. Iron Age, 106.
18. Ann, (1965) Tooling for electrochemical honing. *Tool & Manufacturing Engineering*, 68.
19. N. Given, (1965) Electrochemical honing: Four times faster. *Machinery*, 1965, 72, 119.
20. E. A. Randlett Jr, M. P. Ellis, (1967) Electrochemical honing. American Society of Tools and Manufacturing Engineers (ASTME) Technical Paper MR67-648: 1–13.
21. E. A. RandlettJr, M. P. Ellis, (1968) Electrochemical honing-ECH. American Society of Tools and Manufacturing Engineers (ASTME) Technical Paper MR68-815:1–11.
22. G. Capello, S. Bertoglio, (1979) A new approach by electrochemical finishing of hardened cylindrical gear tooth face. Annals of CIRP 28:103–107.
23. G. Wei, Z. Wang, C. Chen, (1987) Field controlled electrochemical honing of gears. *Precision Engineering*, 9:218–221.
24. F. He, W. Zhang, K. Nezu, (2000) A precision machining of gears: Slow scanning field controlled electrochemical honing. *Japan Society of Mechanical Engineers International Journal (Series C)*, 43:486–491.
25. J. Yi, T. Yang, J. Zhou, (2000) New electrochemical process gear tooth-profile modification. *Manufacturing Technology and Modern Machine*, 9:102–105.
26. J. Yi, J. Zheng, T. Yang, (2002) Solving the control problem for electrochemical gear tooth-profile modification using an artificial neural network. *International Journal of Advanced Manufacturing Technology*, 19:8–13.
27. L. Naik, N. Jain, A. Sharma, (2008) Investigation on precision finishing of spur gears by electrochemical honing. Proceedings of 2nd International and 23rd AIMTDR Conference, Madras, India, 509–514.
28. J. P. Misra, N. K. Jain, P. K. Jain, (2010) Investigations on precision finishing of helical gears by electrochemical honing process. *Proceedings of the Institution of Mechanical Engineers, Part B: Journal of Engineering Manufacture*, 224:1817–1830.
29. J. H. Shaikh, N. K. Jain, V. C. Venkatesh, (2013) Precision finishing of bevel gears by electrochemical honing. *Materials and Manufacturing Processes*, 28:1117–1123.
30. J. H. Shaikh, N. K. Jain, (2014) Modeling of material removal rate and surface roughness in finishing of bevel gears by electrochemical honing process. *Journal of Materials Processing Technology*, 214:200–209.
31. D. Landolt, (1987) Fundamental aspects of electropolishing. *Electrochimica Acta*, 32:1–11.
32. D. Landolt, P.-F. Chauvy, O. Zinger, (2003) Electrochemical micromachining, polishing and surface structuring of metals: Fundamental aspects and new developments. *Electrochimica Acta*, 48:3185–3201.
33. M. Datta, D. Landolt, (1981) Electrochemical machining under pulsed current conditions. *Electrochimica Acta* 29:899–907.
34. X. Zhai, (2013) Analysis of electric field of pulse electrochemical finishing of shaped cathode on gears. *Advanced Materials Research*, 677:207–210.

35. M. Ning, W. Xu, W. Xuyue, W. Zefei, (2011) Mathematical modeling for finishing tooth surfaces of spiral bevel gears using pulse electrochemical dissolution. *International Journal of Advanced Manufacturing Technology*, 54:979–986.
36. G. B. Pang, W. J. Xu, J. J. Zhou, D. M. Li, (2010) Gear finishing and modification compound process by pulse electrochemical finishing with a moving cathode. *Advanced Materials Research*, 126:533–538.
37. J. P. Misra, P. K. Jain, N. K. Jain, H. Singh, (2012) Effects of electrolyte composition and temperature on precision finishing of spur gears by pulse electrochemical honing (PECH). *International Journal of Precision Technology*, 3:37–50.
38. S. Pathak, N. K. Jain, I. A. Palani, (2014) On use of pulsed electrochemical honing to improve micro-geometry of bevel gears. *Materials and Manufacturing Processes*, 29(11–12):1461–1469.
39. S. Pathak, N. K. Jain, I. A. Palani, (2015) Process performance comparison of ECH and PECH for quality enhancement of bevel gears. *Materials and Manufacturing Processes*, 30(7):836–841.
40. J. P. Misra, P. K. Jain, D. K. Dwivedi, N. K. Mehta, (2013) Study of time dependent behavior of ECH of bevel gears. *Procedia Engineering*, 64:1259–1266.
41. J. H. Shaikh, N. K. Jain, S. Pathak, (2016) Investigations on surface quality improvement of straight bevel gears by electrochemical honing process. *Proceedings of the Institution of Mechanical Engineers, Part B: Journal of Engineering Manufacture*, 230(7):1242–1253. (doi: 10.1177/0954405415584899).

Section VI

Miscellaneous

17

Measurement Systems for Characterisation of Micro/Nano-Finished Surfaces

G.L. Samuel

Manufacturing Engineering Section, Department of Mechanical Engineering,
Indian Institute of Technology, Madras, Chennai, India

CONTENTS

17.1 Introduction

Over the recent years, micro/nano-structured surfaces as part of precision manufacturing processes have been taking increasing relevance. Such micro/nano aspects are key in developing the capabilities, functionalisation and performances of future industrial products. The biology, health, electronic, optics, transportation and energy industries are examples of business sectors already benefiting from such processes. However, compared to manufacturing processes at the micrometre and nanometre levels, the ability to control, measure and characterise such features in three dimensions (3D) needs to be further developed. Advanced research in micro- and nano-metrology is important, which will help in understanding the performance of micro/nano-structured surfaces at the micro world.

Surface metrology is a science of measurement that plays a very important role in industrial fields. From an engineering standpoint, it is the measurement of the deviations of a workpiece from its intended size and shape (Whitehouse, 2010). Nowadays, surface metrologists have new challenges, particularly in areas like nano-metrology. Nano-metrology is a subfield of metrology, where the measurements are carried out at the nano-scale level. In nano-manufacturing, nano-metrology has a crucial role in order to produce nanomaterials and devices with a high degree of accuracy and reliability.

The surface properties directly relate to the functionality of the product, which makes it critical to redefine the techniques of micro/nano-metrology to clearly understand the role of the surfaces. For example, the surface topography influences properties like wettability, absorptivity in the microfluidic devices.

The lotus-leaf-like surface is an inspiration to fabricate a hydrophobic surface, where the surface contains structured patterns. Not only that, the surface relates to the performance of micro-electronic devices, optics in defence and satellite imaging and implantable medical devices. Hansen et al. (2006) have described the need and various challenges in micro- and nano-metrology. They also pointed out the difficulty of traceability and calibration of micro/nano-instruments. As the product size decreases to below the micro-level, the more challenging it is for the metrologists.

Precision engineering is the field where micro/nano-structured surfaces find a major application (Stephen et al., 2004). Functional performance, like static and dynamic behaviour, of the micro/nano-surface will determine the overall efficiency of the system. So the characterisation of these micro/nano-scale structured surfaces has great importance, and it is also a challenge for metrologists. Traditional methods cannot be applied for measurement as the structured surfaces are in the micro/nano-level. Conventional contact surface measurement instruments are designed for macro-workpieces with micro-scale topographies. Because its measurement is through the contact mode, it is incapable of meeting the measurement needs of micro/nano-structured surfaces and their evaluation requirements. So there is a need to develop a unique characterisation technique that will provide precise measurement at the nano-level.

Meanwhile, as the size of mircro-feature decreases to the nano-scale, the classical physical principles of the materials are changed. The molecular forces at the surface play a crucial role in determining the functionalization of micro-features. This not only changes the physical properties of materials but also brings about great challenges for the surface geometry measurement and its characterisation. Thus, there is a need to develop new instruments that incorporate the new principles. A unique and precise measurement and characterisation method is needed for evaluating the quality and functional performance of micro/nano-scale structured surfaces. Thus, it is a challenge for metrologists to do research in developing new surface measuring instruments and characterisation methods for the micro/nano-world.

17.2 Challenges in Micro/Nano-Measurements

Micro/nano-structured surface measurements and characterisation techniques require highly sophisticated measuring instruments that can perform precisely at the nano-level. Since the feature size is at the sub-micron level, the dynamic stability of the measuring devices is a great challenge in the performance of the measuring device. Contact and non-contact measuring techniques are available for the characterisation of micro- and nano-features. The real challenge arises when feature size comes down to the nano-scale, as the contact of the measuring instrument will damage the feature. Some of the challenges that arise in the measurement and characterisation of micro/nano-structured surfaces are shown in Figure 17.1.

17.3 Paradigm Shifts in Surface Metrology

The reason for the paradigm shift in surfaces includes the necessity to fabricate surfaces at micro/nanometre ranges and their characterisation to achieve and assess the functional requirements at the micro- and nano-levels. Jiang and Whitehouse (2012) suggested that for optimising functional performance, reducing diffusion filtering and to add the value for manufacture of miniaturised parts, numerical methodology can be adopted. Jiang

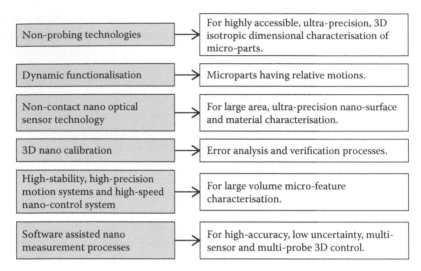

| Non-probing technologies | → | For highly accessible, ultra-precision, 3D isotropic dimensional characterisation of micro-parts. |

| Dynamic functionalisation | → | Microparts having relative motions. |

| Non-contact nano optical sensor technology | → | For large area, ultra-precision nano-surface and material characterisation. |

| 3D nano calibration | → | Error analysis and verification processes. |

| High-stability, high-precision motion systems and high-speed nano-control system | → | For large volume micro-feature characterisation. |

| Software assisted nano measurement processes | → | For high-accuracy, low uncertainty, multi-sensor and multi-probe 3D control. |

FIGURE 17.1
Challenges in measurement and characterisation of micro/nano-structured surfaces.

et al. (2007) described the need for quantification of surface characteristics at all scales such as micrometer, nanometer and atomic scales. This approach had shifted the emphasis of surface metrology from profile to areal characterisation, stochastic to structured surface characterisation and characterisation from simple shapes to complex free form geometries.

17.3.1 Profile to Areal Measurements

The methods for characterising the areal surface texture for the fundamental and functional topographies include the development of the following.

17.3.1.1 Unified Coordinate System for Surface Texture and Form Measurement Surface

Surface texture is changed from profiles to one based on areal surface, meaning there is no consistency requirement for the coordinate system to be related to the lay. A unified coordinate system is being established for both surface texture and form measurement.

17.3.1.2 Scale-Limited Surface

Areal surface characterisation does not require three different groups of surface texture parameters. In areal parameters, Sq alone is defined for the root-mean square (RMS) parameter instead of primary surface Pq, waviness Wq

and roughness Rq in the case of profile. Filters like S-filters, L-filters and form filters were used; now, they have been changed to operators. An SF surface is obtained by an S-filter and an F-operator in combination on a surface and on an SL surface by using an L-filter on an SF surface. SF and SL surfaces are called as scale-limited surfaces (Jiang et al., 2007).

17.3.2 Stochastic and Structured Surfaces

Stochastic surfaces include surfaces produced by procedures like milling, grinding and lapping. The sizes of individual structures have reached the nano-scale in all dimensions, which is a challenging area to characterise. Surfaces with deterministic patterns are economically important in structures like Fresnel lenses and diffraction gratings. There can be different characteristics like height, volume, area and distance between features that have to be characterised. A segmentation procedure is required that identifies the significant and insignificant features defined by the segmentation. There are feature parameters that define these surfaces mentioned in ISO 25178-2. Figure 17.2a shows stochastic surfaces, whereas Figure 17.2b shows structured surfaces.

17.3.3 Simple Geometries to Complex Freeform Geometries

Freeform surfaces include surfaces with steps, edges and facets (Fresnel lenses), surfaces that have a tessellated patterns and smooth surfaces (telescopic mirrors). Traditional techniques are used for measuring freeform surfaces, like scanning electron microscope (SEM)/atomic force microscopy (AFM), contact profilometry, phase shifting interferometry (PSI), geometry measuring machine, etc. The characterisation of freeform surfaces is dependent upon the fitting accuracy of the measured surface and reference template. Fitting methodology has been proposed for the measurement of freeform surfaces. Fractal geometry can be analysed using the AFM imaging technique (Dallaeva et al., 2014).

17.4 Applications of Micro-Structured Surfaces

The number of micro-electro-mechanical systems (MEMS) applications is increasing every year. Due to their unique advantages, MEMS are finding applications in mechanical, electrical, and industrial production areas. Zhu (2012) described various applications of micro-structured surfaces in MEMS. According to him, a major application is in pressure sensors and optical switches.

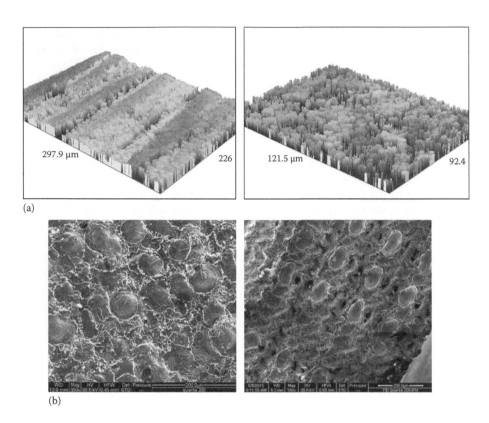

FIGURE 17.2
Stochastic and structured surfaces. (a) Stochastic surfaces obtained by different machining processes. (b) Structured micro-patterns on surfaces.

Micro-structured surfaces are finding more importance in medical applications like critical medical laser delivery systems, surgical systems and in diagnosing the human body. In the defence field, micro-structured surfaces are finding applications in helmet mount displays, long-range surveillance systems, fire control systems, multi-spectra/multi-mission sensor packages, telescopic systems, laser fusion research optics and assemblies, aerial reconnaissance and mirrors, satellite optics and high laser damage threshold assemblies. In the semiconductor industry, micro-surfaces are used for illumination systems, beam expanders, projection lenses, high precision reference mirrors, monolithic glass stage components, polarisers and thin plate fused silica optics and wave plates. In the automotive industry also, micro-surfaces are finding importance in lightweight opto-mechanical systems, single- and multi-axis stage mirrors and reference surfaces, high-precision optical components and coatings.

17.5 Measurement of Micro/Nano-Features

Over the last few years, measurement of micro/nano-scale structured surfaces has been one of the main challenges for surface metrology. Effective measurement of micro/nano-scale structured surfaces is important for understanding and evaluating their static and dynamic properties. Based on the mode of operation and working principle, micro/nano-structure surface measurement can be categorised into three major types, as shown in Figure 17.3. These three categories are

a. Contact-based instruments

b. Optical instruments

c. Scanning probe microscopy (SPM)

Each method has its own benefits and drawbacks, which are given in Figure 17.3.

As micro/nano-structured surfaces are finding great importance in various fields like medical, defence, aerospace, automobile and marine, their measurement and characterisation are a great challenge. The measuring instrument must be fast and reliable to analyse the functional capabilities of the micro/nano-structured surfaces. A unique practical measurement technique is needed that will facilitate the characterisation and functional evaluation of micro/nano-surfaces. Classification and representative instruments for surface metrology are shown in Figure 17.4. Vorburger et al. (2007) studied the comparison of measurement of surface texture using optical and stylus-based techniques. The results suggested that discrepancy is large in the stylus-based method during lower surface roughness value measurement, i.e. 50 to 300 nm, and this can be overcome by the PSI technique.

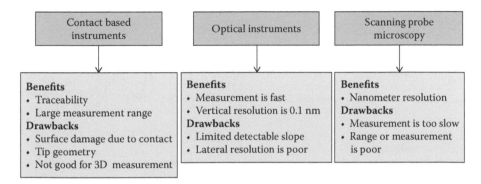

FIGURE 17.3
Classification of different surface measurement techniques.

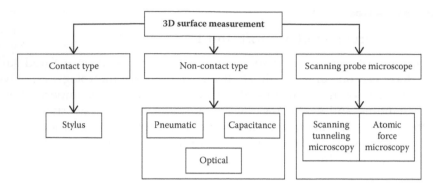

FIGURE 17.4
Classification and representative instruments of surface metrology.

17.6 Typical 2D and 3D Roughness Parameters

To characterise them, surfaces can be broken down into roughness, waviness and form. Surface texture parameters are used to quantify these characteristics. There are different methods for quantitative characterisation of surface texture, which include 2D parameters, 3D parameters, Motif parameters, bearing curve parameters, etc.

17.6.1 Surface Roughness 2D Parameters

The conventional roughness parameters most commonly used are referred as 2D parameters or profile parameters. Surface texture parameters can be divided into three parameters: amplitude, spacing and hybrid. The most commonly used 2D parameters are *Ra*, *Rq*, *Rt* and *Rz*, as shown in Figure 17.5, defined by ISO 4287.

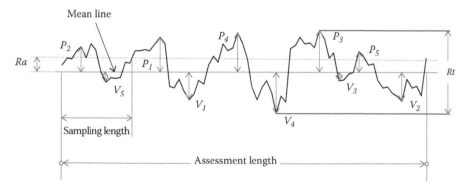

FIGURE 17.5
Representation of 2D surface roughness parameters.

Ra is the arithmetic mean deviation of the assessed profile over the sampling length *L*:

$$Ra = 1/L \int_0^L |Z(x)| \, dx. \tag{17.1}$$

Rq is the RMS average of the profile height deviations from the mean line within the evaluation length:

$$Rq = \left[\frac{1}{L} \int_0^L Z(x)^2 \, dx \right]^{1/2}. \tag{17.2}$$

Rt is the difference between the highest peak and the lowest valley in the sampling length.

Rz is the average of the height difference between the five highest peaks and the five lowest valleys:

$$Rz = \frac{(P_1 + P_2 + P_3 + P_4 + P_5) - (V_1 + V_2 + V_3 + V_4 + V_5)}{5} \tag{17.3}$$

$$= \frac{\sum_{i=1}^{5} P_i - \sum_{i=1}^{5} V_i}{5}.$$

17.6.2 Surface Roughness 3D Parameters

With the development of visualisation and image manipulation techniques, adoption of 3D or areal characterisation is a necessity. According to their geometrical properties, the surface texture areal parameters (3D) can be partitioned into two main classes called 'field' and 'feature' parameters. The term *field* refers to the use of every data point in the evaluation area, and feature parameters take specific points, lines and areas into consideration. The majority of the surface texture parameters are field parameters that allow the characterisation of surface heights, slopes, complexity, wave length, etc. Field parameters are used mainly for areal characterisation and feature parameters are used for structural pattern characterisation. Figure 17.6 shows a profile of a 3D nano-roughness measurement.

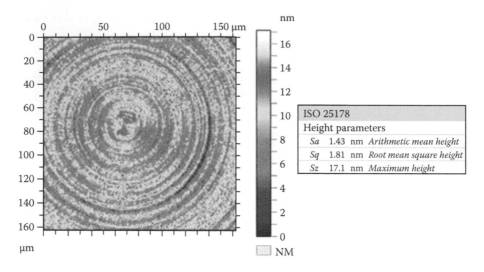

FIGURE 17.6
3D surface profile of a nano-surface. (Courtesy of AMETEK, Precitech Inc.)

17.6.2.1 RMS Height, Sq

The RMS height is defined as the RMSs of the surface departures, $z(x, y)$, within a sampling area A. As explained in Section 17.3.1.2 Scale-Limited Surface, areal parameter Sq alone is defined for the RMS parameter rather than the primary surface Pq, waviness Wq and roughness Rq in the profile case:

$$Sq = \sqrt{\frac{1}{A} \iint_A z^2(x, y)\, dx\, dy}. \tag{17.4}$$

17.6.2.2 Arithmetic Mean Height, Sa

Arithmetic mean height is defined as the arithmetic mean of the absolute value of the height within a sampling area,

$$Sa = \frac{1}{A} \iint_A \left| z(x, y) \right| dx\, dy. \tag{17.5}$$

A simple discrete implementation uses summations as shown. This implementation gives sound results when the data are sufficiently high:

$$Sa = \frac{1}{n_x n_y} \sum_{y=0}^{n_y-1} \sum_{x=0}^{n_x-1} |Z(x,y)|, \qquad (17.6)$$

where n_y is the number of line scans and n_x is the number of points in each line.

17.6.2.3 Maximum Height, Sz

Maximum height is a 3D parameter expanded from the roughness (2D) parameter Rz. It expresses the sum of the maximum value of peak height Zp and the maximum value of valley depth Zy on the surface within the measured area:

$$Sz = Sp + Sv, \qquad (17.7)$$

where Sp is the maximum value peak height Zp on the surface in the measured area and Sv is the maximum value valley depth Zv on the surface in the measured area.

17.7 Measurement and Characterisation Techniques for Micro/Nano-Features

Micro-Feature characterisation is very important in order to achieve the desired performance in the respective area. Measuring the moving components of micro-devices is so critical as they affect the dynamic behaviour of the micro-system. Thus, a proper measurement and characterisation technique is needed for understanding both the static and dynamic behaviour of the micro-system. Figure 17.7 shows an image of a micro-feature component.

17.7.1 Stylus-Based Instruments

The traditional Stylus based measurement techniques where the contact of surfaces comes into play. The contact method gives the stylus-based instruments an advantage in dirty environments. Therefore, this approach is particularly suitable for industrial field of measurements. But the mechanical force generated due to contact has a big influence on the reliability of the system. The geometry of the probe will be disturbed due to the constant contact, which takes place between the surfaces.

For stylus-based instruments, a stylus is first moved to make contact with the specimen and then moved across the surface laterally for a pre-defined

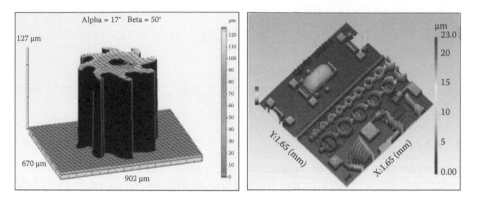

FIGURE 17.7
Photograph showing micro-feature component. (Courtesy of AMETEK, Precitech Inc.)

distance with specified contact force. The small surface variations in terms of peaks and valleys in vertical direction can be sensed by the displacement of the stylus. The transducer will convert the mechanical force into electrical signals. The measuring system is incorporated with a computer to analyse the surface variations and to display the surface profile. Generally, a stylus-based instrument can measure very small features in the vertical direction ranging in height from 10 nm to 1 mm. The radius of a diamond stylus ranges from 20 nm to 25 μm, 60° or 90° cone. The stylus tracking force can range from less than 1 to 50 mg, as shown in Figure 17.8a. A contact-based roughness measurement instrument and the profile obtained are shown in Figure 17.8b and c, respectively.

17.7.2 Non-Contacting Optical Measurement

Surface measurement is carried out without any direct contact of the instrument with the sample. Thus, no mechanical force is generated in the probe surface; hence, damage to the surface and adhesion of the sample are also eliminated in this method. The measurement speed of optical instruments is also much higher than that of stylus-based instruments. An aerial measurement of the surface can be obtained by optical instruments. Optical instruments can measure surfaces though the transparent medium such as glass or plastic film. This feature is ideal for the measurements of some MEMS devices. As MEMS devices are too sensitive, contact mode of measurement is too difficult.

17.7.2.1 Optical Interferometry

Optical interferometry uses the principle of combining two light waves to form interference. It is a non-contact measuring technique of surface topography at high vertical and moderate lateral resolution. Both laser and white

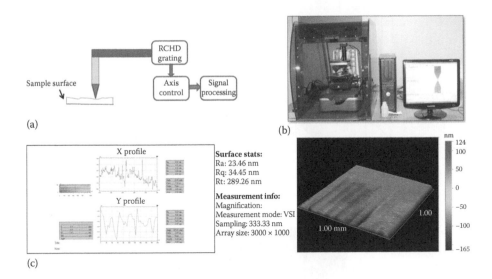

FIGURE 17.8

Stylus based 2D and 3D surface roughness measurements. (a) Schematic diagram of stylus-based instruments. (b) Contact-based roughness measurement instruments. (c) Profile of a component measured using contact-based roughness measurement instruments.

light can be used as the light source of the interferometry. There are two major types of interferometry: PSI and vertical scanning interferometry (VSI).

The basic principle and structures of PSI are shown in Figure 17.9a. In PSI mode, a white-light beam from the light source is filtered by a set of the lenses before it passes through the interferometer objective to the specimen surface. Half of the white-light beam can be reflected by the interferometer beam splitter to the reference surface. Therefore, the inter-ference's fringes can be formed and observed by combining two groups of beams. One is from the specimen surface, and the other is from the reference surface. The fringes appear as dark and bright bands alterna-tively when white light is focused on the surface of the specimen. During the measurement process, a piezoelectric transducer moves the reference surface a known distance linearly to cause a phase shift between the test and reference beams. The instrument records the intensity of the result-ing interference pattern at a number of different relative phase shifts, and then the intensity can be transferred to wave front data by integration. Figure 17.9b and c shows the non-contact-based roughness measurement instrument and profile of a component.

17.7.2.2 Confocal Laser Scanning Microscopy

A confocal microscope is a non-contact measuring instrument. The confo-cal microscope was first invented by Marvin Minsky in 1955 at Harvard

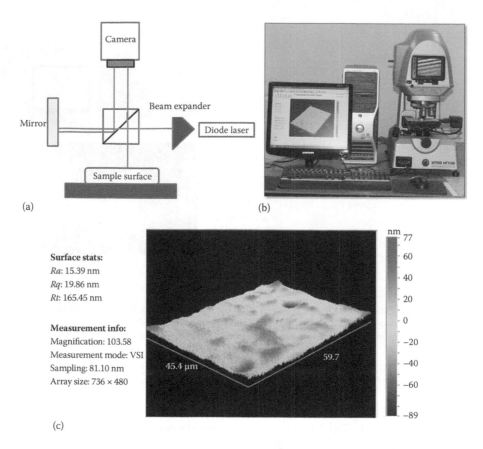

FIGURE 17.9
Measurement of 3D surface topography. (a) The basic principle of PSI. (b) Non-contact-based roughness measurement instruments. (c) Profile of component measured using white light interferometry-based measurement system.

University. Sharp images of the specimen can be obtained in a confocal microscope as it excludes the light that is not from the microscope's focal plane.

The main features of confocal microscopy are

1. Ability to control depth of field;
2. Elimination or reduction of background information away from the focal plane
3. Capability to collect serial optical sections from thick specimens

A confocal microscope is a point-by-point measurement instrument. It consists of two major parts, a stationary optic part and a moving specimen stage. A pin hole aperture is made to focus the light on to the specimen.

The returning rays, which are from the focal plane, are collected back to form the image. The working principle of a confocal microscope is shown in Figure 17.10a. The profile of a surface measured using confocal sensor is shown in Figure 17.10b.

Confocal white-light microscopy is another category that uses white light having a vertical dynamic range from the nanometre level to a large range (Ali, 2012). White light is used to illuminate the surface. After incident with the sample, the surface beam will split and the returning rays are detected by the photodetector, which in turn generates the surface image. The photo detector detects the wavelength of the reflected beam, which in turn gives the information about the height of the surface pattern of the measurement.

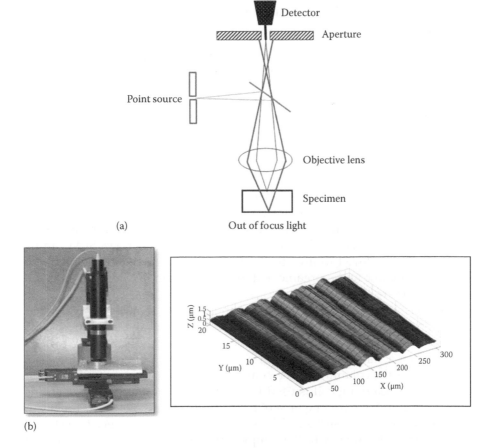

FIGURE 17.10
Confocal Microscopy for surface measurements. (a) Principle of confocal microscopy. (b) Profile of a surface measured using confocal sensor.

17.7.2.3 Fringe Projection Microscopy

The fringe projection microscopy technique helps in generating 3D information on surfaces. This technique has a wide range of applications in optical metrology. Its applications range from measuring the 3D shape of MEMS components to the measurement of the flatness of large panels. Fringe projection microscopy uses the principle of triangulation to analyse the surface profile. A typical fringe projection profilometry system is shown in Figure 17.11.

Gorthi and Rastogi (2010) described the various steps involved in the measurement of shape through fringe projection techniques:

1. Projecting a structured pattern onto the object surface
2. Recording the image of the fringe pattern
3. Calculating the phase modulation
4. Phase unwrapping algorithm to get continuous phase distribution
5. Calibrating the system for mapping the unwrapped phase distribution to real-world 3D coordinates.

Figure 17.12 shows the different steps involved in the measurement of the height distribution of an object using the fringe projection technique and the role of each step.

17.7.2.4 Scanning Electron Microscope

In a scanning electron microscope (SEM), a focused electron beam is used to scan over a surface to create an image. Information regarding surface topography and chemical composition of a sample is obtained by the signals produced by the interaction of electrons with the sample surface. SEM can be successfully employed to investigate the microstructure and chemistry of a range of organic and inorganic materials. The main SEM components are shown in Figure 17.13a.

Electrons are produced at the top of the column, which are accelerated down and passed through a combination of lenses and apertures. This focused beam of electrons will hit the surface of the sample, which is mounted on a stage in the chamber area. The system operates in low vacuum so both the column and the chamber are evacuated by a combination of pumps. A scanning coil is used for controlling the position of electrons with respect to the scanning area of the sample. These coils allow the beam to be scanned over the surface of the sample. Signals are produced due to the interaction of electrons with the sample surface, which in turn are detected by detectors for understanding the surface morphology. When the electron beam hits the surface of the sample, it penetrates the sample to a depth of a few microns, depending on the accelerating voltage and the density of

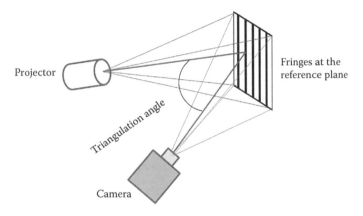

FIGURE 17.11
Fringe projection profilometry system.

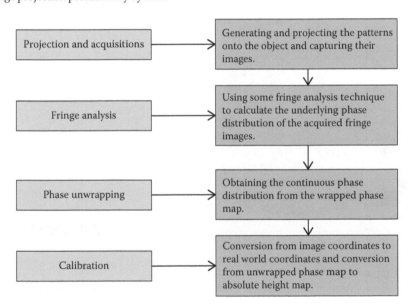

FIGURE 17.12
Steps involved in the measurement of height distribution of an object using the fringe projection technique.

the sample. Secondary electrons, backscattered electrons and characteristic X-rays are produced due to the interaction of electrons with the sample, which in turn are collected by detectors to form images. The surface topography of a diamond turned sample obtained using SEM is shown in Figure 17.13b. The maximum resolution obtained in an SEM depends on multiple factors, like the electron spot size and interaction volume of the electron beam with the sample.

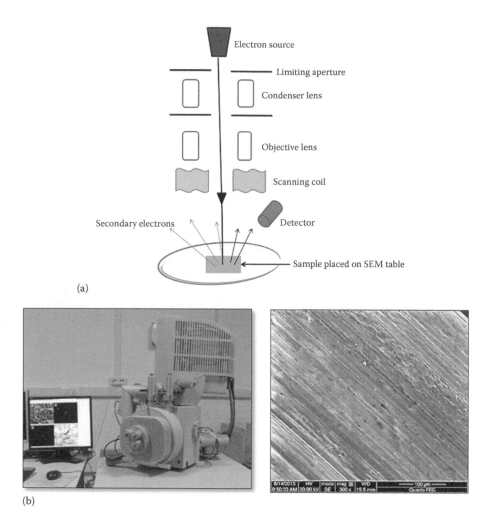

(a)

(b)

FIGURE 17.13
Surface characterisation using SEM. (a) Schematic diagram of an SEM. (b) Surface topography obtained using SEM.

17.7.3 Scanning Probe Microscopy

SPM is another technique used for the characterisation of micro/nano-features. It has a probe scanning system to generate images of sample surfaces. Line-by-line scanning of the probe on the surface in the X and Y directions helps in generating the surface image. It records the probe surface interaction as a function of position. Atomic-level resolution is possible in the case of SPM, which depends mainly on the probe size and volume of surface interaction. Application of SPM is limited due to its lower measuring speed. AFM comes under the category of SPM, which is also a very-high-resolution instrument. It consists of a cantilever and tip used to scan the surface as shown in Figure 17.14a. The cantilever is normally silicon with the very sharp tip made of Si, SiN or diamond.

The working principle of AFM is based on the force between the tip and the surface, which leads to a deflection of the cantilever according to Hooke's law. These forces in AFM include mechanical contact forces, van der Waals forces, electrostatic forces, etc. As the probe traces the surface, the deflection of the cantilever is recorded by a laser spot reflected from the top surface of the cantilever into the photodiode. In order to avoid a collision of the tip with the surface, a feedback mechanism is employed to adjust the tip-to-sample distance to maintain a constant force between

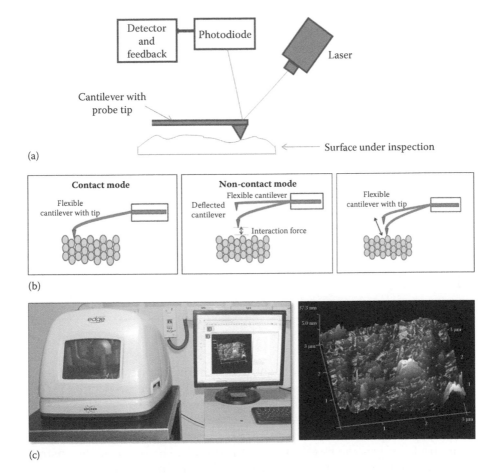

FIGURE 17.14
Measurement of 3D profile of a surface using AFM. (a) Schematic diagram of atomic force microscope. (b) Schematic diagram of the different modes of AFM operation. (c) Profile of a surface measured using AFM.

the tip and the sample. The AFM can be operated in several different modes:

- Contact mode
- Non-contact mode
- Tapping mode

In contact mode, the tip is so close to the specimen surface that the attractive forces can be very strong. The force between the tip and the surface is kept constant during scanning by maintaining a constant deflection. Figure 17.14b shows a schematic diagram of the AFM different mode operations.

In non-contact mode, the distance between the probe tip and the sample surface is kept at a few nanometres. The flexible cantilever that traces the surface oscillates at a resonant frequency. Frequency variation during scanning the surface is compared with that of the resonant frequency, which helps in obtaining the information of the surface. The topographic image of the sample is obtained based on the distance between the AFM tip and the sample surface. As the imagining is done in a non-contact mode, this technique can be applied for soft sample.

Tapping mode is a combination of both contacting and non-contacting modes. In this mode, the oscillating tip is made to lightly touch or tap the surface and then lift up above the surface. The cantilever oscillation will get reduced due to the energy loss at the tip contacting the specimen. This reduction in oscillation amplitude helps in measuring the surface features. In the tapping mode, there is less stiction of the tip from the surface, so it can be used for measuring MEMS components. Figure 17.14c shows the profile of a single-point diamond turned specimen surface measured using AFM.

17.8 Conclusion

For structured surface measurement, the material, fabrication techniques, feature shape and dimensions of the structured surface need to be investigated in association with the measurement range and resolution of surface measurement instruments. Surface measurement instruments based on different principles have their advantages and limitations. Although most structured surfaces can be effectively measured by contacting stylus instruments, non-contacting optical methods and scanning probe microscopes, there are still many challenges in specific samples. Establishing a datum plane is the fundamental analysis for the following microstructure characterisation and evaluation. Geometrical parameters and evaluation methods for micro/nano-scale structured surfaces have been developed associated with experimental analysis for micro fluidics. The methods are practicable to other structured surface applications in the same classification.

Acknowledgements

The facility used for the measurements was sponsored by a grant provided by DST under project no. SR/S3/MERC-68/2004 dated 08-06-2007. I would also like to express special thanks to Indian Institute of Technology Madras, Central Scientific Instruments Organisation (CSIO), Chandigarh, AMETEK and Precitech Inc.

Review Questions

1. Discuss the various roughness parameters for characterising micro/nano-machined surfaces.
2. Explain the importance of measurement and characterisation of micro/nano features.
3. Discuss the importance of roughness measurement in MEMS devices.
4. Compare the different contact methods of surface roughness measurement, explaining the advantages of each method.
5. Explain the basic principle of contact-based 3D roughness measurement techniques.
6. Explain how fringe projection technique helps in finding 3D information about the surfaces.
7. Discuss the different modes of AFM techniques.
8. Explain the working principle of confocal microscopy.
9. Explain how PSI is used for the characterisation of micro/nano features.
10. Discuss various safety precautions while measuring micro/nano components.
11. Compare contact-based roughness measurement with that of a non-contact-based roughness measurement.
12. Discuss the need and importance of metrology of micro/nano features.

Solved Problems

Q1. If the values of the areas shown in the figure are given as follows, estimate the height parameters *Sa* and *Sq*.

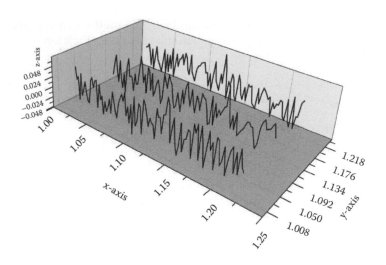

	1.02	1.06	1.1	1.14	1.18	1.22
1.02	0.05	−0.03	0.04	0.07	0.01	0
1.12	−0.01	0.06	0.03	0	0.07	0.03
1.22	0.02	0.04	0.03	0.07	−0.03	−0.04

Solution:

$$\widetilde{Sa} = \frac{1}{n_x n_y} \sum_{y=0}^{n_y-1} \sum_{x=0}^{n_x-1} |z(x,y)|.$$

$Sa = 1/(18 \times 3) \, [0.05 + 0.03 + 0.04 + 0.07 + 0.01 + 0 + 0.01 + 0.06 + 0.03$
$+ 0 + 0.07 + 0.03 + 0.02 + 0.04 + 0.03 + 0.07 + 0.03 + 0.04]$

$$= 0.011 \ \mu m$$

$$Sq = \sqrt{\frac{1}{A} \iint_A z^2(x,y) \, dx \, dy}$$

$Sq = [1/(18 \times 3)[0.05^2 + 0.03^2 + 0.04^2 + 0.07^2 + 0.01^2 + 0 + 0.01^2 + 0.06^2 + 0.03^2$
$+ 0 + 0.07^2 + 0.03^2 + 0.02^2 + 0.04^2 + 0.03^2 + 0.07^2 + 0.03^2 + 0.04^2]]^{1/2}$

$$= 03.023 \ \mu m.$$

Q2. If the values of the profile shown in the Figure 17.5 are given as follows, estimate the surface roughness parameters Ra, Rq, Rt and Rz:

Position	P1	P2	P3	P4	P5	V1	V2	V3	V4	V5
Profile height (μm)	0.03	0.01	0.05	0.02	0.04	0.03	0.02	0.01	0.06	0.01

Solution:

$$Ra = (|Z_1| + |Z_2| + |Z_3| + \ldots\ldots\ldots |Z_N|)/N$$

$$Ra = (0.03 + 0.01 + 0.05 + 0.02 + 0.04 + 0.03 + 0.02 + 0.01 + 0.06 + 0.01)/10$$

$$= 0.28/10$$

$$= 0.028$$

$$Rq = \sqrt{[Z_1^2 + Z_2^2 + Z_3^2 + \ldots\ldots\ldots + Z_N^2)/N]}$$

$$Rq = [(0.03^2 + 0.01^2 + 0.05^2 + 0.02^2 + 0.04^2 + 0.03^2 + 0.02^2 + 0.01^2 + 0.06^2 + 0.01^2)/10]^{1/2}$$

$$= 0.0325$$

$$Rt = Rp + Rv$$

$$= 0.05 + 0.06$$

$$Rt = 0.11$$

$$Rz = [|P_1 + P_2 + P_3 + P_4 + P_5| + |V_1 + V_2 + V_3 + V_4 + V_5|]/5$$

$$Rz = 0.056$$

Exercises

1. The area profiles are given in the following table (all dimensions are in μm). Estimate the surface texture parameters Sa and Sq.

		x-axis								
	0.01	**0.06**	**0.10**	**0.14**	**0.18**	**0.22**	**0.26**	**0.30**	**0.34**	**0.38**
y-axis 0.10	0.023	0.036	0.078	−0.042	0	0.03	0.012	0.014	−0.001	0.020
0.20	0.024	−0.042	−0.069	0.032	0.021	−0.054	0	0.018	0.023	0.021
0.30	−0.038	0.041	0.026	−0.031	−0.052	0.03	0.043	0.041	0.062	0
0.40	0.001	0.051	0.004	0.022	0.03	0.041	0.027	0.039	−0.003	0.04

2. The line profiles are given in the following table. Estimate Ra, Rq and Rt for each line, where P points are peaks and V points are valleys in μm.

	Peak Heights (P)	**Valley Heights (V)**
Line 1	0.011	0.023
	0.042	0.068
	0.065	0.043
	0.05	0.056
	0.039	0.071
	0.072	0.012
	0.048	0.034
Line 2	0.049	0.013
	0.066	0.032
	0.032	0.045
	0.058	0.044
	0.039	0.062
	0.016	0.037

3a. Consider that a peak height value of 0.1 μm and valley height of 0.2 μm are added to line 1 in the previous data in question 2. Comment on the change in surface characteristic by calculating the surface parameters (Ra and Rq).

3b. Consider that a peak height value of 0.1 nm and a valley height of 0.15 nm are added to line 1 in the previous data in question 2. Comment on the change in surface characteristic by calculating the surface parameters (Ra and Rq).

The field parameters and feature parameters are as listed in the following.

Height Parameters (Field Parameter)	
Sq	RMS height (standard deviation of the height distribution)
Ss_k	Skewness (symmetry of the height distribution)
Sku	Kurtosis (spread of the height distribution)
Sa	Arithmetic height
Sz	Maximum height (peak-to-valley)
Sp	Maximum peak height (from reference plane)
Sv	Maximum pit height (from reference plane)

Feature Parameters	
Spd	Density of peaks
Spc	Arithmetic mean curvature of peaks
S_5p	Five-point peak height
S_5v	Five-point pit height
$S_{10}z$	Ten-point height
Sda	Closed dale area
Sha	Closed hill area
Sdv	Closed dale volume
Shv	Closed hill volume

Functional Parameters (Field Parameter)	
Smr(c)	Areal material ratio
Sdc(mr)	Inverse areal material ratio
Sxp	Extreme peak height
Vv(mr)	Void volume
Vm(mr)	Material volume
Vmp	Hill material volume
Vmc	Core material volume
Vvc	Core void volume
Vvv	Dale void volume

Space Parameters (Field Parameter)	
Sal	Autocorrelation length
Str	Texture aspect ratio
Std	Texture direction

Hybrid Parameters (Field Parameter)	
Sdq	Root mean square gradient
Sdr	Developed interfacial area ratio

References

Ali, S.H.R. (2012) Advanced Nanomeasuring Techniques for Surface Characterization. *ISRN Optics*, 2012: 23p.

Dallaeva, D.; Talu, Ş.; Stach, S.; Skarvada, P.; Tomanek, P.; Grmela, L. (2014) AFM Imaging and Fractal Analysis of Surface Roughness of ALN Epilayers on Sapphire Substrates. *Applied Surface Science*, 312: 81–86.

Gorthi, S.S.; Rastogi, P. (2010) Fringe Projection Techniques: Whither We Are? *Optics and Lasers in Engineering*, 48(2): 133–140.

Hansen, H.N.; Carneiro, K.; Haitjema, H.; De Chiffre, L. (2006) Dimensional Micro and Nano Metrology. *CIRP Annals–Manufacturing Technology*, 55(2): 721–743.

Jiang, X.J.; Whitehouse, D.J. (2012) Technological Shifts in Surface Metrology. *CIRP Annals–Manufacturing Technology*, 61(2): 815–836.

Jiang, X.; Scott, P.J.; Whitehouse, D.J.; Blunt, L. (2007) Paradigm Shifts in Surface Metrology. Part II. The Current Shift. *Proceedings of the Royal Society of London A: Mathematical, Physical and Engineering Sciences*, 463(2085): 2071–2099.

Stephen, A.; Sepold, G.; Metev, S.; Vollertsen, F. (2004) Laser-Induced Liquid-Phase Jet-Chemical Etching of Metals. *Journal of Materials Processing Technology*, 149(1–3): 536–540.

Vorburger, T.V.; Rhee, H.G.; Renegar, T.B.; Song, J.F.; Zheng, A. (2007) Comparison of Optical and Stylus Methods for Measurement of Surface Texture. *The International Journal of Advanced Manufacturing Technology*, 33(1–2): 110–118.

Whitehouse, D.J. (2010) *Handbook of Surface and Nanometrology, Second Edition*. CRC Press, Florida.

Zhu, H. (2012) 'Measurement and characterization of micro/nano-scale structured surfaces', Doctoral thesis, University of Huddersfield, UK.

18

Optimisation of Advanced Finishing Processes Using a Teaching-Learning-Based Optimisation Algorithm

R. Venkata Rao and Dhiraj P. Rai

Department of Mechanical Engineering, S.V. National Institute of Technology, Surat Gujarat, India

CONTENTS

18.1 Introduction

Manufacturing industries are experiencing a profound need to manufacture products using materials with extraordinary properties, stringent design requirements, complex geometries, miniature features, improved interchangeability of components, improved quality and control, reduced loss of power due to friction and increased longevity of the product by reducing the wear in sliding components. These requirements are taxing engineers to manufacture parts with micro- and nano-level surface finishes. In order to meet these obligations, advanced finishing processes such as abrasive flow machining (AFM), magnetic abrasive finishing (MAF), magnetorheological finishing (MRF), magnetic float polishing (MFP), magnetorheological abrasive flow finishing (MRAFF), elastic emission machining (EEM) and chemo-mechanical polishing (CMP), etc. have been developed in the recent past. However, these processes are labour intensive, critical and less controllable in nature, as they involve a large number of influential process parameters and require high production time and cost.

It is evident that the performance of any machining process is greatly influenced by its process parameters. Thus, researchers have recognised the need to investigate the effect of process parameters of these finishing processes on performance measures such as surface roughness, material removal rate, cutting forces, etc., keeping abreast of the environmental footprint and sustainability of the process (Rao and Kalyankar, 2014). However, for achieving the best of any machining process, parameter optimisation is necessary, and the advanced finishing processes are no exception.

Most of the real-world machining process optimisation problems involve complex functions and a large number of process parameters. In such problems, traditional optimisation techniques may fail to provide an optimum solution as they may get caught up in local optima. In order to overcome these problems and to search a near-optimum solution for complex problems, many population-based heuristic algorithms have been developed by researchers in the past two decades. These may be classified into two main groups, i.e. evolutionary algorithms and swarm-intelligence-based optimisation algorithms. However, all evolutionary and swarm-intelligence-based optimisation algorithms require common control parameters like population size, number of generations, elite size, etc. Besides the common control parameters, different algorithms require their own algorithm-specific parameters. Improper tuning of algorithm-specific parameters either increases the computational effort or yields to the local optimal solution. In addition to the tuning of algorithm-specific parameters, the common control parameters also need to be tuned, which further enhances the effort.

Considering this fact, Rao et al. (2011) introduced the teaching-learning-based optimisation (TLBO) algorithm, which does not require any algorithm-specific parameters. The TLBO algorithm requires only the common controlling parameters like population size and number of generations for its work. The TLBO algorithm possesses good exploration and exploitation capability, is less complex and has proved its effectiveness in solving single-objective and multi-objective optimisation problems (Rao and Waghmare, 2014). Owing to these characteristics, the TLBO algorithm has gained wide acceptance in the field of engineering optimisation and has been successfully applied by various researchers to solve optimisation problems in different disciplines of engineering and science (Rao, 2015). The TLBO algorithm had also been applied to solve optimisation problems in machining processes (Rao and Kalyankar, 2012, 2013a,b; Pawar and Rao, 2013).

The previous researchers had solved the multi-objective optimisation problem of machining processes using an a priori approach. However, an a priori approach yields a unique optimum solution that corresponds to the weights assigned to the objective. The a priori approach is not suitable in volatile scenarios where weights are subjected to frequent change. This drawback of the a priori approach is eliminated in an a posteriori approach, where it is not required to assign the weights to the objective functions prior to the simulation run. The a posteriori approach does not lead to a unique optimum solution at the end but provides a dense spread of Pareto points (Pareto optimal solutions). The process planner can then select one solution from the set of Pareto optimal solutions based on the order of importance of objectives.

In order to solve multi-objective optimisation problems and to successfully obtain the Pareto set without being trapped in local optima, the following aspects are absolutely indispensable: balance between exploration and exploitation capability and good diversity ensuring mechanism in order to avoid stagnation at the local optima.

Thus, in this work, a parameter-less a posteriori multi-objective optimisation algorithm based on the TLBO algorithm is proposed and named as the non-dominated sorting TLBO (NSTLBO) algorithm. In the NSTLBO algorithm, the teacher phase and learner phase maintain the vital balance between the exploration and exploitation capabilities and the teacher selection is based on the non-dominance rank of the solutions. The crowding distance computation mechanism ensures the selection process towards better solutions with diversity among the solutions in order to obtain a Pareto optimal set of solutions for multi-objective optimisation problems in a single simulation run.

In this chapter, the TLBO and NSTLBO algorithms are applied to solve the single-objective and multi-objective optimisation problems of three nanofinishing processes, namely, rotational-MRAFF (R-MRAFF), the MAF process and magnetorheological-fluid-based finishing (MRFF).

The computer programs for TLBO and NSTLBO algorithms were developed in MATLAB R2009a. A computer system with a 2.93 GHz processor and 4 GB random access memory was used for execution of the programs.

The TLBO and NSTLBO algorithms are described in Sections 18.2 and 18.3, respectively.

18.2 The TLBO Algorithm

The TLBO algorithm emulates the teaching-learning process of a classroom. In each generation, the best candidate solution is considered as the teacher and other candidates' solutions are considered as learners. The learners mostly accept the instructions from the teacher but also learn from each other. In the TLBO algorithm, an academic subject is analogous to an independent variable or candidate solution feature. The TLBO algorithm consists of two important phases, i.e. the teacher phase and the learner phase. In the teacher phase, each independent variable $x_i(s)$ in each candidate solution x_i is modified according to

$$c_i(s) \leftarrow x_i(s) + r(x_t(s) - T_f \bar{x}(s)), \tag{18.1}$$

where

$$\bar{x}(s) = \frac{1}{N} \sum_{k=1}^{N} x_k(s), \tag{18.2}$$

for $i \in [1, N]$ and $s \in$ a $[1, n]$, where N is the population size, n is the problem dimension, x_t is the best individual in the population (i.e. the teacher), r is the random number taken from a uniform distribution on $[0,1]$ and T_f is called the teaching factor and is randomly set equal to either 1 or 2 with equal probability. T_f is not an algorithm-specific parameter. The child c_i replaces the parent x_i if it is better than the parent x_i.

As soon as the teacher phase ends, the learner phase commences. The learner phase mimics the act of knowledge sharing between two randomly selected learners. The learner phase entails updating each learner based on another randomly selected learner as follows:

$$c_i(s) \leftarrow \begin{cases} x_i(s) + r(x_i(s) - x_k(s)) \text{ if } x_i \text{ is better than } x_k \\ x_k(s) + r(x_k(s) - x_i(s)) \text{ otherwise} \end{cases} \tag{18.3}$$

for $i \in [1, N]$ and $s \in [1, n]$, where k is the random integer in $[1, N]$ such that $k \neq i$, and r is a random number taken from a uniform distribution on $[0,1]$. Again, the child c_i replaces the parent x_i if it is better than the parent x_i. The pseudo-code for the TLBO algorithm is given in Figure 18.1. The flowchart of the TLBO algorithm is shown in Figure 18.2.

18.3 The NSTLBO Algorithm

NSTLBO is an extension of the TLBO algorithm. The NSTLBO algorithm is an a posteriori approach for solving multi-objective optimisation problems and maintains a diverse set of solutions. The NSTLBO algorithm consists of teacher phase and a learner phase similar to the TLBO algorithm. However, in order to handle multiple objectives effectively and efficiently, the NSTLBO algorithm is incorporated with the non-dominated sorting approach and crowding distance computation mechanism proposed by Deb et al. (2002).

Initialise a population of candidate solutions $\{x_k\}, k \in [1, N]$
While not (termination criteria)
   Comment: Teacher phase

$$x_t \leftarrow \arg\min_x \left(f(x): x \in \{x_k\}_{k=1}^N \right)$$

   For each individual x_i where $i \in [1, N]$
      For each solution feature $s \in [1, n]$
         $T_f \leftarrow$ random integer $\in \{1, 2\}$

$$\bar{x}(s) \leftarrow \frac{1}{N} \sum_{k=1}^N x_k(s)$$

         $r \leftarrow U[0, 1]$
         $c_i(s) \leftarrow x_i(s) + r(x_t(s) - T_f \bar{x}(s))$

      Next solution feature
      $x_i \leftarrow \arg\min_{x_i c_i} (f(x_i), f(c_i))$

   Next individual
   Comment: Learner phase
   For each individual x_i where $i \in [1, N]$
      $k \leftarrow$ random integer $\in [1, N]: k \neq i$
      if $f(x_i) < f(x_k)$ then
      For each solution feature $s \in [1, n]$
         $r \leftarrow U[0, 1]$
         $c_i(s) \leftarrow x_i(s) + r(x_i(s) - x_k(s))$
      Next solution feature
   else
      For each solution feature $s \in [1, n]$
         $r \leftarrow U[0, 1]$
         $c_i(s) \leftarrow x_i(s) + r(x_i(s) - x_k(s))$
      Next solution feature
   End if
   $x_i \leftarrow \arg\min_{x_i c_i} (f(x_i), f(c_i))$

   Next individual
Next generation

FIGURE 18.1
Pseudo-code for the TLBO algorithm.

The teacher phase and learner phase ensure good exploration and exploitation of the search space, while the non-dominated sorting approach makes certain that the selection process is always towards the good solutions and the population is pushed towards the Pareto front in each iteration. The crowding distance assignment mechanism ensures the selection of teacher from a sparse region of the search space. The pseudo-code for NSTLBO algorithm is given in Figure 18.3.

18.3.1 Non-Dominated Sorting of the Solutions

In this approach, the population is sorted into several ranks (fronts) based on the dominance concept as follows: a solution x_i is said to dominate other

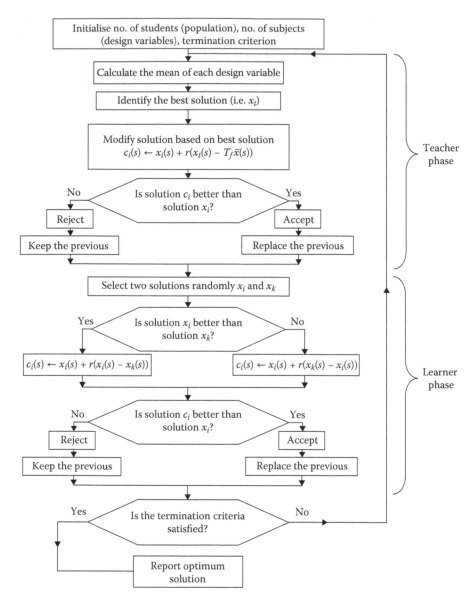

FIGURE 18.2
Flowchart of the TLBO algorithm.

solution x_j if and only if solution x_i is no worse than solution x_j with respect to all the objectives and solution x_i is strictly better than solution x_j in at least one objective. If any of the two conditions are violated, then solution x_i does not dominate solution x_j.

Among a set of solutions P, the non-dominated solutions are those that are not dominated by any solution in the set P. All such non-dominated

Initialise a population of candidate solutions

$P \leftarrow \{x_k\}, k \in [1, N]$ *N* is the population size

While not (termination criteria)

$F \leftarrow$ non-dominated sorting (P) Sort the population *P* based on non-domination concept into fronts $F \leftarrow \{F_1, F_2, F_3, ...\}$

$CD \leftarrow$ crowding distance assignment (F)

$P \leftarrow sort(F, \prec_n)$ Sort *P* front-wise in decreasing order based on crowding-comparison operator

Comment: Teacher phase

$x_t \leftarrow P[1]$ The first solution in the sorted *P* is selected as teacher

For each individual x_i where $x \in P$;
$i \in [1, N]$

For each solution feature $s \in [1, n]$ *n* is the number of dimensions

Tf $\leftarrow$ random integer $\in \{1, 2\}$

$$\bar{x}(s) \leftarrow \frac{1}{N} \sum_{k=1}^{N} x_k(s)$$

$r \leftarrow U[0, 1]$

$c_i(s) \leftarrow x_i(s) + r(x_t(s) - T_f \bar{x}(s))$

Next solution feature

Next individual

Comment: Concatenation

$P \leftarrow merge(x, c)$ Merge initial solutions (x) and new solutions (c) to form combined population P'

$F \leftarrow$ non-dominated sorting (P') Sort the population P' based on non-domination concept into fronts $F \leftarrow \{F_1, F_2, F_3, ...\}$

$CD \leftarrow$ crowding distance assignment (F)

Comment: Selection

$P' \leftarrow sort(F, \prec_n)$ Sort P' front-wise in decreasing order based on crowding-comparison operator

$P \leftarrow P'[1: N]$ Select first *N* solutions from P'

Comment: Learner phase

For each individual x_i where $x \in P$;
$i \in [1, N]$

$k \leftarrow$ random integer $\in [1, N]: k \neq i$

if $x_i \prec_n x_k$ then   Decide superiority among x_i and x_k using crowding-comparison operator

For each solution feature $s \in [1, n]$

$r \leftarrow U[0, 1]$

$c_i(s) \leftarrow x_i(s) + r(x_k(s) - x_i(s))$

FIGURE 18.3

Pseudo-code for the NSTLBO algorithm.   *(Continued)*

Next solution feature

else

For each solution feature $s \in [1, n]$

$r \leftarrow U[0, 1]$

$c_i(s) \leftarrow x_i(s) + r(x_k(s) - x_i(s))$

Next solution feature

End if

Next individual

Comment: Concatenation

$P' \leftarrow merge\ (x, c)$ Merge solutions (x) and new solutions (c) to
form combined population P'

$F \leftarrow non\text{-}dominated\ sorting\ (P')$ Sort the population P' based on non-
domination concept into fronts
$F \leftarrow \{F_1, F_2, F_3, ...\}$

$CD \leftarrow crowding\ distance\ assignment\ (F')$

Comment: Selection

$P' \leftarrow sort\ (F, \prec_n)$ Sort P' front-wise in decreasing order based
on crowding-comparison operator

$P \leftarrow P'\ [1: N]$

Next generation

FIGURE 18.3 (CONTINUED)
Pseudo-code for the NSTLBO algorithm.

solutions that are identified in the first sorting run are assigned rank one (first front) and are deleted from the set P. The remaining solutions in set P are again sorted and the procedure is repeated until all the solutions in set P are sorted and ranked.

18.3.2 Crowding Distance Assignment

Crowding distance is assigned to each solution in the population with the aim to estimate the density of solutions surrounding a particular solution i. Thus, the average distance of two solutions on either side of solution i is measured along each of the objectives m. This quantity is called the crowding distance (CD_i). The following steps may be followed to compute the CD_i for each solution i in the front F.

> Step 1: Determine the number of solutions in front F as $l = |F|$. For each solution i in the set, assign $CD_i = 0$.
>
> Step 2: For each objective function $m = 1, 2,..., M$, sort the set in the worst order of f_m.
>
> Step 3: For $m = 1, 2,..., M$, assign the largest crowding distance to boundary solutions in the sorted list $(CD_1 = CD_l = \infty)$, and for all the other

solutions in the sorted list $j = 2$ to $(l - 1)$, assign crowding distance as follows:

$$CD_j = CD_j + \frac{f_m^{j+1} - f_m^{j-1}}{f_m^{max} - f_m^{min}}, \tag{18.4}$$

where j is a solution in the sorted list, f_m is the objective function value of the mth objective and f_m^{max} and f_m^{min} are the population-maximum and population-minimum values of the mth objective function, respectively.

18.3.3 Crowding-Comparison Operator

The crowded-comparison operator is used to identify the superior solution among two solutions under comparison, based on the two important attributes possessed by every individual i in the population, i.e. non-domination rank $(Rank_i)$ and crowding distance (CD_i). Thus the crowded-comparison operator $(\prec_n)$ is defined as follows:

$$i \prec_n j \text{ if } (Rank_i < Rank_j) \text{ or } ((Rank_i = Rank_j) \text{ and } (CD_i > CD_j)).$$

That is, between two solutions (i and j) with differing ranks, the solution with the lower or better rank is preferred. Otherwise, if both solutions belong to the same front $(Rank_i = Rank_j)$, then the solution located in the lesser crowded region $(CD_i > CD_j)$ is preferred. Figure 18.4 shows the flowchart of the NSTLBO algorithm.

Now, examples of three advanced finishing processes are considered to demonstrate the effectiveness of the TLBO and NSTLBO algorithms for optimisation of parameters.

18.4 Examples

18.4.1 Optimisation of Process Parameters of the R-MRAFF Process

In the R-MRAFF process, magnetorheological polishing medium is extruded through the workpiece surface, and it is given an up-and-down motion in the direction of the tube axis by driving two opposing pistons in the medium cylinders with the help of a hydraulic unit. At the same time, the polishing medium (magnetic in nature) is rotated by imparting a rotational motion to the permanent magnets surrounding the workpiece fixture. By controlling

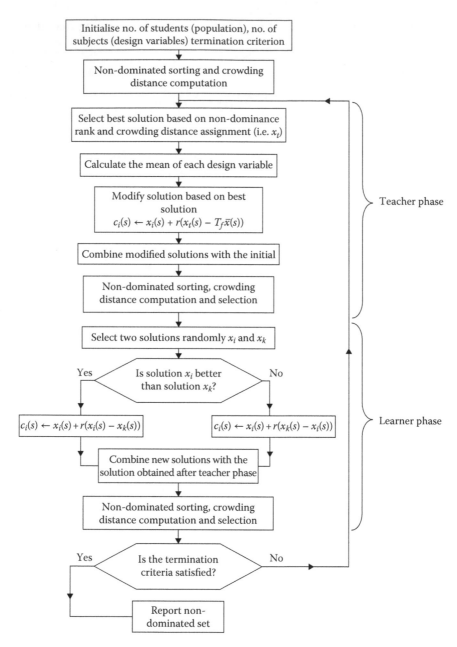

FIGURE 18.4
Flowchart of the NSTLBO algorithm.

these two motions, uniform a smooth mirror-like finished surface in the range of nanometre (nm) can be achieved and out-of-roundness (*OOR*) can be reduced. Additional force other than the axial force acts on the abrasive particle due to the rotational motion provided to it. The combination of these two motions (axial and rotational) in the case of R-MRAFF makes the resultant motion a helical one as compared to the only axial motion in the case of MRAFF. In the case of helical motion, the total length of interaction between an abrasive particle and the workpiece is more than in the case of axial motion alone in MRAFF. Hence, better finish is achieved in case of the R-MRAFF process as compared to the MRAFF process for the same number of finishing cycles. Figure 18.5 shows a schematic diagram of the R-MRAFF process.

The objective of this work is to achieve the maximum improvement in *OOR* at the internal surfaces of the steel tubes. The optimisation problem formulated in this work is based on the empirical models developed by Das et al. (2011) for improvement in *OOR*, ΔOOR, (µm) and improvement in OOR per finishing cycle, $'\overline{\Delta OOR}'$ (nm/cycle), of internal surfaces of stainless steel tubes. The process parameters considered are the same as those considered by Das et al. (2011), and these are hydraulic extrusion pressure P (bar), number of finishing cycles N, rotational speed of magnet S (RPM) and mesh size of abrasive M.

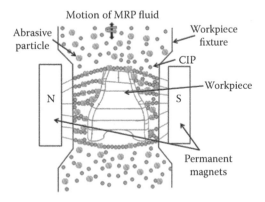

FIGURE 18.5
Schematic diagram of the mechanism of the R-MRAFF process. (Reprinted from *Precision Engineering*, Kumar, S., Jain, V.K., Sidpara, A., Nanofinishing of freeform surfaces [knee joint implant] by rotational–magnetorheological abrasive flow finishing [R-MRAFF] process, Copyright 2015 with permission from Elsevier.)

TABLE 18.1

Coded Levels and Corresponding Actual Values of Process Parameters

Sr. No.	Parameter	Unit	Levels				
			-2	-1	0	1	2
1	Hydraulic extrusion pressure (P)	bar	32.5	35	37.5	40	42.5
2	Number of finishing cycles (N)		600	800	1000	1200	1400
3	Rotational speed of the magnet (S)	RPM	50	100	150	200	250
4	Mesh size of abrasive (M)		90	120	150	180	210

Objective functions are as follows.

The objective functions in terms of coded values of the process parameters are expressed by Equations 18.5 and 18.6. Table 18.1 gives the coded values of the process parameters and their corresponding actual values.

$$\text{maximise } \Delta OOR = 1.24 + 0.18P + 0.18N - 0.083S - 0.11M + 0.068PN - 0.083PS$$

$$- 0.16PM - 3.125E - 3NS - 0.26NM - 0.063SM - 0.11P^2$$

$$- 0.13N^2 - 0.20S^2 - 0.13M^2 \tag{18.5}$$

$$\text{maximise } \overline{\Delta OOR} = 1.24 + 0.17P + 0.041N - 0.086S - 0.074M + 0.013PN - 0.08PS$$

$$- 0.13PM + 0.022NS - 0.26NM - 0.064SM - 0.11P^2 - 0.14N^2$$

$$- 0.20S^2 - 0.13M^2 \tag{18.6}$$

The maximum values of ΔOOR and $\overline{\Delta OOR}$ obtained by Das and Jain (2011) using the desirability function approach are reported in Table 18.2.

The optimisation problem is now solved using the TLBO algorithm considering the individual objective functions, and optimum combinations of process parameters obtained are reported in Table 18.3. It can be observed from Table 18.3 that the TLBO algorithm has achieved improvements of 46.57%

TABLE 18.2

Optimum Combination of Process Parameters for Maximisation of ΔOOR and $\overline{\Delta OOR}$ Obtained Using Desirability Function Approach

Sr. No.	P (bar)	N	S (RPM)	M	ΔOOR (µm)	$\overline{\Delta OOR}$ (nm/cycle)
1	40	1200	137	120	1.8421	–
2	40	1200	140	120	–	1.57

Source: Das, M., Jain, V.K., *Mater. Manuf. Process.*, 2011; 26:1073–1084.

TABLE 18.3

Optimum Combination of Process Parameters for Maximisation of ΔOOR and $\overline{\Delta OOR}$ in the R-MRAFF Process Obtained Using the TLBO Algorithm

Sr. No.	P (bar)	N	S (RPM)	M	ΔOOR (μm)	$\overline{\Delta OOR}$ (nm/cycle)	Computational Time (sec)
1	42.5	1400	133.8	99	**2.7**	–	0.047382
2	42.5	1400	140.7	90	–	**1.9**	0.054587

Note: Items in bold indicate the values of objective functions which are improved by TLBO algorithm.

and 21.02% in $\overline{\Delta OOR}$ and ΔOOR, respectively, as compared to the values obtained by Das et al. (2011) using desirability function approach. Figures 18.6 and 18.7 show the convergence graphs of the TLBO algorithm for ΔOOR and $\overline{\Delta OOR}$, respectively. The numbers of function evaluations required by the TLBO algorithm to obtain the maximum value of ΔOOR and $\overline{\Delta OOR}$ are 360 and 460, respectively.

Furthermore, the NSTLBO algorithm is applied to simultaneously optimise the objective functions of ΔOOR and $\overline{\Delta OOR}$. The non-dominated set of solutions obtained using NSTLBO algorithm is reported in Table 18.4. The NSTLBO algorithm required 2500 function evaluations to achieve the non-dominated set of solutions. Figure 18.8 shows the Pareto front obtained using the NSTLBO algorithm for ΔOOR and $\overline{\Delta OOR}$. The computational time required by the NSTLBO algorithm to obtain the non-dominated set of solutions is 10.389 sec. Table 18.4 and Figure 18.8 act as ready references for the process planner to select the optimum combination of process parameters for the R-MRAFF process.

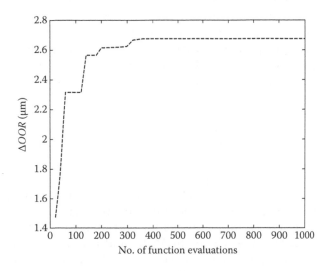

FIGURE 18.6
Convergence graph of the TLBO algorithm for ΔOOR.

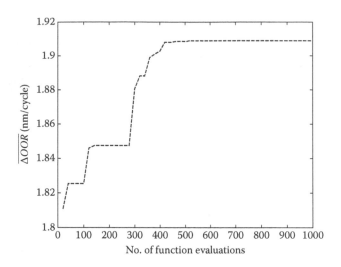

FIGURE 18.7
Convergence graph of the TLBO algorithm for $\overline{\Delta OOR}$.

18.4.2 Optimisation of Process Parameters of the MAF Process

MAF is a process in which the workpiece surface is smoothened by removing the material in the form of microchips by abrasive particles in the presence of magnetic field across the machining gap between the workpiece top surface and bottom face of the rotating electromagnet pole. The working gap between the workpiece and the magnet is filled with a mixture of ferromagnetic and abrasive particles as shown in Figure 18.9. In some cases, bonded ferromagnetic and abrasive particles are also used. These bonded or unbonded blends of ferromagnetic and abrasive particles are referred to as magnetic abrasive particles (MAPs). The MAPs form a flexible magnetic abrasive brush in the presence of magnetic field. The magnetic field retains powder in the machining gap and acts as a binder, causing the powder to be pressed against the surface to be finished. The force due to the magnetic field is responsible for normal force (F_n), causing the abrasive to penetrate inside the workpiece, while rotation of the magnetic abrasive brush intact to the magnetic poles results in tangential force (F_t). The sum of the forces F_t and the tangential component of magnetic force (F_{mt}) is responsible for removing material in the form of tiny chips.

The objective of this work is to achieve the maximum percentage improvement in surface roughness. The optimisation problem formulated in this work is based on the empirical model for percentage improvement in surface roughness ΔRa (%) developed by Mulik and Pandey (2011) for the MAF process. The process parameters considered in this work are same as those considered by Mulik and Pandey (2011), and these are voltage X_1 (volts), mesh

TABLE 18.4

Non-Dominated Set of Solutions Obtained Using the NSTLBO Algorithm
for Optimisation of ΔOOR and $\overline{\Delta OOR}$ in the R-MRAFF Process

Sr. No.	P (bar)	N	S (RPM)	M	ΔOOR (µm)	$\overline{\Delta OOR}$ (nm/cycle)
1	42.5	1400	133.8421	90	2.6729	1.905
2	42.5	1400	140.7277	90	2.6691	1.9088
3	42.5	1400	135.2207	90	2.6727	1.9064
4	42.5	1400	134.1771	90	2.6729	1.9054
5	42.5	1400	134.9625	90	2.6728	1.9062
6	42.5	1400	134.6449	90	2.6728	1.9059
7	42.5	1400	135.3934	90	2.6727	1.9065
8	42.5	1400	135.9532	90	2.6725	1.907
9	42.5	1400	134.8556	90	2.6728	1.9061
10	42.5	1400	134.3398	90	2.6729	1.9056
11	42.5	1400	135.7309	90	2.6726	1.9068
12	42.5	1400	137.1479	90	2.672	1.9078
13	42.5	1400	139.1431	90	2.6706	1.9086
14	42.5	1400	138.2823	90	2.6713	1.9084
15	42.5	1400	135.58	90	2.6726	1.9067
16	42.5	1400	138.8971	90	2.6708	1.9086
17	42.5	1400	136.4938	90	2.6723	1.9074
18	42.5	1400	134.4867	90	2.6728	1.9057
19	42.5	1400	140.2988	90	2.6695	1.9088
20	42.5	1400	134.1824	90	2.6729	1.9054
21	42.5	1400	134.5095	90	2.6728	1.9057
22	42.5	1400	138.4373	90	2.6712	1.9084
23	42.5	1400	136.3457	90	2.6724	1.9073
24	42.5	1400	136.5549	90	2.6723	1.9074
25	42.5	1400	137.2671	90	2.6719	1.9079
26	42.5	1400	137.9876	90	2.6715	1.9082
27	42.5	1400	136.0909	90	2.6725	1.9071
28	42.5	1400	139.5021	90	2.6703	1.9087
29	42.5	1400	136.8912	90	2.6721	1.9077
30	42.5	1400	139.8393	90	2.67	1.9088
31	42.5	1400	139.9245	90	2.6699	1.9088
32	42.5	1400	139.3236	90	2.6705	1.9087
33	42.5	1399.98	140.4188	90	2.6694	1.9088
34	42.5	1400	140.1491	90	2.6697	1.9088
35	42.5	1400	140.0684	90	2.6698	1.9088
36	42.5	1400	136.673	90	2.6722	1.9075
37	42.5	1400	136.9948	90	2.6721	1.9077
38	42.5	1400	135.6582	90	2.6726	1.9068
39	42.5	1400	139.6992	90	2.6701	1.9088

(Continued)

TABLE 18.4 (CONTINUED)

Non-Dominated Set of Solutions Obtained Using the NSTLBO Algorithm
for Optimisation of ΔOOR and $\overline{\Delta OOR}$ in the R-MRAFF Process

Sr. No.	P (bar)	N	S (RPM)	M	ΔOOR (µm)	$\overline{\Delta OOR}$ (nm/cycle)
40	42.5	1400	137.6552	90	2.6717	1.9081
41	42.5	1400	137.5117	90	2.6718	1.908
42	42.5	1400	137.5977	90	2.6718	1.9081
43	42.5	1400	140.539	90	2.6693	1.9088
44	42.5	1400	138.0949	90	2.6714	1.9083
45	42.5	1400	138.1531	90	2.6714	1.9083
46	42.5	1400	137.8973	90	2.6716	1.9082
47	42.5	1400	136.1363	90	2.6725	1.9071
48	42.5	1400	136.3143	90	2.6724	1.9073
49	42.5	1400	136.2325	90	2.6724	1.9072
50	42.5	1400	136.8033	90	2.6722	1.9076

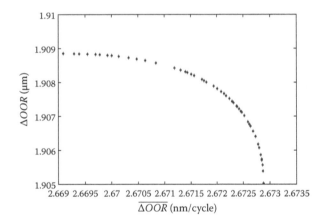

FIGURE 18.8
Pareto front obtained using the NSTLBO algorithm for optimisation of ΔOOR and $\overline{\Delta OOR}$ in
the R-MRAFF process.

number X_2, natural logarithm of rotations per minute of electromagnet X_3
(RPM) and % wt. of abrasives X_4.

The objective functions is as follows.

The objective function is expressed by

$$\text{maximise } \Delta Ra = -674 + 5.48X_1 + 0.0628X_2 + 115X_3 - 15.8X_4 - 0.000018X_2^2$$

$$- 0.976X_1X_3 - 0.00203X_2X_4 - 2.41X_3X_4. \tag{18.7}$$

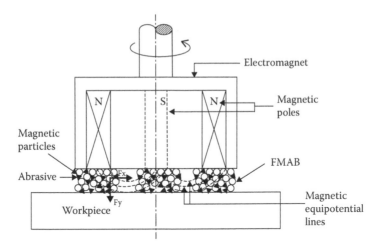

FIGURE 18.9
Schematic diagram of the MAF process. (With kind permission from Springer Science +Business Media: *International Journal of Advanced Manufacturing Technology*, Magnetic abrasive finishing of hardened AISI 52100 steel, 55, 201, 501–515, Mulik, R.S., Pandey, P.M.)

The parameter bounds are

$$50 \leq X_1 \leq 90 \tag{18.8}$$

$$400 \leq X_2 \leq 1200 \tag{18.9}$$

$$5.193 \leq X_3 \leq 6.1092 \tag{18.10}$$

$$15 \leq X_4 \leq 35. \tag{18.11}$$

The optimisation problem is now solved using the TLBO algorithm and the results obtained are reported in Table 18.5. The optimum combinations of process parameters suggested by Mulik and Pandey (2011) are also reported in Table 18.5. The TLBO algorithm achieved a better value of percentage improvement

TABLE 18.5

Optimum Combination of Process Parameters for Maximisation of ΔR_a Obtained Using the TLBO Algorithm

Sr. No.	X_1 (V)	X_2	$ln(X_3)$	X_4	ΔR_a (%)	Computational Time (s)	Methodology
1	70	800	5.6348	15	20.2944	–	Experimentally (Mulik and Pandey, 2011)
2	90	400	5.193	35	**69.0348**	0.078737	TLBO

Note: Items in bold indicate the value of objective function which is improved by TLBO algorithm.

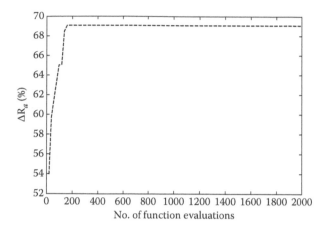

FIGURE 18.10
Convergence graph of the TLBO algorithm for ΔR_a in the MAF process.

in surface roughness as compared to that found experimentally by Mulik and Pandey (2011). Figure 18.10 shows the convergence graph of the TLBO algorithm for ΔR_a (%). It is observed that the number of function evaluations required by the TLBO algorithm to obtain the maximum value of ΔR_a (%) is 260.

It can be seen from Table 18.5 that the TLBO algorithm has achieved a maximum improvement in surface roughness (i.e. ΔR_a = 69.0348%) with optimum process parameters as compared to the ΔR_a value of 20.2944% obtained by Mulik and Pandey (2011) with non-optimum process parameters.

18.4.3 Optimisation of Process Parameters of the MRFF Process

MRFF uses magentorehological (MR) fluid that consists of non-magnetic polishing abrasives and magnetic carbonyl iron particles in water or other carriers. In the absence of a magnetic field, the MR fluid behaves as a Newtonian fluid (viscosity 0.1–1 Pa.s). The magnetic field stiffens the MR fluid ribbon (viscosity 10–20 Pa.s) depending upon the magnetic field strength and behaves like a viscoplastic fluid. Figure 18.11 shows a schematic diagram of the MRFF process. When the MR fluid ribbon is squeezed through the working gap, the workpiece experiences normal force due to compression and tangential force due to rotation of the carrier wheel. Due to the magnetic levitation force, the majority of abrasive particles move away from the magnet or towards the outer periphery of the ribbon and the workpiece surface. Hence, material removal takes place by interaction of abrasive particles (which are gripped by stiffened MR fluid) and workpiece. The normal and tangential forces are responsible for the removal of material from the workpiece surface. Besides the normal and tangential force, an axial force also acts on the workpiece due to the axial feed given to the tool or workpiece.

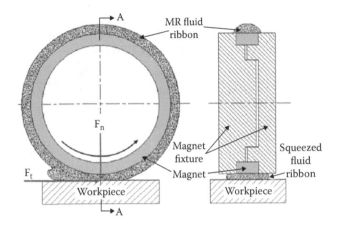

FIGURE 18.11
Schematic diagram of the MRFF process. (Reprinted from *International Journal of Machine Tools and Manufacture*, 69, Sidpara, A., Jain, V.K., Analysis of forces on the freeform surface in magnetorheological fluid based finishing process, 1–10, Copyright 2013, with permission from Elsevier.)

The objective of this work is to achieve the optimum values of normal and tangential forces acting on the workpiece surface. The optimisation problem formulated in this work is based on the empirical models developed by Sidpara and Jain (2013) for the prediction of normal F_n (N) and tangential F_t (N) forces acting on the workpiece during the MRFF process. The process parameters considered are same as those considered by Sidpara and Jain (2013), and these are angle of curvature θ (degree), tool rotational speed S (RPM) and feed rate F (mm/min).

The objective functions are as follows:

$$\text{maximise } F_n = -15.027 - 0.266\,\theta + 0.042\,S + 0.757\,F - 0.0002\,\theta S - 0.0056\,\theta F$$

$$+ 0.0001\,SF + 0.0085\theta^2 - 0.00002S^2 - 0.115F^2 \tag{18.12}$$

$$\text{maximise } F_t = -7.257 - 0.191\,\theta + 0.021\,S + 0.526\,F - 0.00008\,\theta S - 0.008\,\theta F$$

$$+ 0.0001\,SF + 0.006\theta^2 - 0.00001\,S^2 - 0.077F^2 \tag{18.13}$$

$$\text{maximise } F_a = -4.372 - 0.081\,\theta + 0.012\,S + 0.247\,F - 0.00004\,\theta S - 0.0021\,\theta F$$

$$+ 0.00009SF + 0.002\theta^2 - 0.000006S^2 - 0.044F^2. \tag{18.14}$$

TABLE 18.6

Pareto Optimal Solutions Obtained Using the NSTLBO Algorithm for Maximum Normal Tangential and Axial Forces in the MRFF Process

Sr. No.	θ (degrees)	S (RPM)	F (mm/min)	F_n (N)	F_t (N)	F_a (N)
1	5	1011.2484	3.7188	6.3609	3.662	1.6794
2	5	1011.59	3.7151	6.3613	3.6622	1.6794
3	5	1012.0802	3.7228	6.3615	3.6627	1.6794
4	5	1012.3268	3.7152	6.362	3.6628	1.6794
5	5	1013.9747	3.7488	6.3624	3.6643	1.6794
6	5	1013.1812	3.7152	6.3627	3.6634	1.6794
7	5	1047.8376	3.8495	6.3631	3.677	1.6711
8	5	1014.6201	3.7394	6.3632	3.6647	1.6794
9	5	1049.2008	3.8269	6.3634	3.677	1.6707
10	5	1014.0137	3.7095	6.3635	3.6639	1.6794
11	5	1015.9952	3.7483	6.364	3.6657	1.6793
12	5	1015.5636	3.727	6.3643	3.6652	1.6793
13	5	1015.7605	3.731	6.3643	3.6654	1.6793
14	5	1048.0612	3.8147	6.3646	3.677	1.6712
15	5	1045.6634	3.8398	6.3646	3.6769	1.6721
16	5	1017.4313	3.759	6.3647	3.6667	1.6792
17	5	1017.1226	3.7491	6.3648	3.6664	1.6792
18	5	1017.5541	3.7501	6.365	3.6667	1.6792
19	5	1043.1844	3.847	6.3652	3.6767	1.673
20	5	1046.744	3.8074	6.3656	3.6769	1.6718
21	5	1044.412	3.8225	6.366	3.6768	1.6727
22	5	1019.9951	3.7663	6.366	3.6683	1.6789
23	5	1045.1149	3.812	6.3662	3.6768	1.6725
24	5	1020.9165	3.7717	6.3663	3.6689	1.6788
25	5	1019.5586	3.7416	6.3666	3.6678	1.679
26	5	1019.1767	3.7219	6.3669	3.6674	1.6791
27	5	1018.1628	3.6831	6.367	3.6661	1.6791
28	5	1037.447	3.8325	6.3673	3.6757	1.675
29	5	1024.1356	3.7829	6.3674	3.6707	1.6784
30	5	1042.3921	3.8015	6.3676	3.6765	1.6736
31	5	1028.4713	3.7926	6.3685	3.6727	1.6776
32	5	1034.6946	3.8053	6.3687	3.6749	1.676
33	5	1025.0412	3.7562	6.3688	3.6709	1.6783
34	5	1039.8154	3.7796	6.3691	3.676	1.6746
35	5	1025.8243	3.7526	6.3692	3.6712	1.6782
36	5	1042.1433	3.7608	6.3692	3.6761	1.6738
37	5	1040.7126	3.7675	6.3693	3.676	1.6743
38	5	1032.2731	3.7827	6.3695	3.674	1.6767
39	5	1033.7131	3.7799	6.3697	3.6745	1.6764
40	5	1039.0695	3.7592	6.37	3.6756	1.6748

(Continued)

TABLE 18.6 (CONTINUED)

Pareto Optimal Solutions Obtained Using the NSTLBO Algorithm for Maximum Normal Tangential and Axial Forces in the MRFF Process

Sr. No.	θ (degrees)	S (RPM)	F (mm/min)	F_n (N)	F_t (N)	F_a (N)
41	5	1027.7906	3.7429	6.37	3.672	1.6778
42	5	1038.7944	3.7433	6.3705	3.6754	1.6749
43	5	1030.3408	3.7392	6.3707	3.6729	1.6773
44	5	1039.5436	3.7223	6.3709	3.6752	1.6746
45	5	1036.8261	3.7306	6.3711	3.6748	1.6755
46	5	1026.3003	3.615	6.3715	3.6685	1.6774
47	5	1034.7668	3.7088	6.3717	3.6739	1.6761
48	5	1028.0661	3.6122	6.3719	3.6692	1.6771
49	5	1029.7373	3.6064	6.3723	3.6696	1.6766
50	5	1035.6173	3.6296	6.3726	3.6722	1.6753

The parameter bounds are as follows:

$$5 \leq \theta \leq 25 \tag{18.15}$$

$$700 \leq S \leq 1100 \tag{18.16}$$

$$1 \leq F \leq 5. \tag{18.17}$$

The optimisation problem is now solved using the NSTLBO algorithm. The non-dominated set of solutions obtained using the NSTLBO algorithm is reported in Table 18.6, and the same is graphically represented in Figure 18.12.

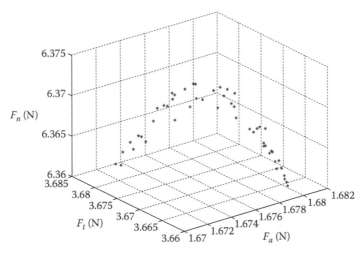

FIGURE 18.12

Pareto front obtained using the NSTLBO algorithm for optimisation of F_n, F_a and F_t in the MRFF process.

The NSTLBO algorithm required 2000 function evaluations and a computational time of 4.2906 sec to achieve the non-dominated set of solutions.

18.5 Conclusions

In the present work, optimisation problems of three advanced finishing processes, namely, R-MRAFF, MAF and MRFF, are solved using the TLBO and NSTLBO algorithms. In the case of the R-MRAFF process, the TLBO algorithm is applied to maximise ΔOOR and $\overline{\Delta OOR}$. The TLBO algorithm has achieved improvements of 46.57% and 21.02% in ΔOOR and $\overline{\Delta OOR}$, respectively, as compared to the values of ΔOOR and $\overline{\Delta OOR}$ obtained using the desirability function approach. Furthermore, the NSTLBO algorithm is applied to simultaneously optimise ΔOOR and $\overline{\Delta OOR}$ and a non-dominated set of solutions is provided to act as a ready reference for the process planner to select the optimum combination of process parameters for the R-MRAFF process.

In the case of the MAF process, the TLBO algorithm is applied to maximise the percentage improvement in surface roughness. The TLBO algorithm has achieved a maximum improvement in surface roughness (i.e. ΔR_a = 69.0348%) as compared to the ΔR_a value of 20.2944% obtained by the previous researchers using experiments with non-optimum process parameters.

In the case of the MRFF process, the NSTLBO algorithm is applied to simultaneously optimise the normal, axial and tangential forces acting on the workpiece. The NSTLBO algorithm has provided a non-dominated set of solutions that will act as a ready reference for the process planner to select the optimum combination of process parameters for the MRFF process.

In the present work, the TLBO and NSTLBO algorithms are applied only to the selected advanced finishing processes, namely, R-MRAFF, MAF and MRFF. However, the TLBO and NSTLBO algorithms can also be applied to solve the optimisation problems pertaining to other advanced finishing processes such as AFM, MRF, MFP, EEM, CMP, etc.

Review Questions

1. What is evolutionary computation? Name three widely used evolutionary algorithms.
2. What is swarm intelligence? Name three swarm intelligence-based optimisation algorithms.
3. What do you understand by non-dominated set of solutions in multi-objective optimisation problems?
4. What is needed to tune the algorithm-specific parameters? In which way is the TLBO algorithm better in this regard?

5. The optimisation of three advanced finishing processes, carried out by the TLBO algorithm, is described in this chapter. Can you use any other optimisation algorithm for the same three processes? Try it and make a comparison of your results with those given by the TLBO algorithm.

References

Das M, Jain VK, Ghoshdastidar PS. The out-of-roundness of the internal surfaces of stainless steel tubes finished by the rotational–magnetorheological abrasive flow finishing process. *Materials and Manufacturing Processes* 2011; 26:1073–1084.

Deb K, Pratap A, Agarwal S, Meyarivan T. A fast and elitist multiobjective genetic algorithm: NSGA-II. *IEEE Transactions on Evolutionary Computation* 2002; 6(2):182–197.

Kumar S, Jain VK, Sidpara A. Nanofinishing of freeform surfaces (knee joint implant) by rotational–magnetorheological abrasive flow finishing (R-MRAFF) process. *Precision Engineering* 2015; 42:165–178. http://dx.doi.org/10.1016/j.precisioneng .2015.04.014.

Mulik RS, Pandey PM. Magnetic abrasive finishing of hardened AISI 52100 steel. *International Journal of Advanced Manufacturing Technology* 2011; 55:501–515.

Pawar PJ, Rao RV. Parameter optimization of machining processes using teaching learning based optimization algorithm. *International Journal of Advanced Manufacturing Technology* 2013; 67:995–1006.

Rao RV. *Teaching-learning-based optimization (TLBO) algorithm and its engineering applications*. Springer-Verlag, London, 2015.

Rao RV, Kalyankar VD. Parameters optimization of machining processes using new optimization algorithm. *Materials and Manufacturing Processes* 2012; 27:978–985.

Rao RV, Kalyankar VD. Multi-pass turning process parameters optimization using teaching learning based optimization algorithm. *Scientia Iranica* 2013a; 20:967–974.

Rao RV, Kalyankar VD. Parameters optimization of modern machining processes using teaching learning based optimization algorithm. *Engineering Applications of Artificial Intelligence* 2013b; 26:524–531.

Rao RV, Kalyankar VD. Optimization of modern machining processes using advanced optimization techniques: A review. *International Journal of Advanced Manufacturing Technology* 2014; 73:1159-1188.

Rao RV, Waghmare GG. A comparative study of teaching learning based optimization algorithm on multi-objective unconstrained and constrained functions. *Journal of King Saud University – Computer and Information Sciences* 2014; 26:332–346.

Rao RV, Savsani VJ, Vakharia DP. Teaching-learning-based optimization: A novel method for constrained mechanical design optimization problems. *Computer-Aided Design* 2011; 43:303–315.

Sidpara A, Jain VK. Analysis of forces on the freeform surface in magnetorheological fluid based finishing process. *International Journal of Machine Tools and Manufacture* 2013; 69:1–10.

19

Molecular Dynamics Simulation (MDS) to Study Nanoscale Cutting Processes

Saurav Goel[1], Saeed Zare Chavoshi[2] and Adrian Murphy[3]

[1]*Precision Engineering Institute, School of Aerospace, Transport and Manufacturing, Cranfield University, Cranfield, Bedfordshire, United Kingdom*

[2]*Mechanical Engineering Department, Imperial College, London, United Kingdom*

[3]*School of Mechanical and Aerospace Engineering, Queen's University, Belfast, United Kingdom*

CONTENTS

Abbreviations

ABOP	Analytical bond order potential
BOP	Bond order potential
d	Uncut chip thickness
EAM	Embedded-atom method
FEA	Finite element analysis
Fx or *Fc*	Tangential cutting force
Fy or *Ft*	Thrust force
LAMMPS	Large-scale atomic/molecular massively parallel simulator
LLNL	Lawrence Livermore National Laboratory
MDS	Molecular dynamics simulation
MPI	Message passing interface
NVE	Microcanonical ensemble
NVT	Canonical ensemble
PBC	Periodic boundary conditions
SDM	Surface defect machining
SiC	Silicon carbide
SPDT	Single point diamond turning

19.1 Introduction

Experimental studies, aside from being expensive, are constrained by the fact that they do not permit direct observation of events occurring at the atomic level, especially at short timescales of a few femtoseconds. For this reason, molecular dynamics simulation (MDS), a versatile numerical analysis tool, has evolved as an appropriate bottom–up simulation approach to investigate atomic scale events. The beauty of MDS is that it is informed by quantum properties, thereby revealing discrete atomistic mechanics that is otherwise impossible to be investigated using the conventional engineering toolbox, e.g. finite element analysis (FEA).

One of the principal differences between FEA and MDS is that the nodes and the distances between the nodes in MDS are not selected on an arbitrary basis but on the basis of more fundamental units of the material, namely the position of the atom as the nodes and inter-atomic distances as the distance between the nodes. Also, the shape and size of the crystal in MDS are dictated by the crystallographic structure of the material and not arbitrarily, such as triangular or rectangular shapes as in FEM. The implementation of MDS was first developed through the pioneering work of Alder and

Wainwright in the late 1950s [1] in their study of the interactions of hard spheres. Since then, MDS has transcended the field of ultra-precision manufacturing, materials science, physics, chemistry and tribology and has contributed significantly towards improved understanding of our knowledge in the inaccessible atomic scale regime.

Ultra-precision manufacturing has emerged as a powerful tool for manipulating optical, electrical and mechanical properties of components by changing their surface and sub-surface structure at the nanometre length scale [2]. This technique can be utilised to achieve nanometre-level tolerances and optical quality finished surfaces, which is required in products used in the defence and security technologies, plastic electronic devices, low-cost photovoltaic cells and next generation displays.

Nanoscale machining in this context is an ultra-precision manufacturing method that is used to fabricate components requiring submicron geometry accuracy and atomically smooth surface finish, such as those employed in the optical, semiconductor and opto-electronics industries. As the nanometric machining process involves removing few atomic layers from the surface, it is extremely difficult to observe the machining process and to measure the process parameters directly from the experiments. Thus, investigation of atomistic processes occurring at such small length scales is more amenable to MDS.

MDS was adapted to study ultra-precision machining at Lawrence Livermore National Laboratory (LLNL), USA, during the late 1980s [3]. Landman et al. [4] and Ikawa et al. [5] pioneered the concept of molecular dynamics (MD) in the framework of contact loading studies, followed by Voter and Kress [6]. Since then, Inamura et al. [7], Rentsch [8] and Komanduri et al. [9] have contributed significantly to this arena and set a sound foundation for the study of nanometric cutting processes using MDS.

Currently, significant work is being undertaken in developing unique algorithms for conducting large-scale atomic simulations with the integrated usage of the 'dislocation extraction algorithm' and a crystal analysis tool [10] for automated identification of crystal defects, dislocation lines and their Burgers Vector from the MDS output data. Further research efforts are being directed to study processes such as hard turning [11–13], thermal spray [14], nanoimpact and nanofatigue [15,16], nanoindentation [17,18] and nanometric cutting [19–23]. The majority of these works are reviewed in two comprehensive review articles [24,25]. For the purpose of brevity, this chapter provides the most relevant information for the scientific community interested in adapting MDS as a tool to study nanoscale machining processes. In this chapter, we will present various aspects of MDSs, including virtual/computer experiments, principles of MDS, potential–energy functions, boundary conditions, MDS procedures and a number of relevant examples of MDS studies focused on engineering problems.

19.2 Principle of MDS

The essence of the MDS method is the numerical solution of Newton's second equation of motion for an ensemble of atoms, where the atoms are assumed to follow Newton's second law as follows:

$$a_{ix} = \frac{F_{ix}}{m_i} = \frac{d^2 x_i}{dt^2}, \quad F_{ix} = -\frac{dV}{dx_i}, \tag{19.1}$$

where a_{ix} represents the ith atom's acceleration in the x direction and m_i is the mass of the ith atom. F_{ix} is the interaction force acting on the ith atom by the jth atom in the x direction, x_i is the ith atom's x-coordinate and V is the potential energy function. In MDS, these equations are integrated by numerical techniques for extremely short time intervals (1–100 pico-seconds [ps]), and equilibrium statistical averages are computed as temporal averages over the observation time. To render atomistic simulation studies practical, an interatomic potential function is necessary. During MDS, the interatomic bonding forces (both attractive and repulsive) are defined by an appropriate empirical potential–energy function. The MDSs can be likened to the dynamic response of numerous non-linear spring-mass (atoms or positive ions) in a system under an applied load, velocity or displacement conditions. From this point of view, MDS is similar to another analysis that engineers routinely undertake, such as the investigation of vibrations of a mechanical system, wherein a series of springless-masses and massless-springs are connected and the response of the system is investigated under given external loading conditions.

Figure 19.1 shows schematically a computer simulation of the deformation of a workpiece being machined with a diamond cutting edge. Every atom shown

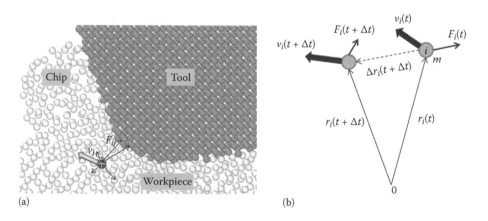

(a)   (b)

FIGURE 19.1
(a) Scheme for performing MDS of nanometric cutting. (b) An atom i having mass m at a distance of r from the other atom exerts force F on other atom, when given a velocity v.

in Figure 19.1 is in motion and interacts with neighbouring atoms in a manner that can be determined from the interatomic potential energy function.

In the simulation, the tool is fed in a stepwise procedure into the workpiece, at time intervals of Δt, which are shorter than the period of lattice vibration. As indicated in Figure 19.1, the position $r_i(t + \Delta t)$ and velocity $v_i(t + \Delta t)$ of the atom i after the tool has been fed in the period of Δt are calculated by employing finite difference method and using Equations 19.2 and 19.3. This calculation is then repeated in order to describe the motion of individual atoms. The behaviour of the collection of atoms that comprise the tool/workpiece model can be then analysed through synthesis of the movement of the individual atoms.

$$r_i(t + \Delta t) = r_i(t) + \Delta t v_i(t) + \frac{(\Delta t)^2}{2m} F_i(t) \tag{19.2}$$

$$v_i(t + \Delta t) = v_i(t) + \frac{\Delta t}{2m} [F_i(t + \Delta t) + F_i(t)] \tag{19.3}$$

In general, temperature assessment in an MDS is done by averaging the velocity of a group of atoms by using the relationship between kinetic energy and temperature as follows:

$$\frac{1}{2} \sum_i m_i v_i^2 = \frac{3}{2} N k_b T, \tag{19.4}$$

where N is the number of atoms, v_i represents the velocity of ith atom, k_b is the Boltzmann constant ($1.3806503 \times 10^{-23}$ J/K) and T represents the atomistic temperature. The instantaneous fluctuations in kinetic energy per atom are usually very high, so this equation is averaged temporally and/or spatially over a number of time steps and reassigned to each atom at every N steps to be converted into equivalent temperature. It may be noted that thermal conduction cannot be analysed correctly in MDS, as the normally used potential functions do not take the electron properties into consideration. Since thermal conduction is mainly governed by electron mobility, in practice, materials exhibit considerably higher thermal conductivities and, consequently, smaller thermal gradients than those estimated by MDS based on lattice vibration. Thus, for a more realistic simulation of nanoscale machining, the gradient in thermal field should be scaled, or adjusted, to coincide with that calculated from continuum thermal conduction theory.

19.2.1 Potential Energy Function

Whenever an atomic-scale problem is treated, we require a constitutive description of the atoms. This interaction is governed by a potential energy function that roughly accounts for quantum interactions between electron shells and represents the physical properties of the atoms being simulated, such as its elastic constants and lattice parameters. In general, there are two ways to construct an interatomic potential. The first way is the artistic route of empirical design using chemical and physical insights together with convenient functional forms. The second alternative is the rigorous approach of a bottom–up derivation of the functional forms from a higher-level theory, in some cases, tight binding approaches [26]. Both of these procedures have been applied in the construction of bond order expressions, and both approaches turned out to be largely successful. Potentials used in chemistry are generally called 'force fields', while those used in materials physics are called 'analytical potentials'. Most force fields in chemistry are empirical and consist of a summation of forces associated with chemical bonds, bond angles, dihedrals, nonbonding forces associated with van der Waals forces and electrostatic forces.

One of the most successful families of interatomic potentials that have been able to stand the demanding requirement of simultaneous efficiency and reliability is the family of models based on the quantum-mechanical concept of bond order.

Empirical potential–energy functions are classified into two-body potentials, three-body potentials and multi-body potentials, depending on the unit of atoms on which those potentials are based. The potentials that fall into the two-body potentials category are Morse, Born-Mayer, Lennard-Jones potentials, etc. The three-body potentials, such as Tersoff and analytical bond order potential (ABOP), are devised to represent covalent bonding that has a directional grip and can be used conveniently for material systems involving silicon, germanium, diamond, etc. Multi-body potentials, such as embedded-atom method (EAM) potentials, are devised to describe the metallic bonding more accurately than the two-body potentials. Adaptive intermolecular reactive empirical bond order (AIREBO) is another multi-body potential developed to accurately describe hydrocarbons.

In a simple pairwise potential, e.g. Morse or Lennard-Jones potential, only the direct interactions between atoms are considered and summed for a certain sphere with a radius that is usually equal to the spacing between four adjacent atoms. They fully determine the total energy without considering any further cohesive terms that arise from the interaction with atoms far away from the particle considered. The atoms are regarded as mass points that have a central interaction with their nearest neighbours. The interaction of any pair of atoms depends only on their spacing. This implies that such potentials are principally radially symmetric and independent of the angular position of other atoms in the vicinity. The Morse potential function is an example of a pair potential that was frequently used in early research work and is used for simulations even now. It may, however, be noted that a common limitation of all the pair potentials is their inability to reproduce the Cauchy pressure, which is a quantity that reflects the nature of the bonding at the atomic level of a material. This was one of the motivations for introducing EAM potential in the year 1984. Unlike Morse potential functions, many of the potentials used in physics, such as those based on bond order formalism, may describe both bond breaking and bond formation (e.g. Tersoff is a three-body potential function, while the AIREBO function is a four-body potential function). Almost all forms of pairwise potentials are empirical due to the approximations necessary to overcome the many-body problem involved in the interaction. The validity of the function, as well as the stability of the crystal for a given material, is checked for various properties, including cohesive energy, the Debye temperature, the lattice constant, the compressibility and the elastic constants as well as the equation of state. The most widely used bond order potential (BOP) formalism was first proposed by Tersoff [27]. Tersoff based his potential on an idea presented by Abell a few years earlier on BOP, which has environmental dependence and no absolute minimum at the tetrahedral angle. While newly developed

TABLE 19.1

List of Potential Functions with Respect to the Time of Introduction

S. No.	Year	Name of the Potential Function	Materials Suited
1	1984	EAM: embedded-atom method [28]	Cu
2	1985	Stillinger-Weber potential [29,30]	Si
3	1987	SPC: simple point charge [31]	H_2O
4	1988	BOP	Si
	1988	• Tersoff-1 variant for silicon [32]	Si
	1989	• Tersoff-2 for better elastic properties of silicon [27]	Si, Ge and C
	1990	• Tersoff-3 for Si, C and germanium [33,34]	Si and C
	1994	• Tersoff-4 for silicon and carbon [35]	SiC
		• Tersoff-5 for amorphous silicon carbide [36]	Si and C
		• Refinements in Tersoff potential function [37–39]	Si and C
		• EDIP [40,41]	
5	1989	MEAM: modified embedded-atom method [42][a]	Universal
6	1990	REBO: reactive empirical bond order [43]	Carbon
7	2000	AIREBO [44] (four body potential function)	Hydrocarbons and carbon
8	2001	ReaxFF: reactive force field [45] (capable of bond breaking and bond formation during the simulation)	Universal
9	2005	ABOP [46] (three-body potential function)	Si and C
10	2007	COMB: charge optimised many-body [47]	SiO_2, Cu and Ti
11	2008	EIM: embedded-ion method [48]	Ionic e.g. NaCl
12	2010	GAP: Gaussian approximation potential [49]	Universal
13	1998–2001	Other important potential functions relevant in contact loading problems [50–52]	Si, B and N
14	2013	Screened potential functions [53,54][b]	Range of materials

[a] Latest modifications (2NN MEAM) are available through https://cmse.postech.ac.kr/home _2nnmeam.

[b] Details available from https://github.com/pastewka/atomistica.

formalisms provide greater accuracy, they are sometimes computationally very expensive, as shown in Table 19.1.

Tersoff functions gained wide popularity in the 1990s for MDSs. However, one key drawback of Tersoff (or similar formulations, e.g. ABOP [46]) is the way in which the next-nearest neighbour atoms are determined, namely, via a narrow distant-dependent cut-off. In their original incarnation, the artificial abrupt change in energy–distance relation enforces the bond breaking forces to be severely overestimated, leading to ductile behaviour in silicon, for example. Consequently, the potential functions proposed by Tersoff and Erhart et al., i.e. BOP and ABOP, fail to reproduce the density–temperature relation of silicon. This suggests that both BOP and ABOP potential functions are not fully reliable to obtain the phase diagram of silicon. In an attempt to address this problem, a recent effort has been made by decoupling the condition for a nearest-neighbour relationship from the range of the potential [54]. Subsequent refinements have led to a formalism, which is developed by using the screening functions to increase the range of these potentials [53]. By changing the cut-off procedure of all the BOP functions, the screening function has been reported to reproduce an improved description of amorphous phases and brittle behaviour of silicon, diamond and silicon carbide (SiC). This improvement, however, has not yet addressed the problem of obtaining the correct melting point of silicon and is still a fertile area of research. Overall, Morse potential functions limit the exploration of interaction within atoms of the workpiece and the cutting tool, while Tersoff potential functions have limitations in accurately describing the thermal aspects of silicon, which might limit the study of some machining processes related to high-temperature applications.

The key message of this section is that a potential energy function is an important consideration for a realistic MDS. There are some shortcomings of the currently used potential functions. For example, the ductile-brittle transition during nanometric cutting of silicon and SiC cannot be described well by the Tersoff potential energy function (that has been a heavily used potential function). Similarly, the mechanism of cleavage on certain crystal orientations of brittle materials is yet another aspect that cannot inherently be captured by current potential energy functions [23,24]. Indeed, in the absence of crystal orientation information, this was perhaps misinterpreted as a ductile–brittle transition in a previously published study [55]. An important consideration for simulating nanometric cutting of some materials is that the surface bonds or the nascent surface of, say, silicon will be reactive and will tend to bond together with the surface of the diamond cutting tool during its approach. In order to avoid such an outcome, it is good practice to saturate the surface bonds by using hydrogen or any other similar material before the start of the simulation. Finally, MD considers the environment as a vacuum; however, the real experimental environment is known to play a key role in influencing the machining outcome, and hence, a more robust potential function would involve representation of the influence of coolants etc.

19.2.2 MDS of Nanometric Cutting

In what follows, the steps involved in an MDS are described briefly. Figure 19.2 provides a general scheme of how an MDS of nanometric cutting is performed. This scheme is generic and can well be extended to study other contact loading processes such as nanoindentation, impact loading, etc. Also, the description is generalised and may be adapted to any software platform.

19.2.2.1 Boundary Conditions and Ensemble

A schematic diagram of the nanometric cutting simulation model is shown in Figure 19.3. In this model, the nano-crystalline workpiece and the cutting tool are modelled as deformable bodies in order to permit tribological interactions between them and to permit the simulation of tool wear. However, if the workpiece is very soft (such as copper and brass), the diamond tool can be modelled as a rigid body since the cutting tool will not wear even after a cutting length of the order of 30 km [56]. The model shown in Figure 19.3 can be seen to have a negative rake angle tool, as this is generally the recommendation for machining of hard and brittle materials.

In Figure 19.3, the atoms of the cutting tool and the workpiece are allocated into one of three different zones: Newton atoms, thermostatic atoms and boundary atoms. The boundary atoms are assumed to remain unaffected and fixed in their initial lattice positions during the simulation, serving to reduce the boundary effects and to maintain the symmetry of the lattice. In conventional machining operations, the energy from plastic deformation in the primary shear zone and friction at the tool–chip interface generate heat, which is carried away by chips and lubricant and by conduction into the tool and workpiece. The nanometric cutting model is, however, extremely small and is not capable of dissipating the heat itself. The velocity of the thermostatic atoms is therefore re-scaled to a desired temperature (300 K) at each step of the computation to dissipate the artificial heat. It may be noted here that a thermostat layer so close to the cutting zone strongly exaggerates the cooling since it forces that zone to have room temperature. In reality, the thermostat area is at a macroscopic distance. Such a problem can be handled by either increasing the size of the simulation model both in the x and in the y directions or by using the multiscale simulation method.

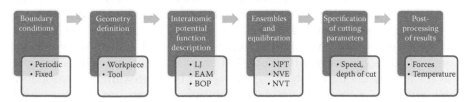

FIGURE 19.2
Generic scheme of performing an MDS of nanometric cutting.

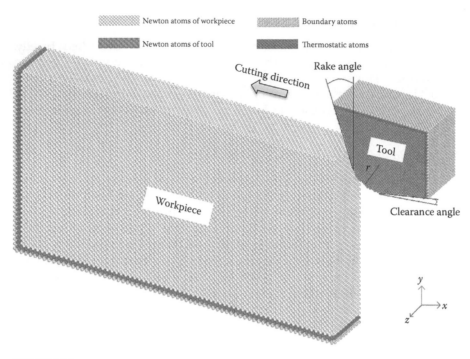

FIGURE 19.3
Schematic of the MDS model. Readers are referred to the web-based version of this work to interpret the correct colour legends (https://www.cranfield.ac.uk/people/Dr-Suarav-Goel-15392730).

MDSs are usually implemented considering a system of N particles in a cubic box of length L. Since N is typically in the range of 100 to 10,000 (very far from the thermodynamic limits), it is necessary to use periodic boundary conditions (PBC) to avoid surface effects. An important consideration for using PBC is to first determine the equilibration lattice parameter [21]. This could be achieved by averaging the lattice constant from the NPT dynamics ran on a small volume of a material at the desired temperature and pressure for a few femtoseconds. An interesting feature to be noticed here is the adjustment of the lattice constants; e.g. nanometric cutting of silicon using a diamond tool involves the use of two different lattice constants, i.e. silicon (0.5432 nm) and diamond (0.356 nm). Care must be taken to choose the periodic cell dimensions in such a way that these two lattice constants are in an integer proportion, e.g. $L_z = n_1 \times a_1 = n_2 \times a_2$, where L_z is the box size (in the z direction), n_1 and n_2 are integers and a_1 and a_2 are the two lattice constants. It is generally difficult to find an exact solution to this equation, but for a large enough system, n_1 and n_2 can be approximated reasonably well. Similarly, a change in crystal orientation also requires an adjustment in the dimension of a periodic boundary. For example, a workpiece may be positioned on the (111) orientation by specifying the basis vectors in the

x direction as (–2 1 1), in the y direction as (1 1 1) and in the z direction as (0 1 –1). An alternative orientation specification could be (–1 1 0), (1 1 1) and (1 1 –2), respectively along the x, y and z directions. In both cases, the z orientation varies, and hence, the simulation box size in the z direction should accordingly be adjusted to accommodate the cutting tool and the workpiece. Inappropriately chosen lattice parameters would lead to a build-up of excessive stress, causing the atoms to explode before the simulation even begins to run. Once the geometry of a model is ready, the velocities to the atoms can be assigned using the Maxwell-Boltzmann distribution. Followed by an energy minimisation, the velocities of all the atoms can be set to a desired temperature. This step is followed by the process of equilibration, wherein the aim is to achieve a desired temperature until a steady state is achieved. The amount of time required for equilibration depends on the system being investigated as well as the initial configuration of the system. Newton atoms are then allowed to follow Newtonian dynamics (LAMMPS NVE dynamics), while atoms in an intermediate thin boundary layer were subjected to a thermostat (large-scale atomic/molecular massively parallel simulator [LAMMPS] canonical ensemble [NVT] dynamics) to dissipate the extra heat generated in the finite simulation volume. This consideration of boundary conditions ensures that the process of deformation is not affected by any artificial dynamics.

19.2.2.2 Calculation of Cutting Forces

Figure 19.4 shows schematically the arrangement of nanometric cutting as well as the two-dimensional (2D) representation of the two coplanar forces (namely, the tangential cutting force [F_c or F_x] and the thrust force [F_t or F_y]) acting on a cutting tool fundamentally governing the cutting action of the tool. The third component, F_z, acts in the direction orthogonal to the x and y planes and mainly influences surface error, as it tends to push the tool away from the workpiece. The tangential force causes displacements in the direction of cut chip thickness, and its variation therefore relates to chatter. These are the reasons why cutting force measurement is an important indicator of tool wear [57]. From the MDS perspective, the calculation of the cutting forces using a diatomic pair potential, such as a Morse or Lennard-Jones function, is relatively simple because the interaction energy will include a pair component that is defined as the pairwise energy between all pairs of atoms where one atom in the pair is in the first group (workpiece) and the other is in the second group (cutting tool). These pair interactions can directly be used to compute the cutting forces. For a many-body potential function, such as EAM, Tersoff, ABOP and AIREBO, in addition to the pair potential, there are other terms that make them computationally expensive. Accounting for these extra terms needs additional computations in addition to those in the pair-wise interactions. Earlier, Cai et al. [58] have reported that ductile mode cutting is achieved when the thrust force acting on the

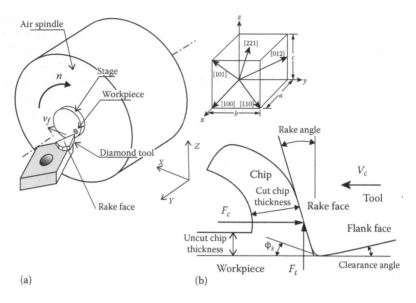

FIGURE 19.4
(a) Schematic diagram of chip formation during nanometric cutting. (From Goel, S. et al., *J. Manuf. Sci. Eng.*, 136, 021015, 2014.) (b) Two-dimensional representation of the machining forces acting on the cutting tool.

cutting tool is larger than the cutting force. While this was found to be true in several experimental studies, this is not the case observed during several nanoscale friction-based simulation studies where cutting forces are found dominant over thrust forces [59,60]. It is therefore yet another important area for future research.

19.2.2.3 Calculation of Machining Stresses

The state of stress acting in the machining zone is shown schematically in Figure 19.5 for both 3D and 2D stress systems. One fundamental problem with the computation of atomic stress is that the volume of an atom does not remain fixed during deformation. To mitigate this problem, the best method is to plot the stresses on the fly by considering an elemental atomic volume in the cutting zone. The total stresses acting on that element could be computed and divided by the pre-calculated total volume of that element to obtain the physical stress tensor. Also, the instantaneous values of the stress calculated from the MDS should always be time averaged. When a stress tensor from the simulation is available, it can readily be used to obtain the Tresca stress, von Mises stress, Octahedral shear stress and principal stresses.

Overall, MDS of nanometric cutting starts with the development of the geometry of the material and a description of the interactions of the atoms with the material using a suitable potential energy function. This is followed

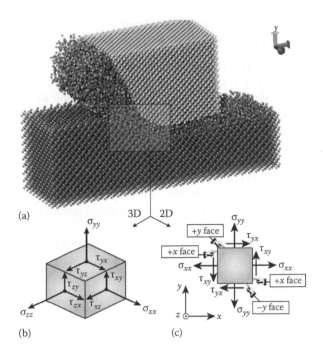

FIGURE 19.5
(a) Schematic representation of the area showing (b) 3D stress tensor in the machining zone and (c) 2D representation of the stress for a plane-stress condition.

by equilibration of the model and simulation in an appropriate ensemble. After the simulation is over, atomic trajectories can be used for post-processing of the results (with or without time averaging depending on the quantity).

19.3 Some Examples of Recent MDS Studies

In this section, we attempt to summarise the recent state-of-the-art studies in the field of MDS of nanometric machining of hard–brittle materials such as silicon and SiC. In a nanometric machining operation, the geometry of the cutting tool and anisotropy of the workpiece play a vital role and substantially influence the cutting behaviour. Hence, first, we present an insight into the effect of the aforementioned variables on the cutting behaviour of the silicon and SiC workpieces. In continuation, we cover MDS of hot machining and a recently developed machining technique known as surface defect machining (SDM) in order to explore their potential capabilities for commercial implementation.

19.3.1 Influence of Cutting Tool Geometry during Nanometric Cutting of Brittle Materials

It has been demonstrated that material removal at extremely fine depths of cut for certain atomic layers involves a high coefficient of friction that is dependent on the rake angle and is independent of the thrust force of the cutting tool [61]. When the uncut chip thickness approaches the size of the cutting edge radius during nanometric cutting, the rake angle of the cutting tool appears to determine both the direction and the magnitude of the resultant cutting force. Also, the use of a negative rake angle tool for single-point diamond turning (SPDT) operations has become somewhat of a conventional practice for the machining of brittle materials [9,62]. A schematic comparison of the cutting process using negative and positive rake angle tools is shown in Figure 19.6.

It can be seen from Figure 19.6b that the tangential force F acts along the wedge of the cutting tool so that the normal force acts on the wedge face. Along these directions, the shear stress and compressive stress on the cutting tool vary during the course of machining (Figure 19.6c). When positive rake angles are used, the normal force exerts a bending stress on

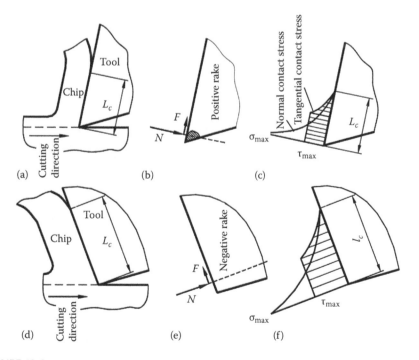

FIGURE 19.6
Difference in the force vector and stress distribution due to positive rake angle (a–c) and negative rake angle (d–f), where L_c is length of contact between cutting tool and chip. (From Astakhov, V.P., *Fundamentals of the Selection of Cutting Tool Geometry Parameters Geometry of Single-point Turning Tools and Drills*, Springer, London, 2010.)

the cutting tip of the tool, under which diamond, being extremely brittle, might eventually chip off. When a negative rake-angled cutting tool is used (Figure 19.6d), this bending effect does not occur because it is replaced by compression on the cutting tool (Figure 19.6e and f). Additionally, a negative rake angle cutting tool is thought to exert a hydrostatic stress state in the workpiece, which inhibits crack propagation and leads to a ductile response from brittle materials during their nanometric cutting [55,64]. Nakasuji et al. [64] noted that the effect of rake angle in cutting is analogous to that of the apex angle of an indenter: low angles of approach result in relatively small hydrostatic stress fields, which, in turn, enable ductile regime machining. Negative rakes of approximately −25° to −45° with clearance angles of approximately 8° to 12° are recommended for improved tool life [65]. The reason for such a selection is that a high clearance angle reduces rubbing, while a corresponding increase in negative rake angle provides mechanical strength to the wedge of the cutting tool [66]. It was also noted that a 0° rake angle (clearance angle of 8°) provided superior performance to a +5° or −5° rake angle for machining electro-less nickel plate die material [67]. However, this was due to the fact that when the depth of cut is smaller than the edge radius, an effective rake angle is presented by the cutting tool [68]. In such cases, a 0° rake angle tool already presents some negative rake, which induces better performance than −25° or −30° rake angle tools. For hard steels, the critical value of the rake angle (the dividing line between efficient and inefficient material removal) is 0° [69]. Table 19.2 summarises a number of key investigations on the effect of the cutting tool rake angle and clearance angle during machining of brittle materials.

It is evident from Table 19.2 that the rake angle and the clearance angle have a significant influence on the critical un-deformed chip thickness and the sub-surface lattice deformation layer depth. However, there is no systematic answer or model available that can be used to determine the best tool geometry for tool longevity. Komanduri et al. [70] used MD and a Tersoff potential function to simulate a wide range of rake angles (Figure 19.7) to observe the mechanism of chip formation during the nanometric cutting of silicon. They proposed a mechanism of material removal in silicon based on extrusion of plastically deformed material ahead of the tool, particularly for large negative rake angle tools, where the space available to accommodate departing chips decreases, causing an increase in chip side flow. From their simulation results, they were able to explain that an increase in the negative rake angle results in a significant increase in the extent of sub-surface deformation.

19.3.2 Influence of Anisotropy of the Workpiece (SiC – 3C Type)

Since advanced ceramics are being used in applications historically reserved for metals, close tolerances and good surface finish are becoming increasingly

TABLE 19.2

Influence of Rake Angle on the Outcome of the SPDT of Brittle Materials

Work Material and Citation	Rake Angle	Clearance Angle	Total Included Angle of the Tool	Remarks/Observations
Germanium [71]	−30°	6°	114°	Better machining conditions (large feed rate) was obtained for a −30° rake tool than a −10° and 0° rake angle tool.
Silicon [72]	−40°	5°	125°	Enabled better plastic deformation of the workpiece than a (−25°) rake angle tool.
Silicon [73]	−40°	10°	120°	A −40° rake angle tool provided a better ductile finished surface than a −20° angle rake tool.
Silicon [74] and SiC [75]	−45°	5°	130°	With an adjustable arrangement for varying rake angle, a −45° rake angle tool was found to provide better response of the workpiece for ductile-regime machining.
Silicon [76]	−25°	10°	105°	Performed better than −15° and −45° rake angle tool; however, the inferior quality of the gem was suspected to be the reason for poor performance of the diamond tool having −45° rake.
Silicon [77]	−25°	10°	105°	Provided a better machined surface finished in comparison to a −15° and 0° rake angle tool.
Silicon [68]	−30°	7°	113°	A rake angle between 0° and 60° was tested by keeping other parameters unchanged, and a 30° rake was found superior by LLNl.
Silicon [78]	0°	Not specified	Not specified	An effective rake angle is presented by the tool when the depth of cut is smaller than the edge radius. In this condition, a 0° rake angle tool already presented some negative rake and was found to provide better finish than a −25° or −30° rake angle tool. However, a 0° rake angle tool permits reduced critical chip thickness and hence low material removal rate.
Silicon [79]	Varying tool rake and clearance		84°	Both tool rake angle and clearance angles were varied from −15° to −45° and from 21° to 51°, respectively. A −30° rake angle tool permitted higher critical chip thickness, while a −45° angle tool enabled to reduce the micro-cracks.

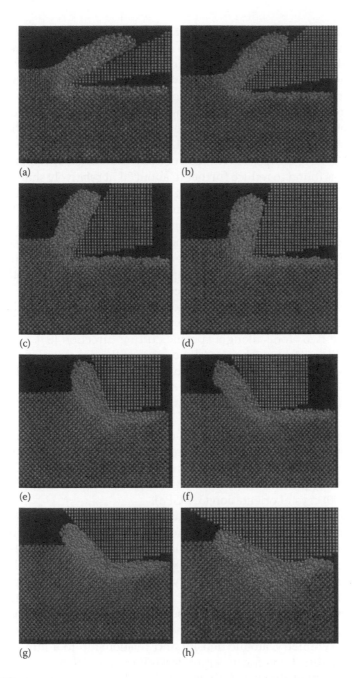

FIGURE 19.7
MDS of turning of silicon over a range of rake angles. (a) +60° tool rake angle, (b) +45° tool rake angle, (c) +30° tool rake angle, (d) +15° tool rake angle, (e) –15° tool rake angle, (f) –30° tool rake angle, (g) –45° tool rake angle and (h) –60° tool rake angle. (From Komanduri, R. et al., Molecular dynamics simulation of the nanometric cutting of silicon. *Philosophical Magazine Part B*, 81(12): 1989–2019, 2001.)

important [80], especially in the field of nanotechnology. SiC is one such non-oxide ceramic exhibiting most of these desirable engineering properties [60]. 3C-SiC is the only polytype of SiC residing in a diamond cubic lattice structure among the 250 various other polytypes of SiC recognised to date [81]. Komanduri et al. [65] cited Tabor, who made a recommendation regarding the selection of the cutting tools, stating that their hardness should be about 5 times that of the workpiece. In the case of SiC as a workpiece and diamond as a cutting tool, the ratio is only about 4:1, owing to the high micro-hardness of SiC (about 28 GPa) compared with that of diamond (100 GPa). Experimentally, this ratio was found to reduce further to about 2:1 at relatively shallow depths of cut [82]. Hence, nanometric cutting of 3C-SiC by a single-point cutting tool is expected to be surrounded by many technical challenges. Strenuous efforts are being made to find a solution to suppress the wear of cutting tools occurring during the process of nanometric cutting [19,21,83]. The crystal orientation of the workpiece significantly affects its cutting behaviour [84], and hence, a properly selected crystallographic setup could benefit both the tool life and the attainable machined surface roughness.

The anisotropic variation in a material originates from the differences in the density of atoms on a particular crystal orientation and the distance between the two atoms along a specific cutting direction [85]. This in turn influences the cutting mechanics (slip occurs on the densest plane and along the shortest direction) and the nature of the plastic deformation of the workpiece material. While the knowledge of the extent of anisotropy of 3C-SiC during nanometric cutting is important, it is also important to know the variation in the Young's modulus of 3C-SiC along three major crystallographic orientations. Table 19.3 provides estimates of the Young's modulus and Poisson's ratio obtained by applying the known analytical solutions. It is evident from Table 19.3 that the maximum Young's modulus of 3C-SiC is on the (111) crystal orientation (557 GPa), while the minimum Young's modulus is on the (100) crystal orientation (314 GPa).

The calculated values of Young's modulus shown in Table 19.3 are also verified experimentally by the National Aeronautics and Space Administration in the United States [86], and its variation proves that 3C-SiC exhibits a high degree of crystal anisotropy (up to 44%). However, the extent of this anisotropy and favourable combination of crystal orientation and cutting direction to cut 3C-SiC and how an individual crystal setup will respond to the cutting process are not known. Figure 19.8 presents a schematic diagram showing the arrangement of nanometric cutting, and Figure 19.9 shows schematically the changes in symmetry, atomic density and relationship to a number of possible cutting directions for different crystallographic planes. As is evident, the arrangement of atoms and, therefore, the behaviour of the material under nanoscale cutting conditions change in conjunction with change in crystal setup (plane and direction) [87]. This is the reason why all the nanoscale mechanical properties, including the plastic response of brittle materials, change with respect to crystal orientations and the direction of applied force.

TABLE 19.3

Direction-Dependent Mechanical Properties in Single-Crystal 3C-SiC

3C-SiC Properties	Reference	Values
Elastic constant C_{11}	Experimental data [46]	390 GPa
Elastic constant C_{12}	Experimental data [46]	142 GPa
Elastic constant C_{44}	Experimental data [46]	256 GPa
Young's modulus (E_{100})	$C_{11} - 2\dfrac{C_{12}}{C_{11}+C_{12}}C_{12}$	314 GPa
Young's modulus (E_{110})	$4\dfrac{\left(C_{11}^2 + C_{12}C_{11} - 2C_{12}^2\right)C_{44}}{4C_{44}C_{11} + C_{11}^2 + C_{12}C_{11} - 2C_{12}^2}$	467 GPa
Young's modulus (E_{111})	$3\dfrac{C_{44}(C_{11}+2C_{12})}{C_{11}+2C_{12}+C_{44}}$	557 GPa
Poisson's ratio	$\dfrac{C_{12}}{C_{11}+C_{12}}$	0.267
Crystal anisotropy	$\dfrac{E_{111}-E_{100}}{E_{111}}$	0.436

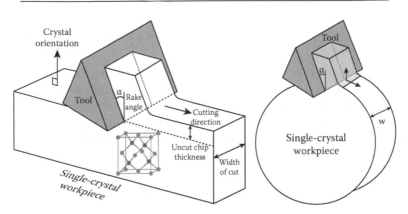

FIGURE 19.8
Schematic diagram showing the arrangement of nanometric cutting.

Based on Figure 19.9, calculations were done, as shown in Table 19.4, to highlight the number of atoms on a crystal plane, atomic density per unit area, minimum distance between atoms on a plane and distance between two adjacent planes of 3C-SiC.

It can be seen from Table 19.4 that the atomic density per unit area in 3C-SiC is highest on the (110) crystal orientation and the distance between the two adjacent planes is farthest compared to the other two orientations. This could be expected to make the (110) plane the weakest plane in comparison to the other two crystal planes. Moreover, the minimum distance between the two atoms

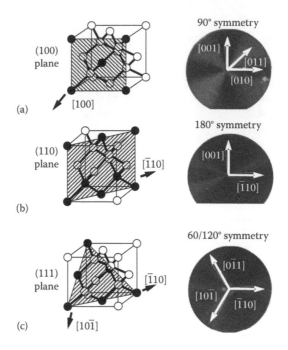

FIGURE 19.9
Schematic illustration of various crystal orientations and cutting directions in 3C-SiC on (a) (100) crystal plane, (b) (110) crystal plane and (c) (111) crystal plane.

on this orientation is shortest, which makes it energetically favourable for the deformation process to occur preferentially on this crystal plane. Table 19.5 provides the details of the two sets of simulation trials performed in this work, including size of the workpiece, uncut chip thickness, cutting tool, nine distinct combinations of the crystal orientations and cutting directions and other relevant parameters.

19.3.2.1 Variation in the Cutting Forces and Coefficient of Kinetic Friction

Quantitative measures of the simulation results, such as thrust forces (F_t or F_y), tangential cutting forces (F_c or F_x), resultant forces and coefficient of kinetic friction (F_c/F_t), are tabulated in Table 19.6 and plotted in Figure 19.10 for comparison purposes. Although different d/r ratios, different tool rake angles and different cutting speeds were used in the two trials, the trend of the variation and the extent of variation in the plots shown in Figure 19.10 were eventually found to exhibit the same pattern. As evident from the results, the thrust forces were found to be highest for the crystal setup (110) <–110> and lowest for (001) <100>. From these values, the extent of anisotropic variation in the thrust forces was observed to be around 45%. Similarly, the tangential cutting forces were found to vary by 30%, whereas the resultant force showed the extent of anisotropy of up to 37%. Contrary to the similar variation of the machining forces in both

TABLE 19.4

Variation in the Properties of 3C-SiC with Respect to Various Crystal Orientations

Orientation	Number of Atoms on the Plane	Atomic Projection Area	Atomic Density per Unit Area	Minimum Distance between the Two Atoms Lying on the Same Plane	Distance between Two Adjacent Planes
Cube (100)	$\frac{4}{4}+1=2$	$a \times a = a^2$	$2/a^2 = 0.1052$	$a/2 = 2.18$ Å	$a = 4.36$ Å
Dodecahedron (110)	$\frac{4}{4}+2+\frac{2}{2}=4$	$\sqrt{2}a \times a = \sqrt{2}a'^2$	$4/\sqrt{2}a^2 = 0.1488$	$a/4 = 1.09$ Å	$1.707a = 6.16$ Å
Octahedron (111)	$\frac{1}{6}\times 3+3\times\frac{1}{2}=2$	$=\frac{\sqrt{3}}{2}a'^2$	$4/\sqrt{3}a^2 = 0.1215$	$\sqrt{3}a/4 = 1.89$ Å	$0.577a = 2.514$ Å

Note: a = lattice constant of 3C-SiC = 4.36 Å.

TABLE 19.5

Process Variables Used in the MDS

Details	Simulation Trial Case 1	Simulation Trial Case 2
Size of the 3C-SiC workpiece	14.262 nm × 4.635 nm	30.1 nm × 11.13 nm
Uncut chip thickness (d) (nm)	1.312	1.9634
Cutting edge radius (r) (nm)	2.297	1.9634
d/r ratio	0.57	1
Cutting tool rake angle	−25°	−30°
Cutting tool clearance angle	10°	
Equilibration temperature	300 Kelvin	
Cutting velocity	10 m/sec	4 m/sec
Potential energy function	ABOP	
Time step	0.5 femtoseconds	
Boundary condition	PBC	
Crystal Setup Combinations	**Crystal Orientation**	**Cutting Direction**
Case 1 or crystal setup 1	(111)	<−110>
Case 2 or crystal setup 2	(111)	<−211>
Case 3 or crystal setup 3	(110)	<−110>
Case 4 or crystal setup 4	(110)	<001>
Case 5 or crystal setup 5	(001)	<−110>
Case 6 or crystal setup 6	(001)	<100>
Case 7 or crystal setup 7	(11−2)	<1−10>
Case 8 or crystal setup 8	(110)	<−11−2>
Case 9 or crystal setup 9	(1−20)	<210>

trials, friction coefficient showed significant variation. For example, a small variation in the friction anisotropy of up to 17% was noticed for a high cutting speed of 10 m/sec and d/r ratio of 0.57 in contrast to a variation of friction anisotropy of up to 35% at a low cutting speed of 4 m/sec and d/r ratio of 1.

19.3.2.2 Variation in the Machining Temperature and Machining Stresses

Figure 19.11 and Table 19.7 show the variation in the temperature in the machining zone of SiC with respect to change in orientations. Figure 19.11a shows the variation in the peak temperature of the workpiece with respect to change in crystal setup and Figure 19.11b shows the variation in the average of the peak temperature of the atoms in the cutting edge radius of the tool. It can be seen from a comparison of Figure 19.11a and b that the trend of plot in both simulation trials is quite similar. It is further evident from Figure 19.11a that the local temperature in the workpiece went up to 2000–2200 K when the cutting was performed on the (111) crystal orientation. It is noteworthy to refer to the earlier discussions on the occurrence of cleavage on (111) crystal orientation. An occurrence of cleavage releases a tremendous amount of elastic energy in the form of heat, which gets transmitted to both the workpiece

TABLE 19.6

Percentage Variation in the Machining Forces and Coefficient of Kinetic Friction

Crystal Orientation		111[a]	111[a]	110	110	001	001	11-2	110	1-20	% Anisotropy Variation
Cutting Direction		-110	-211	-110	001	-110	100	1-10	-11-2	210	
Average thrust force (F_t), nN	Simulation trial 1	1484	1772	2098	1328	1679	1250	1661	2081	1667	40.4%
	Simulation trial 2	1279	1616	1936	1531	1730	1070	1759	1801	1607	44.73%
Average cutting force (F_c), nN	Simulation trial 1	1114	1176	1223	850	1117	885	1099	1225	1097	30.61%
	Simulation trial 2	1194	1389	1285	1004	1316	1066	1198	1316	1282	23.71%
Average resultant force, nN	Simulation trial 1	1855	2126	2428	1576	2016	1531	1991	2414	1995	36.94%
	Simulation trial 2	1749	2130	2323	1830	2185	1510	2128	2230	2055	35%
Coefficient of kinetic friction (F_c/F_t)	Simulation trial 1	0.75	0.66	0.58	0.64	0.67	0.7	0.66	0.588	0.658	17.14%
	Simulation trial 2	0.93	0.86	0.65	0.65	0.77	0.996	0.681	0.73	0.79	34.73%

[a] Mechanism of material removal on (111) plane was cleavage/fracture dominant, that is why cutting forces dropped during the occurrence of fracture. Consequently, the average forces appear to be lower here. Therefore, machining outcomes on (111) crystal orientation in its current format cannot be considered to quantify the anisotropy.

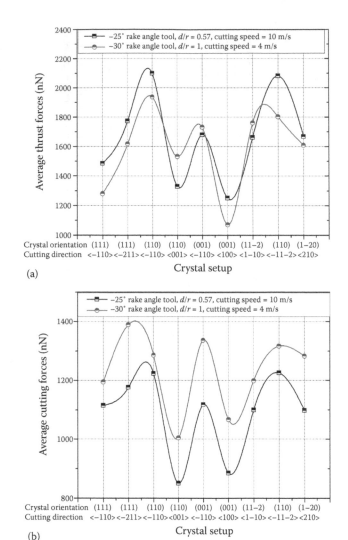

FIGURE 19.10
Variation in forces and friction coefficients: (a) average thrust forces and (b) average cutting
forces. *(Continued)*

and the cutting tool. Consequently, the local temperature of the workpiece
in this particular case increases to about 2200 K, while the cutting tool edge
temperature increases to around 520 K during cutting on the (111) orientation. Interestingly, materials exhibiting a large ratio of covalent bonding do
not get affected at elevated temperatures as much as ionic bonded materials
in terms of their mechanical properties, especially the ease with which they
could mechanically be deformed [80]. This is also the reason why SiC exhibits a lower coefficient of thermal expansion and a high thermal conductivity.

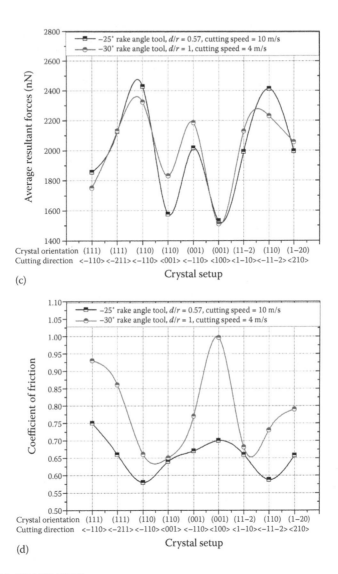

FIGURE 19.10 (CONTINUED)
Variation in forces and friction coefficients: (c) average resultant forces and (d) coefficient of friction.

The ratio of covalent to ionic bonding in SiC is about 1:9, which indicates the fact that it is the concentrated shear rather than the adiabatic shear that drives the ductility in SiC during its nanometric cutting [87]. Along the easy cutting directions, the temperature increase in the workpiece was only up to 600 K. In addition, during cutting on easy cutting directions, i.e. Case 4 and Case 6 in Table 19.7, the lowest temperature on the cutting tool was observed, whereas Case 3, Case 5 and Case 8 showed a somewhat higher temperature on the cutting tool and workpiece.

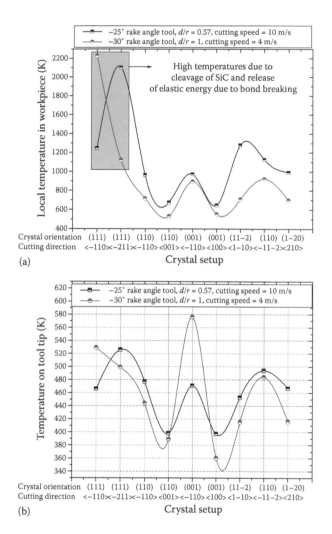

FIGURE 19.11
Variation in the temperature (a) in the machining zone of the workpiece and (b) on the tool tip.

Figure 19.12, in conjunction with Table 19.7, shows the variation in the von Mises stresses in the machining zone. A high magnitude of von Mises stress of up to 112 GPa can be seen when cutting was performed on the (111) orientation, unlike Case 4 (an easy cutting direction), where this magnitude was only a maximum of 58 GPa. Thus, it appears that it is the high stress concentration that is responsible for the occurrence of cleavage on (111) crystal orientation. Knowledge of the stress state on the cutting tool is very promising information. In this context, stresses that acted on the cutting tool during nanometric cutting of 3C-SiC on crystal setups (110) <001> and (001) <100>

TABLE 19.7

Percentage Variation in Temperature and Stresses

		Case 1	Case 2	Case 3	Case 4	Case 5	Case 6	Case 7	Case 8	Case 9
Crystal Orientation		111[a]	111[a]	110	110	001	001	11-2	110	1-20
Cutting Direction		-110	-211	-110	001	-110	100	1-10	-11-2	210
von Mises stress in the machining zone of the workpiece (GPa)	Set 1	100	112[b]	110	65	78	85	89	110[b]	94
	Set 2	110+	96	101	58	90	74	62	86	77
von Mises stress on the cutting edge of the diamond tool	Set 1	343	350	380	322	343	300	322	354	355
	Set 2	400	419	386	386	412	325	337	380	388
Average temperature in the machining zone of the workpiece (K)	Set 1	1247	2102	962	675	973	648	1277	1127	992
	Set 2	2227	1119	715	529	896	553	708	921	699
Average temperature on the cutting edge of the tool (K)	Set 1	466	526	477	398	471	397	453	494	467
	Set 2	529	499	444	389	576	360	415	484	416

[a] Mechanism of material removal on (111) plane was cleavage/fracture dominant, which is the reason that the cleavage energy is transformed to cutting heat and is apparent in the results.

[b] Reflects stress concentration responsible for the occurrence of cleavage.

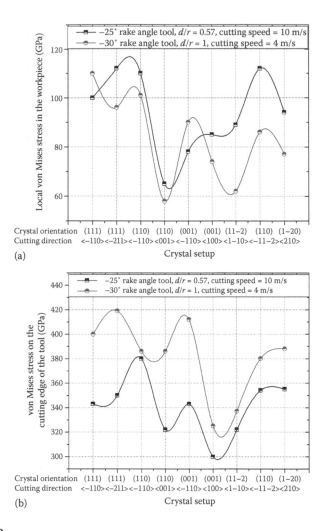

FIGURE 19.12
Variation in the von Mises stress on different crystal setups of (a) workpiece and (b) on the tip of the cutting tool.

were found minimum and similar in magnitude, suggesting that these two setups should be the preferred orientations for cutting 3C-SiC. In common with 3C-SiC, these two crystal setups are once again known to be soft cutting directions for easy material removal in diamond.

An overall analysis of the machining stresses, temperature in the machining zone and cutting forces obtained from the MDS shown previously suggest that the amenability of 3C-SiC of being cut in particular crystal setups can be tabulated, as shown in Table 19.8.

TABLE 19.8

Influence of Anisotropy on the Cutting Behaviour of 3C-SiC Found from the MDS

Crystallographic Planes of 3C-SiC	Least Amenable Cutting Directions	More Amenable Cutting Directions
Cube (001)	<–110>	<100>
Dodecahedron (110)	<–11–2>	<001>
Octahedron (111)	<–211>	<–110>

The anecdote of this MDS study (Table 19.8) is that cutting of 3C-SiC is preferred on the <100> direction on the cubic plane, <001> direction on the Dodec plane and in the <–110> direction on the octahedron plane. These results, along with the previous simulation case study results, demonstrate the value of MDS to understand and improve nanometric machining processes.

19.3.3 SDM and Hot Machining of Nanocrystalline SiC

In this section, the efficiency of SDM and hot machining has been assessed by using MDS. SDM is a recently proposed method of machining that aims to obtain a better quality of a machined product at lower costs. This method utilises pre-defined and machined surface defects on the workpiece to ease material removal. The central idea of this method is to generate surface defects in the form of a series of holes on the top surface of the workpiece prior to the actual machining operation (Figure 19.13). The presence of these defects changes the homogeneity and thus reduces the structural strength of the workpiece, which in turn aids to lower the cutting resistance during machining. Recent experimental trials [13] and numerical simulations [12] on hard steels have shown some very interesting salient features of the SDM method, such as lower machining forces, reduction in overall temperature in the cutting zone, reduced machining stresses and increased chip flow velocity. In the following paragraphs, the applicability of SDM for machining SiC is explored.

This section details the results of a case study made on machining of beta SiC (3C-SiC), but this method is versatile and can readily be applied to other engineering materials that are not readily machinable through conventional approaches, e.g. titanium [88], silicon [89], nickel [90], pretentious low-carbon ferrous alloys [91,92], various polytypes of SiC [75,93], silicon nitride [94] and Al-SiC$_p$ metal matrix composites [95], where rapid tool wear and consequent deterioration of the quality of surface remain major concerns until today. Researchers have also investigated preferential heating and thermal softening of the workpiece by a laser device prior to machining. While this approach has shown promise [96–98], its adverse undesirable effects, such as the lack of control on laser power, sub-surface deformation and transfer of heat to the cutting tool, have

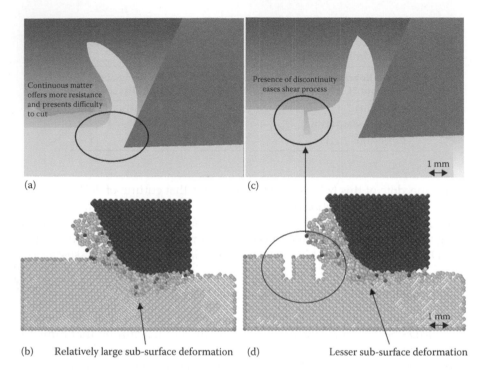

FIGURE 19.13
Schematic diagram indicating the differences between the mode of deformation during conventional machining (a) observed using FEA and (b) observed using MDS and the mechanics of chip formation during SDM (c) observed using FEA and (d) observed using MDS.

impeded its commercial realisation. Overall, the key question explored in this section is this: Can SiC be machined using the SDM method so that the machined surface exhibits improved quality? The results from SDM are compared with (i) hot machining at 1200 K, a temperature where SiC has been reported to gain significant plasticity [99], and (ii) normal machining at 300 K, in order to test the robustness of the proposed method. To describe the interatomic interaction of the atoms, the ABOP [46] potential energy function was employed since it is the only three-body potential energy function developed till date that does not suppress the mechanism of cleavage in 3C-SiC [23]. A comprehensive description, along with the key parameters, of the ABOP potential function is provided elsewhere [46 in its respective reference], and for the purpose of brevity, they are not repeated in this section. Table 19.9 lists the computational parameters, details of the workpiece, uncut chip thickness, details of the cutting tool and other relevant parameters used in this study. The criterion for the selection of these parameters was chosen so that these can readily be linked with the previously performed studies [21,100].

TABLE 19.9

Process Variables Used for Performing the MDS

Workpiece Material	Number of Atoms in the Workpiece	Number of Atoms in the Diamond Cutting Tool
3C-SiC without holes (14.26 nm × 4.6345 nm × 4.278 nm)	28,170	21,192
3C-SiC with surface defects (holes) 14.26 nm × 4.6345 nm × 4.278 nm)	27,782	21,192
Equilibrium lattice parameters for 3C-SiC		$a = 4.36$ Å; $\alpha = \beta = \gamma = 90°$
Details of surface defects (holes)		
Total number		7
Diameter of each hole		0.713 nm
Depth of each hole		1.426 nm
Crystal orientation of the workpiece		(010)
Crystal orientation of diamond tool		Cubic
Cutting direction		<100>
Cutting edge radius (nm)		2.297
Uncut chip thickness/in-feed (nm)		1.3126
Cutting tool rake and clearance angle		−25° and 10°
Equilibration temperature		300 K
Hot machining temperature		1200 K
Cutting velocity		10 m/sec
Time step		0.5 femtoseconds

A snapshot from the MDS after the equilibration process (with surface defects in the 3C-SiC workpiece) is shown in Figure 19.14, where the red and grey colours* correspond to silicon and carbon atoms in the workpiece and yellow colour represents carbon atoms within the diamond cutting tool respectively.

An important constraint of MDS studies is that they are computationally expensive, and therefore, use of high cutting speeds is frequent in MDS studies, e.g. 500–2500 m/sec cutting speed was used by Belak et al. [3,101] and Komanduri et al. [61,102], 150–400 m/sec was used by Wang et al. [103] and Liang et al. [104], 70–100 m/sec was used by Noreyan et al. [105,106], Rentsch and Inasaki [107] and Goel et al. [83,100]. Although these investigations have been successful in capturing the key insights of the cutting process, in the current investigation, high cutting speed could have affected the sensitivity of the results, particularly when cutting of the same configuration is to be compared at 300 K and 1200 K. Therefore, a more realistic cutting speed, 10 m/sec, was employed to perform the MDSs that are presented here. This is accomplished using parallel computing through message passing interface

---

* Readers are referred to the web-based version of this work to interpret the correct colour legends (https://www.cranfield.ac.uk/people/Dr-Suarav-Goel-15392730).

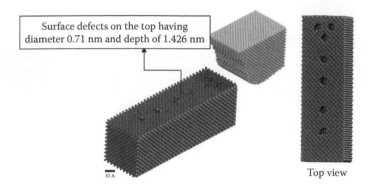

FIGURE 19.14
Snapshot from MDS for 3C-SiC specimen with surface defects on top.

(MPI) interface. The calculation time for each simulation case depends on the model size, cutting speed, cutting distance and the number of CPUs used.

19.3.3.1 Evolution of the Machining Forces

A comparison of the evolution of machining forces (F_c and F_t) obtained from the MDS is shown in Figures 19.15 and 19.16 for all the three cases studied: (i) nanometric cutting at 300 K, (ii) nanometric cutting at 1200 K and (iii) cutting with SDM at 300 K.

It can be seen from Figures 19.15 and 19.16 that the magnitudes of both forces are significantly higher during conventional nanometric cutting performed

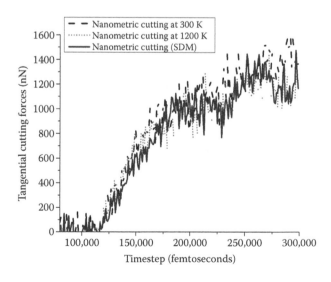

FIGURE 19.15
Tangential cutting forces during nanometric cutting of 3C-SiC in three cases.

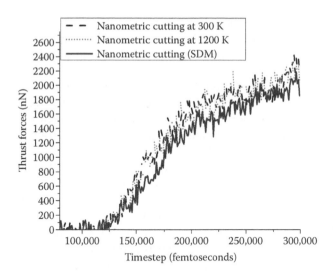

FIGURE 19.16
Thrust forces during nanometric cutting of 3C-SiC in three cases.

at 300 K. However, during nanometric cutting at an elevated temperature of 1200 K, both the tangential cutting force and the thrust force reduce, albeit to a lesser extent. On the other hand, a noticeable reduction, especially in thrust forces, can be seen during nanometric cutting in the case of SDM. It can be noted here that the extent of reduction in cutting force during SDM will depend on various parameters such as the number of holes, dimensions of holes, interspacing between holes and shape of holes. However, since the reduction in the cutting force is of an intermittent nature, the cutting forces and, thus, stresses on the cutting tool would be relieved as soon as the hole is encountered by the cutting tool. Table 19.10 summarises various results obtained from the simulation under different machining conditions, i.e. average cutting force, friction coefficient and resultant cutting force.

TABLE 19.10

Comparison of Cutting Forces and Friction Coefficient during Cutting of 3C-SiC

S.N.	Machining Condition	Average Tangential Cutting Force (F_c)	Average Thrust Force (F_t)	Average Resultant Force $\left(\sqrt{(F_t^2 + F_c^2)}\right)$	Average Friction Coefficient (F_c/F_t)
1	Normal machining (300 K)	835 nN	1185 nN	1449.64 nN	0.7046
2	Hot machining (1200 K)	762 nN	1177 nN	1402.13 nN	0.6474
3	SDM (300 K)	751 nN	1038 nN	1281.19 nN	0.7235

The data in Table 19.10 show that the resultant cutting force reduces from a value of 1449.64 nN to 1402.13 nN when the machining was done at 1200 K instead of at 300 K, signifying a reduction in the cutting resistance of 3C-SiC by 3.27% at 1200 K. However, the extent of this reduction is higher during the SDM process as the resultant forces drop to 1281.19 nN, i.e. a significant reduction of 11.62% compared to normal nanometric cutting at 300 K. It is very interesting to note here that while the cutting force showed a reducing trend, a similar trend is not visible for the coefficient of friction. Compared to the nanometric cutting results at 300 K, the coefficient of friction reduces by 8.11% when the cutting was performed at 1200 K. On the contrary, SDM causes an increase in the friction coefficient by 2.68%. This suggests that a different mechanism of chip formation is associated with the proposed SDM method, which will be discussed here now.

Figure 19.17 shows a superimposed image and a comparison of the chip morphology in all the three cases investigated. On comparing nanometric cutting at 1200 K with conventional nanometric cutting at 300 K, it can be seen that the curliness of the chip has seemingly increased, which is plausible due to the increased plasticity of SiC at a high temperature of 1200 K. However, the shear plane angle remained unchanged, which is not the case with SDM. In the case of SDM, the cut chip thickness has increased, and thus, the shear plane angle has decreased. This was verified using the following equation:

$$\tan\phi = \frac{r\cos\alpha}{1 - r\sin\alpha},\qquad(19.5)$$

where ϕ is shear plane angle, α is rake angle and r is the ratio of uncut and cut chip thicknesses.

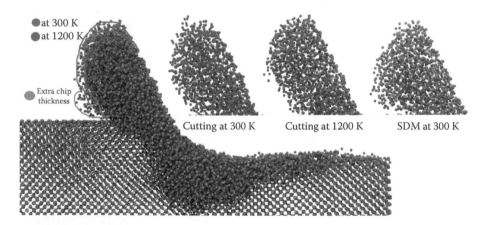

FIGURE 19.17
Chip morphology of 3C-SiC while cutting the workpiece after tool advances to 8.3 nm (see online coloured version for interpreting correct colours, https://www.cranfield.ac.uk/people/Dr-Suarav-Goel-15392730).

TABLE 19.11

Comparison of Chip Morphology and Shear Angle during Machining of 3C-SiC Under Different Machining Conditions

S.N.	Machining Condition	Ratio of Uncut Chip Thickness to Cut Chip Thickness (r)	Shear Plane Angle (φ)
1	300 K	0.525	21.28°
2	1200 K	0.525	21.28°
3	SDM at 300 K	0.505	20.66°

Table 19.11 shows a decrease in shear plane angle from a value of 21.28° to 20.66° using SDM process compared to nanometric cutting at 300 K. A decrease in the value of shear plane angle under the same machining parameters shows the dominance of tangential cutting force over thrust force, which justifies the enhanced cutting action of the tool. This corroborates to the increased force ratio as seen earlier in Table 19.10 during the case of SDM, suggesting the dominance of tangential cutting force to be the reason of the increase in friction coefficient (which improves the cutting action).

Figure 19.18 shows the measure of the cut chip thickness and highlights the variation in the sub-surface crystal deformation lattice layer depth. It is interesting to note that the sub-surface crystal deformation lattice layer becomes wider while cutting at 300 K, while it becomes a little deeper while cutting at 1200 K. Moreover, the extent of the deformation of the crystal layer underneath the finished surface is more pronounced in both these cases compared to that in the SDM operation. It can be postulated that high temperature weakens the bonding forces between the atoms, and hence, the atoms could easily be deformed without having much influence on the neighbour atoms. Therefore, the deformation did not become wider and remained concentrated under the wake of the tool under the influence of high deviatoric stresses. Contrarily, the SDM process shows minimal sub-surface deformation. The waviness of the finished surface also seems to have decreased during SDM. The defects generated manually for the purpose of SDM significantly weaken the material locally, which in turn reduces the bonding strength of the atoms in the area of uncut chip thickness without disturbing the sub-surface. Also, a discontinuity in the material and the consequent lack of resistance to the deformation of the atoms by the adjacent atoms make the shearing process more preferential. Eventually, the material removal is facilitated by local shearing of the material, which is promoted further by stress concentration. This is compounded by the fact that the cutting tool is relieved from high cutting forces intermittently, and this results in minimal sub-surface damage.

19.3.3.2 Temperature and Stress Variation during SDM

A comparison of the evolution of the cutting temperature on the tool cutting edge in all the three cases is shown in Figure 19.19 over the cutting length

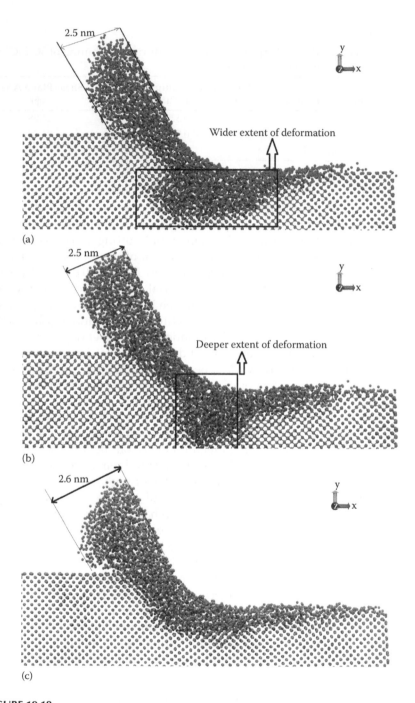

FIGURE 19.18
Sub-surface crystal lattice deformation of 3C-SiC after tool advances to 8.3 nm. (a) Conventional cutting at 300 K. (b) Cutting at 1200 K. (c) Cutting at 300 K with the SDM process.

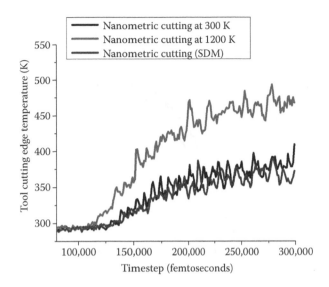

FIGURE 19.19
Variation in the temperature of the cutting edge during nanometric cutting of 3C-SiC.

of 9.35 nm. From this graph, a high temperature of 480 K on the tool cutting edge is evident during nanometric cutting at 1200 K. At higher temperature, graphitisation of the diamond becomes inevitable [100]. The transfer of heat from the bulk of the workpiece to the cutting tool may thus compromise the life of diamond tools [83]. On the contrary, the temperature on the tool cutting edge was observed to be much lower during SDM. This shows that the SDM cutting process could potentially release the cutting tool from high cutting loads and high temperature intermittently, which provides longevity to the tool life.

It will be interesting to analyse the variation in the stresses on the cutting tool during SDM and conventional nanometric cutting to understand the process differences in detail. Figures 19.20 and 19.21 show the evolution of the von Mises stress and shear stress acting on the cutting tool, respectively, over the cutting length of 9.35 nm.

From Figures 19.20 and 19.21, it appears that in all the three cases, the diamond tool cutting edge has undergone a high magnitude of von Mises stress and shear stress during nanometric cutting of 3C-SiC. This can be attributed to the high hardness of 3C-SiC, which offers tremendous cutting resistance in comparison to other relatively softer brittle materials. A high magnitude of von Mises stress of up to 250 GPa having a component of shear stress to the order of 125 GPa is certainly unfavourable for the life of the diamond cutting tool – the ideal strength for diamond prior to graphitisation has been reported to be around 95 GPa [108]. The stress state in all the cases suggests that graphitisation of diamond tool during nanometric cutting of 3C-SiC would be inevitable owing to the high hardness of 3C-SiC. However,

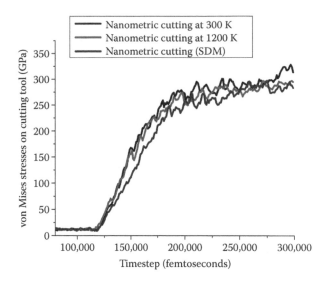

FIGURE 19.20
Variation in von Mises stress acting on the tool during nanometric cutting of 3C-SiC.

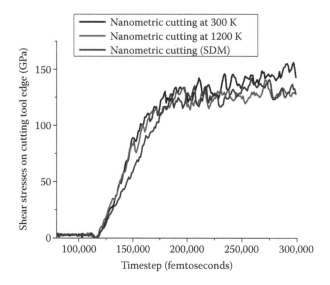

FIGURE 19.21
Variation in shear stress acting on the tool during nanometric cutting of 3C-SiC.

SDM seems to have aided to reduce the stresses on the cutting edge to some extent in comparison to conventional approach of nanometric cutting and hot machining at 1200 K.

In summary, this section has presented a case study on the nanometric machining of silicon and SiC using MDSs. To begin with, the geometry of the cutting tool, particularly the rake angle, has a marked impact in dictating the plasticity from hard-brittle materials. It was also demonstrated that the SiC exhibits strong anisotropy in its ease of deformation, and rigorous analysis shows that the combinations of either (110) <001> or (001) <100> crystal orientations and cutting directions are favourable to cut 3C-SiC. In an effort to overcome the problems of tool wear, a new method, namely SDM, has been presented, which turns out to be advantageous because both cutting as well as polishing mechanisms are realised, improving surface roughness. Further investigations may be directed to realise the optimum shapes of surface defects, pathways to generate such defects in a controlled environment and net saving in energy and cost. Such understanding is only possible through MDS, and these results again demonstrate the value of using MD to understand novel machining processes for the most challenging materials.

19.4 Concluding Remarks

MDSs are constantly being used to make important contributions to our fundamental understanding of material behaviour, at the atomic scale, for a variety of thermodynamic processes. This chapter shows that MDS is a robust numerical analysis tool in addressing a range of complex nanofinishing (machining) problems that are otherwise difficult or impossible to understand using other methods. For example, the mechanism of nanometric cutting of SiC is influenced by a number of variables such as machine tool performance, machining conditions, material properties and cutting tool performance (material microstructure and physical geometry of the contact), and all these variables cannot be monitored online through experimental examination. However, these could suitably be studied using an advanced simulation-based approach such as MDS. This chapter details how MDS can be used as a research and commercial tool to understand key issues of ultra-precision manufacturing research problems, and a specific case was addressed by studying diamond machining of SiC. While this is appreciable, there are a lot of challenges and opportunities in this fertile area. For example, the world of MDSs is dependent on present-day computers and the accuracy and reliability of potential energy functions [109]. This presents a limitation: Real-world scale simulation models are yet to be developed. The simulated length and timescales are far shorter than the experimental ones, which couples further with the fact that contact-loading simulations are typically done in the speed range of a few hundreds of m/sec against the experimental speed of typically about 1 m/sec [17]. Consequently, MDSs suffer from the spurious effects of high cutting speeds and the accuracy of the simulation results has yet to be fully explored. The development of user-friendly software could help facilitate MD as an integral part of computer-aided design and manufacturing to tackle a range of machining problems from all perspectives, including materials science (phase of the material formed due to the sub-surface deformation layer), electronics and optics (properties of the finished machined surface due to the metallurgical transformation in comparison to the bulk material) and mechanical engineering (extent of residual stresses in the machined component) [110]. Overall, this chapter provided key information concerning diamond machining of SiC, which is classed as hard, brittle material. From the analysis presented in the earlier sections, MDS has helped in understanding the effects of crystal anisotropy in nanometric cutting of 3C-SiC by revealing the atomic-level deformation mechanisms for different crystal orientations and cutting directions. In addition to this, the MDS revealed that the material removal mechanism on the (111) surface of 3C-SiC (akin to diamond) is dominated by cleavage. These understandings led to the development of a new approach named the SDM method, which has the potential to be more effective to implement than ductile mode micro-laser-assisted machining or conventional nanometric cutting.

Acknowledgements

The authors would like to acknowledge the use of high-performance computing service of STFC Hartree Centre, UK. SG acknowledges the partial support of the European COST Action 'CA15102' of the Horizon 2020 program in undertaking this work. Also, this research was partially supported by the research funds obtained from the University of Illinois at Urbana Champaign (UIUC).

Solved Problems

Q1. What does MDS stand for?

It stands for molecular dynamics simulation.

Q2. In which scenarios is MDS is favoured over FEA to investigate nano-manufacturing processes?

MDS is informed by quantum properties, thereby revealing discrete atomistic mechanics that is otherwise impossible to be investigated using the traditional FEA.

Q3. Why is it necessary to apply MDS to study nanoscale machining?

As the nanometric machining process involves removing few atomic layers from the surface, it is extremely difficult to observe or to simulate such processes using other simulation methods. Dislocation nucleation, in particular occurring during machining, is best suited to be studied by the MDS.

Q4. What is the generic methodology to carry out an MDS to study nanoscale machining?

The essence of the MDS method is the numerical solution of Newton's second equation of motion for an ensemble of atoms where the atoms are assumed to follow Newton's second law. Newton's second equation of motion is integrated by numerical techniques on short time intervals (typically over 1 to 100 pico-seconds); and equilibrium statistical averages are computed as temporal averages over the observation time. The interatomic forces (both attractive and repulsive) are defined by an appropriate empirical potential–energy function.

Q5. How can the temperature be assessed in atomic simulations?

Temperature is an ensemble property and is not readily measurable. The most suitable way to measure temperature is to convert the kinetic energy of the atoms into equivalent temperature. However, the instantaneous fluctuations in kinetic energy per atom are usually very high, so this equation is averaged temporally and/or spatially over a number of time steps to convert this into equivalent temperature.

Q6. What is a potential function or a force field?

This interaction between atoms is governed by a potential energy function that roughly accounts for quantum interactions between electron shells and represents the physical properties of the atoms being simulated, such as its elastic constants and lattice parameters. It may be noted that the atoms in the MDS are typically assumed to be points.

Q7. What are the limitations of pair potentials over many body potentials?

In a simple pairwise potential, e.g. the Morse or Lennard-Jones potential, only the direct interactions between atoms are considered and summed for a certain sphere with a radius that is usually equal to the spacing between four adjacent atoms. They determine the total energy without considering any further cohesive terms that arise from the interaction with atoms far away from the particle considered. A common limitation of all the pair potentials is their inability to reproduce the correct Cauchy pressure, which became the key motivation for the development of the EAM-type potentials.

Q8. What are the common drawbacks of Tersoff formalism proposed for silicon in the year 1989?

Tersoff formalism fails to reproduce the density–temperature relation of silicon, suggesting that this potential is not fully reliable to obtain the phase diagram of silicon. Also, Tersoff potential functions have limitations in accurately describing the thermal aspects of silicon.

Q9. Explain the boundary conditions used to study MDS of nanoscale machining problem.

Fixed boundaries along the direction of machining forces and PBCs along the direction orthogonal to the cutting forces (where the machining force virtually remains zero) are most commonly used by the researchers.

Q10. Why is it necessary to include a thermostatic zone in MDS of nanometric cutting?

In conventional machining operations, the energy from plastic deformation in the primary shear zone and friction at the tool-chip interface generate heat and are carried away by chips, lubricant and by conduction into the tool and workpiece. The nanometric cutting model is, however, extremely small and is not capable of dissipating the heat itself. The velocity of the thermostatic atoms is therefore re-scaled to a desired temperature (300 K) at each step of the computation to mimic the heat dissipation.

Q11. Explain the von Mises yield criterion.

The von Mises yield criterion is a very common yield criterion used to predict yielding in a material based on the maximum strain energy absorbed by the material before it yields.

Some Unsolved Problems

1. List in chronological order how MDS potential functions have evolved; include which materials each function have been developed for.

2. List the steps involved in an MDS, describing the key input, processes and output for each step.

3. Describe how MDS could be used to understand the influence of rake angle on brittle material nanofinishing.

4. Describe using diagrams the modes of deformation during nanofinishing with and without the use of SDM.

References

1. B. Alder, T. Wainwright. Phase transition for a hard sphere system, *The Journal of Chemical Physics* 27 (1957) 1208.

2. E. Brinksmeier, O. Riemer. Measurement of optical surfaces generated by diamond turning, *International Journal of Machine Tools and Manufacture* 38 (1998) 699–705.

3. J.F. Belak, I.F. Stowers. A Molecular Dynamics model of Orthogonal Cutting process, *Proceedings of American Society Precision Engineering Annual conference* (1990) 76–79.

4. U. Landman, W.D. Luedtke, N.A. Burnham, R.J. Colton. Atomistic Mechanisms and Dynamics of Adhesion, Nanoindentation, and Fracture, *Science* 248 (1990) 454–461.

5. N. Ikawa, S. Shimada, H. Tanaka, Minimum thickness of cut in micromachining, *Nanotechnology* 1 (1992) 6–9.

6. A.F. Voter, J.D. Kress. Atomistic Simulation of Diamond-Tip Machining of Nanoscale Features. Principles of Cutting Mechanics: Applications of Ultra-Precision Machining and Grinding, 1993 Spring Topical Meeting Tucson, AZ, USA: ASPE Proceedings, 1993.

7. T. Inamura, S. Shimada, N. Takezawa, N. Nakahara. Brittle/Ductile Transition Phenomena Observed in Computer Simulations of Machining Defect-Free Monocrystalline Silicon, CIRP *Annals–Manufacturing Technology* 46 (1997) 31–34.

8. R. Rentsch. Influence of Crystal Orientation on the Nanomeric Cutting Process. In *Proceedings of the First International Euspen Conference*. Bremen, Germany, 1999. p.230–233.

9. R. Komanduri, N. Chandrasekaran, L.M. Raff. Effect of tool geometry in nanometric cutting: A molecular dynamics simulation approach, *Wear* 219 (1998) 84–97.

10. S. Goel, B. Beake, C.-W. Chan, N. Haque Faisal, N. Dunne. Twinning anisotropy of tantalum during nanoindentation, *Materials Science and Engineering: A* 627 (2015) 249–261.

11. S. Goel, W.B. Rashid, X. Luo, A. Agrawal, V. Jain. A theoretical assessment of surface defect machining and hot machining of nanocrystalline silicon carbide, *Journal of Manufacturing Science and Engineering* 136 (2014) 021015.

12. W.B. Rashid, S. Goel, X. Luo, J.M. Ritchie. The development of a surface defect machining method for hard turning processes, *Wear* 302 (2013) 1124–1135.

13. W.B. Rashid, S. Goel, X. Luo, J.M. Ritchie. An experimental investigation for the improvement of attainable surface roughness during hard turning process, *Proceedings of the Institution of Mechanical Engineers, Part B: Journal of Engineering Manufacture* 227 (2013) 338–342.

14. S. Goel, N.H. Faisal, V. Ratia, A. Agrawal, A. Stukowski. Atomistic investigation on the structure–property relationship during thermal spray nanoparticle impact, *Computational Materials Science* 84 (2014) 163–174.

15. N.H. Faisal, R. Ahmed, S. Goel, Y.Q. Fu. Influence of test methodology and probe geometry on nanoscale fatigue failure of diamond-like carbon film, *Surface and Coatings Technology* 242 (2014) 42–53.

16. S. Goel, A. Agrawal, N.H. Faisal. Can a carbon nano-coating resist metallic phase transformation in silicon substrate during nanoimpact?, *Wear* 315 (2014) 38–41.

17. S. Goel, N.H. Faisal, X. Luo, J. Yan, A. Agrawal. Nanoindentation of polysilicon and single crystal silicon: Molecular dynamics simulation and experimental validation, *Journal of Physics D: Applied Physics* 47 (2014) 275304.

18. S. Goel, S.S. Joshi, G. Abdelal, A. Agrawal. Molecular dynamics simulation of nanoindentation of Fe3C and Fe4C, *Materials Science and Engineering: A* 597 (2014) 331–341.

19. S. Goel, X. Luo, P. Comley, R.L. Reuben, A. Cox. Brittle–ductile transition during diamond turning of single crystal silicon carbide, *International Journal of Machine Tools and Manufacture* 65 (2013) 15–21.

20. S. Goel, X. Luo, R.L. Reuben. Shear instability of nanocrystalline silicon carbide during nanometric cutting, *Applied Physics Letters* 100 (2012) 231902.

21. S. Goel, X. Luo, R.L. Reuben. Wear mechanism of diamond tools against single crystal silicon in single point diamond turning process, *Tribology International* 57 (2013) 272–281.

22. S. Goel, A. Stukowski, G. Goel, X. Luo, R.L. Reuben. Nanotribology at high temperatures, *Beilstein Journal of Nanotechnology* 3 (2012) 586–588.

23. S. Goel, A. Stukowski, X. Luo, A. Agrawal, R.L. Reuben. Anisotropy of single-crystal 3C–SiC during nanometric cutting, *Modelling and Simulation in Materials Science and Engineering* 21 (2013) 065004.

24. S. Goel. A topical review on 'The current understanding on the diamond machining of silicon carbide', *Journal of Physics D: Applied Physics* 47 (2014) 243001.

25. S. Goel, X. Luo, A. Agrawal, R.L. Reuben. Diamond machining of silicon: A review of advances in molecular dynamics simulation, *International Journal of Machine Tools and Manufacture* 88 (2015) 131–164.

26. L. Pastewka, M. Mrovec, M. Moseler, P. Gumbsch. Bond order potentials for fracture, wear, and plasticity, *MRS Bulletin*-Three decades of many-body potentials in materials research 37 (2012) 493–503.

27. J. Tersoff. Empirical interatomic potential for silicon with improved elastic properties, *Physical Review B* 38 (1988) 9902.

28. M.S. Daw, M.I. Baskes. Embedded-atom method: Derivation and application to impurities, surfaces, and other defects in metals, *Physical Review B* 29 (1984) 6443–6453.

29. F.H. Stillinger, T.A. Weber. Computer simulation of local order in condensed phases of silicon, *Physical Review B* 31 (1985) 5262–5271.

30. F.H. Stillinger, T.A. Weber. Erratum: Computer simulation of local order in condensed phases of silicon [*Physical Review B* 31, 5262 (1985)], *Physical Review B* 33 (1986) 1451–1451.

31. H.J.C. Berendsen, J.R. Grigera, T.P. Straatsma. The missing term in effective pair potentials, *The Journal of Physical Chemistry* 91 (1987) 6269–6271.

32. J. Tersoff. New empirical approach for the structure and energy of covalent systems, *Physical Review B* 37 (1988) 6991.

33. J. Tersoff. Modeling solid-state chemistry: Interatomic potentials for multicomponent systems, *Physical Review B* 39 (1989) 5566.

34. J. Tersoff. Erratum: Modeling solid-state chemistry: Interatomic potentials for multicomponent systems, *Physical Review B* 41 (1990) 3248.

35. J. Tersoff. Carbon defects and defect reactions in silicon, *Physical Review Letters* 64 (1990) 1757.

36. J. Tersoff. Chemical order in amorphous silicon carbide, *Physical Review B* 49 (1994) 16349.

37. P.M. Agrawal, L.M. Raff, R. Komanduri. Monte Carlo simulations of void-nucleated melting of silicon via modification in the Tersoff potential parameters, *Physical Review B* 72 (2005) 125206.

38. R. Devanathan, T. Diaz de la Rubia, W.J. Weber. Displacement threshold energies in β-SiC, *Journal of Nuclear Materials* 253 (1998) 47–52.
39. T. Kumagai, S. Izumi, S. Hara, S. Sakai. Development of bond-order potentials that can reproduce the elastic constants and melting point of silicon for classical molecular dynamics simulation, *Computational Materials Science* 39 (2007) 457–464.
40. M.Z. Bazant, E. Kaxiras, J. Justo. Environment-dependent interatomic potential for bulk silicon, *Physical Review B* 56 (1997) 8542.
41. G. Lucas, M. Bertolus, L. Pizzagalli. An environment-dependent interatomic potential for silicon carbide: Calculation of bulk properties, high-pressure phases, point and extended defects, and amorphous structures, *Journal of Physics: Condensed Matter* 22 (2010) 035802.
42. M.I. Baskes, J.S. Nelson, A.F. Wright. Semiempirical modified embedded-atom potentials for silicon and germanium, *Physical Review B* 40 (1989) 6085–6100.
43. D.W. Brenner. Empirical potential for hydrocarbons for use in simulating the chemical vapor deposition of diamond films, *Physical Review B* 42 (1990) 9458–9471.
44. S.J. Stuart, A.B. Tutein, J.A. Harrison. A reactive potential for hydrocarbons with intermolecular interactions, *The Journal of Chemical Physics* 112 (2000) 6472–6486.
45. A.C.T. van Duin, S. Dasgupta, F. Lorant, W.A. Goddard. ReaxFF: A Reactive Force Field for Hydrocarbons, *The Journal of Physical Chemistry* A 105 (2001) 9396–9409.
46 P. Erhart, K. Albe. Analytical potential for atomistic simulations of silicon, carbon, and silicon carbide, *Physical Review B* 71 (2005) 035211.
47. J. Yu, S.B. Sinnott, S.R. Phillpot. Charge optimized many-body potential for the Si/SiO2 system, *Physical Review B* 75 (2007) 085311.
48. X.W. Zhou, F.P. Doty. Embedded-ion method: An analytical energy-conserving charge-transfer interatomic potential and its application to the La-Br system, *Physical Review B* 78 (2008) 224307.
49. A.P. Bartók, M.C. Payne, R. Kondor, G. Csányi. Gaussian Approximation Potentials: The Accuracy of Quantum Mechanics, without the Electrons, *Physical Review Letters* 104 (2010) 136403.
50. F. de Brito Mota, J.F. Justo, A. Fazzio. Structural properties of amorphous silicon nitride, *Physical Review B* 58 (1998) 8323.
51. K. Matsunaga, Y. Iwamoto. Molecular Dynamics Study of Atomic Structure and Diffusion Behavior in Amorphous Silicon Nitride Containing Boron, *Journal of the American Ceramic Society* 84 (2001) 2213–2219.
52. K. Matsunaga, C. Fisher, H. Matsubara. Tersoff Potential Parameters for Simulating Cubic Boron Carbonitrides, *Japenese Journal of Applied physics* 39 (2000) L48–L51.
53. L. Pastewka, A. Klemenz, P. Gumbsch, M. Moseler. Screened empirical bond-order potentials for Si-C, *Physical Review B* 87 (2013) 205410.
54. L. Pastewka, P. Pou, R. Pérez, P. Gumbsch, M. Moseler. Describing bond-breaking processes by reactive potentials: Importance of an environment-dependent interaction range, *Physical Review B* 78 (2008) 161402.
55. M.B. Cai, X.P. Li, M. Rahman, A.A.O. Tay. Crack initiation in relation to the tool edge radius and cutting conditions in nanoscale cutting of silicon, *International Journal of Machine Tools and Manufacture* 47 (2007) 562–569.

56. E. Brinksmeier, W. Preuss. Micro-machining, *Philosophical Transactions of the Royal Society A: Mathematical, Physical and Engineering Sciences* 370 (2012) 3973–3992.

57. C. Wang, K. Cheng, N. Nelson, W. Sawangsri, R. Rakowski. Cutting force–based analysis and correlative observations on the tool wear in diamond turning of single-crystal silicon, *Proceedings of the Institution of Mechanical Engineers, Part B: Journal of Engineering Manufacture* (2014) 0954405414543316.

58. M.B. Cai, X.P. Li, M. Rahman. Study of the mechanism of nanoscale ductile mode cutting of silicon using molecular dynamics simulation, *International Journal of Machine Tools and Manufacture* 47 (2007) 75–80.

59. T. Zykova-Timan, D. Ceresoli, E. Tosatti. Peak effect versus skating in high-temperature nanofriction, *Nature Materials* 6 (2007) 230–234.

60. X. Luo, S. Goel, R.L. Reuben. A quantitative assessment of nanometric machinability of major polytypes of single crystal silicon carbide, *Journal of the European Ceramic Society* 32 (2012) 3423–3434.

61. R. Komanduri, N. Chandrasekaran, L.M. Raff. Molecular dynamics simulation of atomic-scale friction, *Physical Review B* 61 (2000) 14007–14019.

62. R. Komanduri. Some aspects of machining with negative rake tools simulating grinding, *International Journal of Machine Tool Design and Research* 11 (1971) 223–233.

63. V.P. Astakhov. *Fundamentals of the Selection of Cutting Tool Geometry Parameters Geometry of Single-point Turning Tools and Drills*. Springer, London, 2010. pp. 127–204.

64. T. Nakasuji, S. Kodera, S. Hara, H. Matsunaga, N. Ikawa, S. Shimada. Diamond Turning of Brittle Materials for Optical Components, CIRP *Annals–Manufacturing Technology* 39 (1990) 89–92.

65. R. Komanduri, L. Raff. A review on the molecular dynamics simulation of machining at the atomic scale, *Proceedings of the Institution of Mechanical Engineers, Part B: Journal of Engineering Manufacture* 215 (2001) 1639–1672.

66. R. Komanduri, W.R. Reed Jr. Evaluation of carbide grades and a new cutting geometry for machining titanium alloys, *Wear* 92 (1983) 113–123.

67. A.Q. Biddut, M. Rahman, K.S. Neo, K.M. Rezaur Rahman, M. Sawa, Y. Maeda. Performance of single crystal diamond tools with different rake angles during micro-grooving on electroless nickel plated die materials, *International Journal of Advanced Manufacturing Technology* 33 (2007) 891–899.

68. D. Krulewich, C. Syn, P. Davis, M. Zimmermann, K. Blaedel, J. Carr, J. Haack. An empirical survey on the influence of machining parameters on tool wear in diamond turning of large single crystal silicon optics. Paper prepared for submission to ASPE 14th Annual meeting at Monterey, CA. Livermore, USA: Lawrence Livermore National Laboratory, 1999.

69. Samuels Leonard E. The Mechanisms of Abrasive Machining, *Scientific American* 239 (1978) 132.

70. R. Komanduri, Ch, N. rasekaran, L.M. Raff. Molecular dynamics simulation of the nanometric cutting of silicon, *Philosophical Magazine Part B* 81 (2001) 1989– 2019.

71. W.S. Blackley, R.O. Scattergood. Ductile-regime machining model for diamond turning of brittle materials, *Precision Engineering* 13 (1991) 95–103.

72. M. Wang, W. Wang, Z. Lu. Anisotropy of machined surfaces involved in the ultra-precision turning of single-crystal silicon–a simulation and experimental study, *The International Journal of Advanced Manufacturing Technology* 60 (2012) 473–485.

73. F.S. Shibata Y., E. Makino, M. Ikeda. Ductile-regime turning mechanism of single-crystal silicon, *Precision Engineering* 18 (1996) 129–137.
74. J.A. Patten, W. Gao. Extreme negative rake angle technique for single point diamond nano-cutting of silicon, *Precision Engineering* 25 (2001) 165–167.
75. J. Patten, W. Gao, K. Yasuto. Ductile Regime Nanomachining of Single-Crystal Silicon Carbide, *Journal of Manufacturing Science and Engineering* 127 (2005) 522–532.
76. I. Durazo-Cardenas, P. Shore, X. Luo, T. Jacklin, S.A. Impey, A. Cox. 3D characterisation of tool wear whilst diamond turning silicon, *Wear* 262 (2007) 340–349.
77. T.P. Leung, W.B. Lee, X.M. Lu. Diamond turning of silicon substrates in ductile-regime, *Journal of Materials Processing Technology* 73 (1998) 42–48.
78. F.Z. Fang, V.C. Venkatesh. Diamond Cutting of Silicon with Nanometric Finish, *CIRP Annals–Manufacturing Technology* 47 (1998) 45–49.
79. J. Yan, T. Asami, H. Harada, T. Kuriyagawa. Crystallographic effect on subsurface damage formation in silicon microcutting, *CIRP Annals–Manufacturing Technology* 61 (2012) 131–134.
80. I. Inasaki. Grinding of Hard and Brittle Materials, *CIRP Annals–Manufacturing Technology* 36 (1987) 463–471.
81. S. Goel. An atomistic investigation on the nanometric cutting mechanism of hard, brittle materials. PhD thesis in Mechanical Engineering: Heriot-Watt University, 2013. p. 1–246.
82. D. Ravindra. Ductile mode material removal of ceramics and semiconductors. PhD thesis in the Department of Mechanical and Aeronautical Engineering: Western Michigan University, 2011. p. 312.
83. S. Goel, X. Luo, R.L. Reuben. Molecular dynamics simulation model for the quantitative assessment of tool wear during single point diamond turning of cubic silicon carbide, *Computational Materials Science* 51 (2012) 402–408.
84. S. To, W.B. Lee, C.Y. Chan. Ultraprecision diamond turning of aluminium single crystals, *Journal of Materials Processing Technology* 63 (1997) 157–162.
85. L. Pastewka. Multi-scale simulations of carbon nanomaterials for supercapacitors, actuators and low-friction coatings. Mathematics and Physics Department, vol. PhD. Freiburg, Germany: Fruanhofer IWM, Albert-Ludwigs-Universität, Germany, 2010.
86. J.A. Salem, Z. Li, R.C. Bradt. Thermal expansion and elastic anisotropy in single crystal Al2O3 and SiC Reinforcements, NASA Technical Memorandum 106516 prepared for ASME Symposium on advances in composite materials and structure sponsored by American society of Mechanical Engineers. Anaheim, California, December 10–12, 1986.
87. M.C. Shaw. *Metal Cutting Principles*, Oxford University Press, New York, 2004.
88. A.R. Zareena, S.C. Veldhuis. Tool wear mechanisms and tool life enhancement in ultra-precision machining of titanium, *Journal of Materials Processing Technology* 212 (2012) 560–570.
89. J. Yan, K. Syoji, J. Tamaki. Some observations on the wear of diamond tools in ultra-precision cutting of single-crystal silicon, *Wear* 255 (2003) 1380–1387.
90. M.A. Davies, C.J. Evans, S.R. Patterson, R. Vohra, B.C. Bergner. Application of precision diamond machining to the manufacture of micro-photonics components. In: Kley E.-B., (Ed.). *Lithographic and Micromachining Techniques for Optical Component Fabrication II*, vol. 94. San Diego, USA: SPIE, 2003.

91. T. Fukaya, Y. Kanada, T. Wakabayashi, J. Shiraishi, K. Tomita, T. Nakai. High Precision Cutting of Hardened Steel with Newly Developed PCBN Tools. Accessed from imtp.free.fr/imtp2/C4/Fukaya_Tomohiro.pdf on 10/7/2011 (2011).

92. U. Grimm, C. Muller, W.M. Wolfle. Fabrication of surfaces in optical quality on pretentious tools steels by ultra precision machining. In *Proceedings of the 4th Euspen International Conference*, vol. 4. Glasgow, 2004. p. 193.

93. J. Yan, Z. Zhang, T. Kuriyagawa. Mechanism for material removal in diamond turning of reaction-bonded silicon carbide, *International Journal of Machine Tools and Manufacture* 49 (2009) 366–374.

94. J.A. Patten. Ductile Regime Nanocutting of Silicon Nitride. In *Proceedings ASPE 2000 Annual Meeting*, vol. 22, 2000. p. 106–109.

95. N.P. Hung, C.H. Zhong. Cumulative tool wear in machining metal matrix composites Part I: Modelling, *Journal of Materials Processing Technology* 58 (1996) 109–113.

96. J.A. Patten, J. Jacob, B. Bhattacharya, A. Grevstad, N. Fang, E.R. Marsh. Chapter 2: Numerical simulations and cutting experiments on single point diamond machining of semiconductors and ceramics. In: Yan J., Patten J.A., (Eds.). In *Semiconductor Machining at the Micro-Nano Scale*. Transworld Research Network, Trivandrum-695 023, Kerala, India, 2007. pp. 1–36.

97. D. Ravindra, J.A. Patten. Chapter 4: Ductile regime material removal of silicon carbide(SiC). In: Vanger S.H., (Ed.). In *Silicon Carbide: New Materials, Production Methods and Application*, vol. 1. Nova Publishers, Trivandrum, India, 2011. pp. 141–167.

98. A.R. Shayan, H.B. Poyraz, D. Ravindra, M. Ghantasala, J.A. Patten. Force Analysis, Mechanical Energy and Laser Heating Evaluation of Scratch Tests on Silicon Carbide (4H-SiC) in Micro-Laser Assisted Machining (micro sign -LAM) Process, *ASME Conference Proceedings* 2009 (2009) 827–832.

99. K. Niihara. Slip systems and plastic deformation of silicon carbide single crystals at high temperatures, *Journal of the Less Common Metals* 65 (1979) 155–166.

100. S. Goel, X. Luo, R. Reuben, W. Rashid. Atomistic aspects of ductile responses of cubic silicon carbide during nanometric cutting, *Nanoscale Research Letters* 6 (2011) 589.

101. J. Belak. Nanotribology: Modelling Atoms When Surfaces Collide. In *Energy and Technology Review*, vol. 13: Lawrence Livermore Laboratory, USA, 1994.

102. R. Komanduri, N. Chandrasekaran, L.M. Raff. M.D. Simulation of nanometric cutting of single crystal aluminum–effect of crystal orientation and direction of cutting, *Wear* 242 (2000) 60–88.

103. Y.L. Zhiguo Wang, Mingjun Chen, Zhen Tong, Jiaxuan Chen. *Analysis about Diamond Tool Wear in Nano-Metric Cutting of Single Crystal Silicon Using Molecular Dynamics Method*. In: Li Y., Yoshiharu N., David D.W., Shengyi L., (Eds.), vol. 7655: SPIE, 2010. p.76550O.

104. Y.C. Liang, Y.B. Guo, M.J. Chen, Q.S. Bai. Molecular dynamics simulation of heat distribution during nanometric cutting process. Nanoelectronics Conference, 2008. INEC 2008. 2nd IEEE International, 2008. p.711–715.

105. A. Noreyan, J.G. Amar, I. Marinescu. Molecular dynamics simulations of nanoindentation of beta-SiC with diamond indenter, *Materials Science and Engineering B-Solid* 117 (2005) 235–240.

106. A. Noreyan, J.G. Amar. Molecular dynamics simulations of nanoscratching of 3C SiC, *Wear* 265 (2008) 956–962.

107. R. Rentsch, I. Inasaki. Effects of fluids on the surface generation in material removal processes–Molecular dynamics simulation, *CIRP Annals-Manufacturing Technology* 55 (2006) 601–604.

108. D. Roundy, M.L. Cohen. Ideal strength of diamond, Si, and Ge, *Physical Review B* 64 (2001) 212103.

109. C.L. Kelchner, S.J. Plimpton, J.C. Hamilton. Dislocation nucleation and defect structure during surface indentation, *Physical Review B* 58 (1998) 11085.

110. R. Ghafouri-Azar, J. Mostaghimi, S. Chandra. Modeling development of residual stresses in thermal spray coatings, *Computational Materials Science* 35 (2006) 13–26.

20

Nanofinishing of Biomedical Implants

Naveen Thomas,[1] Ashif Iquebal Sikandar,[2] Satish Bukkapatnam[2] and Arun Srinivasa[1]

[1]Department of Mechanical Engineering, Texas A&M University, College Station, Texas

[2]Department of Industrial and Systems Engineering, Texas A&M University, College Station, Texas

CONTENTS

20.1 Need and Significance of Biomedical Implant Polishing

Biomedical implants are a class of medical devices that are augmented to human (or animal) bones and tissues to support or replace an existing structure and/or enhance an organ's performance. The market for implants is growing at 5% annually and is expected to exceed $35 billion by 2019 [1,2]. Among these, dental and orthotic implants (including those for hip, knee, spine and foot joints) and heart pacemakers share the bulk of the current market. Besides these, artificial heart, cranium and spinal implants are set to experience over 5% market growth rate.

In particular, orthotic implants serve as structural aids when augmented to the bone. They are normally secured to the bones using nuts and bolts, or just placed in the cavity as structural fillers [3]. These are mostly made of materials such as stainless steel (316L SS), cobalt–chromium alloys (CoCrMo), titanium (Ti), titanium alloys (Ti6Al4V, Ti6Al7Nb and TiNi), a variety of ceramic materials with or without acrylic cement [4] and soft polymers, such as ultra-high molecular weight polyethylene (UHMWP) and polyurethane. These materials are chosen to promote bio-acceptance; i.e. (a) implantation is not known to cause undesirable mechanical, chemical or biological stress, and (b) the implant bonds well and promotes the growth of the bone and the nearby tissues.

A collection of different orthotic implants is given in Figure 20.1. Every implant in the figure is composed of two types of surfaces – a textured surface on the regions that bond and mesh with the bone and an ultrasmooth surface (Ra < 100 nm) at the locations where the components slide against each other (i.e. bearing surfaces). Furthermore, these surfaces have complex shapes, sizes and form factors to match a patient's anatomy. Formation of a positive bond between the bone, nearby hard tissue and the implant over the course of the healing time is referred to as bone integration or *osseointegration*. Osseointegration is necessary whenever an implant is inserted into regions where a cavity needs to be filled, a broken bone needs to be restored, or a joint is replaced [16]. The following factors play a primary role in osseointegration: (a) growth of bone tissue on the implant surface [17],

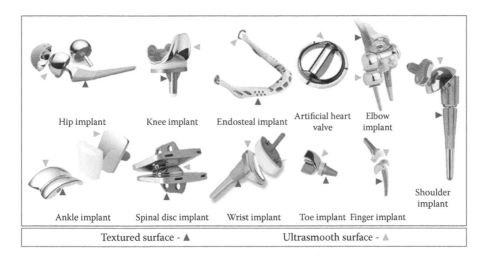

FIGURE 20.1
A montage of important implants and their surface finishing requirements. (Metal on metal hip implants, http://kirkendalldwyer.com/areas-of-practice/medical-devices/metal-on-metal-hip-implants/; Sigma fixed-bearing knees, http://www.kneereplacement.com/DePuy_technology/DePuy_knees/fixed_knee; Custom medical implants – jawbone, layerwise.com, http://www.layerwise.com/medical/custom-implants/; Latitude EV – total elbow prosthesis, http://www.tornier-us.com/upper/elbow/elbrec001/; SMR metal back glenoid in total shoulder arthroplasty, http://www.lima.it/articolo-smr_metal_back_glenoid_in_total_shoulder_arthroplasty-13-53-0.html; Zenith – total ankle replacement, https://www.coringroup.com/medical_professionals/products/extremities/zenith/; Overview of disc arthroplasty – past, present and future, http://www.neurosurgery-blog.com/archives/656; ToeMobile – great toe joint, http://www.hellotrade.com/merete-medical/product.html; Universal total wrist implant system, http://www.medcomtech.es/en/products/orthopedic-surgery-trauma-neurosurgery/upper-limb/wrist/arthroplasty; SR™ PIP (proximal interphalangeal) implants, http://www.totalsmallbone.com/us/products/hand/sr_pip.php4; Pick, A., *Mechanical Heart Valve Replacement Devices*, http://www.heart-valve-surgery.com/mechanical-prosthetic-heart-valve.php.)

(b) adsorption of the proteins onto the surface of the implant (bioactivity) that leads to adhesion [18] and (c) tissue growth through the pores of the implant to form a strong bond with the bone.

Local biological, mechanical and surface material compatibility as well as the finish determine the rate and the strength of osseointegration along the bonding surfaces as well as the durability of the moving parts of the joints. The statistical analysis of Shalabi et al. [17] suggests that higher surface roughness results in a higher bone-to-implant contact area and higher torque resistance. They noted that a roughness Ra of 0.5–8.5 µm was most conducive for osseointegration. This generalises an earlier finding of Wennerberg and Albrektsson [19] that an Ra of 1–1.5 µm is needed for the growth of the bone cells. Götz et al. [20] found that titanium alloy (Ti6Al4V) implants, which were laser-blasted to produce surfaces with pores of the order of 200 µm and a surface roughness of Ra ~7.5 µm, promoted surface integration through deposition and in-growth of the bone tissue. Aniket et al. [21] conducted experiments to study the early osteoblast growth on titanium alloy (Ti6Al4V) implants versus similar alloy implants coated with bioactive silica-calcium phosphate nanocomposite (SCPC50). They observed that the surface roughness of the coating promoted the attachment of the cells to the implant surface but the uncoated surface promoted the growth of the cell. This is because an uncoated surface enhances the concentration of a linear polymer microfilament called F-actin near the surface, which promotes cell growth. The investigations of Lundgren et al. [22] on the growth of bone tissue suggest that the extent of bone growth was not significantly different between a turned smooth surface (Sa = 0.48 ± 0.3 µm) and a titanium oxide blasted surface (Sa = 1.54 ± 0.3 µm), but more mineralised bone was in direct contact with the titanium oxide blasted implant. In this context, Anselme and Bigerelle's [18] investigations with titanium implants and that of Deng et al. [23] with hydroxyapatite-based composites suggest that the adhesion power of the cells were higher on rougher surfaces, and surface texture plays an important role in osseointegration.

As noted in the foregoing, while microscopic roughness is essential for bone growth, ultrasmooth finish Ra < 0.5 µm is necessary on the bearing surfaces (e.g. at hip and knee joints) to ensure performance as well as durability under dynamic loading. As the world population ages, the challenges associated with realising ultrasmooth finish on the bearing surfaces of joints are set to grow. For example, joint replacement surgeries are growing at over 15% annual rates, especially in developed countries, to address issues with arthritic complications and permanent damage to bones and bearing surfaces in old age. Implants for these applications contain surfaces that slide and bear during their functioning.

Joint implants possess components that are in continuous relative motion with each other throughout their life. This relative motion of implant surfaces can lead to surface erosion, chipping and deformation of the geometry of the implants over time. Deformed and worn implants cause debilitation of functions, squeaky noises, stiffening and dislocation of the joints, as well as significant inconvenience and pain to the patient. Smooth finishing of these surfaces is essential to minimise wear. Primary material combinations employed for joint surfaces include polymer linings, metal on polymer, ceramic on polymer, metal on metal and ceramics. Pertinently, polymer liner materials include acrylic, nylon, UHMWP, polyurethane and polypropylene. Metals used for joint implants include medical-grade titanium and cobalt chromium, while ceramics includes alumina, zirconia and diamond thin films [24]. The first-generation implant materials were metallic with appropriate surface treatments. Metal-on-metal components offer strong wear resistance of below 0.05 mm/year. However, the abraded asperities are known to dissolve into and contaminate the bloodstream, causing some health concerns and known hazards such as metallosis, where the worn-out debris deposit and aggregate in the soft tissues causes pain. While polymer linings allow economic generation of ultrasmooth surfaces, they are prone to chipping and create micrometre-scale debris. The issue of debris and high rate of wear (>0.1 mm/year for most common materials) has been noted to be a significant limiter of an implant life. Ceramic and advanced polymeric implants are under investigation for various applications, and many of their variants are in various stages of clinical testing and regulatory approval process.

From a tribological standpoint, these implant surfaces are designed to have a lower coefficient of friction [25–27]. Consequently, very low surface roughness (Ra < 1 μm) is needed for implant bearing surfaces to ensure smooth slipping [26]. The ease of sliding between the surfaces can be measured using a factor known as the lambda ratio (Λ) [28]:

$$\Lambda = \frac{h_{min}}{\sqrt{R_{a(head)}^2 + R_{a(cup)}^2}}, \tag{20.1}$$

where h_{min} is the theoretical film thickness of the lubricating film and $R_{a(head)}$ and $R_{a(cup)}$ are the roughness parameters associated with the two relatively sliding surfaces of the joint. It is suggested that the higher the lambda ratio (Λ) of the bearing surfaces, the lower is the wear rate in the implants.

Thus, finish requirements for biomedical implants tend to be localised – certain parts of the implant would require rough porous surfaces to facilitate osseointegration. Certain other regions, notably the ones whose surfaces are under relative sliding motion, would require a smooth surface finish, in order to reduce friction and, hence, wear. This differential surface finish requirement restricts the use of several traditional as well as advanced finishing processes, including chemical mechanical polishing, electrochemical mechanical polishing and electrochemical polishing to finish. The remainder of this chapter discusses various techniques reported in the literature for polishing biomedical implant surfaces, with emphasis placed on the techniques employed to achieve sub-micrometre scale finish, referred to as ultrasmooth finish in the biomedical implant parlance.

20.2 Classification of Finishing Methods for Biomedical Implants

Surface finishing processes for biomedical implants can be mainly classified into (a) mechanical, (b) chemical, (c) electrochemical, (d) vacuum deposition and (e) laser and thermal treatment techniques [29]. It may be noted that some of these techniques aim to impart specific surface texture on a smooth surface, and others aim to create ultrasmooth surfaces with nanometric scale finish. Conventionally, the industry has been employing handheld polishers for finishing biomedical implants, especially to achieve differential, localised polishing. Such processes tend to be laborious and demand extreme dexterity. Alternatively, localised electro-chemo-mechanical etching methods have been investigated [30,31]. However, they require elaborate masking to achieve localisation of material removal. Many of the methods investigated focus on providing surface texture conducive for osseointegration. Limited techniques have been investigated for nanofinishing of implant surfaces. However, as stated in the foregoing, localised finishing of freeform geometries remains an open issue. The remainder of this section will introduce and provide a critical review of various approaches for polishing implant surfaces.

20.2.1 Mechanical Abrasion-Based Nanofinishing

Finishing methods that use mechanical abrasion as the primary material removal means mainly employ fine abrasive particles to treat and polish the surface of the implant. These techniques, as mentioned earlier, may be used to create specific textures and/or to provide an ultrasmooth finish.

20.2.1.1 Sandblasting

Sandblasting is used primarily to create sub-micrometre textures on ultrasmooth finished surfaces to promote osseointegration. Here, abrasive beads ($\varphi \approx 70$–900 μm) made of alumina, carborundum (SiC) or glass [32–34] are impinged onto the desired regions of the surface. For example, Taga et al. [33] employed this process to create textured surfaces with roughness Ra in the range of 0.5–2.5 μm. Their investigations also suggested that carborundum would provide an effective performance both in terms of achieving texture control as well as improving the structural properties, such as fracture resistance [32]. It was also noted in the literature that the hardness and brittleness (alternatively, toughness) of the abrasive particle, the blast-head power, abrasive powder flow rate, pulse intervals and finishing time are the main parameters that determine the surface characteristics. The process is, however, limited to finishing external surfaces that are in direct line of vision to the blast-heads.

20.2.1.2 Abrasive Flow Finishing

Abrasive flow finishing (AFF) uses a slurry consisting of fine abrasive particles mixed into a viscoelastic medium (henceforth referred to as the medium) to flow under pressure on the target surface to create an ultrasmooth finish [35]. For example, Subramanian et al. [36] employed the AFF technique to polish a hip implant to nano-scale roughness to create ultrasmooth-slipping surfaces. This process can be tuned to polish a variety of rough, porous implants to the required micro-roughness to promote adhesion to the bone.

The key process parameters in this technique are the pressure in the medium, number of cycles, flow rate and rheology of the medium and the abrasive particle size and concentration. Increase in pressure, the number of cycles or the flow rate of the medium increases the material removal rate (MRR) in the process. The medium's viscosity has a high influence on the final roughness and MRR. The viscosity, however, is hard to control during the process as it is dependent on factors such as flow rate and temperature. Generally, a higher viscosity in the medium results in a higher removal rate [35]. The roughness of the workpiece is dependent on the size of the abrasive particles as well. The major limitation in this approach is that local polishing cannot be achieved without the use of masks and elaborately designed confinements. Moreover, as the flow of the abrasive particles on the surface is unidirectional and reciprocative, a uniform surface finish is not assured in all the directions along the surface.

20.2.1.3 Bonnet Polishing

The 'precession' tooling for bonnet polishing was first introduced by Walker et al. [37,38] for polishing glass surfaces. It uses a flexible bonnet-shaped inflated membrane (cf. Figure 20.2) pressed onto a target surface in such a way that the axis of the head is at an angle – called the *precess* angle – to the workpiece surface normal. The bonnet is rotated simultaneously about both the H and the A axes (see Figure 20.2) to achieve the precession motion. The compliance of the bonnet allows it to conform to the freeform shape of the surface being polished. The slurry formed by mixing the fine abrasive particles [39–41] in deionised water with a specific surfactant [42,43] is introduced at the interface. An automated tool motion control is used to ensure that the workpiece is finished to the desired surface quality and form. Pertinently, Zeng et al. [40] employed the bonnet polishing method to produce a multi-radius hip joint implant made of CoCr material using alumina 3 µm abrasives. A roughness level of Sa ≈ 16.1 nm was achieved using the process.

The MRR in bonnet polishing is dependent on the machine parameters, such as precess angle, head speed, tool offset, the down pressure and the slurry composition [39,44]. These parameters bear a non-linear relationship to the surface roughness of the implant. While versatile in terms of the surface forms that can be bonnet polished, the process is limited to polishing only the outer surfaces of the workpiece and it is not suitable to polish

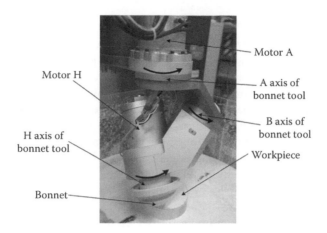

FIGURE 20.2
Bonnet polishing machine for polishing optical surfaces. (From Pan, R., Wang, Z.-Z., Jiang, T., Wang, Z.-S., Guo, Y.-B., *Proc. Inst. Mech. Eng. Part B J. Eng. Manuf.*, 229, 275–285, 2015.)

internal surfaces and narrow pathways. The radius of curvature of concave workpieces that can be polished by using this technique is limited by the size of the bonnet and the form of the target surface on the workpiece.

20.2.1.4 Magnetic Polishing

The most common variants of magnetic polishing processes for biomedical implants use a magnetic or magnetorheological fluid (MR fluid) mixed with abrasive particles. The advantage of using MR fluids is their ability to conform to the shape of the workpiece, which is primarily helpful in polishing free-form surfaces of different materials such as glass, CoCr and titanium alloys. The performance of the process largely depends on the magnetorheological properties of the fluid in terms of the variation of the stiffness and other elastic properties of the fluid with magnetic field strengths. In particular, the elastic modulus of the stiffened fluid determines the abrasion the surface (and hence the MRR), as well as allows conformance to the freeform surface of a workpiece [46]. The magnetic field can also be used to control the relative motion as well as the down force required for polishing. However, in some cases, the down force is provided by pressing a physical tool onto the polishing region. An example of this process is Sidpara et al.'s [47] MR-fluid-based ball end tool to polish freeform surfaces of a knee implant. Other parameters that influence the surface finish include abrasive particle size, fluid concentration and hardness as well as the strength of the magnetic field.

20.2.2 Nanofinishing without Mechanical Abrasion

The main non-traditional finishing methods reported in the literature include chemical, electrochemical and laser finishing techniques. The commonality among these methods is that the main material removal mechanism is not mechanical abrasion or deformation of surface asperities.

20.2.2.1 Chemical Etching

Several chemical methods have been employed to treat and prepare the surface of an implant to make it suitable for insertion into the human body [29]. Chemical etching methods have been employed to achieve varied purposes, including cleaning the surface to remove grease and chemicals, texturing the surface for subsequent finishing or patterning and passivation of the surface to improve in-use corrosion resistance performance of mechanical finishing processes. Acid etching processes are by far the most common chemical etching processes reported for finishing biomedical implants.

Acid etching, as the name suggests, employs a mixture of highly concentrated acids to erode the surface of an implant to achieve the desired texture. This treatment involves washing the surface of the implant with the acids, including nitric, sulphuric and/or phosphoric acids at pH < 0.1 and oftentimes with negative pH to create micro-pits for roughening the surface. For example, Sittig et al. [48] employed concentrated nitric acid with hydrogen fluoride to etch the surface of titanium and titanium alloys to produce a surface roughness of around 3–5 μm. The surface dissolution rate in the workpiece is dependent on key parameters, such as the pH of the acid and the concentration of complexing agent. The choice of chemicals relative to the implant material plays a vital role because of diverse mechanisms involved in an etching process. For example, nitric acid by itself has a negligible effect on the surface roughness of titanium and its alloys, but nitric acid, when mixed with hydrofluoric acid (HNO_3-HF), increases the surface roughness of titanium through grain-structure and lattice orientation-dependent dissolution of the surface. In Ti6Al7Nb and Ti6Al4V alloys, HNO_3-HF increases the surface roughness through selective dissolution of the α-phase alloy.

20.2.2.2 Electrochemical Treatment

Electrochemical polishing methods aim to selectively remove material from surfaces of the electrode or a workpiece via electrolytic reactions [29]. Two variants of this process are electropolishing and anodic oxidation. Electropolishing is primarily used to treat the surface of stainless steel and can be extended to titanium alloys. The workpiece surface is dissolved into the electrolyte in a controlled manner through an electrochemical reaction. It uses mixtures of strong acids as electrolytes. The metallic workpiece (anode) is dissolved into the acid through oxide formation [49]. Most of the surface, with exceptions such as the pits at the grain boundaries, can be smoothened through this process [29]. Latifi et al. [50] demonstrated that when electropolishing is followed by acid dipping, the surface roughness improves to Sa ≈ 0.96 ± 0.29 nm in 316L stainless steel. Larsson et al. [49] showed that when electropolishing of titanium was followed by anodic oxidation, the oxide layer may exhibit the texture conducive for the growth of osteoblasts on the surface. Hryniewicz et al. [51] studied magnetoelectropolishing, a variant of electropolishing. They reported that the dissolution rate and micro-roughness characteristics of the workpiece are dependent on the strength of the magnetic field.

20.2.2.3 Laser Treatment

This process employs pulsed Nd:YAG [20] lasers to create tiny surface pores that serve as nucleation sites for subsequent growth of bone structure that interlocks itself onto the implant. A unique advantage of this technique is that the microscale pores created in this process support the in-growth of bone [20]. Also, as it is a non-contact process, it provides a better control over the patterns desired on the surface. For example, Faeda et al. [52] showed that titanium implants textured using a laser beam (Ra ≈ 1.38 ± 0.23 μm) resulted in better bonding with the bone in terms of the torque required to break the implant free from the bone.

20.2.3 Discussion

While localisation is hard to achieve with free-abrasive methods, many geometric features are inaccessible to handheld polishers. On the other hand, bulk polishing methods based on chemical and/or electrochemical effects need some kind of masks or physical barriers to confine material removal to desired regions. Table 20.1 summarises various techniques for finishing biomedical implants. Among these, Bonnet, AFF and magnetic polishing are the most promising approaches for achieving nano-scale finish on implant surfaces. In particular, recent advances in magneto-viscoelastic fluids offer an interesting opportunity for localised polishing at drastically reduced times [53,54]. These methods are described in more detail in the following sections.

20.3 Advances in AFF

Abrasive-flow-based polishing was first introduced by McCarty in the 1960s [55] as a method to deburr and polish metal products at difficult-to-reach regions. The AFF methods employ a medium made up of fine abrasive particles mixed with polyborosiloxane, commonly referred to as the silly putty [56]. The abrasion occurs when the pressurised medium flows along the surface of the workpiece. These methods can be applied to polish internal as well as external surfaces of a biomedical implant [56].

The AFF technique is normally used in the aerospace and automotive industries to finish complex shapes and extremely small orifices, such as fuel

TABLE 20.1

Comparison of Alternative Nanofinishing Process to Polish Implants

Process	Sandblasting	AFF	Bonnet Polishing	Magnetic Polishing	Chemical Etching	Electrochemical Treatment	Laser Polishing
Mechanism	Large abrasive particles are blasted onto the workpiece surface to cause surface deformation and fracture	Abrasive-laden viscoelastic material is forced along the surface of the workpiece under pressure to polish a smooth surface	A ball-shaped flexible bonnet, rotating at high speed, is used with polishing liquids to locally polish the surface	Abrasion is achieved using an abrasive-laden magnetic fluid in the presence of magnetic field	Strong acids are used to wash the surface of the implant to get a surface with micro-pits	The workpiece is treated as an anode in strong acid electrolyte to dissolve the peaks of the surface in a controlled manner	Laser pulses are used to locally modify the surface of the workpiece
Down pressure	Impinging of the abrasive particles	Pressure in the abrasive medium	Deformation of the bonnet on the surface	Mechanically applied or with the help of magnetic fields	—	—	—
Relative motion	Momentum in the abrasive particles	Flow of the medium	Precession rotations in the bonnet	Rotating or vibrating magnetic fluid or rotating workpiece	Dissolution of the oxides of the metals	Dissolution of the oxide of the metals	—

(Continued)

TABLE 20.1 (CONTINUED)

Comparison of Alternative Nanofinishing Process to Polish Implants

Process	Sandblasting	AFF	Bonnet Polishing	Magnetic Polishing	Chemical Etching	Electrochemical Treatment	Laser Polishing
Key process parameters (MRR)	Blast head power, powder flow rate, finishing time, pulse intervals	Pressure, flow rate, rheology of the medium, abrasive concentration	Precess angle, bonnet tool offset, head speed, tool pressure, dwell time	Strength of magnetic field, concentration of magnetic and abrasive particles, relative motion of abrasives on the surface	pH of the acid, concentration of the complexing agent	Strength of the electrolyte, current density, temperature	Pulse interval, laser power
Key process parameters (finish)	Abrasive hardness and size, blast head power	Abrasive hardness and size, rheology of the medium	Tool path, abrasive particle size and hardness	Abrasive particles size and concentration	–	–	Pulse interval, peak power

injector nozzles, which are not normally amenable to conventional polishing [57]. Recently, these techniques have been further advanced to polish biomedical implants to fine tolerances and finish [36]. The key advantage of AFF is its deformable medium, which conforms to the surface during the finishing process. The MRR and surface finish of the final product depends on different process parameters, including the pressure in the medium, concentration and geometrical properties of the abrasive particles and the rheology of the medium.

20.3.1 Process Setup and Mechanism

An AFF setup consists of one or more hydraulic cylinders connected to hydraulic systems to reciprocatively pump an abrasive medium through the surface to be polished [58]. For internal polishing, the workpiece is mounted between the cylinders such that there is a leak-proof pathway for the medium to flow along the surface of the workpiece between the cylinders. As illustrated in Figure 20.3a, the abrasive medium is reciprocated through the workpiece to achieve material removal. For external polishing, the workpiece is mounted inside a cylinder, which is again connected to the hydraulic system (Figure 20.3b) [36]. The medium primarily consists of a viscoelastic carrier fluid with suspended abrasive particles. The concentration of the abrasive particles in the medium may vary from 5% to 75% by volume [35]. Generally, the viscosity of the polymer medium is chosen based on the viscosity that is desired of the abrasive medium. The composition may vary as per the size and complexity of the surface being polished. The medium is pumped through the workpiece at pressures ranging from 7 to 220 bars, according to the requirements of surface finish and component materials [56].

Three alternative AFF configurations, namely, one-way, two-way and orbital AFF, exist depending on the relative motion desired between the abrasive medium and the target surface. In one-way AFF, the abrasive medium is passed through the workpiece in only one direction and is collected on outflow into a reservoir. Figure 20.3a illustrates the flow of the medium through the workpiece in a typical one-way AFF setup. Here, the finishing process is in a fairly steady state as a fresh, filtered medium devoid of abraded chips or debris from the workpiece, or fractured abrasive particles, is passed through the workpiece every cycle [59]. In a two-way AFF, as shown in Figure 20.3b, the abrasive medium flows back and forth through the workpiece, shearing the surface during both the motions. Here, the same medium with the abraded chips from the workpiece is repeatedly used, thus progressively decreasing its effectiveness. However, a higher number of cycles can be achieved in a short time as the need for a reservoir is avoided. This is the most popular technique of AFF. The orbital finishing method is used for polishing external surfaces and edges. This method differs from the other two methods in that additional small orbital eccentric oscillations are applied on the workpiece

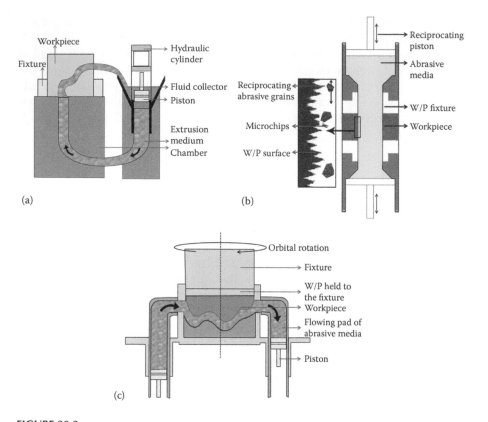

FIGURE 20.3
Schematics of alternative AFF configurations: (a) one-way AFF, (b) two-way AFF and (c) orbital AFF. (From Cheema, M.S., Venkatesh, G., Dvivedi, A., Sharma, A.K, *Proc. Inst. Mech. Eng. Part B J. Eng. Manuf.*, 226, 1951–1962, 2012.)

while the medium is flowing through, as shown in Figure 20.3c [60]. Small vibrations of the order of 0.5–5 mm cause the medium to locally compress very close to the edges, increasing the down pressure generated for finishing the edge [56,61].

As with all finishing processes, the MRR and the finish in AFF depend mainly on the normal down force acting on the surface and the relative motion between the polishing medium and the workpiece. The bulk pressure of the medium is tuned to generate the desired normal force onto the surface for polishing. The relative motion between the medium and the surface is achieved by the flow of the medium. The medium is preferred to have a distinct viscoelastic characteristic as opposed to being purely viscous. The viscoelasticity of the medium restricts the rotation of the abrasive particles during the abrasive action, thereby creating a rubbing action to achieve a finer finish. Gorana et al. [62] stated that this rubbing mechanism of polishing and the concomitant ploughing effect dominate in AFF.

20.3.2 Materials

As discussed in the earlier sections, the abrasive medium consists of polymeric viscoelastic material and fine abrasive particles thoroughly mixed to form a homogenous mixture. Two types of materials are commonly employed to form AFF media, namely, silicon-based polymers and rubber-based polymers. Among these, polyborosiloxane is the most popular silicon-based medium used in AFF processing [63]. The rubber-based polymers reported in the literature include natural rubber, butyl rubber or styrene butadiene derivatives. The most commonly used abrasives in AFF are silicon carbide (SiC) and alumina (Al_2O_3) with an average size of 10–200 μm in concentrations (by volume) of 5%–75% of the medium [35].

20.3.3 Process Parameters

The main process parameters that influence the MRR and the surface quality in the AFF process are (a) the bulk pressure in the medium, (b) the flow rate of the abrasive medium, (c) the rheology of the medium, (d) the abrasive particle size and concentration and (e) the number of flow cycles.

20.3.3.1 Pressure in the Medium

The pumping pressure in AFF relates directly to the normal force applied during lapping and other abrasive finishing process. An average pressure of at least 4 MPa is needed in the medium to observe significant polishing in the workpiece [64]. Figure 20.4a summarises the variation of material removal in a unit time with pressure [65]. Here, we can observe that in all cases, MRR increases with an increase in the bulk extrusion pressure. Although the MRR increases with the medium pressure, the same cannot be deduced for the surface finish. Figure 20.4b summarises the effect of increasing pressure on the surface roughness. Consistent with all mechanical abrasive finishing processes, the surface roughness (Ra and, likely, Sa) initially increases due to the increase in the number of active grains, leading to selective asperity removal. However, after a limit, it increases with further increase in pressure as the tendency for removing materials forms deep valleys, leading to the formation of troughs and pits being enhanced (this condition is also referred to as poor selectivity).

20.3.3.2 Medium Flow Rate

The flow rate of the abrasive medium affects primarily the uniformity of the finish during the process. This is calculated using the volume of medium pumped during any particular cycle and dividing it by the processing time. It is reported that lower flow rates of the medium would reduce MRR but lead to a very uniform finish [57,61]. On the other hand, faster flow rates result in

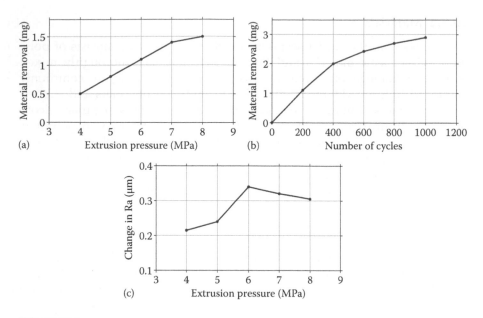

FIGURE 20.4

Variations of (a) MRR with extrusion pressure, (b) MRR with number of cycles and (c) surface roughness Ra with extrusion pressure in abrasive flow finishing. (From Sankar, M.R., Ramkumar, J., Jain, V.K., *Wear*, 266, 688–698, 2009.)

increased MRR. An exception to this effect is observed in brass, where the flow rate of the medium has almost no effect on the material removal [57].

20.3.3.3 Rheology of the Medium

The apparent viscosity of the fluid has a high impact on the polishing effect of the process. It is of primary importance in internal polishing of macro and micro channels. A medium with low viscosity is preferred for long winding channels with small cross-sections. Generally, higher viscosity is preferred for shorter channel lengths. The viscosity of the medium is largely dependent on the size and the concentration of the abrasive particles along with the viscosity of the polymer. It is independent of the pressures acting on them, thereby maintaining a constant viscosity for different working pressures [65]. Rheological studies suggest that an increase in shear rate, induced by the flow rate in the medium, reduces the viscosity of the material [65,66].

20.3.3.4 Abrasive Particle Size and Concentration

The characteristics of the abrasive particles have a direct as well as a secondary effect on the AFF process. The concentration of the abrasive in the medium enhances the MRR and surface finish. However, beyond a certain, material-dependent concentration level, the finish depends on the particle size and the mechanical properties of the medium. A further increase in concentration does not have a significant effect on the finishing process [57]. The size of the abrasive particles affect the finishing process in two ways: it determines the width of cut during polishing and indirectly affects the depth of cut during the finishing process [57]. It is observed that a flexible medium with large particles results in a smooth surface finish as the depth of cut is low.

20.3.3.5 Number of Cycles

The total amount of material removed increases with an increase in the number of cycles the medium is passed through the workpiece. The MRR is normally higher in the initial cycles and decreases in the later cycles, which is evident from the slope of the graph in Figure 20.4c [65]. This trend is partly attributable to the degradation of the medium.

20.3.4 Variants

In order to improve the process efficiency and the final finish, numerous hybrid polishing processes have been developed. Some of the variants of AFF are (a) magnetic AFF, (b) centrifugally assisted AFF and (c) ultrasonic flow polishing. These variants tend to enhance different aspects of the polishing process, and they need to be chosen as per the workpiece and polishing requirements [67–69].

20.3.4.1 Magnetic AFF

In magnetic AFF, carbonyl iron particles (CIPs) are added to the medium. An external magnetic field is applied to increase the down force imparted to the workpiece by the medium or control its flow characteristics. Two variants of magnetic AFF are (a) magnetically assisted AFF and (b) magnetorheological AFF.

Magnetically assisted AFF: Here, the magnetic field is applied to attract the CIP towards the workpiece (see Figure 20.5). This causes an increase in the density of abrasive particles close to the surface to be finished and increases in the normal force onto the surface, resulting in improved MRR. Due to the shielding and reverberation effects, this method is not very effective for workpieces made of ferromagnetic materials but is effective in finishing materials, such as brass and aluminium [67].

Magnetorheological AFF: Here, the magnetic field is instead applied to primarily manipulate the rheological properties of the medium. The application of magnetic field results in the formation of a thick network of CIPs that restricts the rotation of the abrasive particles while flowing. This improves the cutting action of the particles, hence increasing the MRR and, to an extent, the surface finish.

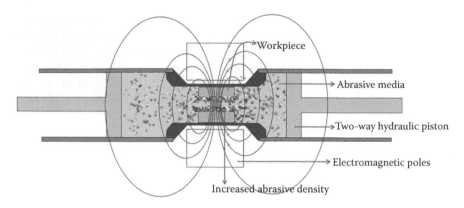

FIGURE 20.5
Schematic of MFA-AFF setup. (From Singh, S., Shan, H., Kumar, P., *J. Mater. Process. Technol.*, 128, 155–161, 2002.)

20.3.4.2 Centrifugally Assisted AFF

In a centrifugally assisted AFF (see Figure 20.6), rods are introduced inside or next to the workpiece in such a way that they are close to the surface to be finished. During the finishing process, these rods are rotated in the medium to introduce a centrifugal force. This helps in building up additional normal force on the workpiece surface, thereby improving MRR. However, this process can only be used to polish simple cylindrical geometries since

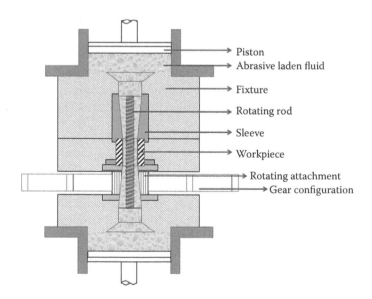

FIGURE 20.6
Schematic of an experimental setup of a centrifugally assisted AFF. (From Reddy, M.K., Sharma, A., Kumar, P., *Proc. Inst. Mech. Eng. Part B J. Eng. Manuf.*, 222, 773–783, 2008.)

developing such setups would be complicated for freeform workpiece geometries [68].

20.3.4.3 Ultrasonic Flow Polishing

This process was invented in 1998 by Cheema and Jones [35,69]. In this technique, the medium flowing through the workpiece during the process is excited using an ultrasonic probe. This excitement of the medium activates the abrasive particles to interact more with the peaks on the surface, resulting in higher and selective removal rates for surface asperities.

20.3.5 Relevance to Biomedical Implants

Biomedical implants are normally made to match the freeform shapes of bones, teeth, etc., that have complex three-dimensional (3D) geometries. Polishing of these freeform surfaces generally requires specialised tools that can adapt to the shape of the implant. AFF is very conducive for polishing these implants as the abrasive medium can assume the shape of the workpiece being polished. For example, Subramanian et al. [36] employed AFF to achieve a surface roughness Ra of 11 nm on a hip prosthesis component.

20.3.6 Limitations

Although the abrasive medium is flexible to reach inaccessible regions in the par and can conform to complex shapes, its flow is generally reciprocative in nature. Because of such a flow pattern, the polishing lay lines at any point are unidirectional in nature. Hence, even though the process yields a fairly smooth finish, one can observe significant differences in the roughness values along different directions. Also, localised polishing is hard to achieve unless the undesired regions are masked. Working pressures of the process also tend to be high (7–220 bars). Consequently, expensive and elaborate setups are needed to run the process.

20.4 Advances in Bonnet Polishing

Bonnet polishing was introduced by Walker et al. [37,70] as a precession motion-based technique for finishing optical surfaces. This process employs a multi-axis computer numerical control (CNC) machine with a polishing bonnet attached to an end. This process is currently widely used for finishing aspheric surfaces of lenses and mirrors and has been investigated for finishing freeform bioimplants.

20.4.1 Process Setup and Mechanism

A bonnet (see Figure 20.7) consists of an inflatable membrane filled with a pressurised fluid (e.g. air) that serves as a flexible polishing pad [70] that can conform to the shape of the surface to be polished. The shank of a bonnet tool consists of two rotating arms connected to the rotating head of a multi-axis CNC machine via an attachment, as shown in Figure 20.2. The arms connected to the CNC machine are oriented at an angle to the spindle head. The adjustability of this angle gives the process the freedom to reach compli-cated surfaces at different angles as required. An abrasive slurry consisting of sub-micrometre scale abrasive particles mixed with surfactants and other ingredients is introduced onto the surface to be bonnet polished [42]. During polishing, the bonnet head is brought in contact with the surface of the work-piece such that the axis of the head (H-axis) is at an angle to the normal to the surface of the workpiece, which is called as precess angle (α in Figure 20.7). While polishing, the bonnet is rotated about these two different axes (H- and A-axes in Figure 20.7) simultaneously. This technique of rotating the bonnet in multiple directions is referred to as continous precession. This method can also be used for polishing as well as very low scale form-correction of work-pieces [71]. As the process is computer controlled, localised polishing to any required roughness level is possible. It also offers a high MRR for fast polish-ing to achieve nanometric surface finish (Sa $\approx$ 16.1 nm) on implants [72].

Mechanism: The mechanism of material removal in bonnet polishing is very similar to that in lapping and other mechanical fine-abrasive finishing processes [73,74]. For conventional polishing processes, MRR ($MRR = \Delta h/\Delta t$, where Δh is a notional reduction in the asperity height after a period of pol-ishing time Δt) is predicted by the Preston equation [75],

$$MRR = C_p\, P\, v, \tag{20.2}$$

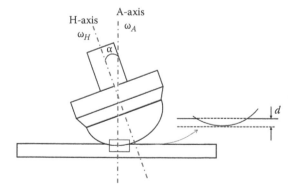

FIGURE 20.7
Schematic referring to the different rotation axes of the bonnet. (From Zeng, S., Blunt, L., Racasan, R., *Int. J. Adv. Manuf. Technol.*, 70, 583–590, 2014.)

where C_p is the Preston coefficient, which includes the effects of the process parameters affecting the interaction between the workpiece and the tool (e.g. pH, slurry, type of abrasive, etc.), P is the down-pressure applied (i.e. normal force, F per unit contact area, A_c, on the workpiece being polished) and v is the relative velocity between the part and the tool. This equation indicates that for conventional polishing, the MRR is controlled by the applied normal load and the relative velocity between the workpiece and the polishing tool/pad.

For estimating MRR, the Preston equation [75] can be modified to accommodate specific process behaviour peculiar to bonnet polishing. The key distinguishing behaviour can be attributed to the spherical (balloon-like) shape of the bonnet, which gives a non-uniform pressure distribution on the surface of the workpiece. Through experimentation under static conditions, this pressure distribution on the contact surface of the workpiece was found to be near-Gaussian, as illustrated in Figure 20.8 [76]; i.e. the height of material removed varies locally at the polishing spot. The material removal at a point B in the polishing spot is given by [73]

$$(M)_B = k\, P_m exp\left(-\frac{1}{2}\left(\frac{r_B}{b}\right)^{\lambda} \right) V_B t_d, \tag{20.3}$$

where M_B is the height of the material removed at B, k is the Preston coefficient, P_m is the maximum pressure observed at the centre of the spot, r_B is the

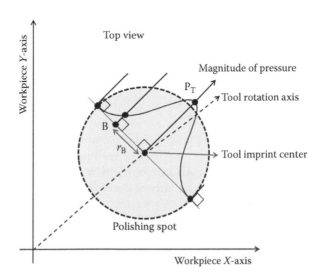

FIGURE 20.8
Distribution of pressure on the surface of the workpiece in static conditions. (From Kim, D.W., Kim, S.-W., *Opt. Exp.*, 13, 910–917, 2005.)

distance from the centre of the spot O to B, b and λ are the parameters of the pressure distribution, V_B is the relative velocity at B and t_d is the dwell time of the tool. Equation 20.3 is also known as the tool influence function (TIF) of the polishing process as it gives the material removal from the polishing area as a function of r_B. This function can be used to predict the shape of the surface for form correction during polishing [44,76]. The integral of TIF over the polishing spot gives the material removal from the region of the workpiece. The MRR can be computed as the material removal per unit dwell time on the surface.

It is important to note that the relative velocity at the point is dependent on the rotational speeds of the bonnet about both the H- and A-axes and the precess angle between them. Precession rotation of the bonnet is needed as polishing cuts from only the rotation about the H-axis are unidirectional in nature. In order to achieve a uniformly polished surface, the bonnet needs to be rotated about the A-axis at the precess angle also to constantly change the direction of cutting in the polishing spot. This process of polishing using both the H- and A-axes rotations was first suggested by Bingham et al. [77]. Pan et al. [45] suggested that a random abrasive path during continuous precessions results in a more uniform surface finish as compared to single precession polishing.

20.4.2 Materials

Bonnet polishing employs a variety of abrasives, such as alumina, diamond powder and cerium oxide, for polishing different types of materials. For example, cerium oxide with a particle size of <1 μm is suitable for polishing glass surfaces such as lenses [42,78]. Metals such as cobalt chromium alloys [40,44] and metal-plated surfaces are usually polished with harder abrasives, such as diamond or alumina with particle sizes in the range of 2–9 μm depending on the finish desired of the polished surface [79–81]. Soft materials that can trap the abrasive particles, such as polyurethane, and polishing 'microcloth' are used for manufacturing bonnets.

20.4.3 Process Parameters

The process parameters such as precess angle, bonnet offset, head speed and tool pressure have a major effect on MRR as well as the quality of the final surface. The surface roughness is also dependent on the size and hardness of the abrasive particles. The following subsections describe the effects of these parameters reported for finishing CoCr surfaces with a GR35 polyurethane bonnet and 3 μm alumina abrasive particles on a seven-axis bonnet polishing machine [39].

20.4.3.1 Precess Angle

In bonnet polishing, precess angles in the range of 5°–30° are generally used for the polishing process. Figure 20.9a shows the variation of the MRR with respect

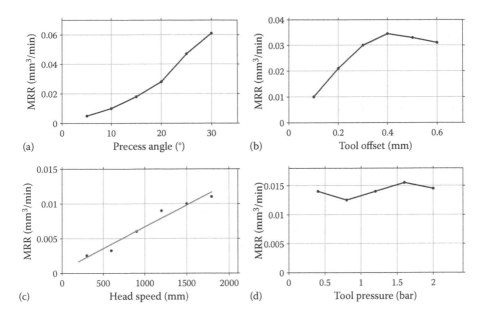

FIGURE 20.9
Variation of the MRR in bonnet polishing with respect to different process parameters such as (a) precess angle, (b) tool offset, (c) head speed and (d) tool pressure. (From Zeng, S.Y., Blunt, L., *Prec. Eng.*, 38, 348–355, 2014.)

to the precess angle of the bonnet. It shows that an increase in precess angle increases the MRR. A six-fold increase in the MRR was observed by increasing the precess angle from 5° to 30°. This increase in MRR can be attributed to the increase in both contact area and the relative speed at the polishing spot.

20.4.3.2 Bonnet Tool Offset

The tool offset is the distance by which the surface of the bonnet is compressed by the workpiece during the polishing process, which is indicated as *d* in Figure 20.7. An increase in the offset of the polishing tool increases the contact area of the polishing tool on the workpiece. It also increases the contact pressure onto the surface up to a limit. However, as the tool offset increases beyond this limit, the contact pressure decreases. Therefore, as shown in Figure 20.9b, MRR increases with an increase in the tool offset up to a limit due to increase in the surface area and down pressure. However, it decreases with further increase in the offset due to a drop in the contact pressure at high bonnet tool offset settings.

20.4.3.3 Head Speed

Head speed refers to the rotational speed of the bonnet head about the H-axis. The effect of the head speed on MRR is the same as that of rotary speed in lapping. As summarised in Figure 20.9c, MRR increases linearly with the head speed. This is in consonance with the Preston equation [75] that suggests a linear relationship between the MRR and the relative velocity, as represented in Equation 20.3.

20.4.3.4 Bonnet Pressure

Bonnet pressure refers to the bulk pressure of the fluid filled inside the bonnet. Under static tool conditions with constant tool offset, an increase in tool pressure would increase the contact pressure on the workpiece [76]. However, under dynamic working conditions, the TIF of the process is not affected significantly with an increase in bonnet pressure. As a result, only a slight increase in MRR is observed when the bonnet pressure is increased. Hence, as summarised in Figure 20.9d, the effect of bonnet pressure on MRR is not significant in comparison with the other parameters such as the precess angle, tool offset and head speed.

20.4.3.5 Tool Path

Tool path plays a vital role in determining the surface roughness and the uniformity of the polished surface. As the mathematical structure of the TIF suggests, the MRR is not constant even within the polishing spot. The spot is generally much smaller in size than the workpiece. This can result in non-uniform local polishing if the bonnet is kept stationary at a spot. Hence, the tool needs to be moved along the surface at a certain feed rate for uniform polishing of the whole surface. The specific path pattern followed by the tool plays a vital role in determining the uniformity of the finish. Dunn and Walker [82] experimented with a randomised unicursal tool path on the surface as opposed to raster tool paths (see Figure 20.10). These images suggest that periodic structures are formed on the surface from raster tool paths. These surfaces have a high Rt with low Ra. On the other hand, randomised tool paths yield a more uniform surface with less periodic structures and Rt values comparable with Ra.

20.4.4 Variants

Bonnet polishing is a secondary finishing process that is best suited for creating ultrasmooth finish on surfaces that are already pre-finished to a certain extent via grinding and other primary finishing processes. This process is not effective with unfinished initial surfaces. Hence, a variant to the basic bonnet polishing method based on using an abrasive jet to pre-polish a machined surface has been investigated [79,80]. This process variant was found to drastically reduce bonnet wear and more efficiently achieve an ultrasmooth surface [79,80].

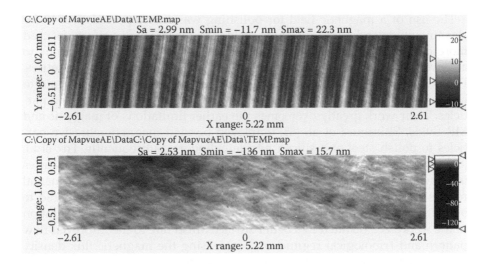

C:\Copy of MapvueAE\Data\TEMP.map
Sa = 2.99 nm Smin = −11.7 nm Smax = 22.3 nm

C:\Copy of MapvueAE\DataC:\Copy of MapvueAE\Data\TEMP.map
Sa = 2.53 nm Smin = −136 nm Smax = 15.7 nm

FIGURE 20.10
Surface images of the workpiece after regular straight path and unicursal path. (From Dunn, C.R., Walker, D.D., *Opt. Exp.*, 16, 18942–18949, 2008.)

20.4.5 Limitations

One of the primary drawbacks with bonnet polishing is that it is a contact polishing process, making it necessary to use mechanical arms and links to reach the polishing spot. This restricts the use of bonnet polishing only to external surfaces as well as certain internal locations of a biomedical implant that allow a bonnet tool to reach. Also, the concavity of the surface that can be efficiently polished is limited by the dimensions of the bonnet and the range of precess angles achievable in the equipment. Small concave features with curvatures higher than that of the bonnet are often difficult to polish using this technique.

20.5 Advances in Magnetic Polishing

Magnetic-field-assisted polishing processes are based on applying a magnetic field on abrasive-mixed magnetic fluid for surface finishing. In the presence of magnetic fields, magnetic particles in the fluid align themselves in the direction of the magnetic field. This increases the apparent stiffness of the Bingham fluid to provide the desired elastic properties to facilitate polishing, expose the abrasive particles to the workpiece surface and/or exert the desired down force.

The use of a magnetic field for polishing was first reported in 1940 for cleaning oxide scales and polishing welded joints [83]. Later efforts focused more on developing improved process designs and fluid compositions to achieve improved polishing results with different workpiece geometries. For example, Shimada et al. [84] developed a magnetic compound fluid to increase the apparent viscosity as well as the stability of the dispersed particles. Their work greatly overcomes the earlier limitations of magnetic and MR fluids. Shinmura et al. [85] employed bonded magnetic abrasive particles to polish steel and silicon nitride cylinders. Subsequently, Fox et al. [86] investigated the effects of using unbounded magnetic abrasive particles in a cylindrical magnetic abrasive finishing process. Their experimental results suggested an increase in MRR and surface roughness compared to bonded abrasives. They have also observed that imparting axial vibration to the workpiece resulted in a better surface finish due to the improved flow pattern and tribological regimes and increasing the magnetic flux density yielded higher MRR and better surface finish [86].

Kim et al. [87] used a pressurised jet of magnetic abrasive particles through a nozzle to finish internal surfaces of a workpiece with non-circular cross-sections. While the magnetic field is used to attract the abrasive particles towards the workpiece surface, they are dragged along the cylinder's surface via the pressurised jet simultaneously to facilitate polishing. More recently, Wang et al. [88,89] employed a gel with magnetorheological properties to finish a cylindrical workpiece. The steel workpiece was treated as a pole of the magnet and was rotated at high speeds to create a relative motion between the asperities and abrasive particles. An axial vibration was introduced to impart a uniform surface finish.

20.5.1 Process Setup and Mechanism

The mechanism of a magnetic polishing process is primarily governed by the behaviour of the magnetic (mostly MR) fluid in the presence of a magnetic field. MR fluids are particularly desired for polishing because they can (a) transport the abrasive particles to the cutting zone and remove abraded chips and heat to prevent scratching and thermal damage, respectively, (b) reach internal surfaces and profiles that are inaccessible to conventional polishing tools, (c) mitigate abrasive and process degradation as the fluid can be continuously recycled, monitored and conditioned, (d) provide optimal stiffness to meet various polishing needs and (e) augment other polishing processes (e.g. abrasive flow/jet polishing) to improve the process efficiency, especially for optical surfaces [90].

20.5.1.1 Setup

Over the past few decades, several magnetic polishing setups have been designed and customised to meet the needs of a variety of workpiece

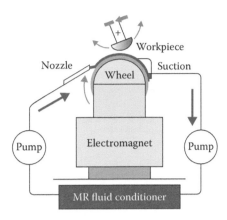

FIGURE 20.11
Magnetorheological finishing. (From Kordonski, W.I., Jacobs, S.D., *Int. J. Modern Phys. B*, 10, 2837–2848, 1996.)

materials and geometries. For example, Kordonski and Jacobs' setup [90] for polishing aspheric surface of lenses [91] is shown in Figure 20.11. In this setup, a nozzle delivers the MR fluid charged with abrasive particles onto the vertical rotating wheel. The rotary wheel carries the fluid to the converging work zone between the lens (workpiece) and the wheel. An electromagnet placed transversely to the carrier wheel magnetises the MR fluid, thus increasing its viscosity and yield stress. Upon the application of magnetic fields, the material stiffens and abrasive particles segregate to the surface of the fluid to promote material removal (along the exposed face of the fluid) [92]. This switching of the rheological properties of the fluid to a higher apparent viscosity and stiffness in the presence of magnetic field is associated with the formation of flexible magnetic abrasive brushes (FMABs) [46]. These brushes change their profile according to the surface form, as well as employ magnetic jigs that form chains of abrasive particles, resulting in material removal from both external [46] as well as internal [93] surfaces. Whereas in cases where bonded abrasives are used, the magnetic forces in these magnetically active bonded abrasive particles provide the required down force and relative motion in the abrasive particles for finishing. As the wheel rotates, the used MR fluid is collected by a suction system, which is then conditioned and recycled for the next cycle [90].

20.5.1.2 Mechanism

The material removal mechanism in this process is illustrated in Figure 20.12. As shown in Figure 20.12a, the abrasive and iron particles are randomly arranged in a non-magnetised MR fluid. Figure 20.12b shows the alignment of CIPs along the magnetic lines of force with abrasive particles suspended on the top. Figure 20.12c shows the sheared fluid with abrasive particles in

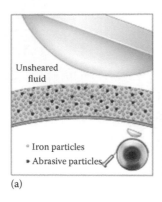
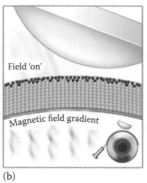
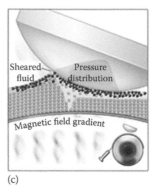

(a)　　　　　　　　　(b)　　　　　　　　　(c)

FIGURE 20.12

Schematic of magnetorheological finishing process showing (a) the distribution of particles in fluid in the absence of a magnetic field or material removal, (b) the preferential disposition of the particles in the fluid in the presence of magnetic field but without material removal and (c) the shearing of the fluid during material removal. (From Jain, V.K., *J. Mater. Process. Technol.*, 209, 6022–6038, 2009.)

contact with the workpiece surface while abrading the surface, thus removing material. Here, the relative motion is caused by the rotation of the carrier wheel and the sweeping motion of the lens. The magnetic force provides the normal component of the force while the tangential component of the force is generated by the rotation of the carrier wheel. The sweeping motion of the lens contributes to both the normal as well as the tangential force components. The relative motion between the workpiece and the abrasive-mixed MR fluid removes the material in the form of micro- and nano-chips [46].

Multiple investigations have addressed different aspects of magnetorheological polishing such as the phenomenology of segregation of the abrasive particles in the fluid and the effect of various process parameters on MRRs. Mori et al. [94] suggested that the down force influences the indentation of the abrasive particle on the surface of the workpiece. The tangential force contributes to restoring particles to their equilibrium chain structures from the disturbances as the workpiece rotates, asperity removal and to spinning the abrasive particles to help climb off the valleys.

Tani et al. [95] described a method that uses a 'battery' of multiple permanent magnets with alternating polarity to set up spatial field gradients. The non-uniform field has a strength of the order of 50–350 kA/m. They employed a magnetic fluid composed of 10–15 nm ferrocolloid-magnetite (Fe_3O_4) in eicosyl naphthalene and 40 vol.% 4 μm SiC. In their setup, the workpiece is rotated over the fluid and the workpiece is polished due to the motion of the part relative to the 'lap' created by the levitational buoyancy force. They demonstrated an MRR of 2 μm/min with an acrylic resin workpiece. Here, it is suggested that the horizontal field gradient prevents the abrasive particles from 'rolling away' along with the workpiece; the vertical gradient causes the magnetic particles to move downward towards the permanent magnets, in turn driving the abrasive particles up to the region between the magnetic particles and the workpiece. The vertical magnetic levitation force that allows the abrasive particles stay afloat is noted to be proportional to the gradient in the magnetic field [96,97], and it is expressed as

$$\frac{F_z}{V} = (\rho_f - \rho_s)g - \mu_o(M\nabla H)_z, \tag{20.4}$$

where F_z is the levitational buoyant force, V is the volume of abrasive particles, ρ_f is the mass density of magnetic fluid, ρ_s is the mass density of the abrasive particles, g is the acceleration due to gravity, M is the ferric induction of magnetic fluid and ∇H is the gradient in the magnetic field. Studies [95] also suggest an increase in MRR with increasing magnetic fields. This is attributed to the increase in buoyancy force resulting from the application of the magnetic field.

Shorey [97] and Miao et al. [98] have also studied the mechanism of material removal in the MR finishing process. Contrary to relating levitation forces to material removal, they stated that the material removal in MR fluids is primarily due to shear forces and not due to normal forces. This was based on the fact that the normal forces due to magnetic levitation (buoyant) (1 × 10^{-9} N) and the normal stress due to bulk deformation and/or hydrodynamic pressure due to flow of MR fluid into converging gap (1 × 10^{-7} N) [97] are much lower in comparison with the forces realised in conventional polishing processes (0.007 N and 0.065 N) [99], while the MRRs are still comparable. Miao et al. [98] reported the following expression for MRR in magnetorheological finishing processes:

$$MRR_{MFF} = C'_{p(MRF(\tau,FOM))} \frac{E}{K_c H_V^2} \cdot \tau v, \tag{20.5}$$

where MRR_{MFF} is the MRR for MR finishing, $C'_{p(MRF(\tau,FOM))}$ is the modified Preston coefficient, which in turn is a function of the shear stress τ and the material's figure of merit FOM, E is the elastic modulus of the workpiece, K_c is the fracture toughness of the workpiece, H_V is the Vicker hardness of the workpiece and v is the relative velocity of the MR fluid with respect to the workpiece. This expression is consistent with hydrodynamic models reported in the literature for material removal in polishing processes.

Other studies on magnetic fluid polishing include finishing of edges based on forming a converging gap between the edge and the magnetised MR fluid [100] and controlling MR fluid flow by adjusting the viscosity and temperature [101]. Augmented abrasive jet finishing has been developed to enhance MRR (the magnetic field is used to pull and collimate the abrasive jet, thus reducing the load on the fluid pump). Studies have also been reported on finishing of deep concave surfaces [102], polishing of elastic workpieces in a condensed medium [103], as well as chemo-mechanical polishing of ceramic balls [104]. The prior investigations have also provided evidence of interesting coupling between hydrodynamics and magnetic effects that affect polishing performance [105].

20.5.2 Materials

A magnetic polishing medium primarily includes three components, namely, the magnetic fluid, the abrasive particles and supplementary ingredients (e.g. surfactants, such as oleic acid, tetramethylammonium hydroxide), that may be employed to increase the colloidal stability of the fluid. In an unbonded magnetic polishing medium, alumina, SiC and CeO_2 particles of the order of 1–25 μm size are mixed with magnetic fluids to form the abrasive medium. The non-magnetic nature of these abrasive particles produces the necessary magnetic levitating force, which results in the formation of the FMAB for polishing. The magnetic fluids used for polishing can belong to one of the following four categories: (a) ferrofluids (at times referred to as magnetic fluids), which are composed of nanometric-sized CIPs suspended in a carrier fluid such as water, oil and eicosyl naphthalene with the help of stabilisers [95], (b) MR fluids that employ ferromagnetic particles (FMPs) of the order of micrometres, which are suspended in oil- or water-based medium using suitable stabilisers or polymers, (c) magnetic compound fluids suggested by Shimada et al. [84] that include both micro- and nano-scale particles where α-cellulose particles are used to stabilise the fluid and (d) magnetic abrasive gels where the magnetic and abrasive particles are suspended in silicone gel, which acts as a viscoelastic medium to suspend the particles [89].

Apart from these, the bonded abrasive particles can also be used with proper solvents and stabilisers. Mori et al. [94,106] reported the application of sintered magnetic abrasives where small abrasive particles (alumina ~5 μm [94]) are attached to the surface of large CIPs (70–170 μm [94]). Bando et al. [107] have also reported the use of bonded magnetic abrasive particles synthesised through electroless plating of diamond abrasives onto ferrous particles for polishing alumina ceramic tubes [108].

20.5.3 Process Parameters

For magnetic ultrasmooth finishing of biomedical implant materials, the MRR and surface finish of magnetic polishing processes are dependent on process parameters, such as the strength of the magnetic field, abrasive particle properties and type of magnetic particles, down force and the relative motion of the particles.

20.5.3.1 Magnetic Field Strength

With an increase in the magnetic field strength, the yield stress (based on the Bingham model) of the magnetic fluid increases [109]. This results in an increase in the pressure exerted by the magnetic particles on the abrasive particles and the workpiece, thereby increasing the MRR. One of the main advantages in magnetic-field-assisted finishing processes is the ability of the method to continuously recycle abrasive particles. However, if the magnetic field strength is increased beyond a critical point, the abrasive particles are not recycled as they are held against the workpiece surface under high pressure, thus resulting in worn-out edges of the active abrasive particles and progressively lower MRR.

20.5.3.2 Abrasive and Magnetic Particle Properties

Geometrical and physical properties of the abrasive and magnetic particles play a major role in influencing the surface finish. For example, when a steel grit is used as the FMP, the surface finish and MRR are better than with an iron grit [110]. This is because a steel grit was able to distribute the pressure more uniformly to the abrasive particles because of its polyhedron shape, compared to a spherical iron grit. Additionally, the hardness of the steel grit was much higher than that of the iron grit. This leads to enhanced surface abrasion and, hence, increased MRR. Also, larger FMP yielded higher MRR and surface roughness [110] because magnetic forces on the particles are directly proportional to their volume. Thus, large-sized FMPs had higher magnetic forces acting on them. This increased the average pressure on each abrasive particle, thereby increasing the MRR [96]. However, an excessively high magnetic force prevented the rolling of the FMP, which resulted in decreasing the pressure on the abrasive particles, thereby lowering the MRR [110].

Also, the larger the abrasive particle size, the higher was the MRR. However, large particle sizes also lead to higher surface roughness. Smaller abrasive particles yielded better surface roughness but low MRR. This is because the smaller the size of the particles is, the lower were the average pressure and shear stress exerted by the FMP on each abrasive particle. Thus, as the average pressure on the abrasive particles decreases, MRR decreases and surface finish improves [110].

20.5.3.3 Down Force

Down force is one of the most crucial parameters for polishing. It affects both the normal and the shear force acting on the workpiece surface. In magnetic-field-assisted polishing, down force on the workpiece can be applied broadly in two ways: (a) using a physical tool to push the magnetic fluid against the workpiece, or vice versa, and (b) using the magnetic field itself to attain the required down force. The first method is mostly used for polishing the external geometries, curved surfaces, etc. For example, the down force in curved and freeform geometries can be applied using a CNC controlled magnetic ball end tool covered with the magnetic fluid [111]. Here, the radius of the ball end tool becomes a limiting factor for polishing small geometrical features.

Magnetic fields, on the other hand, provide an inherent advantage of flexibility in polishing hard-to-reach areas as well as internal geometries, for example, pipes and capillary tubes. An appropriate combination of electromagnets may be needed to adjust the magnetic fields at the target spot to create the required down force. For internal geometries, Yamaguchi et al. [112] developed a magnetic configuration to polish the internal surfaces of a thin capillary tube. The tube was initially filled with the magnetic fluid and then an external magnetic configuration was used to agitate the abrasive particles and provide the down force necessary for polishing.

20.5.3.4 Effect of Relative Motion

The motion of the abrasive particles relative to the surface plays a vital role in material removal. The relative motion between the workpiece and the MR fluid can be achieved through different methods. In most cases, the fluid is agitated relative to the surface of the workpiece either mechanically with the help of wheels, jets, magnetic ball, etc., or using the magnetic field itself. Yamaguchi et al. [106,112] reported the use of magnetic fields to move the abrasive mixed fluid relative to the workpiece.

Since the pressure in this process is very low, an abrasive particle moving along the surface of the workpiece is free to rotate on the target area. But in the presence of the magnetic field, the chains formed by the FMP trap the abrasive particles and stop them from rotating while in contact with the surface of the workpiece. Hence, it is the shearing forces created by the abrasive particles trapped by the magnetic chains that result in finishing of the surface of the workpiece.

The material removal mechanism was further investigated by Mori et al. [94]. They used magnetic abrasive particles made by consolidating and sintering together iron and alumina particles (see Figure 20.13). It was noted that

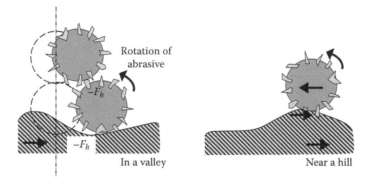

FIGURE 20.13
Schematic showing the motion of abrasive particles and the forces at the interface during polishing at peaks and valleys. (From Mori, T., Hirota, K., Kawashima, Y., *J. Mater. Process. Technol.*, 143, 682–686, 2003.)

the normal force acting on the abrasive particle was responsible for the indentation of the abrasive particle to the surface of the workpiece. To investigate the tangential force, they noted that when the workpiece rotates, the balance in the chain structure of the abrasive particles is disturbed and the abrasive particle at the edge moves a distance δx from its balanced point. As a result, a tangential (frictional) force acts to bring the particle back in the equilibrium position [94]. In Figure 20.13, the dashed lines represent the equilibrium location of the abrasive particle. This equilibrium position of the abrasive particle is perturbed due to the combined effect of friction force between the workpiece surface and the abrasive particle during its relative motion, as well as the repulsive force with other abrasive particles. This results in a return force F_h. If an abrasive particle's equilibrium location is in the vicinity of a crest, the return force F_h results in an asperity material removal. On the other hand, if the abrasive particle's indenting point is in a trough, F_h acts in the centre of the abrasive particle and a reaction force acts at the edge of the abrasive. Thus, a moment acts on the particle and makes it rotate and climb out of the crest. This tendency of the particle to rotate and climb the hill is also due to the fact that the cutting resistance of the material will be larger at the bottom of a crest owing to a larger volume of material to be removed. Hence, the material climbs along the crest with increasing F_h and ultimately cuts material from the top of the hill, where there is minimal cutting resistance.

Among the process parameters, tangential speed v and strength of the magnetic field B (that determines the down force P) are the major determinants of MRR and surface finish [113]. While the magnetic field contributes to an increase in MRR, the surface finish improves with increase in speed v as it promotes 'sloshing' of abrasive particles [113]. Thus, depending on the size, form and type of material being polished and the type and size of the abrasive particles being used, it is critical to choose an optimal magnetic field strength and rotational speed to obtain the desired surface finish.

20.5.4 Variants

Different hybrid methods with magnetic polishing have been developed to improve polishing performance. Prior efforts also investigated the use of electrochemical action (EMAF) to create a passivation layer [114], normal vibrations (VMAF) to enhance the MRR [115] and lubricants to improve fluid flow and finish [116]. The characteristics of FMAB, especially the effects of rotation speeds and magnetic field on the particle distribution, cutting forces and MRR [117], as well as the augmentation of ultrasonic vibration [118], have also received notable attention.

20.5.5 Relevance to Biomedical Implants

The ability of the polishing technique to achieve nano-scale finishes and conform to the shape of the workpiece is particularly useful in finishing of implants to ultrasmooth finish. Sutton [119] used a setup similar to the one described in Figure 20.11 to polish hip implant balls with MR fluids. MAF was also successfully employed by Sidpara and Jain [47] and Yamaguchi and Graziano [120] to polish a knee femoral implant to a surface roughness of Ra ≈ 28 nm and Sa ≈ 5 nm. Both these research groups used a ball-shaped magnetic pole to locally focus the magnetic abrasive fluid to form the FMAB, and the relative motion at the implant surface was enhanced by rotating the magnetic poles.

20.5.6 Limitations

While tremendous progress has been made, especially during the past decade, towards adapting magnetic field and MR fluid to finish various materials and geometric shapes, issues pertaining to localisation, such as confining the magnetic fluid-abrasive mix to polish the desired areas by applying 'optimal' time–space variation of the magnetic field, have received little attention. Recent advances in electro-permanent magnets and magneto-viscoelastic fluidics provide some interesting possibilities for localised polishing.

20.6 Summary

Biomedical implants are medical devices that are surgically placed into the body to enhance and support an organ in its functioning. A variety of metals such as stainless steel (316L SS), titanium alloys (Ti, Ti6Al7Nb, Ti6Al4V, TiNi) and cobalt chromium alloy (CoCrMo); ceramics such as zirconia, sapphire and alumina and polymers such as UHMWP, polyurethane, nylon and polyethylene are used to manufacture orthotic implant components. An implant needs to be chemically, mechanically and biologically accepted by the body without causing any undue stress during its lifetime.

For joint implants, two broad types of surfaces need to be prepared to achieve joint objectives of bio-acceptance and functional durability. A surface with micro-roughness at the interface between the implant and the bone tissue promotes osseointegration, including promotion of growth of bone tissue on the surface, and adsorption of the proteins onto the surface. A surface roughness Ra in the range of 0.5–8.5 μm is known to be most conducive to achieve this objective. An ultrasmooth finish (<50 nm) on the sliding surfaces ensures high durability and smooth movement of the implant at the joints and inhibits debris formation. The smoothness, often quantified by ease of sliding (Λ), varies depending on the material combinations and application.

Two broad categories of processes are employed to simultaneously achieve these differential surface requirements in an implant, namely, textured surface to promote osseointegration and ultrasmooth finished surface for smooth sliding. Texturing of surfaces can be achieved using sandblasting, chemical etching and laser treatment. Nanofinishing processes employed in this context include AFF, bonnet polishing, magnetic polishing and electrochemical treatment. Among these, mechanical methods such as AFF, bonnet polishing and magnetic polishing are best suited to finish complex 3D shapes using free abrasive particles.

In AFF, a homogenous mixture of fine abrasive particles suspended in a viscoelastic medium is pumped through the surface of the workpiece at very high pressures (7–220 bars). The high pressure and the normal stress developed as a result of the flow of the medium cause the abrasive particles to impinge onto the surface to cause asperity removal. In bonnet polishing, a deformable bonnet membrane inflated with a fluid rotates about two different axes along with the abrasive fluid to finely finish the surface of the workpiece. The two rotating axes are oriented at a precess angle. Most magnetic polishing methods are based on creating an FMAB with abrasive particles trapped in the chains of magnetic particles. An FMAB can easily conform to the shape of the implant so that the abrasive particles can polish the surface of the workpiece. External mechanical force and/or the magnetic fields are employed to apply the down force necessary for polishing.

Localisation of the polishing operation is hard to achieve using the current finishing methods. Most of the current techniques are limited to bulk polishing, compelling us to rely on manual methods or the use of elaborate surface masks for polishing local regions. Some techniques are limited by the size constraints of the machine for effective polishing. Significant opportunities exist to advance finishing technologies for polishing local regions of different sizes in implants with complex shapes to nano-scale smoothness.

20.7 Examples

Example 20.1

Synovial fluids present in healthy joints act as a lubricant for smooth sliding of the bones in the joints. In total joint replacement surgeries, periprosthetic synovial fluid is used as the lubricant. Simulator studies that were conducted on a newly designed hip implant for total hip replacement surgery with periprosthetic synovial fluid give a calculated minimum fluid thickness of 24 nm for a bearing load of 1500 N and fluid viscosity of 0.0025 Pa-s. Calculate and discuss on the lambda ratios for the following roughness values for the head and the cup of the hip joint.

Roughness of the Hip Joint Head (Ra_{head}) (μm)	Roughness of the Hip Joint Cup (Ra_{cup}) (μm)
0.1	0.1
0.05	0.05
0.0025	0.1
0.001	0.1

SOLUTION

The lambda ratio (Λ) for the cases discussed previously can be calculated using the following formula:

$$\Lambda = \frac{h_{min}}{\sqrt{Ra_{head}^2 + Ra_{cup}^2}}.$$

The value of the lambda ratio is

$$\Lambda = \frac{h_{min}}{\sqrt{Ra_{head}^2 + Ra_{cup}^2}} = \frac{0.024}{\sqrt{(0.1^2 + 0.1^2)}} = 0.170.$$

The values of lambda ratio for the other cases are given in the following table:

Roughness of the Hip Joint Head (Ra_{head}) (μm)	Roughness of the Hip Joint Cup (Ra_{cup}) (μm)	Lambda Ratio (Λ)
0.1	0.1	0.170
0.05	0.05	0.339
0.0025	0.1	0.233
0.001	0.1	0.240

DISCUSSION

It is interesting to notice that the lambda ratio increases with a decrease in the roughness of the surfaces. But it is also interesting to note that the roughness of both the sliding surfaces needs to be reduced for smooth sliding, and ease of sliding as quantified by the lambda ratio is limited by the rougher surface (compare cases 2 and 3).

Example 20.2

A NiCu-based workpiece is polished using a bonnet made up of Cerium Oxide D'16 polishing cloth with aluminium oxide (1.5 μm) abrasive slurry at a single spot without any feed. The radius of the bonnet is 20 mm. The bonnet is rotated at a rotational speed of 1200 rpm at a precess angle, made by the head axis with the normal to the surface, of 15°

and the offset in the bonnet while polishing is 0.3 mm. The maximum pressure exerted at the polishing spot is observed to be 2 bars.

 a. Calculate the MRR at the centre of the polishing spot, given the Preston coefficient is 3.566×10^{-8} MPa^{-1} and the process parameters are $b = 0.7446$ mm and $\lambda = 3.0432$.
 b. Calculate the height of the material removed at the same spot if the dwelling time is 90 s.
 c. Calculate the MRR at a point A at a distance of 0.4 mm from the centre of the polishing spot. Assume the relative velocity of the bonnet is the same that at the centre.

SOLUTION

Given the following

Precess angle $(\alpha) = 15°$
Radius of the bonnet $(R) = 20$ mm
Rotational speed of the head $(\omega_H) = 1200$ rpm
Maximum pressure at the centre O $(P_m) = 2$ bars
Bonnet offset $(d) = 0.3$ mm
Preston's coefficient $(k) = 3.566 \times 10^{-8}$ MPa^{-1}
$b = 0.7446$ mm
$\lambda = 3.0432$

 a. The distance of the point from O$(r_O) = 0$ mm
   The radius of rotation at O (as shown in Figure 20.14) is

$$r = D\sin \alpha = (R - d) \sin \alpha = (20 - 0.3)\sin(15°) = 5.1 \text{ mm}.$$

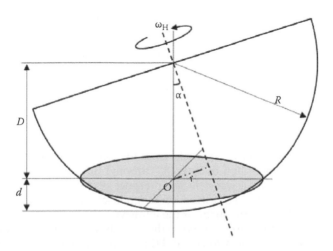

FIGURE 20.14
Schematic of the contact between the bonnet and the workpiece.

The relative velocity at O is

$$v_{rel_O} = r\omega_H = 5.1\times10^{-3}\times1200\times\frac{2\pi}{60} = 1.923\,\text{m/s}.$$

Therefore, the MRR at O is given by

$$MRR_O = kP_m exp\left(-\frac{1}{2}\left(\frac{r_O}{b}\right)^\lambda\right)v_{rel_O} = 3.57\times10^{-8}\times10^{-6}$$

$$\times2\times10^5\times exp\left(-\frac{1}{2}\left(\frac{0}{0.7446}\right)^{3.043}\right)1.923 = 13.7\,\text{nm/s}.$$

b. The value of the material removed at the location after a dwelling time (t_d) of 90 s is

$$M_o = MRR_o \times t_d = 13.7 \times 10^{-3} \times 90 = 1.23\ \mu\text{m}.$$

c.

Distance of the point A from O (r_A) = 0.4 mm

Relative velocity at A (v_{rel_A}) = 1.923 m/s

$$MRR_A = kP_m exp\left(-\frac{1}{2}\left(\frac{r_A}{b}\right)^\lambda\right)v_{rel_A} = 3.57\times10^{-8}\times10^{-6}$$

$$\times2\times10^5\times exp\left(-\frac{1}{2}\left(\frac{0.4}{0.7446}\right)^{3.043}\right)1.923 = 12.7\,\text{nm/s}.$$

Example 20.3

Consider a CeO_2 abrasive particle (size of 5 μm) in an abrasive MR fluid, with a composition of 36 vol.% carbonyl iron and 6 vol.% CeO_2 abrasives mixed in DI water, in the presence of a vertical gradient of field of 90,000 kA/m². The saturation magnetisation of carbonyl iron is 1600 kA/m. Assume the shape of the abrasive to be a sphere. Calculate (a) the levitation force experienced by the particle due to just the magnetic field (ignore the buoyancy forces) and (b) the total levitation force due to the combined mechanical and magnetic effects. Discuss the results.

SOLUTION

a.

$$\text{Gradient of magnetic field } (\nabla H) = -90{,}000 \text{ kA/m}^2$$

$$\text{Saturation magnetisation of carbonyl iron } (M) = 1600 \text{ kA/m}$$

$$\text{Magnetic permeability of free space } (\mu_0) = 1.257 \times 10^{-6} \text{ NA}^{-2}$$

$$\text{Diameter } (d) = 5 \text{ μm}$$

The buoyancy force due to magnetic field is then

$$F_M = -V\mu_0 M\nabla H = -\frac{4}{3}\pi r^3 \mu_0 M\nabla H = -\frac{4}{3}\pi \left(\frac{5}{2} \times 10^{-6}\right)^3$$

$$(1.257 \times 10^{-6})(1600 \times 10^3)(-90{,}000 \times 10^3) = 1.1843 \times 10^{-8} \text{ N.}$$

b.

Consider 1 cu.cm of abrasive MR fluid.

Component	Volume of the Component (cu.cm)	Weight of the Component (g)
Iron	0.36	2.83
Abrasive	0.06	0.459
Water	0.58	0.58
Total	1.00	3.872

Therefore,

$$\text{Density of the abrasive MR fluid} = 3.872 \text{ g/cu.cm}$$

$$F_{mech} = (\rho_f - \rho_s)gV = (3.872 - 7.65)$$

$$\times 10^3 \times 9.81 \times \frac{4}{3}\pi \left(\frac{5}{2} \times 10^{-6}\right)^3 = -2.426 \times 10^{-12} \text{ N}$$

Total force is then

$$F_z = 1.1843 \times 10^{-8} - 2.426 \times 10^{-12} = 1.1841 \times 10^{-8} \ N.$$

Clearly much of the total levitation force emerges due to the magnetic effect compared to the buoyancy effect.

Exercise Questions

1. A commercial MR-fluid-based polishing machine similar to the one shown in Figure 20.11 is considered for polishing a borosilicate glass workpiece. The vertical wheel of diameter of 15 cm is rotated at 1000 rpm to achieve a MRR of 15 μm/min. The modified Preston coefficient for borosilicate glass is 2×10^{-4} MPa m$^{1/2}$. Compute the shear stress acting on the glass during the polishing process (properties of Borosilicate glass are $E = 81$ GPa, $H_v = 6$ GPa, $Kc = 0.8$ MPa.m$^{1/2}$).

2. Describe the different types of AFF processes and compare their relative merits and demerits.

3. On what factors does MRR in bonnet polishing depend and how?

4. Describe the different non-abrasive based nanofinishing processes.

5. How is material removal mechanism in magnetic polishing differ from that is captured in Preston equation?

References

1. G. McGimpsey and T. C. Bradford, *Limb Prosthetics Services and Devices Critical Unmet Need: Market Analysis*, Bioengineering Institute Center for Neuroprosthetics: Worcester Polytechnic Institution, MA, pp. 1–35, 2011.

2. Biomaterials Market for Implantable Devices (Material Type – Metals, Polymers, Ceramics and Natural, Applications – Cardiology, Orthopedics, Dental, Ophthalmology and Others) – Global Industry Analysis, Size, Share, Growth, Trends and Forecast, 2013–2019. http://www.transparencymarketresearch.com, 29 May 2014. [Online]. Available at: http://www.transparencymarketresearch.com/biomaterials-market.html. Accessed 9 October 2015.

3. J. Davis, 'Overview of biomaterials and their use in medical devices', *Handbook of Materials for Medical devices. Illustrated Edition*. J. R. Davis, Ed., Ohio: ASM International, pp. 1–11, 2003.

4. H. Hermawan, D. Ramdan and J. R. Djuansjah, *Metals for Biomedical Applications*. INTECH Open Access Publisher, Rijeka, Croatia, 2011.

5. Metal on metal hip implants. Available at: http://kirkendalldwyer.com/areas-of-practice/medical-devices/metal-on-metal-hip-implants/. Accessed 9 October.

6. Sigma fixed-bearing knees. Available at: https://www.kneereplacement.com/DePuy_technology/DePuy_knees/fixed_knee. Accessed 9 October.

7. Custom medical implants – Jawbone. layerwise.com. Available at: http://www.layerwise.com/medical/custom-implants/. Accessed 9 October.

8. Latitude EV – Total elbow prosthesis. Available at: http://www.tornier-us.com/upper/elbow/elbrec001/. Accessed 9 October.

9. SMR metal back glenoid in total shoulder arthroplasty. Available at: http://www.lima.it/articolo-smr_metal_back_glenoid_in_total_shoulder_arthroplasty-13-53-0.html. Accessed 9 October.

10. Zenith – Total ankle replacement. Available at: https://www.coringroup.com/medical_professionals/products/extremities/zenith/. Accessed 9 October.

11. Overview of disc arthroplasty – Past, present and future. Available at: http://www.neurosurgery-blog.com/archives/656. Accessed 9 October.

12. ToeMobile – Great toe joint. Available at: http://www.hellotrade.com/merete-medical/product.html. Accessed 9 October.

13. Universal total wrist implant system. Available at: http://www.medcomtech.es/en/products/orthopedic-surgery-trauma-neurosurgery/upper-limb/wrist/arthroplasty. Accessed 9 October.

14. SR™ PIP (proximal Interphalangeal) implants. Available at: http://www.totalsmallbone.com/us/products/hand/sr_pip.php4. Accessed 9 October.

15. A. Pick, *Mechanical Heart Valve Replacement Devices*. Available at: http://www.heart-valve-surgery.com/mechanical-prosthetic-heart-valve.php. Accessed 9 October.

16. R. Z. Legeros and R. G. Craig, 'Strategies to affect bone remodeling: Osteointegration', *Journal of Bone and Mineral Research*, vol. 8, pp. S583–S596, 1993.

17. M. Shalabi, A. Gortemaker, M. Van't Hof, J. Jansen and N. Creugers, 'Implant surface roughness and bone healing: A systematic review', *Journal of dental research*, vol. 85, pp. 496–500, 2006.

18. K. Anselme and M. Bigerelle, 'Topography effects of pure titanium substrates on human osteoblast long-term adhesion', *Acta Biomaterialia*, vol. 1, pp. 211–222, 2005.

19. A. Wennerberg and T. Albrektsson, 'Suggested guidelines for the topographic evaluation of implant surfaces', *The International Journal of Oral & Maxillofacial Implants*, vol. 15, pp. 331–344, 1999.

20. H. Götz, M. Müller, A. Emmel, U. Holzwarth, R. Erben and R. Stangl, 'Effect of surface finish on the osseointegration of laser-treated titanium alloy implants', *Biomaterials*, vol. 25, pp. 4057–4064, 2004.

21. Aniket, R. Reid, B. Hall, I. Marriott and A. El-Ghannam, 'Early osteoblast responses to orthopedic implants: Synergy of surface roughness and chemistry of bioactive ceramic coating', *Journal of Biomedical Materials Research Part A*, vol. 103, pp. 1961–1973, 2015.

22. A. Lundgren, D. Lundgren, A. Wennerberg, C. H. F. Hämmerle and S. Nyman, 'Influence of surface roughness of barrier walls on guided bone augmentation: Experimental study in rabbits', *Clinical Implant Dentistry and Related Research*, vol. 1, pp. 41–48, 1999.

23. Y. Deng, X. Liu, A. Xu, L. Wang, Z. Luo, Y. Zheng et al., 'Effect of surface roughness on osteogenesis in vitro and osseointegration in vivo of carbon fiber-reinforced polyetheretherketone–nanohydroxyapatite composite', *International Journal of Nanomedicine*, vol. 10, p. 1425, 2015.

24. K. S. Katti, 'Biomaterials in total joint replacement', *Colloids and Surfaces B: Biointerfaces*, vol. 39, pp. 133–142, 2004.

25. T. Karachalios and G. Karydakis, 'Bearing Surfaces', in *European Instructional Lectures*, vol. 11, G. Bentley, Ed., Dordrecht: Springer, pp. 133–140, 2011.

26. N. Kumar, G. N. C. Arora and B. Datta, 'Bearing surfaces in hip replacement – Evolution and likely future', *Medical Journal, Armed Forces India*, vol. 70, pp. 371–276, 2014 Oct (Epub 2014 Aug 2014).

27. J.-D. Chang, 'Future bearing surfaces in total hip arthroplasty', *Clinics in Orthopedic Surgery*, vol. 6, pp. 110–116, 2014 Mar (Epub 2014 Feb 2014).

28. D. Dowson and Z. Jin, 'Metal-on-metal hip joint tribology', *Proceedings of the Institution of Mechanical Engineers, Part H: Journal of Engineering in Medicine*, vol. 220, pp. 107–118, 2006.

29. R. K. Alla, K. Ginjupalli, N. Upadhya, M. Shammas, R. K. Ravi and R. Sekhar, 'Surface roughness of implants: A review', *Trends in Biomaterials and Artificial Organs*, vol. 25, pp. 112–118, 2011.

30. K.-I. Jang, J. Seok, B.-K. Min and S. J. Lee, 'An electrochemomechanical polishing process using magnetorheological fluid', *International Journal of Machine Tools and Manufacture*, vol. 50, pp. 869–881, 2010.

31. B. M. Basol, 'Method and apparatus for localized material removal by electrochemical polishing', Google Patents, 2006.

32. J. Fischer, A. Schott and S. Märtin, 'Surface micro-structuring of zirconia dental implants', *Clinical Oral Implants Research*, vol. 27.2, pp. 162–166, 2015.

33. Y. Taga, K. Kawai and T. Nokubi, 'New method for divesting cobalt-chromium alloy castings: Sandblasting with a mixed abrasive powder', *The Journal of Prosthetic Dentistry*, vol. 85, pp. 357–362, 2001.

34. C. Aparicio, F. J. Gil, C. Fonseca, M. Barbosa and J. A. Planell, 'Corrosion behaviour of commercially pure titanium shot blasted with different materials and sizes of shot particles for dental implant applications', *Biomaterials*, vol. 24, pp. 263–273, 2003.

35. M. S. Cheema, G. Venkatesh, A. Dvivedi and A. K. Sharma, 'Developments in abrasive flow machining: A review on experimental investigations using abrasive flow machining variants and media', *Proceedings of the Institution of Mechanical Engineers, Part B: Journal of Engineering Manufacture*, vol. 226, pp. 1951–1962, 2012.

36. K. Subramanian, N. Balashanmugam and P. Shashi Kumar, 'Nanometric finishing on biomedical implants by abrasive flow finishing', *Journal of The Institution of Engineers (India): Series C*, vol. 97.1, pp. 55–61, 2016.

37. D. D. Walker, A. T. H. Beaucamp, D. Brooks, R. Freeman, A. King, G. McCavana et al., 'Novel CNC polishing process for control of form and texture on aspheric surfaces', in *Current Developments in Lens Design and Optical Engineering III*, vol. 4767, R. E. Fischer, W. J. Smith, and R. B. Johnson, Eds., Bellingham: Spie-Int Soc Optical Engineering, pp. 99–105, 2002.

38. D. Walker, D. Brooks, A. King, R. Freeman, R. Morton, G. McCavana et al., 'The "Precessions" tooling for polishing and figuring flat, spherical and aspheric surfaces', *Optics Express*, vol. 11, pp. 958–964, 2003.

39. S. Y. Zeng and L. Blunt, 'Experimental investigation and analytical modelling of the effects of process parameters on material removal rate for bonnet polishing of cobalt chrome alloy', *Precision Engineering*, vol. 38, pp. 348–355, 2014.

40. S. Zeng, L. Blunt and R. Racasan, 'An investigation of the viability of bonnet polishing as a possible method to manufacture hip prostheses with multi-radius femoral heads', *International Journal of Advanced Manufacturing Technology*, vol. 70, pp. 583–590, 2014.

41. D. W. Kim and J. H. Burge, 'Rigid conformal polishing tool using non-linear visco-elastic effect', *Optics Express*, vol. 18, pp. 2242–2257, 2010.

42. T. Akahori, T. Ashizawa, K. Hirai, M. Kurihara, M. Yoshida and Y. Kurata, 'CMP abrasive, liquid additive for CMP abrasive and method for polishing substrate', Google Patents, 2004.

43. M. Yoshida, T. Ashizawa, H. Terazaki, Y. Kurata, J. Matsuzawa, K. Tanno et al., 'Cerium oxide abrasive and method of polishing substrates', Google Patents, 2001.

44. S. Zeng, L. Blunt and X. Jiang, 'The investigation of material removal in bonnet polishing of CoCr alloy artificial joints', Proceedings of the 12th International Conference of the European Society for Precision Engineering and Nanotechnology EUSPEN 2012. EUSPEN, Stockholm, Sweden, pp. 352–355, 2012.

45. R. Pan, Z.-Z. Wang, T. Jiang, Z.-S. Wang and Y.-B. Guo, 'A novel method for aspheric polishing based on abrasive trajectories analysis on contact region', *Proceedings of the Institution of Mechanical Engineers, Part B: Journal of Engineering Manufacture*, vol. 229, pp. 275–285, 2015.

46. V. K. Jain, 'Magnetic field assisted abrasive based micro-/nano-finishing', *Journal of Materials Processing Technology*, vol. 209, pp. 6022–6038, 2009.

47. A. M. Sidpara and V. K. Jain, 'Nanofinishing of freeform surfaces of prosthetic knee joint implant', *Proceedings of the Institution of Mechanical Engineers Part B-Journal of Engineering Manufacture*, vol. 226, pp. 1833–1846, Nov 2012.

48. C. Sittig, M. Textor, N. Spencer, M. Wieland and P. Vallotton, 'Surface characterization of implant materials c.p. Ti, Ti–6Al–7Nb and Ti–6Al–4V with different pre-treatments', *Journal of Materials Science: Materials in Medicine*, vol. 10, pp. 35–46, 1999.

49. C. Larsson, P. Thomsen, B. O. Aronsson, M. Rodahl, J. Lausmaa, B. Kasemo et al., 'Bone response to surface-modified titanium implants: Studies on the early tissue response to machined and electropolished implants with different oxide thicknesses', *Biomaterials*, vol. 17, pp. 605–616, 1996.

50. A. Latifi, M. Imani, M. T. Khorasani and M. D. Joupari, 'Electrochemical and chemical methods for improving surface characteristics of 316L stainless steel for biomedical applications', *Surface & Coatings Technology*, vol. 221, pp. 1–12, 2013.

51. T. Hryniewicz, R. Rokicki and K. Rokosz, 'Surface characterization of AISI 316L biomaterials obtained by electropolishing in a magnetic field', *Surface and Coatings Technology*, vol. 202, pp. 1668–1673, 2008.

52. R. S. Faeda, H. S. Tavares, R. Sartori, A. C. Guastaldi and E. Marcantonio Jr, 'Evaluation of titanium implants with surface modification by laser beam: Biomechanical study in rabbit tibias', *Brazilian Oral Research*, vol. 23, pp. 137–143, 2009.

53. D. D. Walker, A. T. Beaucamp, D. Brooks, V. Doubrovski, M. D. Cassie, C. Dunn et al., 'Recent developments of Precessions polishing for larger components and free-form surfaces', in *Optical Science and Technology, the SPIE 49th Annual Meeting*, pp. 281–289, 2004.

54. L. Zhang, J.-L. Zhuang, X.-Z. Ma, J. Tang and Z.-W. Tian, 'Microstructuring of p-Si (100) by localized electrochemical polishing using patterned agarose as a stamp', *Electrochemistry Communications*, vol. 9, pp. 2529–2533, 2007.

55. R. W. McCarty, 'Method of Honing by Extruding', USA Patent 3 521 412, 1968.
56. L. Rhoades, 'Abrasive flow machining: A case study', *Journal of Materials Processing Technology*, vol. 28, pp. 107–116, 1991.
57. H. S. Mali and A. Manna, 'Current status and application of abrasive flow finishing processes: A review', *Proceedings of the Institution of Mechanical Engineers Part B-Journal of Engineering Manufacture*, vol. 223, pp. 809–820, July 2009.
58. R. Minear and N. Nokovich, 'Machine for abrading by extruding', Google Patents, 1974.
59. L. J. Rhoades, T. A. Kohut, N. P. Nokovich and D. W. Yanda, 'Unidirectional abrasive flow machining', Google Patents, 1994.
60. J. R. Gilmore and L. J. Rhoades, 'Self-forming tooling for an orbital polishing machine and method for producing', Google Patents, 2002.
61. L. Rhoades, 'Abrasive flow machining with not-so-silly putty', *Metal Finishing*, vol. 85, pp. 27–29, 1987.
62. V. K. Gorana, V. K. Jain and G. K. Lal, 'Prediction of surface roughness during abrasive flow machining', *The International Journal of Advanced Manufacturing Technology*, vol. 31, pp. 258–267, 2006.
63. W. B. Perry, 'Honing media', Google Patents, 1993.
64. V. K. Gorana, V. K. Jain and G. K. Lal, 'Experimental investigation into cutting forces and active grain density during abrasive flow machining', *International Journal of Machine Tools and Manufacture*, vol. 44, pp. 201–211, 2004.
65. M. R. Sankar, J. Ramkumar and V. K. Jain, 'Experimental investigation and mechanism of material removal in nano finishing of MMCs using abrasive flow finishing (AFF) process', *Wear*, vol. 266, pp. 688–698, 2009.
66. A. C. Wang and S. H. Weng, 'Developing the polymer abrasive gels in AFM process', *Journal of Materials Processing Technology*, vol. 192–193, pp. 486–490, 2007.
67. S. Singh, H. Shan and P. Kumar, 'Wear behavior of materials in magnetically assisted abrasive flow machining', *Journal of Materials Processing Technology*, vol. 128, pp. 155–161, 2002.
68. M. K. Reddy, A. Sharma and P. Kumar, 'Some aspects of centrifugal force assisted abrasive flow machining of 2014 Al alloy', *Proceedings of the Institution of Mechanical Engineers, Part B: Journal of Engineering Manufacture*, vol. 222, pp. 773–783, 2008.
69. A. R. Jones and J. B. Hull, 'Ultrasonic flow polishing', *Ultrasonics*, vol. 36, pp. 97–101, 1998.
70. D. Walker, A. Beaucamp and C. Dunn, 'Computer controlled work tool apparatus and method', Google Patents, 2008.
71. X. Chen, 'Corrective Abrasive Polishing Processes for Freeform Surface', in *Progress in Abrasive and Grinding Technology*. vol. 404, X. Xu, Ed., Stafa-Zurich: Trans Tech Publications Ltd, pp. 103–112, 2009.
72. S. Zeng, L. Blunt and R. Racasan, 'An investigation of the viability of bonnet polishing as a possible method to manufacture hip prostheses with multi-radius femoral heads', *The International Journal of Advanced Manufacturing Technology*, vol. 70, pp. 583–590, 2014.
73. C. F. Cheung, L. B. Kong, L. T. Ho and S. To, 'Modelling and simulation of structure surface generation using computer controlled ultra-precision polishing', *Precision Engineering*, vol. 35, pp. 574–590, Oct 2011.

74. R. Pan, Y. Zhang, C. Cao, M. Sun, Z. Wang and Y. Peng, 'Modeling of material removal in dynamic deterministic polishing', *The International Journal of Advanced Manufacturing Technology*, pp. 1–12, 2015.
75. F. Preston, 'The theory and design of plate glass polishing machines', *Journal of the Society of Glass Technology*, vol. 11, p. 214, 1927.
76. D. W. Kim and S.-W. Kim, 'Static tool influence function for fabrication simulation of hexagonal mirror segments for extremely large telescopes', *Optics Express*, vol. 13, pp. 910–917, 2005.
77. R. G. Bingham, D. D. Walker, D.-H. Kim, D. Brooks, R. Freeman and D. Riley, 'Novel automated process for aspheric surfaces', in *Current Developments in Lens Design and Optical Systems Engineering*, vol. 4093, pp. 445–450, 2000.
78. D. Walker, G. Yu, H. Li, W. Messelink, R. Evans and A. Beaucamp, 'Edges in CNC polishing: From mirror-segments towards semiconductors, paper 1: Edges on processing the global surface', *Optics Express*, vol. 20, pp. 19787–19798, 2012.
79. A. Beaucamp and Y. Namba, 'Super-smooth finishing of diamond turned hard X-ray molding dies by combined fluid jet and bonnet polishing', *CIRP Annals–Manufacturing Technology*, vol. 62, pp. 315–318, 2013.
80. A. T. H. Beaucamp, R. R. Freeman, A. Matsumoto and Y. Namba, 'Fluid jet and bonnet polishing of optical moulds for application from visible to x-ray', in *SPIE Optical Engineering+ Applications*, International Society for Optics and Photonics, San Diego, CA, 2011, pp. 81260U–81260U-8.
81. G. Yu, D. Walker and H. Li, 'Implementing a grolishing process in Zeeko IRP machines', *Applied Optics*, vol. 51, pp. 6637–6640, 2012.
82. C. R. Dunn and D. D. Walker, 'Pseudo-random tool paths for CNC sub-aperture polishing and other applications', *Optics Express*, vol. 16, pp. 18942–18949, 2008.
83. H. P. Coats, 'Method of and Apparatus for Polishing Containers', Google Patents, 1940.
84. K. Shimada, F. Toyohisa, H. Oka, Y. Akagami and S. Kamiyama, 'Hydrodynamic and magnetized characteristics of MCF (Magnetic Compound Fluid)', *Transactions of the Japan Society of Mechanical Engineers Series C*, vol. 67, pp. 3034–3040, 2001.
85. T. Shinmura, K. Takazawa, E. Hatano, M. Matsunaga and T. Matsuo, 'Study on magnetic abrasive finishing', *CIRP Annals–Manufacturing Technology*, vol. 39, pp. 325–328, 1990.
86. M. Fox, K. Agrawal, T. Shinmura and R. Komanduri, 'Magnetic abrasive finishing of rollers', *CIRP Annals–Manufacturing Technology*, vol. 43, pp. 181–184, 1994.
87. J.-D. Kim, Y.-H. Kang, Y.-H. Bae and S.-W. Lee, 'Development of a magnetic abrasive jet machining system for precision internal polishing of circular tubes', *Journal of Materials Processing Technology*, vol. 71, pp. 384–393, 1997.
88. A. C. Wang, L. Tsai, C. H. Liu, K. Z. Liang and S. J. Lee, 'Elucidating the optimal parameters in magnetic finishing with gel abrasive', *Materials and Manufacturing Processes*, vol. 26, pp. 786–791, 2011.
89. A. C. Wang and S. J. Lee, 'Study the characteristics of magnetic finishing with gel abrasive', *International Journal of Machine Tools and Manufacture*, vol. 49, pp. 1063–1069, 2009.
90. W. I. Kordonski and S. D. Jacobs, 'Magnetorheological finishing', *International Journal of Modern Physics B*, vol. 10, pp. 2837–2848, 1996.

91. D. C. Harris, 'History of magnetorheological finishing', in *SPIE Defense, Security, and Sensing*, International Society for Optics and Photonics, San Diego, CA, 2011, pp. 80160N–80160N-22.

92. V. Jain, 'Magnetic field assisted abrasive based micro-/nano-finishing', *Journal of Materials Processing Technology*, vol. 209, pp. 6022–6038, 2009.

93. H. Yamaguchi and T. Shinmura, 'Study of an internal magnetic abrasive finishing using a pole rotation system: Discussion of the characteristic abrasive behavior', *Precision Engineering*, vol. 24, pp. 237–244, 2000.

94. T. Mori, K. Hirota and Y. Kawashima, 'Clarification of magnetic abrasive finishing mechanism', *Journal of Materials Processing Technology*, vol. 143, pp. 682–686, 2003.

95. Y. Tani, K. Kawata and K. Nakayama, 'Development of high-efficient fine finishing process using magnetic fluid', *CIRP Annals – Manufacturing Technology*, vol. 33, pp. 217–220, 1984.

96. A. Sidpara and V. K. Jain, 'Magnetorheological and Allied Finishing Processes', in *Micromanufacturing Processes*, V. K. Jain, Ed., CRC Press (Taylor & Francis), Boca Raton, FL, pp. 133–153, 2013.

97. A. B. Shorey, 'Mechanisms of material removal in magnetorheological finishing (MRF) of glass', Master of Science, University of Rochester, Rochester, MN, 2000.

98. C. Miao, S. N. Shafrir, J. C. Lambropoulos, J. Mici and S. D. Jacobs, 'Shear stress in magnetorheological finishing for glasses', *Applied Optics*, vol. 48, pp. 2585–2594, 2009.

99. V. H. Bulsara, Y. Ahn, S. Chandrasekar and T. N. Farris, 'Mechanics of Polishing', *Journal of Applied Mechanics*, vol. 65, pp. 410–416, 1998.

100. S. D. Jacobs and I. V. Prokhorov, 'Magnetorheological finishing of edges of optical elements', Google Patents, 1997.

101. W. Kordonski, S. Gorodkin and A. Sekeres, 'System for magnetorheological finishing of substrates', Google Patents, 2013.

102. W. I. Kordonski, 'Apparatus and method for abrasive jet finishing of deeply concave surfaces using magnetorheological fluid', Google Patents, 2003.

103. L.-J. Liao, 'Ultra-low temperature magnetic polishing machine', Google Patents, 2013.

104. R. Komanduri, N. Umehara and M. Raghunandan, 'On the possibility of chemo-mechanical action in magnetic float polishing of silicon nitride', *Journal of Tribology*, vol. 118, pp. 721–727, 1996.

105. J. Seok, Y.-J. Kim, K.-I. Jang, B.-K. Min and S. J. Lee, 'A study on the fabrication of curved surfaces using magnetorheological fluid finishing', *International Journal of Machine Tools and Manufacture*, vol. 47, pp. 2077–2090, 2007.

106. H. Yamaguchi and T. Shinmura, 'Study of the surface modification resulting from an internal magnetic abrasive finishing process', *Wear*, vol. 225, pp. 246–255, 1999.

107. S. Bando, A. Tsukada and Y. Kondo, 'A Study on Precision Internal Finishing for Alumina Ceramics Tube', *Journal of the Japan Society for Abrasive Technology*, vol. 45, pp. 46–59, 2001.

108. H. Yamaguchi and T. Shinmura, 'Internal finishing process for alumina ceramic components by a magnetic field assisted finishing process', *Precision Engineering*, vol. 28, pp. 135–142, 2004.

109. J. Huang, J. Zhang and J. Liu, 'Effect of magnetic field on properties of MR fluids', *International Journal of Modern Physics B*, vol. 19, pp. 597–601, 2005.
110. G.-W. Chang, B.-H. Yan and R.-T. Hsu, 'Study on cylindrical magnetic abrasive finishing using unbonded magnetic abrasives', *International Journal of Machine Tools and Manufacture*, vol. 42, pp. 575–583, 2002.
111. A. K. Singh, S. Jha and P. M. Pandey, 'Design and development of nanofinishing process for 3D surfaces using ball end MR finishing tool', *International Journal of Machine Tools & Manufacture*, vol. 51, pp. 142–151, 2011.
112. H. Yamaguchi, T. Shinmura and A. Kobayashi, 'Development of an internal magnetic abrasive finishing process for nonferromagnetic complex shaped tubes', *JSME International Journal Series C*, vol. 44, pp. 275–281, 2001.
113. V. Jain, P. Kumar, P. Behera and S. Jayswal, 'Effect of working gap and circumferential speed on the performance of magnetic abrasive finishing process', *Wear*, vol. 250, pp. 384–390, 2001.
114. B.-H. Yan, G.-W. Chang, T.-J. Cheng and R.-T. Hsu, 'Electrolytic magnetic abrasive finishing', *International Journal of Machine Tools and Manufacture*, vol. 43, pp. 1355–1366, 2003.
115. S. Yin and T. Shinmura, 'Vertical vibration-assisted magnetic abrasive finishing and deburring for magnesium alloy', *International Journal of Machine Tools and Manufacture*, vol. 44, pp. 1297–1303, 2004.
116. H. Yamaguchi and T. Shinmura, 'Internal finishing process for alumina ceramic components by a magnetic field assisted finishing process', *Precision Engineering*, vol. 28, pp. 135–142, 2004.
117. D. K. Singh, V. Jain and V. Raghuram, 'Experimental investigations into forces acting during a magnetic abrasive finishing process', *The International Journal of Advanced Manufacturing Technology*, vol. 30, pp. 652–662, 2006.
118. R. S. Mulik and P. M. Pandey, 'Mechanism of surface finishing in ultrasonic-assisted magnetic abrasive finishing process', *Materials and Manufacturing Processes*, vol. 25, pp. 1418–1427, 2010.
119. J. K. Sutton, 'Orthopaedic component manufacturing method and equipment', Google Patents, 2011.
120. H. Yamaguchi and A. A. Graziano, 'Surface finishing of cobalt chromium alloy femoral knee components', *CIRP Annals – Manufacturing Technology*, vol. 63, pp. 309–312, 2014.

21

Chemical Mechanical Planarisation

David Lee Butler

School of Mechanical and Aerospace Engineering, Nanyang Technological University, Nanyang Avenue, Singapore

CONTENTS

21.1 Introduction

Chemical mechanical polishing or planarisation (CMP) is defined as the process of smoothing surfaces through a combination of chemical and mechanical forces. It is most commonly used in the semiconductor industry, where its introduction in the late 1980s and early 1990s was met with resistance as it can be considered to be a dirty process due to it involving abrasives and subsequently generating particles in the process.

21.1.1 Why Is Planarisation Required?

Planarisation is commonly used for the ultra-precision polishing of bare silicon wafers as well as for device fabrication. In the polishing of silicon wafers both single-sided and double-sided polishing can be used. In the polishing process, silicon wafers are required to be 1 to 2 nm Ry in roughness and free of oxidation, micro-scratches and haze. Polishing is generally carried out in a series of steps, with the first step aimed at efficiently producing planar, mirror surfaces followed by a second process that is aimed at removing oxidation and improving the surface roughness. Subsequent steps are focussed on the production of haze-free and contamination-free surfaces.

Integrated circuits (ICs) typically consist of many layers of circuits which are linked together through multilevel interconnections. Over the years, the number of interconnection layers has greatly increased as IC manufacturers strive to increase performance while at the same time reducing the overall IC size.

As the feature scale is reduced, any surface topographic variation can result in gaps being created, which can cause a number of performance issues. Coupled with a trend for an increasing number of interconnection layers, this problem is compounded. Thus, the challenge in improving step coverage is intrinsically linked to the ability to planarise the wafer to the required specification. Figure 21.1 shows a typical cross-section of a device and the processes used at various stages; the complexity of such a device is clear to see.

Figure 21.2 shows the various planarisation steps used for device wafers. It is ideal to preferentially planarise projected areas only on a rough surface to produce a smooth flat surface, regardless of the substrate conditions. This is termed global planarisation. For the improvement of yield and reliability, it is indispensable to achieve global planarisation across an entire wafer surface.

In addition to the number of layers increasing, the size of the features on each layer is also decreasing. Photolithography is the standard process to create the device features. Any surface irregularities will profoundly influence the resolution and depth of focus (DOF) of the process, making it a challenge to simultaneously focus on both concave and convex surfaces.

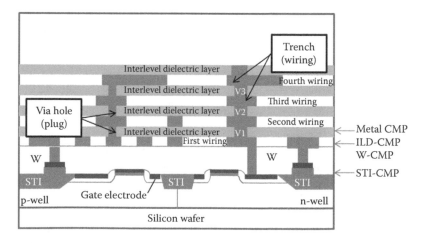

FIGURE 21.1
Sectional diagram of a device and the processes to which planarisation CMP is applied.

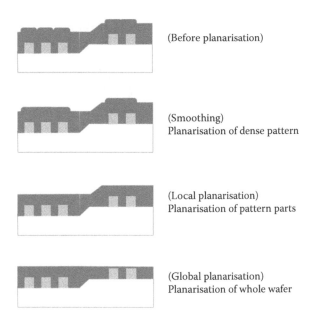

(Before planarisation)

(Smoothing)
Planarisation of dense pattern

(Local planarisation)
Planarisation of pattern parts

(Global planarisation)
Planarisation of whole wafer

FIGURE 21.2
Planarisation mode of a device wafer.

A number of solutions exist such as using a shorter wavelength light or an optical arrangement which is larger than the numerical aperture (NA). However, with a smaller wavelength and a larger NA, the DOF becomes shallow, which creates difficulty for non-flat surfaces. In order to achieve a flat surface, CMP is the preferred process.

21.2 Mechanism

Polishing is carried out without letting fine abrasive particles generate brittle fractures on the work surfaces, while removing these materials in minute steps only by means of plastic deformation, to finally produce a smooth mirror surface. The process detaches material from the surface in a relative motion caused by the protrusion of fixed/free abrasive particles between the opposing surfaces of the polishing pad and workpiece, as shown in Figure 21.3. The polishing mechanism is termed as a two-body abrasion when the abrasive particles are fixed to the polishing pad, while the mechanism is termed as a three-body abrasion when the (loose) abrasive particles are free to rotate between the interfacial surfaces during the material removal. Due to the differences in the removal mechanism, two-body abrasion is considered to produce material removal that is three times more than the three-body

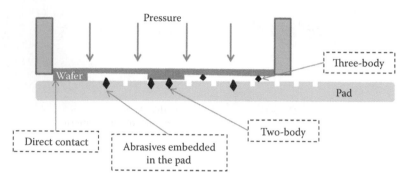

FIGURE 21.3
Material removal mechanisms in CMP.

abrasion under the same loading conditions. Usually, the polishing technique consists of both mechanisms, but the dominance of either of them depends largely on the polishing conditions during the process. During the polishing process, direct chemical actions between the slurry and the workpiece promote the mechanical actions further.

21.3 Polishing Defects

Polishing of patterned wafers typically leads to a number of process defects being generated. Figure 21.4 illustrates some of the more common defects such as dishing, erosion, recess, thinning, micro-scratches and keyhole. The main cause for many of these defects can be attributed to excess polishing, the slurry or the pad. As a consequence, it goes without saying that polishing planarity and uniformity should be secured by optimising polishing conditions and achieving defect-free polished surfaces.

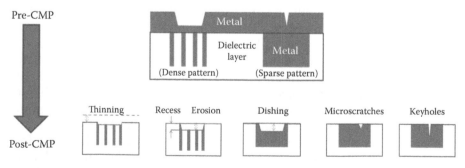

FIGURE 21.4
Typical processing defects and phenomena caused during metal CMP.

21.4 CMP System

There are a number of different tools and consumables used in a complete CMP process. The basic tool used for CMP evolved from the machines that had been used to polish bare silicon substrate surfaces. When the idea for CMP was first conceived at IBM, they purchased silicon polishing tools from Strasbaugh and Westech (now SpeedFam-IPEC). The first tool adapted for CMP had a single wafer carrier. The CMP machine differs from the silicon polishing tool in a number of ways. In silicon polishing, several tens of microns of material are removed, while the CMP process removes only about 0.05–1.0 μm of material. The uniformity of material removal across a wafer in CMP is also much more stringent than that for silicon polishing, which is around ±0.5–0.1 nm for CMP. Furthermore, CMP tools require a higher level of automation, throughput, reliability and critical process-parameter control. Thus, a specific CMP tool was developed for the IC industry. While early CMP tools used a single wafer carrier, more recent designs contain multiple carriers and multiple platens. This increases the throughput of the polisher and allows multi-step CMP processes to be carried out on a single tool. CMP tools that use a platen that moves past the rotating wafer carrier in a linear motion have also been developed and so-called linear CMP tools. These tools offer the potential benefits of higher linear speeds of 400–500 ft/min (122–152 m/min), compared to around 34 ft/min (10 m/min) in rotary CMP tools. However, industry has been reluctant to accept linear CMP tools due to performance issues and cost. A typical CMP system consists of three bodies: a semiconductor wafer, polishing slurry and a polishing pad, as shown in Figure 21.5.

The key components of any CMP tool include the wafer carrier, the platen and slurry and will be described in the following sections.

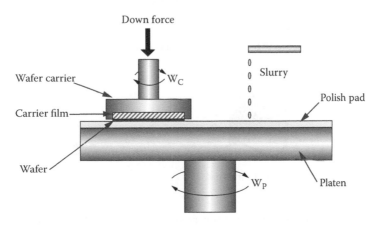

FIGURE 21.5
CMP tool.

21.4.1 Carrier

The basic function of the carrier is to hold the wafer in place while the wafer is processed. Instead of directly holding the wafer on the chuck, a backing film is sandwiched between the wafer and steel chuck. This film provides elasticity between the wafer and chuck. Without a flexible backing film, any defect or particles on the chuck or on the back of the wafer will cause a thin spot in the film being polished. In addition, the possibility of wafer break-age is increased. In conventional wafer carriers, the retaining ring does not make contact with the pad. As a result, there is no pressure applied to the pad by this ring. Thus, the pad pressure at the edge of the wafer is not well controlled and the polishing rate of the film is decreased at the wafer edge. Hence, an edge exclusion ring of 5–7 mm must be enforced for wafers with such carriers, thus reducing the productive space on the wafer.

Carriers used in some CMP tools such as Applied Materials' Mirra (commercially known as Titan Head [1]) are designed with a retaining ring that can apply variable pressure to the pad as shown in Figures 21.6 and 21.7 [2]. An independently controlled pressure is applied onto the retaining

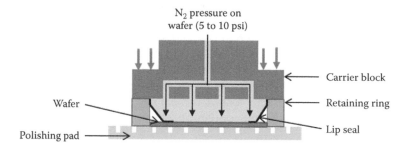

FIGURE 21.6
Illustration of the polishing head showing retainer ring.

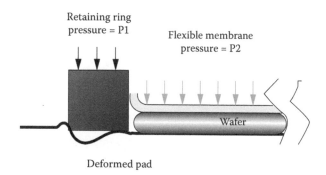

FIGURE 21.7
Schematic diagram and performance of advanced retaining ring of the carrier. (From Wang, S.K., Direct shallow trench isolation CMP with ceria based slurry, PhD Thesis, Nanyang Technological University, 2005.)

ring to absorb pad deformation around the wafer edge during the process. This allows the pad profile at the wafers edge to be controlled. A smaller edge exclusion of around 2–3 mm zone on the wafer can therefore be used. Retaining ring design on the carrier is critical to ensure edge exclusion on the wafer and uniformity of removal rate near the edge of the wafer as well.

21.4.2 Platen

The platen's role in a CMP tool is to support the pad, which is fixed using an adhesive backing. The surface precision of the platen is critical – even more so when polishing bare silicon wafers, which typically require more process control due to the rise in pad temperature during polishing. Typically, bare silicon wafer polishing takes 10 times longer than polishing device wafers. Nonetheless, temperature fluctuations do occur across the pad, which is normally cooled by flowing slurry containing distilled water across the surface.

During polishing, the platen is subjected to high loads. In the case of an 8 in (200 mm) diameter wafer being processed at a pressure of 500 g/cm^2, the load applied to the wafer is around 160 kg. Consequently, a large bearing is required in order to increase the rigidity of the platen, prevent deformation, as well as ensure accuracy during polishing.

21.4.3 Slurry

The goal for each CMP process is to find a slurry that produces high removal rates, good planarity and high selectivity (for the process involved two or more types of thin film materials). Furthermore, it should also be easily cleaned from the wafer surface and cause no defects to the wafer surface such as scratches. Therefore, slurry plays a crucial role in a successful CMP process. Particulates in slurry mechanically abrade the wafer surface and remove surface materials. On the other hand, additives in the slurry solution react with surface materials or the particulates and dissolve the surface material or form other compounds that can be removed by abrasives particles. Slurries consist of small, abrasive particles of specific size in the range from 10 to 1000 nm and specific shape suspended in an aqueous solution. The abrasive particles have roughly the same hardness as the film being polished. Acids or bases are added to the solution, depending on the material to be removed. A number of abrasives are available for CMP slurries and are very much dependent on the specific application. Two of the more common abrasives are fumed and colloidal silica (the former is shown in Figure 21.8), which are used for the polishing of interlayer dielectrics (ILD). Cerium oxide has also found use in the polishing of silicon oxide layers, where it demonstrates a higher material removal rate. Cerium oxides exhibit a phenomenon known as selectivity and remove silicon nitride at a much lower rate.

The transportation medium for the slurry is the distilled water, which also acts as a lubricant and cooling agent between the wafer and polishing pad. There is also a strong correlation between the lubrication and tribological mechanism and regime during the CMP process.

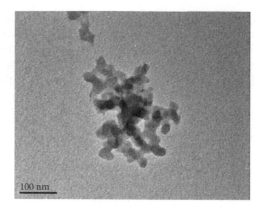

FIGURE 21.8
Transmission electron microscope (TEM) image of ILD 1200 slurry showing particles of 20 to 50 nm diameter.

Abrasive concentration has a direct influence on the coefficient of friction (COF). A lower abrasive concentration would lead to greater contact between the pad and wafer and, thus, a higher COF. Conversely, a higher concentration of abrasive particles allows significant rolling between the pad and wafer, leading to a lower COF.

With respect to the removal rate, researchers have reported inconclusive and contradictory results. The removal rate, being proportional to the applied pressure and relative velocity, has been reported by Cook [3] to be independent of abrasive size. However, Mahajan et al. [4] reported that variation in removal rates was influenced by an increase in abrasive particle size. The size of the abrasive particles in the slurry plays a critical role in the mechanical effects. For abrasives smaller than 0.3 µm, the changes in the removal rate are indistinguishable [5]. The effect of abrasive size on removal rate was modelled by Basim et al. [6] using two different mechanisms. For smaller abrasives, they found that the dominant mode of material removal was the contact area mechanism, whereas for larger abrasives, indentation plays the dominant role. On the other hand, small particles may not be adequate for a sufficient mechanical removal rate, while large particles may cause excessive scratching. Abrasive hardness also affects the potential for scratching.

21.5 Pads

Pads used in the CMP process for IC manufacturing are made of cast and sliced polyurethane. The polyurethane type is resistant to acid and alkaline solutions [7]. While there is no consensus, it is generally agreed that the hardness and porosity of the pad are important parameters and play a crucial role in determining the performance of the process. In many CMP applications, two pads are simultaneously used because a hard pad gives better local planarity but a softer pad gives better uniformity of material removal across the entire wafer. Using a combination of two pads provides a compromise between these extremes. The surface of a pad is either cut with concentric grooves or contains perforations 1 mm wide and 250 µm deep punched into the pad. Grooved pads yield a higher removal rate than perforated pads do, and this can be attributed to the difference in contact area between the two pads, 90.7% for the perforated pad compared to 83.3% for the grooved pad [8]. These grooves or perforations form channels to help transport the slurry between the pad and wafer. Pores present on the pad surface also aid in slurry transport. A comprehensive summary of the effect of polyurethane pad on CMP performance is given in Table 21.1 [9,10].

TABLE 21.1

Impact of Pad Properties on CMP Performances

| | Polishing Scale | | | Condition |
Pad Properties	Wafer	Die	Feature	Ability
Density (porosity)	Removal rate	Defectivity	Conductor dishing	Yes
	Non-uniformity		Oxide loss	
Hardness	Macro-scratch	Defectivity	Defectivity	Yes
			Roughness	
			Conductor dishing	
			Oxide loss	
Tensile properties	Pad life			Yes
Abrasion resistance	Pad life			Yes
Stiffness	Edge effects	Planarisation		
	Non-uniformity			
Modulus				Yes
Thickness	Pad life			
Top pad compressibility		Planarisation	Conductor dishing	
			Oxide loss	
Base pad compressibility	Edge effects	Planarisation		
	Non-uniformity			
Pad texture (grooves and pores)	Pad life			
	Removal rate			
	Edge effects			
	Non-uniformity			
Pad roughness	Removal rate	Planarisation	Conductor dishing	Yes
	Non-uniformity		Oxide loss	
Hydrophilicity	Removal rate			Yes

Source: Lawing, A.S., Rhoades, R., Pad, slurry, conditioning interactions, in Oxide CMP, Vendor's Presentation, Rodel Inc: Phoenix, AZ; James, D., Control of polishing pad physical properties and their relationship to polishing performance, CAMP 5th International CMP Symposium, Clarkson University, 2000.

Figure 21.9 shows a comparison between using a hard pad or soft pad to planarise the surface. With a hard pad, a small amount of stock is removed to produce the planar surface. The removal occurs where features project up. For the soft pad, a larger amount of material is removed to produce the required surface with projecting features aggressively removed as compared to recessed features.

Figure 21.10 shows a topographic plot for a grooved pad, which clearly shows the top contact area and the grooves, which have a depth of 400 μm.

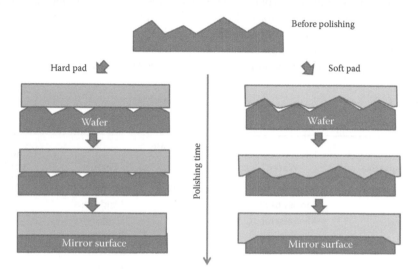

FIGURE 21.9
A comparison of material removal using hard and soft pads.

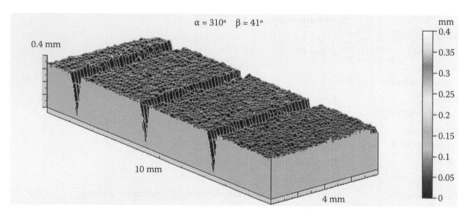

FIGURE 21.10
Topographic plot of a grooved polishing pad.

The alternative to a grooved pad is shown in Figure 21.11, where equal-sized and regularly spaced pores can be clearly seen.

The de facto standard pad for CMP polishing is the IC1000 produced by Rohm and Haas Electronic Materials CMP Inc. (formerly Rodel Inc.). The pores are spherical in nature and range in diameters from 30 to 50 μm. The fraction of the pore is about 35% of the pad total volume but can be changed by manufacturing conditions. The higher the porosity, the lower the pad density. The pad properties are closely related with polishing performance in CMP and the relationship is complex. Polishing performance can be considered at three scale levels: wafer scale, die scale and feature interdependent. The correlation of pad properties with polishing performance is summarised in Table 21.1. Removal rate is dependent on the pad density, texture and hydrophilicity. Non-uniformity depends on pad density, stiffness, texture, roughness and pad compressibility. Pad life depends on pad tensile properties, stiffness, texture and abrasion resistance. Pad density and hardness also affect defectivity (the number of defects that may be produced during the CMP). Planarisation is affected by pad stiffness, compressibility and roughness. Metal interconnect dishing and oxide loss are affected by pad density, hardness, roughness and compressibility [20].

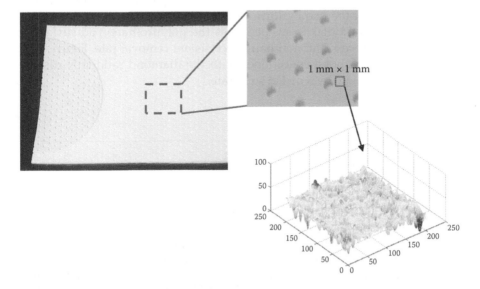

FIGURE 21.11
Partial image of a poromeric pad showing the zoomed images of the larger surface pores and the topographic of the region between these features.

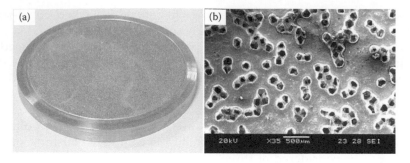

FIGURE 21.12
(a) Photo of a pad conditioner and (b) scanning electron microscope micrograph showing diamond distribution.

21.5.1 Pad Conditioning

As the pad is subjected to the CMP process, removal rates begin to decrease rapidly over time. This occurs due to the surface of the pad rapidly glazing during the planarisation process. The pores of the pad become closed, reducing slurry delivery to the wafer surface and causing an unstable and lower removal rate [12,13]. A consistent removal rate is necessary for a production-worthy CMP process. This can be achieved by pad conditioning, which opens the pores of the pad by forming micro-scratches on the pad surface. Typically, a diamond disk is swept across the polyurethane pad surface to obtain a stable process and maintain a consistent removal rate. Figure 21.12 shows a typical diamond conditioner with the diamond randomly distributed and held in place by a nickel substrate.

21.6 End Point Detection

Knowing when to stop the process is critical in determining the final thickness of the thin film or to ensure that excess material is completely removed to reveal underlying material across the wafer in the case of the inlaid CMP process. Thus, other than estimation made by operator using look-up tables, numerous approaches such as optical, electrical and acoustic sensing means to obtain the optimal process time have been proposed and patented as an in situ end point detection (EPD) system for CMP tool with limited success [14]. This is due to the nature of the CMP process environment and the presence of various patterns on the wafer surface, which add a great deal of complexity to the signals. Furthermore, the material removal rate in the CMP process is more inclined to be erratic as the result of complex interactions between the various input parameters, thus making it difficult to integrate a robust and reliable real-time measurement of film thickness into the CMP tool. Typically, ex situ measurement of thin film thickness is required to characterise the process in IC fabrication.

21.7 Material Removal Rate in CMP

One of the most important and basic performances measured in the CMP process is the material removal rate and is usually expressed in terms of Å/min in IC fabrication. A blanket wafer (i.e. a wafer void of any device features) is used to characterise the material removal rate and non-uniformity across the wafer for a specific configuration of the inputs, as shown in Figure 21.13. It serves as a fundamental understanding for patterned wafers (die scale, feature scale, particle scale). Variation of the material removal rate across the wafer scale results in material thickness variation known as *within-wafer non-uniformity*, whereas variation of material removal rate across the die scale causing the post-thickness variation is termed as *within-die non-uniformity*.

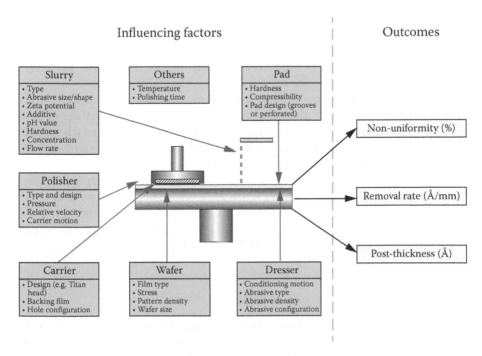

FIGURE 21.13
Influencing factors on the CMP process.

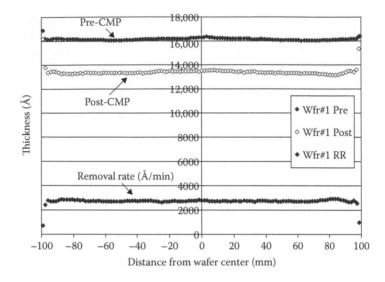

FIGURE 21.14
Diameter scan for pre- and post-thickness and removal rate. (From Wang, S.K., Direct shallow trench isolation CMP with ceria based slurry, PhD Thesis, Nanyang Technological University, 2005.)

Figure 21.14 illustrates the diameter scan for a 200-mm bare silicon wafer. It can be seen that there is some deviation in the thickness near the edge, which can be attributed to the carrier design and retaining ring.

In addition to profile measurements across the wafer, polar scans with either 49 or 121 points are also employed to get a clearer three-dimensional picture of thickness variation. Figure 21.15 shows the polishing of a bare silicon wafer indicating a variation in removal rate across the wafer surface and non-uniformity values ranging from 1% to 4%.

The planarisation performance is usually measured by feature scale (step height reduction), as indicated in Figure 21.3, whereas particle scale is used to measured surface roughness. Blanket material removal rate models provide an important starting point for elucidating the other aspects of the CMP process. A reliable model will facilitate the development of process recipes.

21.7.1 Material Removal Rate Models

Preston presented the first phenomenological model in 1927 based on results of glass polishing, and it retains a significant influence on CMP experimental and modelling work [15]. To date, the model still remains the standard reference for most detailed models that are used to describe the wafer scale, die scale, feature scale and surface finish in the CMP process of IC fabrication.

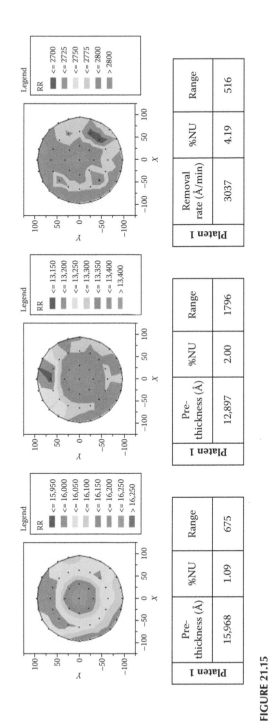

FIGURE 21.15

Polar map with 49 pts scan. (From Wang, S.K., Direct shallow trench isolation CMP with ceria based slurry, PhD Thesis, Nanyang Technological University, 2005.)

According to Preston, the material removal rate at any position across the wafer surface can be described as

$$\frac{\Delta H}{\Delta t} = K_p * \left(\frac{L}{A}\right) * \left(\frac{\Delta s}{\Delta t}\right), \tag{21.1}$$

where $\frac{\Delta H}{\Delta t}$ is the change in thickness or height (H) over time (t), L is the applied load or total load and A is the surface area on which wear occurs, $\frac{\Delta s}{\Delta t}$ is the relative velocity between the wafer and the pad and K_p is the Preston coefficient, which is process dependent with units of cm^2/dyn [10]. Commonly, it is rewritten or simplified to $RR = K_pPV$, where P is the applied pressure, which is calculated by first determining the apparent wafer surface area, and V is the average relative velocity. Despite the fact that the model was developed for glass polishing and is not physical based, it provided the foundation for the dielectric planarisation in the semiconductor industry and this empirical model serves as a convenient storehouse for experimental data. In the literature, some works have also showed that Preston's equation is also valid for the polishing of certain metal thin films [16–18].

Besides Preston, another well-known reference for CMP process was published by Cook [10] in 1990, which focuses on a chemical explanation of the process. Cook describes the polishing mechanics by a Hertzian indenter model as abrasive penetration into the wafer surface under uniform applied load, which can be given by

$$\frac{\Delta H}{\Delta t} = (2E)^{-1} * P * \left(\frac{\Delta s}{\Delta t}\right). \tag{21.2}$$

This is equivalent to the Preston equation (Equation 21.1), but with the Preston constant expressed as the inverse of twice Young's modulus, E. Cook also introduced the term of 'chemical tooth', which implies that hydroxyl groups played a key role in the material removal process. Cook explained that material removal during glass polishing is determined by the relative rates of five processes: (i) the rate of molecular water diffusion into the glass surface; (ii) subsequent glass dissolution under the load imposed by the polishing particle; (iii) the adsorption rate of dissolution products onto the surface of the polishing grain; (iv) the rate of silica redeposition back onto the glass surface and (v) the aqueous corrosion rate between particle impacts. He concluded that the major factors influencing these processes are the load and velocity of the polishing particles, the elastic properties of both glass surface and particles, the chemical durability of the glass, the surface charge of the glass and the surface charge and ion-exchange capacity of the particle.

In 1994, Runnels and Eyman [19] presented a tribological analysis to demonstrate that hydroplaning is possible in the CMP process and concluded that the fluid layer between the pad and wafer is profound. An analogy to Preston's equation was proposed:

$$R = \hat{k}\sigma\tau, \tag{21.3}$$

where $\hat{k}$ is a Preston-like constant and σ and τ are the magnitudes of the normal and shear stresses, respectively, on the wafer surface. The study of Runnels and Eyman was based on fluid mechanics to the stresses in a solid–liquid–solid system. The assumption they made in the simplification of flow modelling was that the wafer and pad are both rigid and smooth. No experimental data were used to verify the model in their study.

Other researchers such as Liu et al. [20] used a combination of statistical methods and elastic theory to describe the wear mechanism of the silicon wafer surface. Tseng and Wang [21] revisited Preston's equation and modified it to take into consideration the dependence between pressure and speed for CMP. Wang et al. [22] developed a model for stress distribution across a wafer during CMP by considering the carrier film and pad compressibility on removal rate non-uniformity. They concluded that the non-uniformity across the wafer surface was not due to the normal contact stress and uniformity could be improved by decreasing the compressibility of the polyurethane pad and carrier film. Maury et al. [23] further modified Preston's equation to account for non-zero material removal intercept when the product of pressure and velocity is high. The model also demonstrated a better fit to experimental data for oxide CMP than the models proposed by Tseng and Wang [21] and Shi and Zhao [24].

The interaction of the particle with the wafer was the focus for researchers such as Liu et al. [20], who developed a CMP model based on the elastic theory and the kinematics of an abrasive particle moving in the air gap of a pair of surfaces in rolling contact. Fu et al. [25] made the assumption of perfect plastic deformation occurring when particles abraded the softer hydrated layer on the SiO_2 surface. One of the most highly cited work on CMP modelling was that of Luo and Dornfeld [26], who based their investigations on the following assumptions: plastic contact between the wafer-abrasive and the pad-abrasive interfaces, periodic roughness of the polishing pad and a normal distribution of particle size. The synergetic effect of chemical and mechanical forces was represented by a dynamic hardness of the wafer surface.

Although many models have been developed over the years, each has focused on only a few aspects of the CMP process, and there have been very few attempts to combine the effects of the material properties of the particles, wafer and polishing pad and process conditions such as pressure, relative velocity and slurry flow rate into one model.

21.8 Summary

Over the past 30 years, CMP has established itself as a key process in the semiconductor fabrication industry. With many different applications, there has been extensive development work to optimise the slurry, pad and process parameters to achieve the required productivity. In the last 10 years, significant effort has been made to scaling up the process to accommodate the anticipated next generation of wafers of 450 mm diameter; however, this has now become quite unlikely to be implemented as the investment cost and perceived economy of scales do not seem to be realisable. Hence, current research seems to focus on developing improved process control as well as new pads that have fixed abrasives and do not require the need for slurry.

Questions

1. Since the beginning of the use of CMP in semiconductor applications, there have been concerns about the creation of debris in the process. As debris can be formed through the scratching of the surface, where do you think would be the most likely sources of such particles to undertake such an action?
2. Chemical mechanical planarisation, as the name suggests, includes both the actions of the chemicals and mechanical mechanisms to planarise the wafer. Do you foresee that chemical planarisation could replace CMP? Justify your answer.

References

1. Osterheld, T.H., Zuniga, S., Huey, S., McKeever, P., Garretson, C., Bonner, B., Bennett, D., and Jin, R.R., A novel retaining ring in advanced polishing head design for significantly improved CMP performance. *Proceedings of Materials Research Society Spring Meeting*, 1999. 566: pp. 63–68.
2. Wang, S.K., Direct Shallow Trench Isolation CMP with Ceria Based Slurry, PhD Thesis, Nanyang Technological University, 2005.
3. Cook, L.M., Wang, J.F., James, D.B., and Sethuraman, A.R., *Semiconductor International*, November 1995: p. 141.
4. Mahajan, U., Bielmann, M., and Singh, R., Abrasive effects in oxide chemical mechanical polishing. *Proceedings of Materials Research Society Spring Meeting*, 1999. 566: pp. 27–32.
5. Luo, Q., Ramarajan, S., and Babu, S.V., Modification of the Preston equation for the chemical–mechanical polishing of copper. *Thin Solid Films*, 1998. 335(1–2): pp. 160–167.

6. Basim, G.B., Adler, J.J., and Mahajan, U., Effect of particle size of CMP slurries for enhanced polishing with minimal Defects. *Journal of the Electrochemical Society*, 2000(147): p. 3523.

7. Sax, N.I. and Lewis, R.J., Eds., *Hawley's Condensed Chemical Dictionary*, New York: Van Nostrand Reinhold Co., 2007. p. 944.

8. Clark, A.J., Witt, K.B., and Rhoades, R.L., Oxide removal rate interactions between slurry, pad, downforce and conditioning, CMP-MIC conference, Santa Clara, CA, 1999. pp. 401–404.

9. Lawing, A.S. and Rhoades, R., Pad, slurry, conditioning interactions, in Oxide CMP, Vendor's Presentation, Rodel Inc: Phoenix, AZ, 2000. p. 22.

10. Cook, L.M., Chemical processes in glass polishing. *Journal of Non-Crystalline Solids*, 1990. 120: pp. 152–171.

11. James, D., Control of polishing pad physical properties and their relationship to polishing performance, CAMP 5th International CMP Symposium, Clarkson University, 2000.

12. Ali, I. and Roy, S.R., Pad conditioning in interlayer dielectric CMP, in *Solid State Technology*, 1997. pp. 185–187.

13. Breivogel, J.R., Blanchard, L.R., and Prince, M.J., *Polishing pad conditioning apparatus for wafer planarization process*. US Patent No. 5,216,843. 1993, Intel Corporation.

14. Li, S.H. and Miller, R.O., *Chemical Mechanical Polishing in Silicon Processing*, Academic Press, New York, 2000. p. 263.

15. Preston, F.W., The theory and design of plate glass polishing machine. *Journal of the Society of Glass*, 1927. 11: pp. 214–256.

16. Cooper, K., Cooper, J., Groschopf, J., Flake, J., Solomentsev, Y., and Farkas, J., Effects of particle concentration on chemical mechanical planarization. *Electrochemical and Solid-State Letters*, 2002. 5(12): pp. 109–112.

17. Steigerwald, J.M., Zirpoli, R., Murarka, S.P., Price, D., and Gutmann, R.J., Pattern geometry effects in the chemical-mechanical polishing of inlaid copper structures. *Journal of the Electrochemical Society*, 1994. 141(10): pp. 2842–2848.

18. Stavreva, Z., Zeidler., D., Plötner, M., and Drescher, K., Chemical mechanical polishing of copper for multilevel metallization. Applied Surface Science MAM 1995. First European Workshop on Materials for Advanced Metallization, 19–22 March 1995. 91: pp. 192–196.

19. Runnels, S.R. and Eyman, L.M., Tribology analysis of chemical-mechanical polishing. *Journal of the Electrochemical Society*, 1994. 141(6): pp. 1698–1701.

20. Liu, C-W., Dai, B-T., Tseng, W-T., and Yeh, C-F., Modeling of the wear mechanism during chemical-mechanical polishing. *Journal of the Electrochemical Society*, 1996. 143(2): pp. 716–721.

21. Tseng, W.T. and Wang, Y-L., Re-examination of pressure and speed dependences of removal rate during chemical-mechanical polishing processes. *Journal of the Electrochemical Society*, 1997. 144(2): pp. 15–17.

22. Wang, D., Lee, J., Holland, K., Bibby, T., Beudoin, S., and Cale, T., Von Mises stress in chemical-mechanical polishing processes. *Journal of the Electrochemical Society*, 1997. 144(3): pp. 1121–1127.

23. Maury, A., Ouma, D., Boning, D., and Chung, J., A modification to Preston's equation and impact on pattern density effect modeling. Program Abstracts, Advanced Metallization and Interconnect Systems for ULSI Applications, 1997.

24. Shi, F.G. and Zhao, B., Modeling of chemical-mechanical polishing with soft pads. *Applied Physics A*, 1998. A67(2): pp. 249–252.
25. Fu, G., Chandra, A., Guha, S., and Subhash, G., Plasticity-based model of material removal in chemical–mechanical polishing (CMP). *IEEE Transactions on Semiconductor Manufacturing*, 2001. 14: pp. 406–417.
26. Luo, J. and Dornfeld, D.A., Material removal mechanism in chemical mechanical polishing: Theory and modeling. *IEEE Transactions on Semiconductor Manufacturing*, 2001. 14: pp. 112–133.

Author Index

A

Adler, T. A., 374
Adsul, S. G., 99
Agarwal, S. S., 385
Agrawal, K., 135, 240
Akbari, J., 54
Albrektsson, 552
Alcock, S., 128, 129
Alisha, S., 152
Ali Tavoli, M., 103
Amero, J. J., 331
Anderson, R., 167
Ando, M., 128
Ang, Y. J., 88, 89
Ann, 428
Anselme, K., 552
Aoki, S., 129
Aoyama, T., 385
Appleton, B. R., 159
Arima, K., 128
Arora, G., 88
Arrasmith, S. A., 15, 217, 218
Arrasmith, S. R., 231
Asari, M., 164
Aziz, M. J., 164

B

Baets, R., 156
Bahre, D., 39
Bai, Y. J., 37
Bakuzis, A. F., 301, 302
Balasubramaniam, R., 201
Balos, S., 44, 51
Bando, S., 578
Baseri, H., 206
Basim, G. B., 604
Behera, P. K., 198, 199
Beier, M., 231
Belak, J. F., 529
Belyshkin, D. V., 135
Bertoglio, S., 429

Berujon, S., 128, 129
Bhagavatula, S. R., 272
Bhowmik, P., 88
Bidanda, B., 102
Bigerelle, M., 552
Billo R. E., 102
Bingham, R. G., 570
Blunt, L., 568, 571
Bronsvoort, W. F., 236
Brown, S. R., 168
Btyumenfel'd, L. A., 342

C

Cabanettes, F., 49
Cai, M. B., 509
Callister, W. D., 289
Campbell, R. J., 236
Cao, G., 296
Capello, G., 429
Carlson, D. J., 300
Cen, Q., 94, 95
Center for Optics Manufacturing
        (COM), 240, 252
Chandrasekaran, N., 21
Chang, A., 159
Chao, P. Y., 121, 122, 123, 124, 126
Chartier, T., 224
Chattopadhyay, A. B., 63, 69, 70
Chattopadhyay, A. K., 63, 69, 70
Cheema, M. S., 562, 567
Chen, C., 414
Chen, C. H., 163
Chen, K. Y., 87, 88
Chen, S.-H., 278
Cheng, H. B., 228
Cheng, K. C., 87, 88
Childs, T. H. C., 270, 273, 274, 275
Chiou, Y., 273, 274, 276
Choi, M. S., 187, 188, 365
Choudhary, S. K., 385
Chow, H. M., 100, 181, 203

Subject Index

Page numbers followed by f and t indicate figures and tables, respectively.